T0227950

MARINE NUTRACEUTICALS

Prospects and Perspectives

Nutraceuticals: Basic Research/Clinical Applications
Series Editor: Yashwant Pathak, PhD

Marine Nutraceuticals: Prospects and Perspectives
Se-Kwon Kim

NUTRACEUTICALS Basic Research/Clinical Applications

MARINE NUTRACEUTICALS

Prospects and Perspectives

Edited by **Se-Kwon Kim**

CRC Press
Taylor & Francis Group
Boca Raton London New York

CRC Press is an imprint of the
Taylor & Francis Group, an **informa** business

CRC Press
Taylor & Francis Group
6000 Broken Sound Parkway NW, Suite 300
Boca Raton, FL 33487-2742

First issued in paperback 2017

© 2013 by Taylor & Francis Group, LLC
CRC Press is an imprint of Taylor & Francis Group, an Informa business

No claim to original U.S. Government works

ISBN 13: 978-1-138-19996-5 (pbk)
ISBN 13: 978-1-4665-1351-8 (hbk)

Library of Congress Cataloging-in-Publication Data

Marine nutraceuticals : prospects and perspectives / editor, Se-Kwon Kim.
      p. ; cm. -- (Nutraceuticals : basic research/clinical applications)
   Includes bibliographical references and index.
   ISBN 978-1-4665-1351-8 (hardcover : alk. paper)
   I. Kim, Se-Kwon. II. Series: Nutraceuticals : basic research/clinical applications.
   [DNLM: 1. Dietary Supplements. 2. Aquatic Organisms. 3. Seafood. QU 145.5]

615.3'2--dc23   2012037586

Visit the Taylor & Francis Web site at
http://www.taylorandfrancis.com

and the CRC Press Web site at
http://www.crcpress.com

Contents

Series Preface

Nutraceuticals: Basic Research/Clinical Applications

Nutraceuticals and functional food industries have grown significantly in the last two decades. Foods that promote health beyond providing basic nutrition are termed "functional foods." These foods have the potential to promote health in ways not anticipated by traditional nutrition science. The acceptance of these products by larger populations, especially in the West, is ever increasing. The United States has dominated the scene of the nutraceuticals and functional food market recently.

Nutraceuticals provide health benefits, facilitate the healing process, and prevent diseases, thus attracting a large clientele. Health conscious citizens worldwide now look at nutraceuticals as one of the major options for their health care. This is also reflected by the fact that nutraceuticals and alternative medicine have been incorporated in the curriculum in all health-care professional education.

Nutraceutical products provide people with a safe and healthy lifestyle. This may contribute to overall market growth and reach more than $90 billion by the end of 2015 in the United States alone, while it may reach $180 billion worldwide. This has been attributed principally to the aging population worldwide, increased prevalence of serious diseases due to changes in lifestyle, and enhanced focus on alternative medicines globally.

One of the challenges nutraceuticals, alternative medicines, dietary supplements, and functional foods face is proper characterization processes, reproducible activity, and clinical evidence to support claims of their application in prevention or treatment. Looking ahead, it is likely that tougher Food and Drug Administration USA (FDA) guidelines and Good Manufacturing Practices (GMPs) will be imposed in the next few years. In the near future, the FDA may ask for some sort of clinical evidence to get nutraceuticals and functional food products in to the market.

It is most appropriate to address these issues at this time, and CRC Press has given utmost importance to these issues by starting a new book series entitled "Nutraceuticals: Basic Research/Clinical Applications." The aim of this new series is to publish a range of books edited by distinguished scientists and researchers who have significant experience in scientific pursuit and critical analysis. This series will address various aspects of nutraceutical products, including the historical perspective, traditional knowledge base, analytical evaluations, and green food processing and applications. It will be useful not only for researchers and academicians but will be valuable reference books for personnel in the nutraceutical and food industries.

This series will encourage editors to come up with books within the scope of the series and in turn help nutraceutical manufacturers and academicians to address the issues in a timely manner.

We are happy to provide the first book of the series entitled *Marine Nutraceuticals: Prospects and Perspectives*. This book has been edited by Dr. Se-Kwon Kim (Pukyong National University, Busan, South Korea), who is a well-known scientist in the field. Dr. Kim has significant experience dealing with marine sciences, and this book will kick-start the series with many more to follow.

The marine environment, rich in biological as well as chemical diversity, has been the source of potential molecules used as pharmaceuticals, nutraceuticals (nutritional supplements), cosmeceuticals, molecular probes, fine chemicals, and agrochemicals. Macro- and microorganisms of marine habitat display a wide array of secondary metabolites, including terpenes, steroids, polyketides, peptides, alkaloids, polysaccharide, proteins, and porphyrins. It is well established that marine organisms contain various unique bioactive substances that can be transformed into active substances by bioconversion technology (chemical and enzymatic) through membrane bioreactors.

The book discusses recent advances in marine nutraceutical research and includes a wide range of applications, from both traditional Chinese medicinal practice, which employs more than 200 species of marine organisms, and contemporary uses, which employ only 10–20 species for medicinal purposes.

This book will clarify historic misconceptions of marine-derived medicines (traditional and oriental), provide an insight into current trends and approaches, and include future applications of marine nutraceuticals. It will also explore marine habitat for novel nutraceutical compounds, cover the current status and future of natural marine compounds, examine chitosan and body weight management, and discuss the benefits of marine-based nutraceuticals for obesity and diabetes.

Yashwant Pathak, PhD
Tampa, Florida

Preface

Recently, consumers have shown a great deal of interest in natural bioactive compounds such as nutraceuticals or functional ingredients in food products due to their various health benefits. The ongoing research on nutraceuticals worldwide will lead to a new generation of foods, which will certainly cause the interface between foods and drugs to become increasingly permeable.

Marine-derived nutraceuticals are alternative sources for synthetic ingredients that can contribute to consumers' well-being and play a vital role in human health and nutrition. They offer an abundant source of nutritionally as well as pharmacologically active agents with great chemical diversity and complexity, and the potential to produce valuable health or medicinal foods. The growing need for novel bioactives for the treatment of chronic conditions such as cancer, microbial infections, and inflammatory processes, combined with the recognition that marine nutraceuticals provide new pharmaceutically active functional ingredients. Marine macro-algae, micro-algae, blue-green algae, invertebrates, vertebrates, and marine-derived microorganisms are rich sources of active nutraceuticals and have been recognized since ancient times.

This book focuses on various types of marine-derived nutraceuticals, such as secondary metabolites like phlorotannins, phenolics, fucoxanthin, and astaxanthin as well as carotenoid pigments, polysaccharides, proteins, bioactive peptides, chito-oligosaccharide derivatives from chitin and chitosan, fucosterol, polyunsaturated fatty acids, etc., and presents an overview of their nutraceutical benefits. With the valued contributions of leading experts from Korea, Japan, Brazil, Turkey, Sri Lanka, Vietnam, Malaysia, Thailand, and Indonesia, this book provides a comprehensive account of marine-derived nutraceuticals and their potential health benefits such as antioxidant, anticancer, antiviral, anticoagulant, anti-diabetic, antiallergic, anti-inflammatory, antihypertensive, antibacterial, and radioprotective properties. It also discusses the sources, isolation and purification, chemistry, functionality interactions, applications, and industrial perspectives of a variety of marine-derived nutraceuticals. The book may be used as a text or reference for students in food science and technology, seafood science, food chemistry, and health and nutritional sciences at the senior undergraduate and graduate levels. Scientists in academia, research laboratories, marine biochemistry, natural products sciences, and industry will also find it useful.

I am grateful to the experts who have provided state-of-the-art, valued contributions to this book. I am also grateful to CRC Press for successful production of this book.

Se-Kwon Kim, PhD
Pukyong National University
Busan, Republic of Korea

Editor

Professor Se-Kwon Kim, PhD, currently works as a senior professor of marine biochemistry in the Department of Chemistry and is the director of the Marine Bioprocess Research Center (MBPRC) at Pukyong National University in the Republic of Korea. He received his BS, MS, and PhD from Pukyong National University and joined as a faculty member. He then worked as a research professor at the Bioprocess Laboratory, Department of Food Science and Technology, University of Illinois, Urbana-Champaign, Illinois (1988–1989). He later became a visiting scientist at the Memorial University of Newfoundland in Canada (1999–2000).

Professor Kim served as president of the Korean Society of Chitin and Chitosan (1986–1990) and the Korean Society of Marine Biotechnology (2006–2007). He was also the chairman for the Seventh Asia-Pacific Chitin and Chitosan Symposium, which was held in South Korea in 2006. He is one of the board members of the International Society of Marine Biotechnology and the International Society for Nutraceuticals and Functional Foods and has served as the editor in chief of the *Korean Journal of Life Sciences* (1995–1997), the *Korean Journal of Fisheries Science and Technology* (2006–2007), and the *Korean Journal of Marine Bioscience and Biotechnology* (2006–present). He has won several awards, including the Best Paper Award from the American Oil Chemist's Society (AOCS) and the Korean Society of Fisheries Science and Technology in 2002.

His major research interests include investigation and development of bioactive substances derived from marine organisms and their application in oriental medicine, nutraceuticals, and cosmeceuticals via marine bioprocessing and mass-production technologies. To date, he has authored over 500 research papers and holds 110 patents. In addition, he has written or edited more than 40 books.

Contributors

Yasantha Athukorala
School of Nutrition
Ryerson University
Toronto, Ontario, Canada

Mahanama De Zoysa
Laboratory of Aquatic Animal
    Diseases
College of Veterinary Medicine
Chungnam National
    University
Daejeon, Republic of Korea

Pradeep Dewapriya
Department of Chemistry
Pukyong National
    University
Busan, Republic of Korea

Ariyanti Suhita Dewi
Research and Development
    Center for Marine and Fisheries
    Product Processing
    and Biotechnology
Jakarta, Indonesia

Bianca F. Glauser
Institute of Medical
    Biochemistry
University Hospital Clementino
    Fraga Filho
Federal University of Rio de
    Janeiro
Rio de Janeiro, Brazil

Yoshihiko Hayashi
Department of Cardiology
Graduate School of Biomedical
    Sciences
Nagasaki University
Nagasaki, Japan

S.W.A. Himaya
Department of Chemistry
Pukyong National University
Busan, Republic of Korea

Takeshi Ikeda
Department of Cardiology
Nagasaki University Hospital
Nagasaki, Japan

Hari Eko Irianto
Research and Development
　Center for Marine and Fisheries
　Product Processing
　and Biotechnology
Jakarta, Indonesia

Jae-Young Je
School of Food Technology and
　Nutrition
Chonnam National University
Yeosu, Republic of Korea

You-Jin Jeon
School of Marine Biomedical
　Sciences
Jeju National University
Jeju, Republic of Korea

Kaipeng Jing
Department of Biochemistry
College of Medicine
and
Infection Signaling Network
　Research Center
Chungnam National University
Daejeon, Republic of Korea

Won-Kyo Jung
Department of Marine Life Science
Marine Bio Research and
　Education Center
Chosun University
Gwangju, Republic of Korea

Fatih Karadeniz
Marine Bioprocess Research Center
Pukyong National University
Busan, Republic of Korea

Mi Eun Kim
Department of Biology
College of Natural Science
Chosun University
Gwangju, Republic of Korea

Se-Kwon Kim
Department of Chemistry
Marine Bioprocess Research
　Center
Pukyong National
　University
Busan, Republic of Korea

Seok-Chun Ko
School of Marine Biomedical
　Sciences
Jeju National University
Jeju, Republic of Korea

Zenya Koyama
Department of Cardiology
Nagasaki University Hospital
Nagasaki, Japan

Maheshika S. Kurukulasuriya
Faculty of Agriculture
Department of Animal Science
University of Peradeniya
Peradeniya, Sri Lanka

Jun Sik Lee
Department of Biology
College of Natural Science
Chosun University
Gwangju, Republic of Korea

Kyu Lim
Department of Biochemistry
College of Medicine
and
Infection Signaling Network
　Research Center
and
Cancer Research Institute
Chungnam National University
Daejeon, Republic of Korea

Masahiro Matsumiya
Department of Marine Science
and Resources
College of Bioresource Sciences
Nihon University
Fujisawa, Japan

Paulo A.S. Mourão
Institute of Medical
Biochemistry
University Hospital Clementino
Fraga Filho
Federal University of Rio de
Janeiro
Rio de Janeiro, Brazil

Ngo Dang Nghia
Institute of Biotechnology and
Environment
Nha Trang University
Nha Trang, Vietnam

Dai-Hung Ngo
Department of Chemistry
Marine Bioprocess Research Center
Pukyong National University
Busan, Republic of Korea

Ratih Pangestuti
Department of Chemistry
Marine Bioprocess Research Center
Pukyong National University
Busan, Republic of Korea

Vitor H. Pomin
Institute of Medical
Biochemistry
University Hospital Clementino
Fraga Filho
Federal University of Rio de
Janeiro
Rio de Janeiro, Brazil

Siswa Setyahadi
Biocatalyst Production Technology
Division
Center for Bioindustrial Technology
Agency for the Assessment and
Application of Technology
Jakarta, Indonesia

Willem Frans Stevens
Center of Excellence for Shrimp
Biotechnology
Mahidol University
Bangkok, Thailand

Quang Van Ta
Department of Chemistry
Marine Bioprocess Research Center
Pukyong National University
Busan, Republic of Korea

Trang Si Trung
Department of Biochemistry
and Microbiology
Nha Trang University
Nha Trang, Vietnam

Janak K. Vidanarachchi
Faculty of Agriculture
Department of Animal Science
University of Peradeniya
Peradeniya, Sri Lanka

Thanh-Sang Vo
Department of Chemistry
Marine Bioprocess Research Center
Pukyong National University
Busan, Republic of Korea

Isuru Wijesekara
Department of Chemistry
Marine Bioprocess Research Center
Pukyong National University
Busan, Republic of Korea

W.M. Niluni Wijesundera
Faculty of Agriculture
Department of Animal Science
University of Peradeniya
Peradeniya, Sri Lanka

Tin Wui Wong
Faculty of Pharmacy
Non-Destructive Biomedical
    and Pharmaceutical Research
    Centre
MARA University
    of Technology
Selangor, Malaysia

Shizuka Yamada
Department of Cardiology
Graduate School of Biomedical
    Sciences
Nagasaki University
Nagasaki, Japan

Kajiro Yanagiguchi
Department of Cardiology
Nagasaki University Hospital
Nagasaki, Japan

Soon Kong Yong
Faculty of Applied Sciences
MARA University
    of Technology
Selangor, Malaysia

and

Centre for Environment Risk
    Assessment and Remediation
University of South Australia
Adelaide, South Australia,
    Australia

Yvonne V. Yuan
School of Nutrition
Ryerson University
Toronto, Ontario, Canada

Marine-Derived Nutraceuticals
Trends and Prospects

Se-Kwon Kim and Isuru Wijesekara

Contents

1.1 Introduction

Recently, a great deal of interest has been paid by the consumers toward natural bioactive compounds as functional ingredients or nutraceuticals. Especially, bioactive compounds derived from marine organisms have served as a rich source of health-promoting components. As more than 70% of the world's surface is covered by oceans, the wide diversity of marine organisms offer a rich source of natural products. Marine environment contains a source of functional materials, including polyunsaturated fatty acids (PUFA), polysaccharides, essential minerals and vitamins, phenolic phlorotannins, enzymes, and bioactive peptides (Barrow and Shahidi, 2008; Kim and Wijesekara, 2010; Ngo et al., 2011; Wijesekara and Kim, 2010; Wijesekara et al., 2010, 2011).

The term "nutraceutical" was coined in 1989 by the Foundation for Innovation in Medicine (New York) to provide a name for this rapidly growing area of biomedical research. A nutraceutical was defined as any substance that may be considered a food or a part of a food and provides medical or health benefits including the prevention and treatment of disease (Andlauer and Furst, 2002). Nutraceuticals possess pertinent physiological functions and valuable biological activities. Interestingly, during the last 2000 years, from the time of Hippocrates (460–377 BC) to the dawn of modern medicine, little distinction was made between food and drugs. Furthermore, Hippocrates clearly recognized the essential relationship between food and health and emphasized that "... differences of diseases depend on nutriment." Nutraceuticals, as defined by Zeisel (1999), are dietary supplements that deliver a concentrated form of a presumed bioactive agent from a food, presented in a nonfood matrix, and used with the purpose of enhancing health in dosages that exceed those that could be obtained from normal foods. The ongoing researches of nutraceuticals in the world will lead to a new generation of foods, which will certainly cause the interface between food and drug to become interestingly permeable. The present accumulated knowledge about nutraceuticals represents undoubtedly a great challenge for nutritionists, physicians, food technologists, and food chemists. Functional foods are those that when consumed regularly produce a specific beneficial health effect beyond their nutritional properties. The boundary between nutraceuticals and functional foods is not always clear, the main difference being the format in which they are consumed: nutraceuticals are consumed as capsules, pills, tablets, etc., while functional foods are always consumed as ordinary foods. Thus, when a phytochemical is included, a food formulation is considered a functional food. If the same phytochemical is included in a capsule, it will constitute a nutraceutical. There is a great potential in the marine bioprocess industry to convert and utilize most of the marine food products and marine food by-products as valuable functional ingredients. Apparently, there has been an increasing interest in the utilization of marine products, and novel bioprocessing technologies are being developed for isolation of some bioactive substances with antioxidative property from marine food products to be used as functional foods and nutraceuticals. This chapter focuses on marine-derived nutraceuticals and presents their sources, potential health benefits, and industrial perspectives.

1.2 Types of marine nutraceuticals

Marine-derived bioactive peptides have been obtained widely by enzymatic hydrolysis of marine proteins (Kim and Wijesekara, 2010) and have shown to possess many physiological functions, including antioxidant, antihypertensive or ACE inhibition, anticoagulant, and antimicrobial activities. In fermented marine food sauces, such as blue mussel sauce and oyster sauce, enzymatic hydrolysis has already been done by microorganisms, and bioactive peptides can be purified without further hydrolysis. In addition, marine processing by-products contain bioactive peptides with valuable functional properties.

Sulfated polysaccharides (SPs) not only are chemically anionic and widespread in marine algae but also occur in animals, such as mammals and invertebrates. Marine algae are the most important source of nonanimal SPs, and their chemical structures vary according to the species of algae, such as fucoidan in brown algae (Phaeophyceae), carrageenan in red algae (Rhodophyceae), and ulvan in green algae (Chlorophyceae).

Phlorotannins are phenolic compounds formed by the polymerization of phloroglucinol or defined as 1,3,5-trihydroxybenzene monomer units and biosynthesized through the acetate–malonate pathway. They are highly hydrophilic components with a wide range of molecular sizes between 126 and 650,000 Da. Marine brown algae accumulate a variety of phloro-glucinol-based polyphenols, as phlorotannins could be used as functional ingredients in nutraceuticals with potential health effects (Wijesekara et al., 2010).

Marine lipids provide unique health benefits to consumers and are highly prone to oxidation. Fucosterol, a phytosterol found in brown seaweeds, is well recognized for its health-beneficial biological activities, such as antioxidative, cholesterol-reducing, and antidiabetic properties. Fucosterol obtained from the n-hexane fraction of *Pelvetia siliquosa* (Phaeophyceae) is effective against free radical and CCl_4-induced hepatotoxicity in vivo.

1.3 Health benefits of marine nutraceuticals

Marine nutraceuticals might have a positive effect on human health as they can protect human body against damage by reactive oxygen species (ROS), which attack macromolecules such as membrane lipids, proteins, and DNA and lead to many health disorders such as cancer, diabetes mellitus, neuro-degenerative and inflammatory diseases with severe tissue injuries. Recently, chito oligosaccharides (COS) have been the subject of increased attention in terms of their pharmaceutical and medicinal applications (Kim and Mendis, 2006), due to their missing toxicity and high solubility, as well as their positive physiological effects such as antioxidant, ACE enzyme inhibition, antimicrobial, anticancer, antidiabetic, hypocholesterolemic, hypoglycemic, anti-Alzheimer's, anticoagulant properties, and adipogenesis inhibition.

Carotenoids are thought to be responsible for the beneficial properties in preventing human diseases, including cardiovascular diseases, cancer, and other chronic diseases. Moreover, marine-derived sterols have received much attention in the last few years because of their cholesterol-lowering prop-erties. Further, marine algal-derived SPs exhibited various health-beneficial biological activities such as anti-HIV-1, anticoagulant, immunomodulating, and anticancer activities (Wijesekara et al., 2011). Moreover, some bioactive peptides from marine organisms have been identified to possess nutraceutical potentials for human health promotion and disease risk reduction (Shahidi and Zhong, 2008), and recently the possible roles of food-derived bioactive peptides in reducing the risk of cardiovascular diseases have been demon-strated (Erdmann et al., 2008). In addition, saringosterol, a derivative of fucos-terol, discovered in several brown algae (Phaeophyceae), such as *Lessonia nigrescens* and *Sargassum ringgoldianum*, has been shown to inhibit the growth of *Mycobacterium tuberculosis*.

1.4 Conclusions

With so many new species of marine resources still to be discovered, the potential for new marine-derived bioactive nutraceuticals is immense with beneficial effects on human health, and the food industry is poised for accelerated development in the near future. Marine resources have been well recognized for their biologically active substances with a great potential to be used as nutraceuticals. Moreover, much attention has been paid recently by the consumers toward healthy lifestyle with natural bioactive ingredients. Recent studies have provided evidence that marine-derived bioactive nutraceuticals play a vital role in human health.

This chapter discusses the recent trends, findings, and prospects of marine-derived potential nutraceuticals.

References

Andlauer, W. and Furst, P. (2002). Nutraceuticals: A piece of history, present status and outlook. *Food Research International*, 35, 171–176.

Barrow, C. and Shahidi, F. (2008). *Marine Nutraceuticals and Functional Foods*, CRC Press, Boca Raton, FL.

Erdmann, K., Cheung, B. W. Y., and Schroder, H. (2008). The possible roles of food-derived bioactive peptides in reducing the risk of cardiovascular disease. *Journal of Nutritional Biochemistry*, 19, 643–654.

Kim, S. K. and Mendis, E. (2006). Bioactive compounds from marine processing byproducts—A review. *Food Research International*, 39, 383–393.

Kim, S. K. and Wijesekara, I. (2010). Development and biological activities of marine-derived bioactive peptides: A review. *Journal of Functional Foods*, 2, 1–9.

Ngo, D. H., Wijesekara, I., Vo, T. S., Ta, Q. V., and Kim, S. K. (2011). Marine food-derived functional ingredients as potential antioxidants in the food industry: An overview. *Food Research International*, 44, 523–529.

Shahidi, F. and Zhong, Y. (2008). Bioactive peptides. *Journal of AOAC International*, 91, 914–931.

Wijesekara, I. and Kim, S. K. (2010). Angiotensin-I-converting enzyme (ACE) inhibitors from marine resources: Prospects in the pharmaceutical industry. *Marine Drugs*, 8, 1080–1093.

Wijesekara, I., Pangestuti, R., and Kim, S. K. (2011). Biological activities and potential health benefits of sulfated polysaccharides derived from marine algae. *Carbohydrate Polymers*, 84, 14–21.

Wijesekara, I., Yoon, N. Y., and Kim, S. K. (2010). Phlorotannins from *Ecklonia cava* (Phaeophyceae): Biological activities and potential health benefits. *Biofactors*, 36, 408–414.

Zeisel, S. H. (1999). Regulation of 'nutraceuticals'. *Science*, 285, 1853–1855.

2

Nutritional Value of Sea Lettuces

Se-Kwon Kim and Ratih Pangestuti

Contents

2.1 Introduction

Seaweeds represent one of the most nutritious plant foods, and general utilization of seaweeds in food products has grown steadily since the early 1980s (Besada et al., 2009). The ancient tradition as well as daily intake of seaweeds has made possible a large number of epidemiological studies showing the health benefits of seaweed consumption (Pisani et al., 2002; Yuan and Walsh, 2006). Furthermore, Jorm and Jolley (1998) reported that neurodegenerative cases in East Asian countries were lower than that in Europe ($p < 0.0004$). Studies showed that less number of cancer and neurodegenerative disease cases in the Eastern hemisphere is associated with large amounts of seaweed consumption. In Asian culture, seaweeds have always been of particular interest as food sources (Khan et al., 2010). In recent years, consumers in developed countries are turning toward more natural and nutritional products such as seaweeds (Van Netten et al., 2000). Seaweeds have recently been approved in France for human consumption, thus providing an opening for the food and fisheries industries. During 2003, it was estimated that about 1 million ton of seaweeds was harvested in 35 countries mainly as food sources (Garcia-Casal et al., 2009).

Several seaweed species are consumed by humans directly after only minor preprocessing such as drying. *Porphyra* sp. which is commercially known as nori or laver is most widely consumed among edible red seaweeds worldwide (Watanabe et al., 1999). Among green seaweeds, sea lettuces are most common, ubiquitous, and environmentally important genera (Tan et al., 1999). Sea lettuces comprise the genus *Ulva*, a group of edible green seaweeds which is widely distributed along the coasts of the world's oceans and often found in the mid and upper tidal zones. Sea lettuces or sometimes termed as green lavers are found in tidal and near tidal seawater worldwide, generally anchored to rocks or other algae. They are easily identified by their paper-thin, semitranslucent, and vibrant green color. Most sea lettuces are gathered wild as they grow prolifically wherever there are sufficient nutrients, but some are farmed. Many species of sea lettuces are reported to be tolerant to organic and metal pollution; hence, if we consume them, we need to make sure they are collected far from any potential sources of pollution.

There are a number of reviews available on the pharmaceuticals and medicinal bioactive compounds derived from marine algae. This chapter focuses specifically on the culinary use, nutritional value of sea lettuces, emphasizing their associated health-promoting effects. Furthermore, it is important to acknowledge that there are gaps in our knowledge of local names for sea lettuces, some of which also lack common names in English. Hence, in this chapter we present several local names for sea lettuces in several countries.

2.2 Nutritional value of sea lettuces

2.2.1 Polysaccharides

Sea lettuces contain large amounts of polysaccharides, which constitute around 38%–54% of the dry matter. These include four polysaccharide families in sea lettuces: two major ones, the water-soluble ulvan and insoluble cellulose, and two minor ones, xyloglucan and glucuronan (Lahaye and Robic, 2007). When faced with the human intestinal bacteria, most of these polysaccharides are not digested by humans, and hence, can be regarded as dietary fibers. Water-soluble and insoluble fibers have been associated with different biological activities and health-promoting effects. Soluble polysaccharides are primarily associated with hypocholesterolemic and hypoglycemic effects (Panlasigui et al., 2003). Furthermore, insoluble fibers, such as cellulose are associated with excretion of bile acids, increased fecal bulk, and reduced intestinal transit time (Burtin, 2003; Moore et al., 1998).

Among the polysaccharides isolated from sea lettuces, ulvan has attracted greater attention as it displays several physicochemical and biological features. The name ulvan was actually derived from the terms ulvin and ulvacin, which refer to different fractions of *Ulva lactuca* water-soluble polysaccharides. Presently, it is being used to refer to the polysaccharides from the members of the Ulvales, mainly sea lettuces. Researchers have revealed that ulvan exhibits various biological activities, such as anticoagulant, antiviral, antioxidant, antiallergic, anticancer, antiinflammatory, antihyperlipidemia, etc. For example, Qi et al. (2005) have prepared ulvans of different molecular weights from *Ulva pertusa* using sulfur trioxide/*N,N*-dimethylformamide (SO_3–DMF) in formamide, and their antioxidant activities were investigated. The results showed that low molecular weight ulvans have a strong antioxidant activity. The rationale for this is that the low molecular weight of ulvan may incorporate into the cells more efficiently and donate proton effectively compared to the high-molecular weight one (Qi et al., 2005). Ulvan may also modulate lipid metabolism in rats and mice. A decrease of serum high-density lipoprotein cholesterol (HDL-cholesterol) and an increase of low-density lipoprotein cholesterol (LDL-cholesterol) and triglyceride are considered to be significant risk factors in cardiovascular diseases. Ulvan or ulvan-derived oligosaccharides significantly lowered the level of serum total cholesterol, LDL-cholesterol, and reduced triglyceride, while they increased the levels of serum HDL-cholesterol (Pengzhan et al., 2003). In addition, Mao et al. (2006) found that the anticoagulant activity of ulvan from *Ulva conglobata* mainly consisted of rhamnose, with variable contents of glucose and fucose, trace amounts of xylose, galactose, and mannose. Anticoagulant activity of ulvan has also been reported from *U. lactuca* (AhdEl-Baky et al., 2009). In comparison, *Ulva neumatoidea* extracts have higher anticoagulant activity compared to other seaweed species, such as *Egregia menziesii, Silvetia compressa*, and *Codium fragile* (Guerra-Rivas et al., 2010).

These biological properties of ulvan open up a wide field of potential applications in food, pharmaceutical, agricultural, cosmetic, and chemical applications. Some of these potential applications are already the subject of patents.

2.2.2 Protein and amino acids

The protein content of sea lettuces varies with the species, but is generally present in high amounts. For example, protein content in *Ulva reticulata* is 21.06% of the dry weight, whereas higher protein contents (27.2% of the dry weight) are recorded in *U. lactuca* (Ortiz et al., 2006; Ratana-arporn and Chirapart, 2006). These levels are comparable to those found in high-protein terrestrial vegetables, such as soybeans, in which protein makes up 40% of the dry mass (Murata and Nakazoe, 2001).

The general levels of some amino acids in sea lettuce proteins are higher than those found in terrestrial plants. Eight essential amino acids (cysteine, isoleucine, leucine, lysine, methionine, phenylalanine, tyrosine, and valine), which cannot be synthesized by our body are present in a high level in sea lettuces. The amino acid compositions of sea lettuces are presented in **Table 2.1**.

TABLE 2.1 Comparison of Amino Acid Composition of Some Sea Lettuce Species (g/100g Dry Basis)

Amino Acids	Ulva lactuca[a]	Ulva reticulata[b]	Ulva pertusa[c]	Ulva armoricana[d]
Histidine	1.8	0.23	4.00	2.10
Isoleucine	6.1	0.90	3.50	2.99
Leucine	9.2	1.68	6.90	5.92
Methionine	1.8	—	1.60	2.58
Phenylalanine	6.3	1.12	3.90	7.10
Threonine	4.6	1.15	3.10	6.88
Tryptophan	—	—	0.30	—
Valine	7.7	1.34	4.90	5.01
Lysine	6.3	1.28	4.50	4.01
Alanine	8.5	1.72	6.10	7.05
Arginine	5.1	1.84	14.9	6.28
Aspartic acid	9.2	2.66	6.50	6.09
Cysteine	2.2	—	—	—
Glycine	7	1.38	5.20	6.34
Glutamic acid	10	2.76	6.90	18.24
Proline	5.2	1.08	4.90	6.92
Hydroxyproline	—	—	—	1.89
Serine	4	1.36	3.0	5.92
Tyrosine	—	0.77	1.40	4.76

[a] Adapted from Mai et al. (1994).
[b] Adapted from Ratana-arporn and Chirapart (2006).
[c] Adapted from Fujiwara-Arasaki et al. (1984).
[d] Adapted from Fleurence et al. (1999).

Several sea lettuce species, such as *U. lactuca*, *U. pertusa*, and *Ulva armoricana* are rich in leucine. Leucine is one of the building blocks for protein, and recent studies have reported that a diet rich in the amino acid leucine might help prevent the muscle loss that typically comes with aging (Anthony et al., 2000). French researchers found that a leucine-supplemented diet restored a more youthful pattern of muscle-protein breakdown and synthesis in elderly rats. In addition to leucine, other amino acids that are found in high amounts in sea lettuces are threonine, arginine, alanine, aspartic acid, and glutamic acid (Fleurence et al., 1999; Fujiwara-Arasaki et al., 1984; Mai et al., 1994; Ratana-arporn and Chirapart, 2006; Wong and Cheung, 2000). The proteins from *U. reticulata* and *U. armoricana* exhibit an amino acid composition close to that of soybean protein (Fleurence et al., 1999; Ratana-arporn and Chirapart, 2006). Furthermore, *U. reticulata* proteins are of high quality since the essential amino acids represented almost 40% of total amino acids (Ratana-arporn and Chirapart, 2006).

Bioactive lectins are found in sea lettuces (Sampaio et al., 1998; Wang et al., 2004). However, lectins derived from sea lettuces are relatively the recent ones compared to other seaweeds. In human body, lectins are involved in numerous biological processes, such as cell–cell communication, induction of apoptosis, host–pathogen interaction, cancer metastasis and differentiation, recognizing and binding carbohydrates, increasing the agglutination of blood cells (erythrocytes), detection of disease-related alterations of glycan synthesis, including infectious agents, such as viruses (Holdt and Kraan, 2011).

2.2.3 Lipids and fatty acids

Lipids are a broad group of naturally occurring molecules, which includes fats, waxes, sterols, fat-soluble vitamins, mono-, di-, and triacylglycerol, diglycerides, phospholipids, etc. The literature has established that in general, the lipid content in sea lettuces is less than 4% (Ortiz et al., 2006). Total lipid contents in *U. lactuca*, *U. reticulata*, *Ulva fasciata* were 1.64, 0.75, 3.6 g/100 g, respectively (McDermid and Stuercke, 2003; Ratana-arporn and Chirapart, 2006; Wong and Cheung, 2000).

Furthermore, sea lettuces show an interesting polyunsaturated fatty acid (PUFA) composition (**Table 2.2**). There are two major families of dietary PUFA, the (ω6) and (ω3) families. The (ω6) PUFA are derived from the parent compound linoleic acid [LA; 18:2(ω6)]. They are fatty acids containing at least two double bonds where the first double bond is located six carbons from the methyl end of the molecule (Whelan and McEntee, 2004). Meanwhile, the (ω3) PUFA have the first double bond located at the third carbon from the methyl terminus and contain up to six double bonds. Sea lettuces are particularly rich in ω3 fatty acids (Ortiz et al., 2006). Eicosapentanoic acid (EPA; 20:5) and docosahexanoic acid (DHA; 22:6) are the two important fatty acids of sea lettuces, along with the precursor α-linolenic acid (ALA; 18:3). Both EPA and DHA are basically derived from ALA through elongation and desaturation (Alamsjah et al., 2008; Ortiz et al., 2006; Ratana-arporn and Chirapart, 2006). The ω3 fatty acids have been demonstrated to play significant role in human body. In human body, the beneficial effect of ω3 fatty acids can be classified into two main areas. First, these fatty acids sustain normal healthy life through the reduction of blood pressure, plasma triglycerides, and cholesterol, together with increased blood coagulation time. Both, EPA and DHA are important for the maintenance of normal blood flow as they lower fibrinogen levels and also prevent platelet from sticking to each other. Second, they alleviate certain diseases, such as blood vessel disorders and inflammatory conditions. Deficiency of ω3 fatty acids causes several disorders, such as restrictive growth, abnormality of the skin and hair, damage of reproductive system, and abnormal composition of serum and tissue fatty acids. The human body cannot synthesize ω3 fatty acids de novo; hence, to obtain their potential health-promoting effects, ω3 fatty acids should be introduced in

TABLE 2.2 Fatty Acid Profiles of Some Sea Lettuce Species

Fatty Acids	Ulva lactuca[a]	Ulva reticulata[b]	Ulva fasciata[c]	Ulva pertusa[c]	Ulva arasakii[c]	Ulva conglobata[c]
C10:0 (decanoic acid)	—	—	0.78	0.96	0.29	0.55
C12:0 (lauric acid)	0.14	—	—	—	—	—
C14:0 (myristic acid)	1.14	—	0.70	0.68	0.44	1.03
C14:1 (myristoleic acid)	—	—	1.81	2.07	0.87	3.35
C15:0 (pentadecanoic acid)	0.2	—	—	—	—	—
C16:0 (palmitic acid)	14	1.43	29.32	27.36	25.43	34.16
C16:1 (palmitoleic acid)	0.69	0.32	—	—	—	—
C18:0 (stearic acid)	8.39	0.92	0.91	1.03	—	2.39
C18:1ω9 (oleic acid)	0.37	0.13	5.12	5.15	7.4	6.31
C18:2ω6 (linoleic acid)	8.31	0.14	7.87	8.09	21	6.81
C18:3ω3 (linolenic acid)	4.38	0.19	17.25	17.96	22.98	14.37
C18:4ω3 (stearidonic acid)	0.41	—	—	—	—	—
C20:0 (arachidate)	0.19	0.11	—	—	—	—
C20:1 (eicosanoate)	4.21	0.06	1.98	1.35	0.42	1.77
C20:4ω6 (arachidonic acid)	0.34	0.04	—	—	—	—
C20:5ω3 (eicosapentanoic acid)	1.01	0.03	—	—	—	—
C22:0 (behanate)	0.27	0.03	0.80	0.65	2.87	2.5
C22:1 (erucate)	0.79	0.003	2.46	2.88	1.42	2.57
C22:6ω3 (docosahexaenoic acid)	0.8	0.04	—	—	—	—

[a] Adapted from Ortiz et al. (2006).
[b] Adapted from Ratana-arporn and Chirapart (2006).
[c] Adapted from Alamsjah et al. (2008).

human diet. One of the potential sources of EPA, DHA, and ALA is sea lettuces. In addition, sea lettuces also contain significant quantities of C18:4ω3 (stearidonic acid), which has recently been demonstrated to possess several biological activities similar to EPA (Whelan, 2009). *U. lactuca* is the best source of EPA and DHA among several sea lettuce species tested in several studies. Meanwhile, *Ulva arasakii* is a better source of palmitic acid. Sea lettuces are therefore good sources of ω3 fatty acids and also important sources of supply of ω3 fatty acids for homeostasis and in promoting human health.

2.2.4 Vitamins

Sea lettuces contain considerable amounts of vitamins. These include both water- and fat-soluble vitamins, such as vitamins A, B, D, and E (**Table 2.3**).

2.2.4.1 Vitamin B complex

Sea lettuces are sources of vitamins from group B (MacArtain et al., 2007; McDermid and Stuercke, 2003). For instance, *U. lactuca* contains large amounts of cobalamin or vitamin B_{12}. Vitamin B_{12} plays a key role in homeostasis of the brain and nervous system, and in the formation of blood (Scalabrino, 2009). Daily ingestion of 1.4 g of *U. lactuca* will be enough to meet the daily requirements of vitamin B_{12} (MacArtain et al., 2007). One of the most important vitamin B occurring in *U. reticulata* is riboflavin (vitamin B_2). Vitamin B_2 deficiency is often endemic in human populations that subsist on diets poor in dairy products and meat. Vitamin B_2 cannot be synthesized by mammals, and there is only limited, short-term storage capacity for this vitamin in the liver. Humans are vulnerable to develop a vitamin B_2 deficiency during periods of dietary deprivation or stress, and this may lead to a variety of clinical abnormalities, such as growth retardation, anemia, skin lesions, and degenerative changes in the nervous system. Therefore, this water-soluble vitamin should be added to the diet on a daily basis (Van Herwaarden et al., 2007).

2.2.4.2 Vitamin C

Water-soluble vitamins such as vitamin C are present in large amounts in sea lettuces. The levels of vitamin C in sea lettuces average from 500 to 3000 mg/kg

TABLE 2.3 Vitamin Contents of Some Sea Lettuce Species (mg/100 g Edible Portion *Except for Vitamins A and D in IU)

Vitamins	Ulva lactuca[a,b]	Ulva reticulata[c]	Ulva fasciata[d]
A*	6050	—	—
B_1	0.42	10	—
B_2	0.03	13	0.1
B_3	8	—	6.6
B_9	0.01	—	—
B_{12}	6.3	—	—
C	10	—	22
D*	848	—	—
E	13.7	—	—

[a] Adapted from Briand and Morand (1997).
[b] Adapted from MacArtain et al. (2007).
[c] Adapted from Ratana-arporn and Chirapart (2006).
[d] Adapted from McDermid and Stuercke (2003).

of dry matter. These levels of vitamin C are comparable with that of parsley, blackcurrant, and peppers. In sea lettuces, the highest level of vitamin C was found in *U. fasciata* (22 mg/100 g) (McDermid and Stuercke, 2003). Vitamin C is of interest for many reasons. First, it strengthens the immune defense system, activates the intestinal absorption of iron, acts as a reversible reductant and antioxidant in the aqueous fluid and tissue compartments. Furthermore, this vitamin is specifically required for the activity of eight human enzymes involved in collagen, hormone, amino acid, and carnitine synthesis or metabolism (Jacob and Sotoudeh, 2002).

2.2.4.3 Vitamin E

Burtin (2003) reported that green seaweeds contain low amounts of vitamin E (Burtin, 2003). In contrast, Ortiz et al. (2006) demonstrated that *U. lactuca* showed high levels of vitamin E. In accordance, daily ingestion of 109.5 g/day of *U. lactuca* will be enough to meet the daily requirements of vitamin E (Briand and Morand, 1997). The determined levels of vitamin E show a good nutritional complement that confirms the importance of sea lettuces in the normal diets for humans.

2.2.5 Minerals

Minerals are inorganic elements that retain their chemical identity in foods. It can be classified into two groups: macro (calcium, phosphorus, potassium, sulfur, sodium, chloride, and magnesium) and trace minerals. Sea lettuces, which are drawn from the sea contain wealth of mineral elements (**Table 2.4**). Calcium, one of the most important minerals essential for human body is accumulated in sea lettuces at a higher level when compared to milk, brown rice, spinach, peanuts, and lentils (MacArtain et al., 2007). Calcium contents in *U. lactuca*, *U. reticulata*, *U. fasciata* were 32.5, 147, and 0.47 mg/100 g edible portion, respectively. Moreover, potassium and sodium are known as electrolytes because of their ability to dissociate into positively and negatively charged ions when dissolved in water. Potassium is the major cation of intracellular fluid. Together with sodium, it maintains normal water balance. In addition, potassium also promotes cellular growth and maintains normal blood pressure. Potential sources of potassium are *U. reticulata*, which contains 1540 mg potassium per 100 g edible portion (Ratana-arporn and Chirapart, 2006).

Iodine is an important mineral in metabolic regulation and growth patterns. The recommended daily intake of iodine for adults is 150 μg/day. During pregnancy and lactation, additional doses of 25 and 50 μg/day, respectively are recommended. Notably, the iodine deficiency is prevalent worldwide, which corresponds to the worldwide phenomena of brain damage and mental retardation. During pregnancy, infancy, and childhood it may lead to endemic

TABLE 2.4 Mineral Contents of Some Sea Lettuce Species (mg/100 g Edible Portion)

Minerals	Ulva lactuca[a]	Ulva reticulata[b]	Ulva fasciata[c]
Ca	32.5	140	0.47
K	24.5	1540	2.87
Mg	46.5	140	2.19
Na	34	—	—
Cu	0.03	0.6	5
Fe	1.53	174.8	86
I	0.16	1.124	—
Zn	0.09	3.3	9
N	—	—	3.62
P	—	180	0.22
S	—		5.24
B	—	—	77
Mn	—	48.1	12

[a] Adapted from MacArtain et al. (2007).
[b] Adapted from Ratana-arporn and Chirapart (2006).
[c] Adapted from McDermid and Stuercke (2003).

and irreversible cretinism in infants or children. *U. reticulata* and *U. lactuca* have been described as good sources of iodine. Hence, considering the high mineral contents, sea lettuces could be used as food supplements to meet the daily intake of essential minerals.

2.3 Conclusions

Sea lettuces are rich in nutrients with medicinal and health-promoting effects. From a nutritional standpoint, the main properties of sea lettuces are their richness in polysaccharides, proteins and amino acids, fatty acids, minerals, and vitamins. Therefore, their nutritional values make them valuable food supplements. Furthermore, sea lettuces may be used to fortify processed foods. Food preparation from sea lettuces worldwide may be studied to increase sea lettuce utilization. Moreover, recognition of sea lettuces as sources of diverse bioactive principles may open the medicinal potential of sea lettuces, and there is a great potential to be used in pharmaceuticals. Therefore, combination between culinary use and research on bioactive compounds may revitalize the use of sea lettuces in the new health-conscious consumers. Sea lettuce products could be used for food fortification, enrichment, and multipurpose applications.

References

AhdEl-Baky, H.H., F.K.E. Gamai, and S. ElBaroîy. 2009. Potential biological properties of sulphated polysaccharides extracted from the macroalgae *Ulva lactuca* L. *Academic Journal of Cancer Research* 2:1–11.

Alamsjah, M.A., S. Hirao, F. Ishibashi, T. Oda, and Y. Fujita. 2008. Algicidal activity of polyunsaturated fatty acids derived from *Ulva fasciata* and *U. pertusa* (Ulvaceae, Chlorophyta) on phytoplankton. *Journal of Applied Phycology* 20:713–720.

Anthony, J.C., T.G. Anthony, S.R. Kimball, T.C. Vary, and L.S. Jefferson. 2000. Orally administered leucine stimulates protein synthesis in skeletal muscle of postabsorptive rats in association with increased eIF4F formation. *Journal of Nutrition* 130:139.

Besada, V., J.M. Andrade, F. Schultze, and J.J. González. 2009. Heavy metals in edible seaweeds commercialised for human consumption. *Journal of Marine Systems* 75:305–313.

Briand, X. and P. Morand. 1997. Anaerobic digestion of *Ulva* sp. 1. Relationship between *Ulva* composition and methanisation. *Journal of Applied Phycology* 9(6):511–524.

Burtin, P. 2003. Nutritional value of seaweeds. *Electronic Journal of Environmental, Agricultural and Food Chemistry* 2:498–503.

Fleurence, J., E. Chenard, and M. Luçcon. 1999. Determination of the nutritional value of proteins obtained from *Ulva armoricana*. *Journal of Applied Phycology* 11:231–239.

Fujiwara-Arasaki, T., N. Mino, and M. Kuroda. 1984. The protein value in human nutrition of edible marine algae in Japan. *Hydrobiologia* 116:513–516.

Garcia-Casal, M.N., J. Ramirez, I. Leets, A.C. Pereira, and M.F. Quiroga. 2009. Antioxidant capacity, polyphenol content and iron bioavailability from algae (*Ulva* sp., *Sargassum* sp. and *Porphyra* sp.) in human subjects. *British Journal of Nutrition* 101:79–85.

Guerra-Rivas, G., C.M. Gómez-Gutiérrez, G. Alarcón-Arteaga, I.E. Soria-Mercado, and N.E. Ayala-Sánchez. 2010. Screening for anticoagulant activity in marine algae from the Northwest Mexican Pacific coast. *Journal of Applied Phycology* 23:495–503.

Holdt, S.L. and S. Kraan. 2011. Bioactive compounds in seaweed: Functional food applications and legislation. *Journal of Applied Phycology* 3: 543–597.

Jacob, R.A. and G. Sotoudeh. 2002. Vitamin C function and status in chronic disease. *Nutrition in Clinical Care* 5:66–74.

Jorm, A.F. and D. Jolley. 1998. The incidence of dementia: A meta-analysis. *Neurology* 51:728–733.

Khan, S.B., C.S. Kong, J.A. Kim, and S.K. Kim. 2010. Protective effect of Amphiroa dilatata on ROS induced oxidative damage and MMP expressions in HT1080 cells. *Biotechnology and Bioprocess Engineering* 15:191–198.

Lahaye, M. and A. Robic. 2007. Structure and functional properties of ulvan, a polysaccharide from green seaweeds. *Biomacromolecules* 8:1765–1774.

MacArtain, P., C.I.R. Gill, M. Brooks, R. Campbell, and I.R. Rowland. 2007. Nutritional value of edible seaweeds. *Nutrition Reviews* 65:535–543.

Mai, K., J.P. Mercer, and J. Donlon. 1994. Comparative studies on the nutrition of two species of abalone, *Haliotis tuberculata* L. and *Haliotis discus hannai* Ino: II. Amino acid composition of abalone and six species of macroalgae with an assessment of their nutritional value. *Aquaculture* 128:115–130.

Mao, W., X. Zang, Y. Li, and H. Zhang. 2006. Sulfated polysaccharides from marine green algae *Ulva conglobata* and their anticoagulant activity. *Journal of Applied Phycology* 18:9–14.

McDermid, K.J. and B. Stuercke. 2003. Nutritional composition of edible Hawaiian seaweeds. *Journal of Applied Phycology* 15:513–524.

Moore, M.A., C.B. Park, and H. Tsuda. 1998. Soluble and insoluble fiber influences on cancer development. *Critical Reviews in Oncology/Hematology* 27:229–242.

Murata, M. and J. Nakazoe. 2001. Production and use of marine algae in Japan. *Japan Agricultural Research Quarterly* 35:281–290.

Ortiz, J., N. Romero, P. Robert et al. 2006. Dietary fiber, amino acid, fatty acid and tocopherol contents of the edible seaweeds *Ulva lactuca* and *Durvillaea antarctica*. *Food Chemistry* 99:98–104.

Panlasigui, L.N., O.Q. Baello, J.M. Dimatangal, and B.D. Dumelod. 2003. Blood cholesterol and lipid-lowering effects of carrageenan on human volunteers. *Asia Pacific Journal of Clinical Nutrition* 12:209–214.

Pengzhan, Y., Z. Quanbin, L. Ning, X. Zuhong, W. Yanmei, and L. Zhi'en. 2003. Polysaccharides from *Ulva pertusa* (Chlorophyta) and preliminary studies on their antihyperlipidemia activity. *Journal of Applied Phycology* 15:21–27.

Pisani, P., F. Bray, and D.M. Parkin. 2002. Estimates of the world-wide prevalence of cancer for 25 sites in the adult population. *International Journal of Cancer* 97:72–81.

Qi, H., Q. Zhang, T. Zhao et al. 2005. Antioxidant activity of different sulfate content derivatives of polysaccharide extracted from *Ulva pertusa* (Chlorophyta) in vitro. *International Journal of Biological Macromolecules* 37:195–199.

Ratana-arporn, P. and A. Chirapart. 2006. Nutritional evaluation of tropical green seaweeds *Caulerpa lentillifera* and *Ulva reticulata*. *Kasetsart Journal: Natural Sciences* 40:75–83.

Sampaio, A.H., D.J. Rogers, and C.J. Barwell. 1998. Isolation and characterization of the lectin from the green marine alga *Ulva lactuca* L. *Botanica Marina* 41:427–434.

Scalabrino, G. 2009. The multi-faceted basis of vitamin B12 (cobalamin) neurotrophism in adult central nervous system: Lessons learned from its deficiency. *Progress in Neurobiology* 88:203–220.

Tan, I.H., J. Blomster, G. Hansen et al. 1999. Molecular phylogenetic evidence for a reversible morphogenetic switch controlling the gross morphology of two common genera of green seaweeds, Ulva and Enteromorpha. *Molecular Biology and Evolution* 16:1011–1018.

Van Herwaarden, A.E., E. Wagenaar, G. Merino et al. 2007. Multidrug transporter ABCG2/breast cancer resistance protein secretes riboflavin (vitamin B2) into milk. *Molecular and Cellular Biology* 27:1247–1253.

Van Netten, C., S.A. Hoption Cann, D.R. Morley, and J.P. Van Netten. 2000. Elemental and radioactive analysis of commercially available seaweed. *Science of the Total Environment* 255:169–175.

Wang, S., F.D. Zhong, Y.J. Zhang, Z.J. Wu, Q.Y. Lin, and L.H. Xie. 2004. Molecular characterization of a new lectin from the marine alga *Ulva pertusa*. *Acta Biochimica et Biophysica Sinica* 36:111–117.

Watanabe, F., S. Takenaka, H. Katsura et al. 1999. Dried green and purple lavers (Nori) contain substantial amounts of biologically active vitamin B12 but less of dietary iodine relative to other edible seaweeds. *Journal of Agricultural and Food Chemistry* 47:2341–2343.

Whelan, J. 2009. Dietary stearidonic acid is a long chain (n-3) polyunsaturated fatty acid with potential health benefits. *Journal of Nutrition* 139:5–10.

Whelan, J. and M.F. McEntee. 2004. Dietary (n-6) PUFA and intestinal tumorigenesis. *Journal of Nutrition* 134:3421S.

Wong, K.H. and P.C.K. Cheung. 2000. Nutritional evaluation of some subtropical red and green seaweeds: Part I—Proximate composition, amino acid profiles and some physico-chemical properties. *Food Chemistry* 71:475–482.

Yuan, Y.V. and N.A. Walsh. 2006. Antioxidant and antiproliferative activities of extracts from a variety of edible seaweeds. *Food and Chemical Toxicology* 44:1144–1150.

3

Prospects and Potential Applications of Seaweeds as Neuroprotective Agents

Se-Kwon Kim and Ratih Pangestuti

Contents

3.1 Introduction

The oceans cover more than 70% of the Earth's surface with marine species comprising approximately half of the total global biodiversity (Kim and Wijesekara, 2010; Swing, 2003). Hence, the wide diversity of marine organisms is being recognized as a rich source of functional materials, including polyunsaturated fatty acids (PUFA), polysaccharides, natural pigments (NPs), essential minerals, vitamins, enzymes, and bioactive peptides (Shahidi, 2008; Shahidi and Alasalvar, 2010; Shahidi and Janak Kamil, 2001). Among marine organisms, seaweeds are still identified as underexploited plant resources although they have long been used in food diets as well as traditional remedies in Eastern hemisphere (Heo et al., 2009).

The ancient tradition and everyday consumption of seaweeds have made possible a large number of epidemiological studies showing their health benefits. When considering together with international diet-related chronic disease incidences, significant environmental factors including dietary difference between populations varying in seaweed consumption have been revealed. As an example, 1 year prevalence case of breast cancer rates per 100,000 people in Japan and China is 42.2 and 13.1, respectively, versus 125.9 and 106.2 cases in North America and Europe, respectively; and 1 year prevalence case of prostate cancer rates is 10.4 and 0.7 in Japan and China, respectively, versus 117.2 and 53.1 cases in North America and Europe, respectively (Pisani et al., 2002; Yuan and Walsh, 2006). Furthermore, migrants from Japan, Korea, and China to the United States progressively alter their risk of these cancer toward that of native-born populations. Thus, results of epidemiological studies have shown that low breast and prostate cancer cases in Japan and China are associated with environmental factors, such as food diets (Lacey et al., 2002). One of the main dietary differences between Eastern and Western hemispheres is the higher seafood consumption, such as fish and seaweeds by the former (Terry et al., 2003). In accordance with the epidemiological studies, in vivo studies using rodent model demonstrate protective effects of dietary kelps and other red and green seaweeds against mammary carcinogenesis. In East Asian culture, seaweeds have always been of particular interest (Khan et al., 2010). Seaweeds accounted for more than 10% of Japanese diet with average consumption of 1.4 kg/person/year (Burtin, 2003).

In recent years, seaweeds have served as important sources of bioactive natural substances. Moreover, many metabolites isolated from seaweeds have shown to possess biological activities and potential health benefits. Furthermore, several scientific studies have provided insight into neuroprotective properties of seaweeds. Many species of seaweeds have long been used in food diets as well as traditional remedies in Eastern countries, and more recently in Europe and America. Hence, seaweeds have great potential to be used in neuroprotection (Zarros, 2009).

Biological activities, nutritional value, and potential health benefits of seaweeds have been intensively investigated and reviewed. This chapter, however, focuses specifically on the neuroprotective effects of seaweeds and emphasizes their potential application as future functional foods and nutraceutical candidates to prevent neurodegenerative diseases.

3.2 Prospects of seaweeds as neuroprotective agents in nutraceuticals

Neurodegenerative diseases are estimated to surpass cancer as the second most common cause of death among elderly by the 2040s (Ansari et al., 2010; Bjarkam et al., 2001). For this reason, a great deal of attention has been paid by scientists regarding safe and effective neuroprotective agents. Many categories of natural and synthetic neuroprotective agents have been reported. However, synthetic neuroprotective agents are believed to have certain side effects, such as dry mouth, tiredness, drowsiness, sleepiness, anxiety or nervousness, difficulty to balance, etc. (Narang et al., 2008). Hence, nowadays researchers have an interest in studying natural bioactive compounds that can act as neuroprotective agents.

Marine environment, due to its phenomenal biodiversity, is very attractive; not only as nutritious food source, but also as a treasure of novel, biologically active compounds. Numerous studies have documented the health benefit effects of consuming seaweeds. In addition, seaweeds represent one potential candidate neuroprotective agent. More recently, there has been growing interest in seaweeds and their constituents as functional foods and nutraceuticals with potential health benefit effects as sources of antioxidants to reduce the risk of neurodegenerative diseases. For example, various NPs isolated from seaweeds have attracted attention in the fields of food, cosmetic, and pharmacology. Novel extraction and separation techniques, such as supercritical CO_2 extraction, centrifugal partition chromatography, and pressurized liquid method have recently been applied in the development NPs derived from seaweeds (Kim et al., 2011; Roh et al., 2008; Shang et al., 2010). However, development of seaweeds as neuroprotective agents still faces several challenges. The rationale for seaweeds neuroprotective effects treatment in the CNS is based on established observations and experiments in vitro or in animal models only. Until now, none of the marine algal neuroprotective effects have been examined in human subjects. Therefore, small clinical studies and further large-scale controlled studies are needed. Another important challenge in the development of seaweeds as neuroprotective agents is that many drugs failed to provide real neuroprotection in practice. Potential reasons for this failure include inappropriate use of specific neuroprotection(s) for a given disease or stage of disease progression or the use of suboptimal doses (Gilgun-Sherki et al., 2001). Hence, future studies are needed focusing on the synergistic benefits of consuming different seaweed species, recommended doses and timing of intake, and preparation methods for marine algal bioactive compounds in order to maximize the desired protective effect in the prevention of neurodegenerative diseases.

It has been reported that neurodegenerative diseases in East Asian countries were lower than in Europe ($p < 0.0004$) (Jorm and Jolley, 1998; Mishra and Palanivelu, 2008). Many studies have indicated potential health benefits of seaweed consumption (Burtin, 2003; Smit, 2004). Thus, lower incidence of neurodegenerative diseases in East Asia may correlate to high fish and seaweed consumption by East Asian populations. More recently, there has been growing interest in seaweeds and their constituents as functional foods and nutraceuticals with potential health benefit effects as sources of antioxidant to reduce the risk of neurodegenerative diseases. Seaweeds are important sources of bioactive ingredients that can be applied to many aspects of processing healthier foods and developing functional neuroprotective foods.

In addition, the wide diversity of seaweeds and numerous undiscovered unique metabolites present in seaweeds are interesting sources to increase the number of novel drugs against neurodegenerative diseases. However, large-scale human studies are required to identify the prophylactic and therapeutic neuroprotective effects of seaweeds.

3.3 Neuroprotective effects of seaweeds

3.3.1 Antioxidant activity

Oxidative stress is the result of an imbalance between pro-oxidant and anti-oxidant homeostases that leads to the generation of toxic reactive oxygen species (ROS) (Barnham et al., 2004). When compared to other parts of our body, the central nervous system (CNS) is more sensitive to oxidative stress due to its high oxygen consumption and lipid content. Increased oxidative stress in the CNS will further lead to lipid peroxidation, DNA and protein damage (Akyol et al., 2002). Oxidative stress in the CNS has been demonstrated to involve excitotoxicity and apoptosis, the two main causes of neuronal death. Furthermore, oxidative stress has also implicated the progression of Alzheimer's disease (AD), Parkinson's disease (PD), multiple sclerosis (MS), and other neurodegenerative diseases (Behl and Moosmann, 2002; Migliore and Coppedè, 2009). Antioxidants might have a positive effect in the CNS and seem to be a promising approach to neuroprotection therapy, as they can protect the CNS against free radical-mediated oxidative damage (Andersen, 2004). However, our endogenous antioxidant defenses are not always completely effective, and exposure to damaging environmental factors is increasing; therefore, it seems reasonable to propose that exogenous antioxidants could be effective in diminishing the cumulative effects of oxidative damage. Presently, antioxidants constitute a major component of clinical and experimental drugs that are currently considered for the prevention of neurodegenerative diseases and therapy (Moosmann and Behl, 2002).

Antioxidant activities of seaweeds have been determined by various methods such as 1,1-diphenyl-2-picryl hydrazyl (DPPH) radical scavenging, 2,2'-azinobis-3-ethylbenzothiazoline-6-sulfonate (ABTS) radical scavenging, singlet oxygen quenching activity, lipid peroxide inhibition, superoxide and hydroxyl radical scavenging assays. Lim et al. demonstrated that *Neorhodomela aculeate*, which is also known as *Rhodomela confervoides*, was able to scavenge DPPH with an IC50 = 90 µg/mL, and at a concentration of 20 µg/mL completely suppressed H_2O_2-induced lipid peroxidation in rat brain homogenate. Furthermore, Fallarero et al. (2003) showed that *Halimeda incrassata* and *Bryothamnion triquetrum* are potent ROS scavengers in mouse hypothalamic (GT1–7) cells. Vidal Novoa et al. (2001) reported that the antioxidant and ROS scavenging activities of *B. triquetrum* are related to their high phenolic contents. Dieckol, a phenolic compound isolated from brown seaweeds has been shown to scavenge ROS production in murine microglia (BV2) cells (Jung et al., 2009b). Wijesekara et al. (2010b) reported that most phenolic compounds which were purified from marine seaweeds are responsible for marine algal antioxidant activities and protective effects against oxidative stress-induced cell damage. Phenolic compounds act as free radical scavengers, reducing agents, and metal chelators, and thus effectively inhibit lipid oxidation. In addition, Yan et al. demonstrated that carotenoids have a strong radical scavenging activity and are found as major antioxidants in marine seaweeds (Nomura et al., 1997; Yan et al., 1999). Young and Lowe (2001) indicated that the structure, physical form, location or site of action, potential interaction with another antioxidant, concentration, and partial pressure of oxygen may affect the antioxidant activities of carotenoids in biological systems. Fucoxanthin, obtained from *Padina tetrastromatica*, has shown higher potential to be used as an antioxidant than β-carotene in modulating the antioxidant enzyme in the plasma and liver of retinol-deficient rats. However, the exact mechanisms of how fucoxanthin exerts antioxidative effect in rat induced by retinol deficiency are not yet completely understood. Moreover, the cytoprotective effect of fucoxanthin against ROS formation induced by H_2O_2 in monkey kidney fibroblast (Vero) cells has been observed (Heo et al., 2008). Two hydroxyl groups present in the ring structure of fucoxanthin may correlate to the inhibition of ROS formation. Indeed, it has been reported that the number of hydroxyl groups on the ring structure is correlated with the effects of ROS suppression. Moreover, it has also been shown that some marine algal sulfated polysaccharides (SPs) can be used as potent antioxidants (Jiao et al., 2011; Wijesekara et al., 2010a). Antioxidant activity of marine algal SPs depends on their structural features such as degree of sulfating, molecular weight, type of the major sugar, and glycosidic branching (Qi et al., 2005; Zhang et al., 2003). However, bioactivities of marine algal carotenoids and SPs against oxidative stress in the CNS have not been demonstrated yet.

Collectively, it can be suggested that among various organisms in the marine environment, marine seaweeds prove to be one of the useful candidates that can protect the CNS against oxidative degradation. Hence, developing novel molecules derived from marine seaweeds which promote antioxidant activity in the CNS may lead to the development of effective neuroprotective agents. Furthermore, it is also important to determine whether antioxidants derived from marine seaweeds can be used as prophylactic neuroprotective agents in order to slow down the progression of neurodegenerative diseases in populations that are at high risk, such as the elderly. Additionally, antioxidant activities of marine algal carotenoids, SPs, and other bioactive compounds in the CNS warrant further investigations.

3.3.2 Antineurotoxic activity

Neurotoxins are varied groups of compounds, whose highly specific effects on the nervous system of animals and humans, are possible by interfering with nerve impulse transmission (Patockaa and Stredab, 2002). They are able to produce neuronal damage or neurodegeneration when administered in vivo or in vitro (Segura-Aguilar and Kostrzewa, 2004). As an example, β-amyloid (Aβ) peptides have been demonstrated to possess neurotoxic effect on neuron and glial cells, although the precise mechanisms by which this occurs are yet to be elucidated (Allan Butterfield, 2002). Excessive accumulation of Aβ in the brain has been characterized as a major pathological hallmark of AD, and recently, fucoidan has been reported to block Aβ neurotoxicity in neuronal cells (Jhamandas et al., 2005). Fucoidan treatment abolished the inhibitory effect of Aβ on the phosphorylation of protein kinase C (PKC), which has been demonstrated to stimulate the survival of neurons and prevents Aβ neurotoxicity. PKC causes GSK-3β inactivation, and this inactivation in turn leads to the accumulation of cytoplasmic β-catenin and the subsequent translocation of β-catenin to the nucleus, causing TCF/LEF-1-dependent transcriptional activation of growth- and differentiation-related genes, which is required to stimulate neuronal survival (Garrido et al., 2002). In addition, Luo et al. (2009) showed that fucoidan isolated from *Laminaria japonica* was able to protect against 1-methyl-4-phenyl-1,2,3,6-tetrahydropyridine (MPTP)-induced neurotoxicity in animal model of Parkinsonism (C57/BL mice) and dopaminergic (MN9D) cells. The mechanisms of protection provided by fucoidan may partly relate to its antioxidative activity. Furthermore, the results of these studies suggest potential application of fucoidan for PD prevention and/or treatment. Moreover, the possible roles of alginates to protect human neuronal (NT2) cells against H_2O_2-induced neurotoxicity have previously been demonstrated (Eftekharzadeh et al., 2010). *H. incrassata* and *B. triquetrum* at a concentration of 0.2 mg/mL have been shown to protect methyl mercury-induced neurotoxicity in GT1–7 cells (Fallarero et al., 2003). Collectively, seaweeds and their bioactive compounds can be used for the development of new-generation therapeutic neuroprotective agents against neurotoxins in the CNS.

3.3.3 Anti-neuroinflammatory activity

Inflammation has been found to be the pathophysiological mechanism underlying many chronic diseases, such as cardiovascular diseases, diabetes, certain cancers, arthritis, and neurodegenerative diseases (Allen and Barres, 2009). Recent studies demonstrated that the resulting production of inflammatory responses and neurotoxic factors in the CNS is sufficient to induce neurodegeneration in a rat model (Liu et al., 2002). Several cell types have been demonstrated as contributors in inflammation-mediated neurodegeneration, yet microglia are implicated as critical components of the immunological insult to neurons (Block et al., 2007). Microglia are the immune cells in the CNS; they enter the system through the blood circulation early in an organism's development and play a role in immune surveillance (Allen and Barres, 2009). Ramified or resting microglia constitute 5%–20% of glial populations in the CNS (Kim and de Vellis, 2005). Recent study demonstrated that activation of microglia and the resulting production of pro-inflammatory and neurotoxic factors are sufficient to induce neurodegeneration in a rat model. Furthermore, activation of microglia and excessive amounts of pro-inflammatory mediators released by microglia have been observed during the pathogenesis of PD, AD, MS, AIDS dementia complex, as well as post-neuronal death in cerebral stroke and traumatic brain injury (Liu et al., 2002; Lull and Block, 2010). Therefore, a mechanism to regulate inflammatory response release by microglia may have important therapeutic potential for the treatment of neurodegenerative diseases.

Numerous studies have documented anti-inflammatory activities of seaweeds in vitro and in vivo (Abad et al., 2008). However, scientific analysis of anti-neuroinflammatory activity of seaweeds has been poorly carried out, and until now only few studies were reported. *Ecklonia cava* (Phaeophyceae; Laminariaceae), also known as "sea trumpet," has been reported to possess anti-inflammatory activity (Jung et al., 2009a; Maegawa et al., 1987; Serisawa et al., 2001). *E. cava* was able to suppress the levels of pro-inflammatory mediators such as nitric oxide (NO), prostaglandine-E_2 (PGE_2), and pro-inflammatory cytokines (tumor necrosis factor-α [TNF-α], interleukin-6 [IL-6], and interleukin-1β [IL-1β]) in lipopolysaccharides (LPS)-stimulated BV2 cells by blocking nuclear factor-κB (NF-κB) and mitogen-activated protein kinases (MAPKs) activation (Jung et al., 2009a,b). Furthermore, *N. aculeate* decreased NO production, inhibiting inducible NO synthase (iNOS) expression in inter-feron-gamma (IFN-γ)-stimulated BV2 cells (Lim et al., 2006). A number of bromophenols have been previously isolated from *N. aculeate* and may be potential anti-neuroinflammatory candidates (Fan et al., 2003; Ma et al., 2006; Xu et al., 2003; Zhao et al., 2004). Another study conducted by Cui et al. (2010) provides the first evidence that fucoidan isolated from *L. japonica* has a potent inhibitory effect against LPS-induced NO production in BV2 cells. In their study, the average molecular weight of fucoidan was 7000 Da, consisting of 48% total sugar (including 28% fucose) and 29% sulfate. Fucoidan, at a concentration of 125 μg/mL, significantly inhibited NO production to 75% (Cui et al., 2010). NO is a cytotoxic, short-lived highly diffusible signaling molecule. A number of studies demonstrated that NO generated by iNOS causes injury and cell death of neuron and oligodendrocytes in the CNS; hence, NO is implicated in the pathogenesis of various neurodegenerative diseases (Heales et al., 1999; Lee et al., 2000). Anti-neuroinflammatory activity of another seaweed species, *Ulva conglobata* has been reported. Methanolic extracts of *U. conglobata* were able to suppress the expression of pro-inflammatory enzymes, iNOS and cyclooxygenase-2 (COX-2), which accounted for the large production of NO and PGE_2, respectively (Jin et al., 2006; Salvemini et al., 1995). Among other mediators released by microglia, NO and PGE_2 are the main cytotoxic mediators participating in the innate response in the CNS (Boscá et al., 2005; Vane and Botting, 1995). Pro-inflammatory mediators have been found to be elevated in the brain of early AD (Blasko et al., 2004). For these reasons, agents that inhibit the production of pro-inflammatory mediators have been previously considered as potential candidates for the treatment of neurodegenerative diseases.

Epidemiological studies show that application of nonsteroidal anti-inflammatory drugs (NSAIDs) reduces the risk and delays the onset of inflammation in the CNS which further participates in the pathogenesis of some neurodegenerative diseases. NSAIDs mainly act by inhibiting the production of pro-inflammatory mediators. Hence, attenuation of pro-inflammatory mediators in microglia by seaweeds demonstrates its potential neuroprotective activity. Furthermore, seaweeds as potential anti-neuroinflammatory agents have a great potential application in the pharmaceutical area as well as the food industry. There are numerous advantages of seaweed use in pharmaceuticals and functional foods, such as relatively low production costs, low cytotoxicity, safety, and wide acceptability. However, further studies are needed with clinical trials for marine algal anti-neuroinflammatory activity.

3.3.4 Cholinesterase inhibitory activity

Alzheimer's disease (AD) is an irreversible, progressive neurodegenerative disease, which results in memory loss, behavior disturbances, personality changes, and a decline in cognitive abilities (Pietrini et al., 2000). It was stated in the cholinergic hypothesis, that a serious loss of cholinergic function in the CNS contributes significantly to the cognitive symptoms associated with AD (Bartus, 2000). Accordingly, neuropathological studies demonstrated that AD was associated with deficiency in the brain neurotransmitter acetylcholine (ACh) (Tabet, 2006). The inhibition of acetylcholinesterase (AChE) enzyme, which catalyzes the breakdown of ACh, may be one of the most realistic approaches to the symptomatic treatment of AD (Pangestuti and Kim, 2010). Recently, a variety of plants has been reported to possess AChE inhibitory activity. *Huperzia serrata*, a Chinese terrestrial herb has been demonstrated to be a potent AChE inhibitor (Cheng et al., 1996). In addition, Houghton et al. reported cholinesterase (ChE) inhibitory activity of *Crinum jagus* and *Crinum glaucum*, two Nigerian *Crinum* species (Houghton et al., 2004). A number of studies have recently shown AChE inhibitory activity of several seaweed species. A list of seaweeds reported to have significant AChE inhibitory activity is presented in **Table 3.1**.

Recently, Myung et al. (2005) reported that dieckol and phlorofucofluoroeckol possess memory-enhancing and AChE inhibitory activity. Furthermore, Yoon et al. (2008) screened ethanolic extracts of 27 Korean seaweeds, for inhibitory activity on AChE, and found that extracts from *Ecklonia stolonifera* showed significant inhibitory activity. Two sterols and eight phlorotannins were isolated from *E. stolonifera*. Eckol, dieckol, 2-phloroeckol, and 7-phloroeckol demonstrated selective dose-dependent inhibitory activities toward AChE, whereas, eckstolonol and phlorofucofuroeckol-A exhibited inhibitory activities toward both AChE and butyrylcholinesterase (BChE). However, phloroglucinol, which is a monomer, and triphlorethol-A, the open-chain trimer of phloroglucinol, did not inhibit cholinesterase (ChE) at the concentrations tested. The exact mechanisms underlying this phenomenon have not yet been identified. However, the possible relation between structure of phlorotannins and AChE inhibitory activity has been reported; it is suggested that phlorotannins as polymers

Table 3.1 Acetylcholinesterase Inhibitory Activities of Several Seaweed Species

Seaweeds	Extracts/Compounds	IC$_{50}$
Caulerpa racemosa	MeOH extracts	5.5 mg/mL
Codium capitatum	MeOH extracts	7.8 mg/mL
Ulva fasciata	MeOH extracts	4.8 mg/mL
Halimeda cuneata	MeOH extracts	5.7 mg/mL
Amphiora ephedraea	MeOH extracts	5.1 mg/mL
Amphiora bowerbankii	MeOH extracts	5.3 mg/mL
Dictyota humifusa	MeOH extracts	4.8 mg/mL
Hypnea valentiae	MeOH extracts	2.6 mg/mL
Padina gymnospora	MeOH extracts	3.5 mg/mL
Ulva reticulate	MeOH extracts	10 mg/mL
Gracilaria edulis	MeOH extracts	3 mg/mL
Ecklonia stolonifera	EtOH extracts	108.11 μg/mL
Ecklonia stolonifera	24-Hydroperoxy-24-vinylcholesterol	389.1 μM
Ecklonia stolonifera	Eckstolonol	42.66 μM
Ecklonia stolonifera	Eckol	20.56 μM
Ecklonia stolonifera	Phlorofucofluoroeckol-A	4.89 μM
Ecklonia stolonifera	Dieckol	17.11 μM
Ecklonia stolonifera	2-Phloroeckol	38.13 μM
Ecklonia stolonifera	7-Phloroeckol	21.11 μM
Ishige okamurae	MeOH extracts	163.07 μM
Ishige okamurae	EtOAc extracts	137.25 μM
Ishige okamurae	6,6′-Bieckol	46.42 μM

IC$_{50}$ values for eserine and galanthamine were 0.004 and 0.0007 mg mL, respectively.

of phloroglucinol have appropriately bulky structures, which are then able to mask the ChE and prevent the binding of the substrates. Moreover, as the phloroglucinol monomer and open-chain trimer of phloroglucinol were not able to inhibit the ChE activity, it may be suggested that the degree of polymerization and closed-ring structure of phlorotannins play key roles in the inhibitory potential of phlorotannins toward the ChE (Yoon et al., 2008). In addition, *Hypnea valentiae* and *Ulva reticulate*, two seaweed species from Tamil Nadu, India, also have been reported to inhibit both AChE and BChE activities (Suganthy et al., 2010). A good balance between AChE and BChE activities has been reported to result in higher efficacy for the treatment of AD (Greig et al., 2002). BChE are considered to play a minor role in regulating brain AChE levels. Notably, mixed inhibition of AChE and BChE has been found in tacrine and physostigmine, which are the licensed drugs used in the treatment of AD.

Taken together, marine red, brown, and green seaweeds have the potential to be used as functional neuroprotective agents due to their effectiveness in inhibiting ChE activity. Furthermore, some compounds derived from seaweeds provided mixed-type ChE (AChE and BChE) inhibitory activities, which have been considered to be more effective in the treatment of AD. Some AChE synthetic commercial drugs are known to produce side effects. Hence, researchers have a great interest to study natural herbs that can act as AChE inhibitors. Many kinds of seaweeds, consumed for centuries in East Asian countries, are well tolerated and lack harmful side effects. Interestingly, several seaweed species have also been demonstrated as potential AChE inhibitors. Hence, AChE inhibitory activity of seaweeds should be screened, and further studies with clinical trials are also needed.

3.3.5 Inhibition of neuronal death

A common pathological hallmark of various neurodegenerative diseases is the loss of particular subsets of neurons (Mattson, 2000). Neurodegeneration of these neural subsets may be a consequence of various forms of neural cell death, including necrosis and apoptosis (Bains and Shaw, 1997). A study carried out by Jhamandas et al. (2005) successfully showed that fucoidan isolated from *Fucus vesiculosus* was able to protect rat cholinergic neuronal death induced by $A\beta_{1-42}$. Fucoidan pretreatment blocked the activation of caspase-9 and caspase-3. Caspase-9 and caspase-3 have been suggested to mediate the terminal stages of neuronal apoptosis (Cowan et al., 2001). Caspase-9 and caspase-3 are two of several central components of the machinery responsible for apoptosis. Therefore, the ability of fucoidan to block the activation of caspase-9 and caspase-3 suggest that inhibition of neuronal death by fucoidan mainly occurs through apoptotic inhibition. In neurodegenerative diseases, apoptosis might be pathogenic, and targeting this process might mitigate neurodegenerative diseases (Vila and Przedborski, 2003). Furthermore, aqueous extracts of *B. triquetrum* have been demonstrated to protect GT1–7 cells' death produced by severe (180 min) chemical hypoxia/aglycemia insult, which further reduced the cytotoxicity and early production of free radicals. The protection exerted by *B. triquetrum* extract seems to be linked to its ability to reduce free-radical generation (Fallarero et al., 2006). The authors suggest that the protective effects of *B. triquetrum* extract are partially related to the presence of ferulic acid (Fallarero et al., 2006).

3.3.6 Other neuroprotective effects

Neurite outgrowth is a fundamental neuronal feature and plays an important role in neuronal development during embryogenesis and in the adult brain (Khodosevich and Monyer, 2010). *Sargassum macrocarpum* and its two active components, sargaquinoic acid and sargachromenol, have been shown to promote neurite outgrowth in rat pheochromocytoma (PC12) cells (Kamei and Sagara, 2002; Tsang and Kamei, 2004; Tsang et al., 2005). Structure and neurite outgrowth-promoting relationship of sargaquinoic acid has been reported by Tsang et al. (2001). They reported that quinone is the structural moiety of the sargaquinoic acid molecule which is responsible for the neurite outgrowth-promoting activity. Notably, the hydroxyl group bonded to quinone had a significant effect on neuritogenic activity. In addition, pheophytin a, a chlorophyll-related compound and its analog, vitamin B_{12} derived from *Sargassum fulvellum* also have potential neurite outgrowth-promoting activity (Ina and Kamei, 2006; Ina et al., 2007).

Phlorotannins derived from *Eisenia bicyclis* have been demonstrated to inhibit β-amyloid cleavage enzyme (BACE-1) activity (Jung et al., 2010). BACE-1 represents candidate biomarkers of AD, since it initiates the formation of Aβ (Tang et al., 2006). When considering that almost all currently available medications for AD are AChE inhibitors, suppression of BACE-1 by phlorotannins will enhance the medications and/or therapy for AD patients.

In addition, Lee et al. (2007) demonstrated that fucoidan treatment resulted in an increase in cell proliferation of human neuroblastoma (SH-SY5Y) cell induced by Aβ. Hence, it may be suggested that fucoidan has potential neuroprotective effects.

3.4 Conclusions

Seaweeds are valuable sources of neuroprotective agents and could be introduced for the preparation of novel functional ingredients in pharmaceuticals and functional foods as a good approach for the treatment and/or prevention of neurodegenerative diseases. Functional ingredients derived from seaweeds can be suggested as alternative sources to synthetic ingredients that can contribute to neuroprotection by being a part of pharmaceuticals and functional foods. Moreover, the wide range of biological activities associated with natural compounds derived from seaweeds increases the potential to expand the neuroprotective effects and health beneficial value of marine organisms in the pharmaceutical industry. However, until now, most of the biological and neuroprotective activities of functional ingredients derived from seaweeds have been observed in vitro or in mouse model systems. Therefore, further research studies are needed in order to investigate neuroprotective activities of functional ingredients derived from seaweeds in human subjects and further in large-scale controlled studies.

References

Abad, MJ, LM Bedoya, and P Bermejo. 2008. Natural marine anti-inflammatory products. *Mini Reviews in Medicinal Chemistry* 8:740–754.

Akyol, Ö, H Herken, E Uz et al. 2002. The indices of endogenous oxidative and antioxidative processes in plasma from schizophrenic patients: The possible role of oxidant/antioxidant imbalance. *Progress in Neuro-Psychopharmacology and Biological Psychiatry* 26:995–1005.

Allan Butterfield, D. 2002. Amyloid-peptide (1–42)-induced oxidative stress and neurotoxicity: Implications for neurodegeneration in Alzheimer's disease brain. A review. *Free Radical Research* 36:1307–1313.

Allen, NJ and BA Barres. 2009. Neuroscience: Glia—More than just brain glue. *Nature* 457:675–677.

Andersen, JK. 2004. Oxidative stress in neurodegeneration: Cause or consequence? *Nature Reviews Neuroscience* 5:S18-S25.

Ansari, JA, A. Siraj, and NN Inamdar. 2010. Pharmacotherapeutic approaches of Parkinson's disease. *International Journal of Pharmacology* 6:584–590.

Bains, JS and CA Shaw. 1997. Neurodegenerative disorders in humans: The role of glutathione in oxidative stress-mediated neuronal death. *Brain Research Reviews* 25:335–358.

Barnham, KJ, CL Masters, and AI Bush. 2004. Neurodegenerative diseases and oxidative stress. *Nature Reviews Drug Discovery* 3:205–214.

Bartus, RT 2000. On neurodegenerative diseases, models, and treatment strategies: Lessons learned and lessons forgotten a generation following the cholinergic hypothesis. *Experimental Neurology* 163:495–529.

Behl, C and B Moosmann. 2002. Antioxidant neuroprotection in Alzheimer's disease as preventive and therapeutic approach. *Free Radical Biology and Medicine* 33:182–191.

Bjarkam, CR, JC Sørensen, NÅ Sunde, FA Geneser, and K Østergaard. 2001. New strategies for the treatment of Parkinson's disease hold considerable promise for the future management of neurodegenerative disorders. *Biogerontology* 2:193–207.

Blasko, I, M Stampfer-Kountchev, P Robatscher, R Veerhuis, P Eikelenboom, and B Grubeck-Loebenstein. 2004. How chronic inflammation can affect the brain and support the development of Alzheimer's disease in old age: The role of microglia and astrocytes. *Aging Cell* 3:169–176.

Block, ML, L Zecca, and JS Hong. 2007. Microglia-mediated neurotoxicity: Uncovering the molecular mechanisms. *Nature Reviews Neuroscience* 8:57–69.

Boscá, L, M Zeini, PG Través, and S Hortelano. 2005. Nitric oxide and cell viability in inflammatory cells: A role for NO in macrophage function and fate. *Toxicology* 208:249–258.

Burtin, P 2003. Nutritional value of seaweeds. *Electronic Journal of Environmental, Agricultural and Food Chemistry* 2:498–503.

Cheng, DH, H Ren, and XC Tang. 1996. Huperzine A, a novel promising acetylcholinesterase inhibitor. *Neuroreport* 8:97–101.

Cowan, CM, J Thai, S Krajewski et al. 2001. Caspases 3 and 9 send a pro-apoptotic signal from synapse to cell body in olfactory receptor neurons. *Journal of Neurosciences* 21:7099–7109.

Cui, YQ, LJ Zhang, T Zhang et al. 2010. Inhibitory effect of fucoidan on nitric oxide production in lipopolysaccharide activated primary microglia. *Clinical and Experimental Pharmacology and Physiology* 37:422–428.

Eftekharzadeh, B, F Khodagholi, A Abdi, and N Maghsoudi. 2010. Alginate protects NT2 neurons against H_2O_2-induced neurotoxicity. *Carbohydrate Polymers* 79:1063–1072.

Fallarero, A, JJ Loikkanen, PT Männistö, O Castañeda, and A Vidal. 2003. Effects of aqueous extracts of Halimeda incrassata (Ellis) Lamouroux and Bryothamnion triquetrum (S.G. Gmelim) Howe on hydrogen peroxide and methyl mercury-induced oxidative stress in GT1–7 mouse hypothalamic immortalized cells. *Phytomedicine* 10:39–47.

Fallarero, A, A Peltoketo, J Loikkanen, P Tammela, A Vidal, and P Vuorela. 2006. Effects of the aqueous extract of *Bryothamnion triquetrum* on chemical hypoxia and aglycemia-induced damage in GT1–7 mouse hypothalamic immortalized cells. *Phytomedicine* 13:240–245.

Fan, X, NJ Xu, and JG Shi. 2003. Bromophenols from the red alga Rhodomela confervoides. *Journal of Natural Products* 66:455–458.

Garrido, JL, JA Godoy, A Alvarez, M Bronfman, and NC Inestrosa. 2002. Protein kinase C inhibits amyloid {beta} peptide neurotoxicity by acting on members of the Wnt pathway. *FASEB Journal* 16:1982–1984.

Gilgun-Sherki, Y, E Melamed, and D Offen. 2001. Oxidative stress induced-neurodegenerative diseases: The need for antioxidants that penetrate the blood brain barrier. *Neuropharmacology* 40:959–975.

Greig, NH, DK Lahiri, and K Sambamurti. 2002. Butyrylcholinesterase: An important new target in Alzheimer's disease therapy. *International Psychogeriatrics* 14:77–91.

Heales, SJR, JP Bolaños, VC Stewart, PS Brookes, JM Land, and JB Clark. 1999. Nitric oxide, mitochondria and neurological disease. *Biochimica et Biophysica Acta (BBA)— Bioenergetics* 1410:215–228.

Heo, SJ, JY Hwang, JI Choi, JS Han, HJ Kim, and YJ Jeon. 2009. Diphlorethohydroxycarmalol isolated from Ishige okamurae, a brown algae, a potent [alpha]-glucosidase and [alpha]-amylase inhibitor, alleviates postprandial hyperglycemia in diabetic mice. *European Journal of Pharmacology* 615:252–256.

Heo, SJ, SC Ko, SM Kang et al. 2008. Cytoprotective effect of fucoxanthin isolated from brown algae Sargassum siliquastrum against H_2O_2-induced cell damage. *European Food Research and Technology A* 228:145–151.

Houghton, PJ, JM Agbedahunsi, and A Adegbulugbe. 2004. Choline esterase inhibitory properties of alkaloids from two Nigerian Crinum species. *Phytochemistry* 65:2893–2896.

Ina, A, KI Hayashi, H Nozaki, and Y Kamei. 2007. Pheophytin a, a low molecular weight compound found in the marine brown alga *Sargassum fulvellum*, promotes the differentiation of PC12 cells. *International Journal of Developmental Neuroscience* 25:63–68.

Ina, A and Y Kamei. 2006. Vitamin B 12, a chlorophyll-related analog to pheophytin a from marine brown algae, promotes neurite outgrowth and stimulates differentiation in PC12 cells. *Cytotechnology* 52:181–187.

Jhamandas, JH, MB Wie, K Harris, D MacTavish, and S Kar. 2005. Fucoidan inhibits cellular and neurotoxic effects of β-amyloid (Aβ) in rat cholinergic basal forebrain neurons. *European Journal of Neuroscience* 21:2649–2659.

Jiao, G, G Yu, J Zhang, and H Ewart. 2011. Chemical structures and bioactivities of sulfated polysaccharides from marine algae. *Marine Drugs* 9:196–223.

Jin, DQ, CS Lim, JY Sung, HG Choi, I Ha, and JS Han. 2006. *Ulva conglobata*, a marine algae, has neuroprotective and anti-inflammatory effects in murine hippocampal and microglial cells. *Neuroscience Letters* 402:154–158.

Jorm, AF and D Jolley. 1998. The incidence of dementia: A meta-analysis. *Neurology* 51:728–733.

Jung, W-K, Y-W Ahn, S-H Lee et al. 2009a. Ecklonia cava ethanolic extracts inhibit lipopolysaccharide-induced cyclooxygenase-2 and inducible nitric oxide synthase expression in BV2 microglia via the MAP kinase and NF-[kappa]B pathways. *Food and Chemical Toxicology* 47:410–417.

Jung, WK, SJ Heo, YJ Jeon et al. 2009b. Inhibitory effects and molecular mechanism of Dieckol isolated from marine brown alga on COX-2 and iNOS in microglial cells. *Journal of Agricultural and Food Chemistry* 57:4439–4446.

Jung, HA, SH Oh, and JS Choi. 2010. Molecular docking studies of phlorotannins from Eisenia bicyclis with BACE1 inhibitory activity. *Bioorganic & Medicinal Chemistry Letters* 20:3211–3215.

Kamei, Y and A Sagara. 2002. Neurite outgrowth promoting activity of marine algae from Japan against rat adrenal medulla pheochromocytoma cell line, PC12D. *Cytotechnology* 40:99–106.

Khan, SB, CS Kong, JA Kim, and SK Kim. 2010. Protective effect of Amphiroa dilatata on ROS induced oxidative damage and MMP expressions in HT1080 cells. *Biotechnology and Bioprocess Engineering* 15:191–198.

Khodosevich, K and H Monyer. 2010. Signaling involved in neurite outgrowth of postnatally born subventricular zone neurons in vitro. *BMC Neuroscience* 11:1–18.

Kim, SM, YF Shang, and B-H Um. 2011. A preparative method for isolation of fucoxanthin from *Eisenia bicyclis* by centrifugal partition chromatography. *Phytochemical Analysis* 22:322–329.

Kim, SU and J de Vellis. 2005. Microglia in health and disease. *Journal of Neuroscience Research* 81:302–313.

Kim, SK and I Wijesekara. 2010. Development and biological activities of marine-derived bioactive peptides: A review. *Journal of Functional Foods* 2:1–9.

Lacey Jr, JV, SS Devesa, and LA Brinton. 2002. Recent trends in breast cancer incidence and mortality. *Environmental and Molecular Mutagenesis* 39:82–88.

Lee, HR, H Do, SR Lee, ES Sohn, S Pyo, and E Son. 2007. Effects of fucoidan on neuronal cell proliferation-association with NO production through the iNOS pathway. *Journal of Food Science and Nutrition* 12:74–78.

Lee, JM, MC Grabb, GJ Zipfel, and DW Choi. 2000. Brain tissue responses to ischemia. *Journal of Clinical Investigation* 106:723–731.

Lim, CS, DQ Jin, JY Sung et al. 2006. Antioxidant and anti-inflammatory activities of the methanolic extract of *Neorhodomela aculeate* in hippocampal and microglial cells. *Biological & Pharmaceutical Bulletin* 29:1212–1216.

Liu, BIN, H-M Gao, J-Y Wang, G-H Jeohn, CL Cooper, and J-S Hong. 2002. Role of nitric oxide in inflammation-mediated neurodegeneration. *Annals of the New York Academy of Sciences* 962:318–331.

Lull, ME and ML Block. 2010. Microglial activation and chronic neurodegeneration. *Neurotherapeutics* 7:354–365.

Luo, D, Q Zhang, H Wang et al. 2009. Fucoidan protects against dopaminergic neuron death in vivo and in vitro. *European Journal of Pharmacology* 617:33–40.

Ma, M, J Zhao, S Wang et al. 2006. Bromophenols coupled with methyl γ-ureidobutyrate and bromophenol sulfates from the red alga *Rhodomela confervoides*. *Journal of Natural Products* 69:206–210.

Maegawa, M, Y Yokohama, and Y Aruga. 1987. Critical light conditions for young *Ecklonia cava* and *Eisenia bicyclis* with reference to photosynthesis. *Hydrobiologia* 151:447–455.

Mattson, MP. 2000. Apoptosis in neurodegenerative disorders. *Nature Reviews Molecular Cell Biology* 1:120–130.

Migliore, L and F Coppedè. 2009. Environmental-induced oxidative stress in neurodegenerative disorders and aging. *Mutation Research/Genetic Toxicology and Environmental Mutagenesis* 674:73–84.

Mishra, S and K. Palanivelu. 2008. The effect of curcumin (turmeric) on Alzheimer's disease: An overview. *Annals of Indian Academy of Neurology* 11:13–19.

Moosmann, B and C Behl. 2002. Antioxidants as treatment for neurodegenerative disorders. *Expert Opinion on Investigational Drugs* 11:1407–1435.

Myung, CS, HC Shin, HY Bao, SJ Yeo, BH Lee, and JS Kang. 2005. Improvement of memory by dieckol and phlorofucofuroeckol in ethanol-treated mice: Possible involvement of the inhibition of acetylcholinesterase. *Archives of Pharmacal Research* 28:691–698.

Narang, S, D Gibson, AD Wasan et al. 2008. Efficacy of dronabinol as an adjuvant treatment for chronic pain patients on opioid therapy. *Journal of Pain* 9:254–264.

Nomura, T, M Kikuchi, A Kubodera, and Y Kawakami. 1997. Proton-donative antioxidant activity of fucoxanthin with 1, 1-diphenyl-2-picrylhydrazyl (DPPH). *IUBMB Life* 42:361–370.

Pangestuti, R and S-K Kim. 2010. Neuroprotective properties of chitosan and its derivatives. *Marine Drugs* 8:2117–2128.

Patockaa, J and L Stredab. 2002. Brief review of natural nonprotein neurotoxins. *ASA Newsletter* 89:16–24.

Pietrini, P, GE Alexander, ML Furey, H Hampel, and M Guazzelli. 2000. The neurometabolic landscape of cognitive decline: In vivo studies with positron emission tomography in Alzheimer's disease. *International Journal of Psychophysiology* 37:87–98.

Pisani, P, F Bray, and DM Parkin. 2002. Estimates of the world-wide prevalence of cancer for 25 sites in the adult population. *International Journal of Cancer* 97:72–81.

Qi, H, Q Zhang, T Zhao et al. 2005. Antioxidant activity of different sulfate content derivatives of polysaccharide extracted from *Ulva pertusa* (Chlorophyta) in vitro. *International Journal of Biological Macromolecules* 37:195–199.

Roh, MK, MS Uddin, and BS Chun. 2008. Extraction of fucoxanthin and polyphenol from Undaria pinnatifida using supercritical carbon dioxide with co-solvent. *Biotechnology and Bioprocess Engineering* 13:724–729.

Salvemini, D, PT Manning, BS Zweifel et al. 1995. Dual inhibition of nitric oxide and prostaglandin production contributes to the antiinflammatory properties of nitric oxide synthase inhibitors. *Journal of Clinical Investigation* 96:301–308.

Segura-Aguilar, J and R Kostrzewa. 2004. Neurotoxins and neurotoxic species implicated in neurodegeneration. *Neurotoxicity Research* 6:615–630.

Serisawa, Y, Y Yokohama, Y Aruga, and J Tanaka. 2001. Photosynthesis and respiration in bladelets of *Ecklonia cava* Kjellman (Laminariales, Phaeophyta) in two localities with different temperature conditions. *Phycological Research* 49:1–11.

Shahidi, F 2008. *Bioactives from Marine Resources*, ACS Symposium Series, ACS Publications, Oxford University Press, Cary, NC, pp. 24–34.

Shahidi, F and C Alasalvar, eds. 2010. Marine oils and other marine nutraceuticals, Chapter 36. In *Handbook of Seafood Quality, Safety and Health Applications* (eds C. Alasalvar, F. Shahidi, K. Miyashita, and U. Wanasundara),Wiley-Blackwell, Oxford, U.K., pp. 444–454.

Shahidi, F and YVA Janak Kamil. 2001. Enzymes from fish and aquatic invertebrates and their application in the food industry. *Trends in Food Science & Technology* 12:435–464.

Shang, YF, SM Kim, WJ Lee, and BH Um. 2010. Pressurized liquid method for fucoxanthin extraction from *Eisenia bicyclis* (Kjellman) Setchell. *Journal of Bioscience and Bioengineering* 111:237–241.

Smit, AJ 2004. Medicinal and pharmaceutical uses of seaweed natural products: A review. *Journal of Applied Phycology* 16:245–262.

Suganthy, N, S Karutha Pandian, and K Pandima Devi. 2010. Neuroprotective effect of seaweeds inhabiting South Indian coastal area (Hare Island, Gulf of Mannar marine biosphere reserve): Cholinesterase inhibitory effect of *Hypnea valentiae* and *Ulva reticulata*. *Neuroscience Letters* 468:216–219.

Swing, JT. 2003. What future for the oceans? *Foreign Affairs* 82:139–152.

Tabet, N. 2006. Acetylcholinesterase inhibitors for Alzheimer's disease: Anti-inflammatories in acetylcholine clothing! *Age and Ageing* 35:336–338.

Tang, K, LS Hynan, F Baskin, and RN Rosenberg. 2006. Platelet amyloid precursor protein processing: A bio-marker for Alzheimer's disease. *Journal of the Neurological Sciences* 240:53–58.

Terry, PD, TE Rohan, and A Wolk. 2003. Intakes of fish and marine fatty acids and the risks of cancers of the breast and prostate and of other hormone-related cancers: A review of the epidemiologic evidence. *American Journal of Clinical Nutrition* 77:532–543.

Tsang, CK, A Ina, T Goto, and Y Kamei. 2005. Sargachromenol, a novel nerve growth factor-potentiating substance isolated from *Sargassum macrocarpum*, promotes neurite outgrowth and survival via distinct signaling pathways in PC12D cells. *Neuroscience* 132:633–643.

Tsang, CK and Y Kamei. 2004. Sargaquinoic acid supports the survival of neuronal PC12D cells in a nerve growth factor-independent manner. *European Journal of Pharmacology* 488:11–18.

Tsang, CK, A Sagara, and Y Kamei. 2001. Structure-activity relationship of a neurite outgrowth-promoting substance purified from the brown alga, *Sargassum macrocarpum*, and its analogues on PC12D cells. *Journal of Applied Phycology* 13:349–357.

Vane, JR and RM Botting. 1995. New insights into the mode of action of anti-inflammatory drugs. *Inflammation Research* 44:1–10.

Vidal Novoa, A, M Motidome, J Mancini Filho et al. 2001. Actividad antioxidante y ácidos fenólicos del alga marina Bryothamnion triquetrum (SG Gmelim) Howe; Antioxidant activity related to phenolic acids in the aqueous extract of the marine seaweed Bryothamnin triquetrum (SG Gmelim) Howe. *RBCF, Rev. bras. ciênc. farm.(Impr.)* 37:373–382.

Vila, M and S Przedborski. 2003. Targeting programmed cell death in neurodegenerative diseases. *Nature Reviews Neuroscience* 4:365–375.

Wijesekara, I, R Pangestuti, and SK Kim. 2010a. Biological activities and potential health benefits of sulfated polysaccharides derived from marine algae. *Carbohydrate Polymers* 84:14–21.

Wijesekara, I, NY Yoon, and SK Kim. 2010b. Phlorotannins from *Ecklonia cava* (Phaeophyceae): Biological activities and potential health benefits. *Biofactors* 306:408–414.

Xu, N, X Fan, X Yan, X Li, R Niu, and CK Tseng. 2003. Antibacterial bromophenols from the marine red alga *Rhodomela confervoides*. *Phytochemistry* 62:1221–1224.

Yan, X, Y Chuda, M Suzuki, and T Nagata. 1999. Fucoxanthin as the major antioxidant in Hijikia fusiformis, a common edible seaweed. *Bioscience, Biotechnology, and Biochemistry* 63:605–607.

Yoon, NY, HY Chung, HR Kim, and JS Choi. 2008. Acetyl and butyrylcholinesterase inhibitory activities of sterols and phlorotannins from *Ecklonia stolonifera*. *Fisheries Science* 74:200–207.

Young, AJ and GM Lowe. 2001. Antioxidant and prooxidant properties of carotenoids. *Archives of Biochemistry and Biophysics* 385:20–27.

Yuan, YV and NA Walsh. 2006. Antioxidant and antiproliferative activities of extracts from a variety of edible seaweeds. *Food and Chemical Toxicology* 44:1144–1150.

Zarros, A 2009. In which cases is neuroprotection useful. *Advances and Alternative Thinking in Neuroscience* 1:3–5.

Zhang, Q, N Li, G Zhou, X Lu, Z Xu, and Z Li. 2003. In vivo antioxidant activity of polysaccharide fraction from *Porphyra haitanesis* (Rhodephyta) in aging mice. *Pharmacological Research* 48:151–155.

Zhao, J, X Fan, S Wang et al. 2004. Bromophenol derivatives from the red alga *Rhodomela confervoides*. *Journal of Natural Products* 67:1032–1035.

4

Chitosan-Based Biomaterials against Diabetes and Related Complications

Se-Kwon Kim and Fatih Karadeniz

Contents

4.1 Introduction

4.1.1 Chitin and chitosan

Chitin is an abundant natural polysaccharide, which can be found in the exoskeleton of crustaceans, cuticle of the insects, and cell wall of some microorganisms. Chitosan is a common derivative of chitin and gained by N-deacetylation in the presence of alkaline. Chitosan is reported to be a functional and basic linear polysaccharide. Generally, deacetylation cannot be completely achieved even under harsh treatment. The degree of deacetylation usually ranges from 70% to 95%, depending on the method used. Chitosan is widely available with different molecular weights and functionalities due to differences regarding the deacetylation process which results in different degrees of deacetylation ranging from 70% to 95% and filters used for obtaining the final compound. The industrial production and application fields of chitosan have been steadily increasing since 1970s. Chitosan is soluble in most organic acids and mildly acidic when the pH of the environment is below 6. However, aquatic conditions are not able to solubilize chitosan, which is a limiting condition for chitosan to be utilized as a biofunctional agent. Therefore, chitosan was mainly used as wastewater treatment agent, adsorbent for heavy metals, helping compound for biological experiments, food preserving material, and animal feed. On the other hand, interest in chitosan increased in recent decades which resulted in different insights for bioactivity of chitosan and generating derivatives with more effective utilization. As a result, chitosan earned a lead role in several industrial processes and found itself at the center of attention as a promising cosmetic, drug carrier, and pharmaceutical and nutraceutical agent. This attention and numerous researches have made chitin and chitosan to be known to exhibit important bioactivities, such as antitumor, antibacterial, hypocholesterolemic, and antihypertensive activities (Fei Liu et al. 2001; Kim and Rajapakse 2005; Qin et al. 2002). Natural abundance, biodegradability, and nontoxic nature of the compounds were the main motives behind the development of further applications of chitosan. Against all promising improvements in this field, there is evidence about the nill or indigent absorption of chitin and chitosan in the human intestine. This drawback is mostly due to the lack of enzymes to cleave the β-glucosidic linkage in chitosan, which results in poor absorption in mammalian body.

In recent years, chitosan and its monomer glucosamine were subjected to various derivatizations to obtain new natural compounds with improved bioactivity than their predecessors (Fenton et al. 2000; Jiang et al. 2007; Prabaharan 2008). The main aim of the first derivatization was to produce the soluble forms of chitosan, which makes it more biofriendly and easily absorbed by the intestines for enhanced and rapid biological effects (Hai et al. 2003; Il'ina and Varlamov 2004; Kuroiwa et al. 2002; Mao et al. 2004). Since effectiveness of the compound is based on its absorption rate by the body, this kind of derivation opened up new angles for chitosan derivation toward novel bioactive compounds. In this respect, enzymatic hydrolysis of chitosan to obtain oligomers with higher water solubility is of great interest recently (Jeon and Kim 2000). Expectedly, this interest promoted a new derivative with much improved solubility and promising bioactivities compared to that of chitosan itself. In this manner, the obtained oligosaccharide (COS) is the hydrolyzed derivative of chitosan composed of β-(1 → 4) D-glucosamine units. It has been reported to possess better properties such as relatively smaller molecular size in comparison to chitosan and high solubility in aqueous solutions. Moreover, the research conducted also suggested that chitosan oligosaccharides (COS) is effective agents for lowering blood cholesterol and pressure, controlling arthritis, and enhancing antitumor properties (Kim and Rajapakse 2005). Besides oligomerization, another main derivation for chitosan and its monomer glucosamine is adding negative and/or positively charged side chains. In this manner, glucosamine, chitin, chitosan, and COS were reformed under chemical conditions to give sulfated-, phosphorylated-, carboxymethyl-, deoxymethyl-derivatives, and so on (Cho et al. 2011; Huang et al. 2005; Je and Kim 2006; Kim et al. 2005, 2010). Since oligosaccharides and many other derivatives are biodegradable, water-soluble, and nontoxic compounds (Qin et al. 2006), they might be beneficial biomaterials for diseases such as diabetes and obesity with increasing morbidity and mortality rates.

4.1.2 Diabetes

As a chronic metabolic syndrome, diabetes is characterized by deterioration in both secretion and action of insulin, which eventually result in raised levels of blood glucose. Increased blood glucose levels (hyperglycemia) are dangerous complications of diabetes, which can damage many crucial body parts such as blood vessels and nerves. Surveys of global mortality report that 5% of the global population is diagnosed with diabetes and that it is the fifth leading cause of death (Roglic et al. 2005). Despite such risky numbers, diabetes is still surprisingly prevalent and could not be cured. The only treatment for the disorder aims in improved life quality and is limited to control the disease and the improvements in the diabetic complications. Most of the approved treatments of diabetes are directed to take control over hyperglycemia which is evidently known as the mediator of most of the diabetic complications and impaired body systems (Ceriello 2005; Williamson et al. 1993). Blood sugar levels can lead to serious cell damage with increasing levels of reducing sugars resulting in high levels of reactive oxygen species after a set of reactions. Under hyperglycemic conditions, these increased levels of reactive oxygen species, as a result of processing high blood sugar levels, directly damage the tissue inducing an oxidative stress (Poitout and Robertson 2002; Robertson and Harmon 2006). Patients who are diagnosed with type 2 diabetes can retain their insulin production by healthy pancreatic β-cells for many years after the onset. However, untreated or poorly controlled hyperglycemia will impair insulin secretion function in the later stages of the disorder. Impaired β-cell function leads to increased cellular damage and less controllable complications throughout the body (Ihara et al. 1999; Robertson et al. 2003).

4.2 Chitosan and its derivatives against diabetes

Chitosan-based products are known to have many biological activities, such as antitumor, anti-HIV, antifungal, antibiotic, and act against oxidative stress (Artan et al. 2010; Kendra and Hadwiger 1984; Kim et al. 2008; Nishimura et al. 1998; Xie et al. 1999). Activities can be grouped into two according to the use of chitin-based products. These products are highly used as indirect helping agents to enhance the effectiveness of other active compounds through chemical modification or nonchemical linkage against diabetes and obesity. On the other hand, the main role of chitin-based products is to act as therapeutic nutraceutical agents directly against diabetes and obesity. In both cases, derivatives of these natural products express a high and significant potential in the light of searching bioactive pharmaceuticals against obesity and obesity-related diabetes.

4.2.1 Compounds with complementary utilization

The preferred route of drug administration for patients is mostly the oral route in chronic therapy of diseases and complications. However, delivery of many therapeutic peptides and proteins through the digestion system is still an unsolved problem, basically because of the size, hydrophilicity, and unstable conditions of these molecules. Thus, several chitosan derivatives have been developed over the years with improved properties for enhanced applicability (Fernandez-Urrusuno et al. 1999; Thanou et al. 2001). Therefore, recent studies focused on carrier products for administration of insulin efficiently in pre- or post-diabetic patients, and lately, one of these products is chitosan derivative. It has been reported by Portero et al. (2007) that chitosan sponges are quite successful in buccal administration of insulin. Moreover, up-to-date studies presumed that chitosan-derived particles are intensely usable for insulin administration orally with their high protective effect and harmless structure (Hari et al. 1996; Krauland et al. 2004, 2006). Results of some related studies have suggested that the observed drug delivery activity of chitosan is highly promising in the case of insulin. For example, studies showed that chitosan–insulin nanoparticles have a strong affinity to rat intestinal epithelium after 3 h post-oral administration (Ma et al. 2005). This suggests that chitosan as a cofactor for drug delivery makes insulin absorption safe and rapid. Carboxymethyl–hexanoyl chitosan is an amphiphilic chitosan derivative with important swelling ability and water solubility under natural conditions, and studies showed that these hydrogels can be used for encapsulating the poorly water-soluble drugs for effective drug delivery (Liu and Lin 2010), which lightens up the way for efficient insulin delivery by chitosan derivatives. Furthermore, Mao et al. (2005) showed that PEG–trimethyl chitosan complexes are efficiently coupled with insulin and easily taken up by Caco-2 cells.

Besides drug delivery activity for insulin, studies have shown that chitosan complexes can be efficiently used for gene delivery in gene therapy (Koping-Hoggard et al. 2001). Therefore, it can be easily adduced that chitosan complex derivatives are potent gene delivery targets for high prevalent diseases such as diabetes. Furthermore, it has been reported that these chitosan complexes possess relatively higher uptake and transfection efficiency than that of other polysaccharide complexes used for both drug and gene delivery (Huang et al. 2004). Several researches were conducted to prove chitosan as a nontoxic alternative to other cationic polymers, and results demonstrated a prominent potential for further studies on chitosan-based gene delivery systems (Sato et al. 2001). All these results suggest that chitosan and chitosan-based derivatives shed light upon the search for a harmless agent for drug and gene delivery, which is extremely crucial for diabetic patients' improved life standards.

Moreover, studies on streptozotocin (STZ)-induced diabetic rats expressed that chitosan-based sponges are highly effective at healing diabetic wounds in addition to treatment of diabetic patients. Wang et al. (2008) suggest that application of chitosan–collagen complex is an ideal wound-healing cover to enhance recovery from wounds such as diabetic skin wound, which provides a great potential for chitosan and its derivatives to be used clinically for diabetic patients.

To conclude, chitosan-based polymers show great potential for treatment of diabetes therapeutically with their high efficient drug and gene delivery properties as well as effectiveness on diabetic wound healing.

4.2.2 Compounds with direct effect

Overweight and obesity, two common health-threatening conditions, are considered to result in diabetes worldwide although there are no considerably enough treatments. Therefore, studies of chitosan are focused on its fat-lowering and fat-preventing activities. Several researchers have demonstrated that chitosan tends to bond with the ingested dietary fat and carry it out in the stool while preventing their absorption through the gut (Kanauchi et al. 1995). Relevant researches about fat-lowering activity of chitosan also have shown that chitosan is capable of absorbing fat up to five times of its weight. In respect to these results, there are several studies showing that chitosan derivatives lower the levels of low-density lipoprotein (LDL) while increasing the high-density lipoprotein (HDL) levels. Studies of chitosan and its fat-lowering activity have expressed chitosan and its derivatives as highly effective hypocholesterolemic agents with the ability of decreasing blood cholesterol level up to as much as 50% (Jameela et al. 1994; Maezaki et al. 1993). Moreover, diabetic patient-based studies clearly showed that daily administration of chitosan could drop the blood cholesterol levels by 6% with an increased level of HDL. Additionally, COS, the oligomerized derivatives of chitosan show high activity in regulating blood cholesterol levels. Especially, studies reported that COS are capable of regulating cholesterol levels even in liver. COS prevent the development of fatty liver caused by the action of hepatotropic poisons. Despite few studies on the action mechanism of COS in regulating the serum cholesterol level, several of them suggested possible mechanism of COS lowering the LDL levels. As Remunan-Lopez et al. (1998) suggested, the ionic structure of COS binds bile salts and acid, which inhibits lipid digestion through micelle formation. However, Tanaka et al. (1997) suggest a different mechanism of action for chitosan and COS where lipids and fatty acids are directly bonded by chitosan.

In addition to fat-lowering mechanisms of chitosan and its derivatives, studies have also proven that chitosan administration can lead to increased insulin sensitivity in animal models (Neyrinck et al. 2009). It has been shown that 3 month administration of chitosan significantly increased insulin sensitivity in obese patients and expressed a highly notable decrease in body weight and triglyceride levels (Hernandez-Gonzalez et al. 2010).

On the other hand, glucosamine and its derivatives are reported to be highly effective at inhibiting adipogenesis in vitro. Recent studies showed that phosphorylated derivative of glucosamine inhibited adipogenesis as well as fat accumulation of 3T3-L1 cells (Kim et al. 2010). Several researches suggested that acetylated chitin treatment causes adipocytes to break down fats and lower their triglyceride accumulation as much as half of control cells (Kong et al. 2011).

Kong et al. (2009) demonstrated clearly that sulfated derivative of glucosamine inhibited the proliferation and adipogenesis mechanism through 5′ adenosine monophosphate-activated protein kinase (AMPK) pathways in 3T3-L1 cells. Glucosamine, acetylated-, sulfated-, and phosphorylated-glucosamine derivatives are reported as successful adipogenic inhibitors with intense potential to prevent weight gain by adipogenesis in patients who are at risk for diabetes. Further, it has been reported that COS inhibit fat accumulation and adipogenesis in 3T3-L1 cell line (Cho et al. 2008). In addition, studies have shown that treatment with glucosamines reduced the triglyceride content of adipocytes and enhanced glycerol secretion as a lipid-lowering effect. Most of these studies have expressed better activity of chitosan-based compounds such as COS and glucosamines, after derivation by adding a charged side chain by phosphorylation and sulfation. Therefore, it can be suggested that cationic power of glucosamine and COS plays the main role in its anti-obesity effect. Further, a selective synthesis of phosphorylated or sulfated derivatives of chitosan and glucosamine will open up the way to a better understanding behind the structure–mechanism relation. However, up-to-date researchers have strong proofs to show that chitosan displays its anti-obesity effect through the PPAR-γ pathway of adipogenic differentiation, which results in fewer adipocytes and lipid accumulation. Collectively, chitosan and its derivatives such as glucosamine and COS successfully inhibit the differentiation of cells into adipocytes, and also enhance adipocytes to hydrolyze the triglycerides which show a significant effect against lipid accumulation in the body. This effect of chitosan and its derivatives demonstrates an important impact against obesity in the way of diabetes progression. Hence, they display a greater potential to be used as pharmaceutical agents.

Furthermore, chitosan and its oligosaccharides act as antidiabetic agents for treatment of diabetes for protecting pancreatic β-cells. In type 2 diabetes, although patients can retain healthy pancreatic β-cells for many years after the disease onset, chronic exposure to high glucose will impair β-cell function in later stages. Impaired β-cell functionality leads to cellular damage in type 2 diabetic patients (Ihara et al. 1999). Therefore, the protection of β-cells is quite important for elevated insulin secretion as a part of diabetes treatment. Recent studies reported COS as protective agents for pancreatic β-cells against high glucose-dependent cell deterioration (Karadeniz et al. 2010). It is suggested that at the same time COS could effectively accelerate the proliferation of pancreatic islet cells with elevated insulin secretion in the aid of lowering blood glucose levels. Liu et al. (2007) reported that COS treatment could improve the general situation and diabetic symptoms of rats, decrease the blood glucose levels, and normalize the impaired insulin sensitivity. Moreover, COS were reported as agents preventing the nonobese diabetic mice from developing type 1 diabetes, which might be related to several bioactivities of COS (Cao et al. 2004). These results supported the hypothesis that COS can protect pancreatic β-cells of diabetic patients and normalize the crucial insulin secretion. The mechanism behind this protection is studied and suggested as related to immunopotentiation and antioxidation activities of COS.

Renal failure is one of the most common diseases caused by diabetes mellitus. The metal cross-linked complex of chitosan, chitosan-iron (III), has been recently reported to be highly active in reducing phosphorus serum levels to treat chronic renal failure (Schoninger et al. 2010). This relatively new derivative of chitosan is significantly capable of adsorbing serum phosphorus in alloxan diabetes-induced rats with symptoms of renal failure progression.

Moreover, recent studies indicate that diabetics may be at higher risk for blood coagulation than nondiabetics. This life-threatening condition urges treatment for diabetic patients. Therefore, sulfated derivative of chitosan has been shown to possess anticoagulant potency (Vongchan et al. 2002). Furthermore, studies have reported that sulfated chitosan does not show antiplatelet activity unlike heparin, which is an effective anticoagulant agent. Collectively, results proved that sulfated chitosan is a more efficient agent than heparin, although heparin has been used for a long time in blood coagulation treatment.

In addition to COS, chitosan has also been reported to prevent the development and symptoms of noninsulin-dependent diabetes in rats as well as the complications of STZ inducement (Kondo et al. 2000). Briefly, reports suggest that chitosan products protect pancreatic cells and insulin secretion mechanism in diabetic conditions. Furthermore, these compounds can decrease the progression and complication rate of diabetes onset in animal models, demonstrating a great potential for chitosan products to be used as nutraceuticals in the treatment of diabetes.

4.3 Conclusions

The number of patients diagnosed with diabetes is rapidly increasing in recent years. There is no cure for diabetes, and controlling the diabetic complications is not at the desired level. On the other hand, high mortality and morbidity of diabetes urge effective preventing and treatment methods to this disorder. Environmental and genetic factors, which lead to diabetes and impaired pancreatic functions in later stages, also have to be kept under control for improved prevention of diabetes onset. On this matter, chitosan and its derivatives are gaining deserved interest for this purpose due to possessing various biological activities and having a remarkable potential to be used in numerous therapeutic applications. Many researches have been conducted in order to promote a way to use chitosan-based compounds for diabetic complication treatment as well as preventing them. Several studies reported that chitosan and its various derivatives are indeed promising lead compounds for efficient disorder control. Many of these studies exhibit valuable insights to provide acting mechanisms of compounds and proper utilization of them for prevention and/or treatment of diabetes-related disorders. Chitosan, its monomer glucosamine, oligomeric derivative chitooligosaccharides, and other reported derivatives express highly efficient activity in a manner of lowering lipid accumulation and cholesterol as well as pancreatic β-cell protection.

In addition, chemical modification of these materials prove to be an efficient way to improve their activity and broaden their utilization range as well as helping to understand the mechanism behind their antidiabetic effect. With further studies directed to enhance the potential and effectiveness of chitosan-based biomaterials, novel compounds with higher and more successful treatment efficiency and lower side effects are about to be achieved. In conclusion, reported evidences suggest that chitosan and its derivatives are promising lead compounds with highly potent utilization as nutraceuticals for treatment and prevention of diabetes and diabetes-related complications.

References

Artan, M., F. Karadeniz, M. Z. Karagozlu, M. M. Kim, and S. K. Kim. 2010. Anti-HIV-1 activity of low molecular weight sulfated chitooligosaccharides. *Carbohydrate Research* 345 (5):656–662.

Cao Z., B. Li, and X. Qiao. 2004. The effect of chitooligosaccharides on preventing the on set of diabetes in NOD mice. *The Journal of Medical Theory and Practice* 12:25–31.

Ceriello, A. 2005. Postprandial hyperglycemia and diabetes complications: Is it time to treat? *Diabetes* 54 (1):1–7.

Cho, Y. S., S. H. Lee, S. K. Kim, C. B. Ahn, and J. Y. Je. 2011. Aminoethyl-chitosan inhibits LPS-induced inflammatory mediators, INOS and COX-2 expression in RAW264.7 mouse macrophages. *Process Biochemistry* 46 (2):465–470.

Cho, E. J., M. A. Rahman, S. W. Kim, Y. M. Baek, H. J. Hwang, J. Y. Oh, H. S. Hwang, S. H. Lee, and J. W. Yun. 2008. Chitosan oligosaccharides inhibit adipogenesis in 3T3-L1 adipocytes. *Journal of Microbiology and Biotechnology* 18 (1):80–87.

Fei Liu, X., Y. Lin Guan, D. Zhi Yang, Z. Li, and K. De Yao. 2001. Antibacterial action of chitosan and carboxymethylated chitosan. *Journal of Applied Polymer Science* 79 (7):1324–1335.

Fenton, J. I., K. A. Chlebek-Brown, T. L. Peters, J. P. Caron, and M. W. Orth. 2000. The effects of glucosamine derivatives on equine articular cartilage degradation in explant culture. *Osteoarthritis and Cartilage* 8 (6):444–451.

Fernandez-Urrusuno, R., P. Calvo, C. Remunan-Lopez, J. L. Vila-Jato, and M. J. Alonso. 1999. Enhancement of nasal absorption of insulin using chitosan nanoparticles. *Pharmaceutical Research* 16 (10):1576–1581.

Hai, L., T. Bang Diep, N. Nagasawa, F. Yoshii, and T. Kume. 2003. Radiation depolymerization of chitosan to prepare oligomers. *Nuclear Instruments and Methods in Physics Research Section B: Beam Interactions with Materials and Atoms* 208:466–470.

Hari, P. R., T. Chandy, and C. P. Sharma. 1996. Chitosan/calcium–alginate beads for oral delivery of insulin. *Journal of Applied Polymer Science* 59 (11):1795–1801.

Hernandez-Gonzalez, S. O., M. Gonzalez-Ortiz, E. Martinez-Abundis, and J. A. Robles-Cervantes. 2010. Chitosan improves insulin sensitivity as determined by the euglycemic-hyperinsulinemic clamp technique in obese subjects. *Nutrition Research* 30 (6):392–395.

Huang, M., E. Khor, and L. Y. Lim. 2004. Uptake and cytotoxicity of chitosan molecules and nanoparticles: Effects of molecular weight and degree of deacetylation. *Pharmaceutical Research* 21 (2):344–353.

Huang, R., E. Mendis, and S. K. Kim. 2005. Factors affecting the free radical scavenging behavior of chitosan sulfate. *International Journal of Biological Macromolecules* 36 (1–2):120–127.

Ihara, Y., S. Toyokuni, K. Uchida, H. Odaka, T. Tanaka, H. Ikeda, H. Hiai, Y. Seino, and Y. Yamada. 1999. Hyperglycemia causes oxidative stress in pancreatic beta-cells of GK rats, a model of type 2 diabetes. *Diabetes* 48 (4):927–932.

Il'ina, A. V. and V. P. Varlamov. 2004. Hydrolysis of chitosan in lactic acid. *Applied Biochemistry and Microbiology* 40 (3):300–303.

Jameela, S. R., A. Misra, and A. Jayakrishnan. 1994. Cross-linked chitosan microspheres as carriers for prolonged delivery of macromolecular drugs. *Journal of Biomaterials Science. Polymer Edition* 6 (7):621–632.

Je, J. Y. and S. K. Kim. 2006. Chitosan derivatives killed bacteria by disrupting the outer and inner membrane. *Journal of Agricultural and Food Chemistry* 54 (18):6629–6633.

Jeon, Y. J. and S. K. Kim. 2000. Production of chitooligosaccharides using an ultrafiltration membrane reactor and their antibacterial activity. *Carbohydrate Polymers* 41 (2):133–141.

Jiang, L., F. Qian, X. He, F. Wang, D. Ren, Y. He, K. Li, S. Sun, and C. Yin. 2007. Novel chitosan derivative nanoparticles enhance the immunogenicity of a DNA vaccine encoding hepatitis B virus core antigen in mice. *Journal of Gene Medicine* 9 (4):253–264.

Kanauchi, O., K. Deuchi, Y. Imasato, M. Shizukuishi, and E. Kobayashi. 1995. Mechanism for the inhibition of fat digestion by chitosan and for the synergistic effect of ascorbate. *Bioscience, Biotechnology, and Biochemistry* 59 (5):786–790.

Karadeniz, F., M. Artan, C. S. Kong, and S. K. Kim. 2010. Chitooligosaccharides protect pancreatic β-cells from hydrogen peroxide-induced deterioration. *Carbohydrate Polymers* 82 (1):143–147.

Kendra, D. F. and L. A. Hadwiger. 1984. Characterization of the smallest chitosan oligomer that is maximally antifungal tofusarium solani and elicits pisatin formation inpisum sativum. *Experimental Mycology* 8 (3):276–281.

Kim, J. H., Y. S. Kim, K. Park, S. Lee, H. Y. Nam, K. H. Min, H. G. Jo, J. H. Park, K. Choi, S. Y. Jeong, R. W. Park, I. S. Kim, K. Kim, and I. C. Kwon. 2008. Antitumor efficacy of cisplatin-loaded glycol chitosan nanoparticles in tumor-bearing mice. *Journal of Controlled Release* 127 (1):41–49.

Kim, J. A., C. S. Kong, S. Y. Pyun, and S. K. Kim. 2010. Phosphorylated glucosamine inhibits the inflammatory response in LPS-stimulated PMA-differentiated THP-1 cells. *Carbohydrate Research* 345 (13):1851–1855.

Kim, S. K., P. J. Park, W. K. Jung, H. G. Byun, E. Mendis, and Y. I. Cho. 2005. Inhibitory activity of phosphorylated chitooligosaccharides on the formation of calcium phosphate. *Carbohydrate Polymers* 60 (4):483–487.

Kim, S. K. and N. Rajapakse. 2005. Enzymatic production and biological activities of chitosan oligosaccharides (COS): A review. *Carbohydrate Polymers* 62 (4):357–368.

Kondo, Y., A. Nakatani, K. Hayashi, and M. Ito. 2000. Low molecular weight chitosan prevents the progression of low dose streptozotocin-induced slowly progressive diabetes mellitus in mice. *Biological & Pharmaceutical Bulletin* 23 (12):1458–1464.

Kong, C. S., J. A. Kim, S. S. Bak, H. G. Byun, and S. K. Kim. 2011. Anti-obesity effect of carboxymethyl chitin by AMPK and aquaporin-7 pathways in 3T3-L1 adipocytes. *Journal of Nutritional Biochemistry* 22 (3):276–281.

Kong, C. S., J. A. Kim, and S. K. Kim. 2009. Anti-obesity effect of sulfated glucosamine by AMPK signal pathway in 3T3-L1 adipocytes. *Food and Chemical Toxicology* 47 (10):2401–2406.

Koping-Hoggard, M., I. Tubulekas, H. Guan, K. Edwards, M. Nilsson, K. M. Varum, and P. Artursson. 2001. Chitosan as a nonviral gene delivery system. Structure-property relationships and characteristics compared with polyethylenimine in vitro and after lung administration in vivo. *Gene Therapy* 8 (14):1108–1121.

Krauland, A. H., D. Guggi, and A. Bernkop-Schnurch. 2004. Oral insulin delivery: The potential of thiolated chitosan-insulin tablets on non-diabetic rats. *Journal of Controlled Release* 95 (3):547–555.

Krauland, A. H., D. Guggi, and A. Bernkop-Schnurch. 2006. Thiolated chitosan microparticles: A vehicle for nasal peptide drug delivery. *International Journal of Pharmaceutics* 307 (2):270–277.

Kuroiwa, T., S. Ichikawa, O. Hiruta, S. Sato, and S. Mukataka. 2002. Factors affecting the composition of oligosaccharides produced in chitosan hydrolysis using immobilized chitosanases. *Biotechnology Progress* 18 (5):969–974.

Liu, T. Y. and Y. L. Lin. 2010. Novel pH-sensitive chitosan-based hydrogel for encapsulating poorly water-soluble drugs. *Acta Biomaterialia* 6 (4):1423–1429.

Liu, B., W. S. Liu, B. Q. Han, and Y. Y. Sun. 2007. Antidiabetic effects of chitooligosaccharides on pancreatic islet cells in streptozotocin-induced diabetic rats. *World Journal of Gastroenterology* 13 (5):725–731.

Ma, Z., T. M. Lim, and L. Y. Lim. 2005. Pharmacological activity of peroral chitosan-insulin nanoparticles in diabetic rats. *International Journal of Pharmaceutics* 293 (1–2):271–280.

Maezaki, Y., K. Tsuji, Y. Nakagawa, Y. Kawai, M. Akimoto, T. Tsugita, W. Takekawa, A. Terada, H. Hara, and T. Mitsuoka. 1993. Hypocholesterolemic effect of chitosan in adult males. *Bioscience, Biotechnology, and Biochemistry* 57 (9):1439–1444.

Mao, S., O. Germershaus, D. Fischer, T. Linn, R. Schnepf, and T. Kissel. 2005. Uptake and transport of PEG-graft-trimethyl-chitosan copolymer-insulin nanocomplexes by epithelial cells. *Pharmaceutical Research* 22 (12):2058–2068.

Mao, S., X. Shuai, F. Unger, M. Simon, D. Bi, and T. Kissel. 2004. The depolymerization of chitosan: Effects on physicochemical and biological properties. *International Journal of Pharmaceutics* 281 (1–2):45–54.

Neyrinck, A. M., L. B. Bindels, F. De Backer, B. D. Pachikian, P. D. Cani, and N. M. Delzenne. 2009. Dietary supplementation with chitosan derived from mushrooms changes adipocytokine profile in diet-induced obese mice, a phenomenon linked to its lipid-lowering action. *International Immunopharmacology* 9 (6):767–773.

Nishimura, S. I., H. Kai, K. Shinada, T. Yoshida, S. Tokura, K. Kurita, H. Nakashima, N. Yamamoto, and T. Uryu. 1998. Regioselective syntheses of sulfated polysaccharides: Specific anti-HIV-1 activity of novel chitin sulfates. *Carbohydrate Research* 306 (3):427–433.

Poitout, V. and R. P. Robertson. 2002. Minireview: Secondary beta-cell failure in type 2 diabetes-a convergence of glucotoxicity and lipotoxicity. *Endocrinology* 143 (2):339–342.

Portero, A., D. Teijeiro-Osorio, M. J. Alonso, and C. Remuñán-López. 2007. Development of chitosan sponges for buccal administration of insulin. *Carbohydrate Polymers* 68 (4):617–625.

Prabaharan, M. 2008. Chitosan derivatives as promising materials for controlled drug delivery. *Journal of Biomaterials Applications* 23 (1):5–36.

Qin, C., Y. Du, L. Xiao, Z. Li, and X. Gao. 2002. Enzymic preparation of water-soluble chitosan and their antitumor activity. *International Journal of Biological Macromolecules* 31 (1–3):111–117.

Qin, C., H. Li, Q. Xiao, Y. Liu, J. Zhu, and Y. Du. 2006. Water-solubility of chitosan and its antimicrobial activity. *Carbohydrate Polymers* 63 (3):367–374.

Remunan-Lopez, C., A. Portero, J. L. Vila-Jato, and M. J. Alonso. 1998. Design and evaluation of chitosan/ethylcellulose mucoadhesive bilayered devices for buccal drug delivery. *Journal of Controlled Release* 55 (2–3):143–152.

Robertson, R. P. and J. S. Harmon. 2006. Diabetes, glucose toxicity, and oxidative stress: A case of double jeopardy for the pancreatic islet beta cell. *Free Radical Biology & Medicine* 41 (2):177–184.

Robertson, R. P., J. Harmon, P. O. Tran, Y. Tanaka, and H. Takahashi. 2003. Glucose toxicity in beta-cells: Type 2 diabetes, good radicals gone bad, and the glutathione connection. *Diabetes* 52 (3):581–587.

Roglic, G., N. Unwin, P. H. Bennett, C. Mathers, J. Tuomilehto, and S. Nag. 2005. The burden of mortality attributable to diabetes: Realistic estimates for the year 2000. *Diabetes Care* 28: 2130–2135.

Sato, T., T. Ishii, and Y. Okahata. 2001. In vitro gene delivery mediated by chitosan. Effect of pH, serum, and molecular mass of chitosan on the transfection efficiency. *Biomaterials* 22 (15):2075–2080.

Schoninger, L. M., R. C. Dall'Oglio, S. Sandri, C. A. Rodrigues, and C. Burger. 2010. Chitosan iron(III) reduces phosphorus levels in alloxan diabetes-induced rats with signs of renal failure development. *Basic & Clinical Pharmacology & Toxicology* 106 (6):467–471.

Tanaka, Y., S. Tanioka, M. Tanaka, T. Tanigawa, Y. Kitamura, S. Minami, Y. Okamoto, M. Miyashita, and M. Nanno. 1997. Effects of chitin and chitosan particles on Balb/c mice by oral and parenteral administration. *Biomaterials* 18 (8):591–595.

Thanou, M., J. C. Verhoef, and H. E. Junginger. 2001. Oral drug absorption enhancement by chitosan and its derivatives. *Advanced Drug Delivery Reviews* 52 (2):117–126.

Vongchan, P., W. Sajomsang, D. Subyen, and P. Kongtawelert. 2002. Anticoagulant activity of a sulfated chitosan. *Carbohydrate Research* 337 (13):1239–1242.

Wang, W., S. Lin, Y. Xiao, Y. Huang, Y. Tan, L. Cai, and X. Li. 2008. Acceleration of diabetic wound healing with chitosan-crosslinked collagen sponge containing recombinant human acidic fibroblast growth factor in healing-impaired stz diabetic rats. *Life Sciences* 82 (3–4):190–204.

Williamson, J. R., K. Chang, M. Frangos, K. S. Hasan, Y. Ido, T. Kawamura, J. R. Nyengaard, M. van den Enden, C. Kilo, and R. G. Tilton. 1993. Hyperglycemic pseudohypoxia and diabetic complications. *Diabetes* 42 (6):801–813.

Xie, W., P. Xu, and Q. Liu. 2001. Antioxidant activity of water-soluble chitosan derivatives. *Bioorganic & Medicinal Chemistry Letters* 11 (13):1699–1701.

5

Nutraceutical Benefits of Marine Sterols Derivatives

Se-Kwon Kim and Quang Van Ta

Contents

5.1 Introduction

Marine organisms are rich sources of structurally diverse bioactive compounds with various beneficial biological activities that promote good health. The importance of marine organisms as a source of novel bioactive substances is growing rapidly. With marine species comprising approximately one half of the total global biodiversity, the sea offers an enormous resource for novel compounds. Moreover, different kinds of substances have been procured from marine organisms, because they live in an exigent, competitive, and aggressive surrounding, which is different in many aspects from the terrestrial environment, a situation that demands the production of specific and potent active molecules. Among them, sterols had been reported as potential bioactive compounds.

Sterols are an important family of lipids, present in the majority of eukaryotic cells. In addition, sterols are categorized as the steroids group, and also contain the same fused four-ring core structure and have different biological roles as hormones and signaling molecules. Sterols are essential components of the membranes of all eukaryotic organisms, controlling membrane fluidity and permeability. Sterols are highly diverse in nature (Fahy et al., 2005; Gaulin et al., 2010; Kamenarska et al., 2006). Plant sterols have been shown to inhibit uptake of both dietary and endogenously produced (biliary) cholesterol from the intestine. Clinical studies have demonstrated that sterols have the ability of lowering "bad" low-density lipoprotein cholesterol (LDL-C) levels. In addition, plant sterols did not have effect on "healthy" high-density lipoprotein cholesterol (HDL-C) and triacylglycerol levels. Clinical studies have shown that plant sterols could reduce the risk of heart disease by prevention and reduction of hypercholesterolemia. Several studies have indicated that phytosterols might have health-promoting effects such as anticancer activity. Additionally, plant sterols have been suggested to have anti-inflammatory, antibacterial, antifungal activities. Furthermore, long-term studies on animal models and humans did not show toxicity effect of plant sterols (Moreau et al., 2002). Sterols also can be found in marine algae. It has been reported that brown algae (Phaeophyceae) contain mainly fucosterol and fucosterol derivatives; that red algae (Rhodophyceae) contain mainly cholesterol and cholesterol derivatives; and that green algae (Chlorophyceae) contain mainly ergosterol and 24-ethylcholesterol (Sánchez-Machado et al., 2004). Moreover, marine animals and microbes also contain various bioactive substances.

Hence, the search for natural bioactive sterols as safe alternatives from marine organisms is important in the food industry. This chapter focuses on the potential of marine-derived sterols on human health benefits.

5.2 Biological activities of marine sterols

5.2.1 Antituberculosis activity

Tuberculosis is the second commonest cause of death worldwide. Thirty-two percent of the world's populations are infected with *Mycobacterium tuberculosis*, the main cause of tuberculosis. Most forms of active tuberculosis can be treated with 6 months of medication. Unfortunately, outbreaks of multidrug-resistant tuberculosis have been occurring since 1990s. Natural products form one avenue in the search for new antituberculosis agents. Recently, marine algae have become targets for screening programmers in search of novel compounds of potential medical value. It had been reported that saringosterol had been isolated from brown algae *Sargassum ringgoldianum* and had antitubercular activity (Ikekawa et al., 1968). Its minimum inhibition concentration (MIC) versus drug susceptance *M. tuberculosis* $H_{37}Rv$ was determined to be 0.25 µg/mL, which is the lowest value found for plant-derived natural products, compared to the tuberculosis drug rifambicin that also determined a MIC of 0.25 µg/mL in the same assay. In addition, low concentration of saringosterol showed no toxicity against mammalian cells in the monkey kidney epithelial (Vero) cells assay. In this assay, sargosterol showed half-maximal inhibitory concentration (IC_{50}) on Vero cells higher than $128\,\mu g/mL$ of concentration. The sargosterol isolated from *Lessonia nigrescens* is a 1:1 mixture of both 24*S* and 24*R* epimers. Individual isomers were isolated by normal-phase high performance liquid chromatography. In the antitubercular assay, the 24*R* isomer was found eight times more active against *M. tuberculosis* $H_{37}Rv$ with MIC of $0.125\,\mu g/mL$ than the 24*S* isomer, which had a MIC of $1\,\mu g/mL$. Saringosterol may be considered as an excellent lead compound due to its activity, specificity, and low toxicity (Wächter et al., 2001). Recently, screening extracts from 15 algae on antimycobacterial activity, it was found that extracts from *Isochrysis galbana* inhibited multidrug-resistant (MDR) *M. tuberculosis*. Screening on seven isolated MDR *M. tuberculosis*, which resisted more than three antibacterial drugs, extracts from *I. galbana* showed the MIC of $50\,\mu g/mL$ compared to the tuberculosis drugs rifambicin ($40\,\mu g/mL$), amikacin ($700\,\mu g/mL$), streptomycin ($4\,\mu g/mL$), ρ-amino salicylic acid ($2.5\,\mu g/mL$), isoniazide ($0.2\,\mu g/mL$). Extracts were purified and chromatographed by gas chromatography. The sterol composition of eluted compounds showed 13 unsaturated sterols with three major sterols such as 24-oxocholesterol acetate, ergost-5-en-3 β-ol, and Cholest-5–24–1,3-(acetyloxy)-,3β-ol. This finding indicated that the presence of sterols might have an effect on MDR *M. tuberculosis* (Prakash et al., 2010). Parguesterols A and B from Caribbean Sea sponge *Svenzea zeai* had been reported to show potent antituberculosis activity. Antituberculosis assay on *M. tuberculosis* $H_{37}Rv$ showed that MICs for the antitubercular activity of parguesterols A and B were determined as 7.8 and $11.2\,\mu g/mL$, respectively. Moreover, they showed low toxicity against Vero cells ($IC_{50} = 52\,\mu g/mL$) (Wei et al., 2007). Therefore, marine sterols have the potential for the development of new antituberculosis agents.

5.2.2 Antifungal activity

Candida fungus has been reported to cause superficial infections such as oropharyngeal candidiasis and vulvovaginal candidiasis. In debilitated or immunocompromised patients, if introduced intravenously, candidiasis may become a systemic disease producing abscess, thrombophlebitis, endocarditis, or infections of the eyes or other organs (Darwazeh et al., 1990). It has been reported that isolated sterol from sponge *Dysidea arenaria*, 9α,11α-epoxycholest-7-ene-3β,5α,6α,19-tetrol 6-acetate (ECTA), showed enhancement in antifungal activity when combined with fluconazole, one common antifungal drug. Fluconazole alone showed $IC_{50} = 300\,\mu M$ on drug-resistant *Candida albicans*; however, IC_{50} of fluconazole is decreased to $8.5\,\mu M$ when combined with $3.8\,\mu M$ of ECTA (Jacob et al., 2003). Sterols from the Red Sea marine sponge *Lamellodysidea herbacea*, Cholesta-8,24-dien-3β,5α,6α-triol and Cholesta-8(14),24-dien-3β,5α,6α-triol, also showed antifungal activity against *Candida tropicalis* (Sauleau and Bourguet-Kondracki, 2005). Collectively, sterols from marine resources have potential for applications in antifungal activities.

5.2.3 Antiviral activity

Human immunodeficiency virus type-1 (HIV-1) is identified as the causative agent for acquired immunodeficiency syndrome (AIDS), which is one of the most important diseases. The first generation anti-HIV drugs have been developed to treat AIDS patients after the infection of AIDS in early 1980s. However, failure in anti-AIDS treatment is observed in most of the patients infected with HIV as a result of the drug-resistant strains of the virus. Therefore, the search for potential drug candidates containing higher inhibitory activity against various HIV strains is increasing in the pharmaceutical industry. In this regard, natural bioactive compounds and their derivatives are great sources for the development of new-generation anti-HIV therapeutics, which are more effective with minor side effects.

Marine organisms are valuable resources of anti-HIV compounds. The sulfated sterol ibisterol was isolated from the deep-water Caribbean sponge *Topsentia* sp. that was investigated as an anti-HIV agent (McKee et al., 1993). Sun et al. (1991) reported that Weinbersterol disulfates A was isolated from the sponge *Petrosia weinbergi*, which is active in vitro against HIV. Studies on sterols isolated from marine sponges have shown that Ibisterol sulfate from *Topsentia* sp., halistanol sulfate from *Halichondria* cf. *moorei*, 26-methylhalistanol sulfate and 25-demethylhalistanol sulfate from *Pseudaxinyssa digitata* showed essentially complete protection against the cytopathic effects of HIV-1 infection at half-maximal effective concentration (EC_{50}) of 13, 6, 3, and 6 µM, respectively. The 10 sulfated sterols, isolated from sea stars *Tremaster novaecaledoniae, Asterias amurensis, Styracaster caroli*, and *Echinaster brasiliensis*, showed inactivation against HIV at the concentration of 100 µg/mL (McKee et al., 1994). Moreover, clathsterol isolated from the Red Sea sponge *Clathria* sp., inhibits HIV-1 reverse transcriptase activity at a concentration of 10 µM (Rudi et al., 2001). Hence, marine sterols have the potential for the development of new antiviral agents.

5.2.4 Antioxidant activity

The deterioration of some foods has been identified due to oxidation of lipids or rancidity and formation of undesirable secondary lipid peroxidation products. Lipid oxidation by reactive oxygen species (ROS) such as superoxide anion, hydroxyl radicals, and H_2O_2 also causes a decrease in the nutritional value of lipid foods, and affects their safety and appearance. In the food and pharmaceutical industries, many synthetic commercial antioxidants, such as butylated hydroxytoluene, butylated hydroxyanisole, tert-butylhydroquinone, and propyl gallate have been used to retard the oxidation and peroxidation processes. However, the use of these synthetic antioxidants must be under strict regulation due to potential health hazards (Hettiarachchy et al., 1996; Park et al., 2001). Hence, the search for natural antioxidants as safe alternatives from natural resources such as marine algae is important in the food industry.

Fortunately, it was reported that fucosterol isolated from marine algae *Pelvetia siliquosa* had antioxidant activity. Rats were treated with fucosterol at a dose of 30 mg/kg/day for 7 days, prior to the administration of carbon tetrachloride (CCl_4). Fucosterol causes a significant elevation of free radical scavenging enzyme activities such as superoxide dismutase (SOD), catalase, and glutathione peroxidase (GSH-px). Increase in the catalase activity with respect to CCl_4 treatment indicated that fucosterol could play an important role in scavenging hydrogen peroxide. Elevation of SOD activity indicated that fucosterol could help in cellular defense mechanism by preventing cell membrane oxidation. In addition, the increase in glutathione peroxidase activity indicated that fucosterol also helped in the restoration of vital molecules such as cytochrome and glutathione. The results showed that the sterol has an antioxidant activity on the rat model (Lee et al., 2003). Hence, fucosterol can be used as potential antioxidants in the food industry.

5.2.5 Anticancer activity

Some of the marine algae and their secondary metabolites have shown promising anticancer activities, and hence they are important sources in the manufacture of novel anticancer drugs. Sheu et al. had shown that sterols from brown alga *Turbinaria ornata* show cytotoxicity effect on P-388 (mouse lymphocytic leukemia) cells, KB (human mouth epidermoid carcinoma) cells, and HT-29 (human colon carcinoma) cells (Sheu et al., 1997). Oxygenated fucosterol from *Turbinaria conoides* also shows cytotoxic effect on cancer cells (Sheu et al., 1999). Moreover, new sterols from *Sargassum carpophyllum* showed cytotoxic effect on HL-60 (human promyelocytic leukemia) cells (Tang et al., 2002). It was also reported that oxysterol from red alga *Jania rubens* showed an ID_{50} value of 0.5 µg/mL toward KB cells (Ktari et al., 2000). Recently, two sterol glycosides isolated from red alga *Peyssonnelia* sp. displayed moderate activity toward human cancer cell lines. These compounds displayed good cytotoxicity toward human breast cancer MDA-MB-468 with IC_{50} = 0.71 and 0.86 µM, respectively; and human lung cancer cell line A549, which is often more resistant to cytotoxins than other common cancer cells, was inhibited with IC_{50} = 0.93 and 0.97 µM, respectively (Lin et al., 2010).

Marine sponges are the sources of numerous novel sterols also. Their sterols have shown promising anticancer activities, and hence, they are important sources in the manufacture of novel anticancer drugs. It had been reported that 5α,6α-epoxy-24*R**-ethylcholest-8(14)-en-3β,7α-diol, 5α,6α-epoxy-24*R**-ethylcholest-8-en-3β,7α-diol had been purified from the marine sponge *Polymastia tenax*. These sterols showed antiproliferative activity toward lung (A549), colon (HT-29 and H-116), mice endothelial (MS-1), and human prostate carcinoma (PC-3) cell lines (Santafé et al., 2002). Orostanal, the purified sterol from the marine sponge *Stellatta hiwasaensis*, inhibited proliferation in human leukemia (HL-60) cell line at IC_{50} = 1.7 µM (Miyamoto et al., 2001). Moreover, sterols from the marine sponge *Lanthella* sp., petrosterol-3,6-dione,

5α,6α-epoxy-petrosterol, and petrosterol, showed the cytotoxicity activity against A549, HT-29, breast (MCF-7), ovary (SK-OV-3), and two types of leukemia (HL-60 and U937) human cancer cell lines with IC_{50} values from 8.4 to 69.9 μM. Petrosterol-3,6-dione showed growth-inhibitory effects with IC_{50} values of 8.4, 19.9, 17.8, 16.2, and 22.1 μM against A549, HL-60, MCF-7, SK-OV-3, and U937 cell lines, respectively. 5α,6α-epoxy-petrosterol showed cytotoxic effects with IC_{50} values of 9.8, 21.3, 19.4, 22.6, and 19.9 μM. The case of petrosterol showed growth-inhibitory activities with 11.5, 21.5, 16.4, 19.8, and 18.7 μM, respectively (Nguyen et al., 2009).

In addition, sterols can also be found in other marine organisms such as corals and starfish. Lobophytosterol from the soft coral *Lobophytum laevigatum* showed cytotoxicity effect on A549 and HL-60 with IC_{50} values of 4.5 and 5.6 μM, respectively. Treatment with lobophytosterol induced apoptosis, which is evident by chromatin condensation in the treated cells (Tran et al., 2011). Sterols from the Caribbean gorgonian *Plexaurella grisea* showed that cholest-3β,5α,6β,7β-tetrol and cholest-3β,5α,6β-triol exhibited cytotoxicity activity against the A549 and HT 29 cell lines with effective dose (ED_{50}) values of 1 μg/mL. The new compounds 5β,6β-epoxyergost-24(28)-ene-3β,7β-diol, together with the known compound ergost-24(28)-ene-3α,5β,6β-triol, exhibited strong and selective cytotoxicity against the HT-29 cell line with an ED_{50} = 0.1 μg/mL, and an ED_{50} = 0.25 μg/mL for ergost-24(28)-ene-3β,5α,6β,7β-tetrol (Rueda et al., 2001). Purified armartnol A from soft coral *Nephthea armata* exhibited cytotoxicity against A549, HT-29, and P-388 (mouse lymphocytic leukemia) cell lines with IC_{50} value of 7.6, 6.5, and 6.1 μM, respectively. Moreover, armartnol B showed cytotoxicity against P-388 and HT-29 cells with IC_{50} values of 3.2 and 3.1 μM, respectively (El-Gamal et al., 2004). Wang et al. (2004) reported that certonardosterol Q6 from starfish *Certonardoa semiregularis* exhibited cytotoxicity effect on A549, SK-OV-3 (human ovarian cancer), SK-MEL-2 (human skin cancer), and HCT 15 (human colon cancer) cell lines at ED_{50} values of 0.43, 0.22, 0.17, and 0.48 μg/mL, respectively. Collectively, sterols from marine resources have the potential for the development of anticancer agents.

5.2.6 Anti-inflammation activity

Inflammation is the response of the body to injury and danger. It was characterized by persistently activated immune cells. The sources of inflammation include infectious agents, physical and chemical agents, and dysfunction of immune system. Moreover, chronic inflammation is proposed to be at the root of approximately one fifth of all human cancers (De Marzo et al., 2007). Sustained cellular injuries can cause inflammation. Various inflammation and innate immune cells are often recruited at the site of infection or damage. In response to proinflammatory stimuli, activated immune cells generate ROS and reactive nitrogen species (RNS), which can function as chemical effectors in inflammation (Halliday, 2005). In addition, it has been reported that cyclooxygenase 2 (COX-2) is a key mediator of inflammation.

COX-2 mediates prostaglandin synthesis, which in turn promotes cell proliferation, cytokine synthesis, and suppresses immune surveillance (Maitra et al., 2002). Another factor related to inflammation is nuclear factor kappa-B (NF-κB). The NF-κB protein complex is activated in response to inflammation (Karin, 2006), which is evident from the fact that COX-2 inhibitors reduce the incidence of carcinoma and reduce the risk of colorectal cancer in patients that overexpress COX-2 (McKay et al., 2008). Hence, finding anti-inflammatory agents is required for the prevention of carcinogenesis. Fortunately, marine resources provide many bioactive products. Among them, sterols are abundant in marine microbes. It was reported that new sterol Bendigole F, isolated from marine sponge-derived *Actinomadura* sp. SBMs009, showed translocation inhibition of NF-κB at IC_{50} of 71 μM. The inhibition of NF-κB translocation suggested that Bendigole F has potential anti-inflammation activity (Simmons et al., 2011). Sterol glycoside Carijoside A, isolated from the octocorals belonging to the genus *Carijoa* (=*Telesto*), had shown the anti-inflammatory activity in the significant inhibition of superoxide anion generation (IC_{50} = 1.8 μg/mL) and elastase release (IC_{50} = 6.8 μg/mL) by human neutrophils (Liu et al., 2010). 3-Hydroxy-26-norcampest-5-en-25-oic acid from *Euryspongian* sp. was reported to have an effect on cyclooxygenase pathway, and thus could be a potential anti-inflammatory natural compound derived from the marine sponge (Mandeaua et al., 2005). Huang et al. had reported that eight new sterols were isolated from the soft coral *Nephthea chabroli* Audouin (Nephtheidae), and was named Nebrosteroids A–H. Studies on LPS-stimulated RAW 264.7 cells showed that Nebrosteroids D, E, G at a concentration of 10 μM significantly reduced the level of iNOS and COX-2 protein expression. Nebrosteroids A, B, C, and H significantly reduced the iNOS protein expression, but did not inhibit the COX-2 protein expression (Huang et al., 2008). In addition, fractionations of the CH_2Cl_2 extracts of the Formosan soft coral *Clavularia viridis* Quoy and Gaimard (class Anthozoa, subclass Octocorallia, order Stolonifera) resulted in the isolation of four new steroids, Stoloniferones R, S, T, and (25S)-24-methylenecholestane-3, 5, 6-triol-26-acetate. Among them, Stoloniferones S and (25S)-24-methylenecholestane-3, 5, 6-triol-26-acetate significantly reduced the levels of iNOS protein at a concentration of 10 μM. Stoloniferones T and (25S)-24-methylenecholestane-3, 5, 6-triol-26-acetate at a concentration of 10 μM, significantly reduced COX-2 protein expression (Chang et al., 2008). Collectively, these results suggest that sterols derived from marine soft corals and sponges have a promising potential to be used as valuable chemopreventive agents in inflammation therapy.

5.2.7 Antidiabetic activity

Diabetes mellitus is a chronic metabolic disorder characterized by high blood glucose levels. Diabetes without proper treatments can cause many complications. Acute complications include hypoglycemia, diabetic ketoacidosis, or nonketonic hyperosmolar coma. Serious long-term complications include cardiovascular disease, chronic renal failure, and retinal damage.

Hence, antidiabetic agents are urgently required. Luckily, it was reported that fucosterol from *P. siliquosa* has antidiabetic activity. Fucosterol at doses of 100 and 300 mg/kg reduced hyperglycemic effect by 25%–33% in epinephrine-induced diabetes mouse model. Moreover, fucosterol at doses of 100 and 300 mg/kg showed a decrease of 23% and 29% in glycogen degradation of mouse liver. It was suggested that fucosterol from marine algae *P. siliquosa* has the potential in development of antidiabetic agents (Lee et al., 2004).

5.2.8 Antihypertensive activity and prevention of cardiovascular disease

Angiotensin-I-converting enzyme (ACE) plays an important physiological role in the regulation of blood pressure by converting angiotensin I to angiotensin II, a potent vasoconstrictor. Further, ACE is implicated in cell oxidative stress, augmenting the generation of ROS and peroxynitrite, and also in thrombosis, during which ACE induces platelet activation, aggregation, and adhesion (McFarlane et al., 2003). Inhibition of ACE is considered to be a useful therapeutic approach in the treatment of hypertension. Therefore, in the development of drugs to control high blood pressure, ACE inhibition has become an important activity. Many studies have been attempted in the synthesis of ACE inhibitors such as captopril, enalapril, alcacepril, and lisinopril, which are currently used in the treatment of essential hypertension and heart failure in humans. However, these synthetic drugs are believed to have certain side effects such as cough, taste disturbances, skin rashes, or angioneurotic edema all of which might be intrinsically linked to synthetic ACE inhibitors. The search for natural ACE inhibitors as alternatives to synthetic drugs is of great interest to prevent several side effects (Atkinson and Robertson, 1979). Fucosterol was reported as a safety agent on animal models. The modulation of ACE levels was studied using fucosterol in cultured bovine carotid endothelial cells. Dexamethasone was treated to elevate the levels of ACE in the cells. After adding fucosterol to the culture medium, the activity of ACE in endothelial cells was decreased; however, fucosterol did not directly inhibit ACE activity. It was found that fucosterol lowers the ACE levels on endothelial cells by inhibiting the synthesis of glucocorticoid receptors involved in the regulation of ACE levels (Hagiwara et al., 1986).

LDL cholesterol is called "bad" cholesterol, because elevated levels of LDL cholesterol are associated with an increased risk of coronary heart disease. LDL lipoprotein deposits cholesterol on the artery walls, causing the formation of a hard, thick substance called cholesterol plaque. Over time, cholesterol plaque causes thickening of the artery walls and narrowing of the arteries, a process called atherosclerosis. By contrast, HDL cholesterol is called the "good cholesterol," because HDL cholesterol particles prevent atherosclerosis by extracting cholesterol from the artery walls and disposing them through the liver. Higher HDL-C level provides greater capacity to remove cholesterol and prevent dangerous blockages from developing in blood vessels. HDL-C helps to keep blood vessels widened (dilated), thereby promoting better blood flow. HDL-C also reduces blood vessel injury through its antioxidant and anti-inflammatory functions, among other effects (Toth, 2005). Plant sterols have been reported as agents that can reduce the risk of heart disease by lowering LDL-C level. Therefore, it has been suggested that marine algal sterols could be used to prevent cardiovascular diseases. According to Plaza et al., sterols from several edible marine algae, such as *Himanthalia elongata, Undaria pinnatifida, Porphyra* spp., *Chondrus crispus, Cystoseira* spp., *Ulva* spp., have potential effects in reducing total and LDL-C levels (Plaza et al., 2008). In addition, 4-methylsterols from *Crypthecodinium cohnii* had no effect on any serum or liver lipid parameter. However, the percentage of serum HDL-C level was increased by 25%. Addition of bile salt to a cholesterol-containing diet will raise serum and liver cholesterol levels. Rats fed cholesterol diet and cholesterol plus bile salt showed significant increase in total serum cholesterol and decrease in HDL-C level. Moreover, cholesterol–bile salt plus 4-methyls-terols raised the amount of HDL-C and triglyceride levels significantly, but no other serum or liver lipid parameters were affected (Kritchesky et al., 1999). These effects of marine sterols were considered potential in preventing the risk of cardiovascular diseases.

5.3 Conclusions

Marine organisms showed potential role in application for human health and nutrition. Furthermore, increasing knowledge on sterols derived from marine organisms has raised the awareness and demand for novel functional food ingredients and pharmaceuticals. Hence, sterol compounds from marine resources can be used as functional ingredients to reduce infectious diseases and cancer in human body. However, there are fewer studies on marine organism sterols compared with plant sterols. Therefore, future studies on marine sterols would give more beneficial sterols as bioactive compounds. Collectively, the sterols, derived from marine organisms, have the potential for expanding its health-beneficial value not only in the food industry but also in the pharmaceutical industries.

References

Atkinson, A. B. and Robertson, J. I. S. 1979. Captopril in the treatment of clinical hypertension and cardiac failure. *Lancet* **2**:836–839.

Chang, C. H., Wen, Z. H., Wang, S. K., and Duh, C. Y. 2008. New anti-inflammatory steroids from the Formosan soft coral *Clavularia viridis*. *Steroids* **73**:562–567.

Darwazeh, A., Lamey, P., Samaranayake, L. et al. 1990. The relationship between colonisation, secretor status and in-vitro adhesion of *Candida albicans* to buccal epithelial cells from diabetics. *J Med Microbiol* **33**:43–49.

De Marzo, A. M., Platz, E. A., Sutcliffe, S. et al. 2007. Inflammation in prostate carcinogenesis. *Nat Rev Cancer* **7**:256–269.

El-Gamal, A. A. H., Wang, S. K., Dai, C. F., and Duh, C. Y. 2004. New nardosinanes and 19-oxygenated ergosterols from the soft coral *Nephthea armata* collected in Taiwan. *J Nat Prod* **67**:1455–1458.

Fahy, E., Subramaniam, S., Brown, H. A. et al. 2005. A comprehensive classification system for lipids. *J Lipid Res* **46**:839–861.

Gaulin, E., Bottin, A., and Dumas, B. 2010. Sterol biosynthesis in oomycete pathogens. *Plant Signal Behav* **5**:258–260.

Hagiwara, H., Wakita, K., Inada, Y., and Hirose, S. 1986. Fucosterol decreases angiotensin converting enzyme levels with reduction of glucocorticoid receptors in endothelial cells. *Biochem Biophys Res Commun* **139**:348–352.

Halliday, G. M. 2005. Inflammation, gene mutation and photoimmunosuppression in response to UVR-induced oxidative damage contributes to photocarcinogenesis. *Mutat Res* **571**:107–120.

Hettiarachchy, N. S., Glenn, K. C., Gnanasambandan, R., and Johnson, M. G. 1996. Natural antioxidant extract from fenugreek (*Trigonella foenumgraecum*) for ground beef patties. *J Food Sci* **61**:516–519.

Huang, Y. C., Wen, Z. H., Wang, S. K., Hsua, C. H., and Duh, C. Y. 2008. New anti-inflammatory 4-methylated steroids from the Formosan soft coral *Nephthea chabroli*. *Steroids* **7**:1181–1186.

Ikekawa, N., Morisaki, N., Tsuda, K., and Yoshida, T. 1968. Sterol compositions in some green algae and brown algae. *Steroids* **12**:41–48.

Jacob, M. R., Hossain, C. F., Mohammed, K. A. et al. 2003. Reversal of fluconazole resistance in multidrug efflux-resistant fungi by the *Dysidea arenaria* sponge sterol $9\alpha,11\alpha$-epoxycholest-7-ene-$3\beta,5\alpha,6\alpha,19$-tetrol 6-acetate. *J Nat Prod* **66**:1618–1622.

Kamenarska, Z., Stefanov, K., Stancheva, R., Dimitrova-Konaklieva, S., and Popov, S. 2006. Comparative investigation on sterols from some Black Sea red algae. *Nat Prod Res* **20**:113–118.

Karin, M. 2006. Nuclear factor-kappa B in cancer development and progression. *Nature* **441**:431–436.

Kritchevsky, D., Tepper, S.A., Czarnecki, S. K., and Kyle, D. J. 1999. Effects of 4-methylsterols from algae and of β sitosterol on cholesterol metabolism in rats. *Nutr Res* **19**:1649–1654.

Ktari, L., Blond, A., and Guyot, M. 2000. 16β-Hydroxy-5α-cholestane-3,6-dione, a novel cytotoxic oxysterol from the red alga *Jania rubens*. *Bioorg Med Chem Lett* **10**:2563–2565.

Lee, S., Lee, Y. S., Jung, S. H., Kang, S. S., and Shin, K. H. 2003. Anti-oxidant activities of fucosterol from the marine algae *Pelvetia siliquosa*. *Arch Pharm Res* **26**:719–722.

Lee, Y. S., Shin, K. H., Kim, B. K., and Lee, S. 2004. Anti-diabetic activities of fucosterol from *Pelvetia siliquosa*. *Arch Pharm Res* **27**:1120–1122.

Lin, A. S., Engel, S., Smith, B. A. et al. 2010. Structure and biological evaluation of novel cytotoxic sterol glycosides from the marine red alga *Peyssonnelia* sp. *Bioorg Med Chem* **18**:8264–8269.

Liu, C. Y., Hwang, T. L., Lin, M. R. et al. 2010. Carijoside A, a bioactive sterol glycoside from an octocoral *Carijoa* sp. (Clavulariidae). *Mar Drugs* **8**:2014–2020

Maitra, A., Ashfaq, R., and Gunn, C. R. 2002. Cyclooxygenase 2 expression in pancreatic adenocarcinoma and pancreatic intraepithelial neoplasia—An immunohistochemical analysis with automated cellular imaging. *Am J Clin Pathol* **118**:194–201.

Mandeaua, A., Debitus, C., Aries, M. F., and David, B. 2005. Isolation and absolute configuration of new bioactive marine steroids from *Euryspongian* sp. *Steroids* **70**:873–878.

McFarlane, S. I., Kumar, A., and Sowers, J. R. 2003. Mechanisms by which angiotensin-converting enzyme inhibitors prevent diabetes and cardiovascular disease. *Am J Cardiol* **91**(Suppl):30H–37H.

McKay, C. J., Glen, P., and McMillan, D. C. 2008. Chronic inflammation and pancreatic cancer. *Best Pract Res Clin Gastroenterol* **22**:65–73.

McKee, T. C., Cardellina, J. H., Riccio, R. et al. 1994. HIV-inhibitory natural products. 11. Comparative studies of sulfated sterols from marine invertebrates. *J Med Chem* **37**:793–797.

McKee, T. C., Cardellina, J. H., Tischler, M., Snader, K. M., and Boyd, M. R. 1993. Ibisterol sulfate, a novel HIV-inhibitory sulfated sterol from the deep water sponge *Topsentia* sp. *Tetrahedron Lett* **34**:389–392.

Miyamoto, T., Kodama, K., Aramaki, Y., Higuchia, R., and Van Soest, R. W. M. 2001. Orostanal, a novel abeo-sterol inducing apoptosis in leukemia cell from a marine sponge, *Stelletta hiwasaensis*. *Tetrahedron Lett* **42**:6349–6351.

Moreau, R. A., Whitaker, B. D., and Hicks, K. B. 2002. Phytosterols, phytostanols, and their conjugates in foods: Structural diversity, quantitative analysis, and health-promoting uses. *Progr Lipid Res* **41**:457–500.

Nguyen, H. T., Chau, V. M., Tran, T. H. et al. 2009. C29 sterols with a cyclopropane ring at C-25 and 26 from the Vietnamese marine sponge *Ianthella* sp. and their anticancer properties. *Bioorg Med Chem Lett* **19**:4584–4588.

Park, P. J., Jung, W. K., Nam, K. D., Shahidi, F., and Kim, S. K. 2001. Purification and characterization of antioxidative peptides from protein hydrolysate of lecithin-free egg yolk. *J Am Oil Chem Soc* **78**:651–656.

Plaza, M., Cifuentes, A., and Ibáñez, E. 2008. In the search of new functional food ingredients from algae. *Trends Food Sci Technol* **19**:31–39.

Prakash, S., Sasikala, S. L., and Huxley, J. A. V. 2010. Isolation and identification of MDR-*Mycobacterium tuberculosis* and screening of partially characterised antimycobacterial compounds from chosen marine micro algae. *Asian Pacif J Trop Med* **3**:655–661.

Rudi, A., Yosief, T., Loya, S., Hizi, A., Schleyer, M., and Kashman, Y. 2001. Clathsterol, a novel anti-HIV-1 RT sulfated sterol from the sponge *Clathria* species. *J Nat Prod* **64**:1451–1453.

Rueda, A., Zubía, E., Ortega, M. J., and Salva, J. 2001. Structure and cytotoxicity of new polyhydroxylated sterols from the Caribbean gorgonian *Plexaurella grisea*. *Steroids* **66**:897–904.

Sánchez-Machado, D. I., López-Hernández, J., Paseiro-Losada, P., and López-Cervantes, J. 2004. An HPLC method for the quantification of sterols in edible seaweeds. *Biomed Chromatogr* **18**:183–190.

Santafé, G., Paz, V., Rodriguez, J., and Jimenez, C. 2002. Novel cytotoxic oxygenated C29 sterols from the Colombian marine sponge *Polymastia tenax*. *J Nat Prod* **65**:1161–1164.

Sauleau, P. and Bourguet-Kondracki, M. L. 2005. Novel polyhydroxysterols from the Red Sea marine sponge *Lamellodysidea herbacea*. *Steroids* **70**:954–959.

Sheu, J. H., Wang, G. H., Sung, P. J., Chiu, Y. H., and Duh, C. Y. 1997. Cytotoxic sterols from the formosan brown alga *Turbinaria ornata*. *Plant Med* **63**:571–572.

Sheu, J. H., Wang, G. H., Sung, P. J., and Duh, C. Y. 1999. New cytotoxic oxygenated fucosterols from the brown alga *Turbinaria conoides*. *J Nat Prod* **62**:224–227.

Simmons, L., Kaufmann, K., Garcia, R., Schwär, G., Huch, V., and Müller, R. 2011. Bendigoles D-F, bioactive sterols from the marine sponge-derived *Actinomadura* sp. SBMs009. *Bioorg Med Chem* **19**:6570–6575.

Sun, H. H., Gross, S. S., Gunasekera, M., and Koehn, F. E. 1991. Weinbersterol disulfates A and B, antiviral steroid sulfates from the sponge *Petrosia weinbergi*. *Tetrahedron* **47**:1185–1190.

Tang, H. F., Yang-Hua, Y., Yao, X. S., Xu, Q. Z., Zhang, S. Y., and Lin, H. W. 2002. Bioactive steroids from the brown alga *Sargassum carpophyllum*. *J Asian Nat Prod Res* **4**:95–101.

Toth, P. P. 2005. The "good cholesterol": High-density lipoprotein. *Circulation* **111**:e89–e91.

Tran, H. Q., Tran, T. H., Chau, V. M. et al. 2011. Cytotoxic and PPARs transcriptional activities of sterols from the Vietnamese soft coral *Lobophytum laevigatum*. *Bioorg Med Chem Lett* **21**:2845–2849.

Wächter, G. A., Franzblau, S. G., Montenegro, G., Hoffmann, J. J., Maiese, W. M., and Timmermann, B. N. 2001. Inhibition of *Mycobacterium tuberculosis* growth by saringosterol from *Lessonia nigrescens*. *J Nat Prod* **64**:1463–1464.

Wang, W. H., Jang, H. J., Hong, J. K. et al. 2004. Additional cytotoxic sterols and saponins from the starfish *Certonardoa semiregularis*. *J Nat Prod* **67**:1654–1660.

Wei, X. M., Rodriguez, A. D., Wang, Y. H., and Franzblau, S. G. 2007. Novel ring B abeo-sterols as growth inhibitors of *Mycobacterium tuberculosis* isolated from a Caribbean sea sponge, *Svenzea zeai*. *Tetrahedron Lett* **48**:8851–8854.

6

Nutritional Value, Bioactive Compounds, and Health-Promoting Properties of Abalone

Mahanama De Zoysa

Contents

6.1 Introduction

Maintaining a healthy lifestyle is a basic need for most people. It is well known that good health is strongly associated with diet and many other factors such as genetics, environment, lifestyle habits, and physical activity. People are highly concerned about selecting healthy foods with a wide range of medicinal values to reduce their risk of chronic diseases such as coronary heart diseases, hypertension, obesity, cancer, diabetes, and osteoporosis. Numerous reports have shown that marine-based food products have remarkably higher benefits in maintaining good health and well-being (Jha and Zi-rong, 2004; Rajasekaran et al., 2008). There is a wide range of marine food sources, such as carbohydrates, fats, proteins, vitamins, and minerals, to fulfill the basic nutritional requirements, but consumers are increasingly interested in marine foods with additional health benefits (Larsen et al., 2011). Such foods that aid specific functions for promoting health benefits are called functional foods. Any food, either whole or a partial ingredient, becomes a nutraceutical product when it contains specific health-enhancing compounds with capacities of preventing and treating diseases (Brower, 1990). At present, introducing a new functional food with higher nutraceutical value is a priority, with the aim of improving human health as well as reducing medicinal cost and supporting related industries and people who culture or grow such food sources.

Abalone is a commercially important marine "archeogastropod" mollusk with characteristic single auriform shell under the family of Haliotidae (Lee and Vaequier, 1995). As a food, abalone has been in demand for a long time due to its rich nutritional value, superior taste, and various other benefits to human health among other mollusk species; hence, it is known as "the emperor of the seashells," "mother of shellfish," or "ginseng in the ocean" (Kim et al., 2006; Lee et al., 2010). There are 56 recognized abalone species that belong to the genus *Haliotis*; however, it is believed that there could be more than 100 species of abalones (Geiger, 2000). Abalones are distributed worldwide in temperate and tropical marine waters. Naturally, abalones are found along reefs and rocky shores from sea level to a depth of 30 m (Degnan et al., 2006). Aquaculture of abalone begun in the late 1950s in Japan and China. At present, China is the largest producer of cultured abalone, but almost full production is consumed domestically. The remaining abalone production is shared mainly by Taiwan, Chile, South Africa, Australia, New Zealand, South Korea, France, Ireland, and the United States (Freeman, 2001; Troell et al., 2006). Also, several abalone species (*Haliotis discus hannai, Haliotis discus discus, Haliotis diversicolor supertexta*) are popular in commercial aquaculture, depending on the area where they are cultured. The commercial value (as a food) of these animals is mainly due to the presence of an abductor and foot muscles, digestive tract, and other body masses.

As with other marine-derived foods, abalones also contribute to health-promoting effects, mainly through nutrients and bioactive components (Benkendorff, 2010). However, most of the available information related to health benefits of abalone has not been well documented. Moreover, therapeutic properties are reported mainly from crude extracts of abalone and are required to isolate and identify the active compounds or agents. Also, molecular mechanisms related to bioactive compounds and their structure–function relationships are not understood in most of the identified samples. To highlight the nutraceutical potential and medicinal value of abalone, establishing and updating available data should be prioritized in a systematic way. That could be important for transferring such information to the research community or to product development professionals to focus on new directions in producing abalone-based nutraceuticals. In general, nutraceuticals are considered as biomolecules or agents, which have health-promoting effects on cardiovascular, anticoagulant, antiobese, antidiabetes, anticancer, antimicrobial, immune enhancer, chronic inflammatory, and degenerative diseases (Rajasekaran et al., 2008). According to available information, abalone is a safe marine organism as a food source with much nutraceutical and therapeutic benefit. The aim of this chapter is to discuss the evidence of nutritional benefits, available bioactive compounds, and the health-promoting effects of abalone and their potentials in nutraceutical developments.

6.2 Nutritional composition and related health-promoting properties

Although shellfish like abalones contain varied nutrient content, which often can be changed depending on temperature (season), extraction time (growth stage), and reproductive stage, the consumption of abalone provides multiple benefits, including rich nutrition diet and protection against various diseases (Yoo and Chung, 2007; Zhou et al., 2012). Abalone contains lower amount of calories compared with other animal foods. Current dietary recommendations indicate that reducing total calories in the diet results in maintaining body weight and decreasing the risk of various disorders, especially the cardiovascular diseases (Larsen et al., 2011). Numerous studies have shown that abalone-like mollusks carry balanced nutritional characteristics that enhance human health (Benkendorff, 2010). Abalone is a rich source of protein that helps to maintain all kinds of health benefits. Analysis of nutritional composition shows that the edible portion of 100 g abalone consists of 20 g of protein (www.health-benefits-of.org). Collagen is the most abundant structural protein in the animal body. Two types of fibril-forming collagens have been identified from *Haliotis discus* foot muscle (Yoneda et al., 1999). Also, a significantly higher quantity of taurine is present in abalone that aids in protecting the intestine, reducing the oxidative damage via antioxidant function, maintaining a healthy liver, quick recovery from fatigue, preventing myocardial infection, and minimizing allergic reactions (Kim et al., 2006, 2007). Seven and eight kinds of nonvolatile organic acids have been identified from dried–boiled and fresh abalone, respectively. The most abundant organic acids are reported as succinic, lactic, and pyroglutamic acids (Jo and Park, 1985).

Different isolation techniques have been applied to screen polysaccharide profile in abalone. Most commonly applied methods are water, methanol, ethanol, and alkaline protease extractions (Zhu et al., 2008). A polysaccharide has been isolated from *Haliotis diversicolor* Reeve that consists of glucose, galactose, mannose, xylose, fucose, and galactofuranosiduronic acid (She et al., 2002). Most of the isolated polysaccharides from abalone are combined with sulfate groups as sulfated polysaccharides, which are associated with various bioactivities such as antioxidant and anticoagulant activities (Zhu et al., 2008). Bioactivities of various abalone products will be discussed in detail at the later part of this chapter.

Zhou et al. (2012) have determined the fatty acid composition of *Haliotis discus hannai* Ino, using enzyme-assisted organic solvent extraction technique.

It further describes that gonads of abalone contain high level of lipids (20.87% on dry basis), which is mainly contributed by 89.05% of triglycerides, 7.34% of phospholipids, 1.46% of cholesterol, and 2.15% of free fatty acids. More importantly, abalone has higher amount of beneficial fats of omega ($\acute{\omega}$)-3 polyunsaturated fatty acids ($\acute{\omega}$-3PUSFs), particularly $\acute{\omega}$-3 eicosapentaenoic acid (EPA) and $\acute{\omega}$-3 docosahexaenoic acid (DHA). EPA and DHA are considered as marine fatty acids, and mostly seafood contain naturally higher amounts of these fatty acids. Marine phytoplankton is the main source of these fatty acids, which are accumulated in abalone-like animals via food chain (Larsen et al., 2011). Consumption of foods with EPA and DHA has been associated with positive effects on numerous diseases, including obesity, metabolic syndrome, diabetes, and anti-inflammatory diseases (Nettleton and Katz, 2010). Moreover, a daily intake of $\acute{\omega}$-3PUSFs up to 400–500 mg with EPA and DHA is generally recommended (Oh et al., 2010). Hence, consuming abalone-derived foods could be of benefit to every person since it contains considerably higher amounts of natural EPA and DHA. Also, the amount as well as the type of fat has a great influence on health. Foods with higher saturated fatty acids are linked with great risk of diseases. Total lipid content (0.76 g/100 g) in abalone is lower than that in other seafood such as shrimp (1.73 g/100 g), trout (3.46 g/100 g), and salmon (5.93 g/100 g) (http://www. health-benefits-of.org). Therefore, considerably lower amounts of total fat and higher $\acute{\omega}$-3PUSFs in abalone can have beneficial effects in reducing the risk of such diseases. Some micronutrients are abundantly found in marine organisms (Larsen et al., 2011). Vitamin and mineral profile of abalone shows that it constitutes higher amounts of vitamin E, B_{12}, magnesium, sodium, and selenium. Additionally, it is an excellent source of iron; a 100 g serving of abalone can cover the 21% of daily iron intake. However, abalone has high level of cholesterol than other seafood, which may have negative effect on health (http://www.health-benefits-of.org).

Abalone viscera accounts for 15%–25% of the total body weight of abalone, which are by-products and generally discarded (Zhou et al., 2012). Abalone viscera also contains higher levels of nutrients such as polysaccharides, protein, fat, and certain minerals (Li et al., 2007). In freeze-dried abalone (*H. discus hannai* Ino), viscera consists of 53.04% protein, 19.28% sugar, and 18.09% polysaccharides (Sun et al., 2010). It is important to determine the real facts that are associated with high level of nutritional components and other bioactive substances in abalone viscera. The biodiversity of marine ecosystem allows abalones to have different strategies on maintaining nutritional features. Abalones largely consume the marine macroalgae (phytoplankton) such as *Ulva rigida*, Ecklonia, Laminaria, and Undaria, which contain excessive amounts of complex carbohydrates, dietary fiber, EPA, and DHA (Hwang et al., 2009). It is easy to assume that the visceral components of abalone may contain these molecules originated from algae as well as can be accumulated in abalone tissues via the food chain. Moreover, nutritional components derived from algae could be further modified into various types of bioactive molecules.

Nonedible abalone shell has been used in traditional medicine from a long time. The chemical and physical properties of abalone shell hydrolysate have been analyzed, and results showed that it is a complicated system with water, minerals, polysaccharides, proteins, glycoproteins, β-charatin, and lactic acid radical (Wang et al., 2003; Weiss at al., 2001). The main ingredient of abalone shell is calcium carbonate, and it contains a variety of other elements including iron, magnesium, sodium, silicon, and some amino acids such as aspartic acid and glutamic acid (Lee, 1994; Xu, 1999). In Chinese medicine, powdered abalone shell that is mixed with water or other ingredients like ginger (*Zingiber officinale*) has been used to make liquid form (tonic) to improve visual activity and for treating cataract, hypertension, eye-related pain, red sore eyes, dizziness, and headache (Yeung, 1983). Also, a patent has been taken for burn remedial composition and related methods, which is produced by mixing the powder of the high heated (>3000° C) abalone shells with sesame oil (Lee, 1994). Applying abalone shell-derived burn remedial mixture can completely cure the heat and fire burns within 3–5 days.

6.3 Value-added health-promoting products from abalone

Generally, abalone products are marketed in raw/fresh, dried, smoked, and cooked or various other forms, such as making pickles, drinks, fermented products with oriented herbs (Koh et al., 2009; Shin et al., 2008). In Korea, "Jeonbok-juk" is a rice porridge cooked with mince abalone, which is given as a stamina food to women who have given birth, elderly, weak or sick people due to its higher protein, mineral, and vitamin content (Kim et al., 2006). Abalone source "Jeonbokjang" is another popular and value-added item that is made with hot pepper paste, soy sauce, and oriental herbs (Shin et al., 2008). Moon et al. (2011) have determined that herbal "Jeonbokjang" has 106.22 mg of antioxidant capacity (mg vitamin C equivalent/100 g sample) and 120.25 mg of total phenolic content (mg garlic acid equivalent/100 g). It is further revealed that antioxidant capacity and total phenolic content of "Jeonbokjang" can be enhanced by cold storage at 5°C. Abalone intestine as well as the viscera portion is commonly used for preparing sashimi or pickle. In Chinese traditional medicine, differently prepared abalone such as dried body mass and shell powders is given as a drug to reduce high blood pressure, prevent anemia, and enhance blood circulation. It is given to increase the appetite of the patients. Also, the dried abalone mixed with turnip (*Brassica rapa*) has been used for treating diabetes (Yeung, 1983).

At present, few companies have developed several value-added commercial products using abalone as the main ingredient. Abalone (*H. discus hannai*)-based food drink named as "abalone essence healthy liquid" has been developed in China, especially for treating weak-health and after-sickness persons (Dalian Haibao Fishery Company Ltd., China). It contains an anti-cancer substance named as "abalone essence," polysaccharides, and peptides, the individual components of which have not been specifically described in the product information. Producers of abalone essence healthy liquid® stated that it can significantly boost immunity, suppress tumor by activating macrophages and T cells, reduce blood sugar, resist inflammation, and provide benefiting effects to lung, stomach, and intestine. "Abalone active element capsule" has been developed using concentrated nutritious extract from 5- to 10-year-old abalones living in Japanese water (Qingdao King Industry Co., China). It is being recommended for different groups, including children; high intensive brain workers; elderly or weak people; patients with high blood pressure, cholesterol, and sugar; and breastfeeding women. Abalone capsule as a dietary supplement is being developed using freeze-dried abalone powder that is mixed with silicon dioxide and magnesium citrate (HealthChemist, New Zealand). It contains highly concentrated ingredients, with more than 25 vitamins and minerals, especially selenium, magnesium, and vitamin B_{12}. This capsule is supposed to enhance overall immunity and antifatigue effects. Abalone *Haliotis iris* powder is another dietary supplement product, which contains rich glycosaminoglycans, proteins, minerals, and vitamins. It has shown various health-promoting effects such as enhancement of liver, supporting joints and connective tissues, proper functioning of heart and blood circulation, preventing anemia, and supporting immune system and sexual life (Aroma, New Zealand). Few abalone-based cosmetics have been reported. Abalone *H. iris* (Paua) is an endemic species in New Zealand and is highly valued for its nutritional and therapeutic properties (Hooker and Creese, 1995). Skin care products like "Paua facial scrub" are made from *H. iris* extract that can be used as deep cleanser for removing excess oil on the skin (http://www.skincarenz.com). Further isolation and functional characterization of bioactive materials are required to develop nutraceutical products from abalone.

6.4 Bioactive compounds and health-promoting properties

Availability of a wide range of bioactive molecules in particular foods could lead to support health benefits via modulating molecular mechanisms in cells. It is considered that marine environment is the best source for unique natural products, which are mainly accumulated in algae and abalone-like marine organisms (Jha and Zi-rong, 2004). Various studies have shown that marine-derived natural bioactive molecules are applied in biomedical applications due to their pharmacological activities (Mayer et al., 2009). At present, abalones are being studied for developing different bioactive molecules for biomedical applications. However, studies dealing with the structural characterization of these abalone-derived bioactive compounds are limited and mostly identified as crude samples or at macromolecular level. Polysaccharides or their derivatives are macromolecules abundantly available and play an important role in various cellular functions including signal transduction, regulation of immune responses, cell proliferation and differentiation, metabolic pathways, and other direct bioactive functions (e.g., killing of tumor cells). Next to cellulose, chitin is the most abundant polysaccharide in nature. Chitin can be converted to biologically active chitosan by chemical deacetylation. Chitosan exhibits a wide range of health-promoting properties that include antioxidant, wound healing, antibacterial, and anticoagulant (Prashanth and Tharanathan, 2007). Zenth et al. (2001) have extracted chitosan from the nacre of abalone *Haliotis tuberculata* and it may have potential health-promoting properties. However, the amount of chitosan is low when compared with sources like crustaceans.

Carotenoids are a group of phytochemicals that are involved in various protective effects against cancer, oxidative stress, and cardiovascular diseases (Larsen et al., 2011; Rajasekaran et al., 2008). Presence of β-carotene, lutein, fucoxanthin, neofucoxanthin, and other carotenoids have been found in shell, muscles, ovary, and testis of algal-fed abalones (Kantha, 1989). Eleven carotenoids (α-carotene, carotene, lutein, zeaxanthin, siponaxanthin, siponein, fucoxanthin, loroxanthin-like, and fucoxanthinnol-like) have been identified in the viscera of abalone *H. discus hannai*, and some of them were reported for the first time in mollusks (Ahn, 1974). A recent study describes the structural analysis of series of carotenoid compounds isolated from the viscera of abalone *Haliotis diversicolor aquatilis* (Maoka et al., 2011). These carotenoids are related to fucoxanthin pyropheophorbide A ester, halocynthiaxanthin 3′-acetate pyropheophorbide A ester, lutein pyropheophorbide A ester, and mutatoxanthin pyropheophorbide A ester. Previous data show that fucoxanthin is able to reduce obesity by inhibiting intestinal lipase activity (Matsumoto et al., 2010). Even though there are no reports related to the bioactivities of carotenoids isolated from abalone, we assume that these compounds may play an important role in human health when abalones are consumed.

The following section of this chapter presents the bioactivities and new pharmacological findings on the antioxidant, anticoagulant, anticancer, antimicrobial, and immune-modulatory area reported from abalone.

6.4.1 Antioxidant activities

Antioxidants play an important role in minimizing the oxidative stress via detoxifying the reactive oxygen species (ROS). The most common ROS are superoxide, hydroxyl anion, and nonradical oxygen derivatives (H_2O_2 and singlet oxygen). In general, cells produce ROS during their various cell activities. Uncontrolled generation of ROS can damage cellular protein, DNA, and lipids (Seifried et al., 2007). Foods containing higher dietary antioxidants can greatly support the human antioxidant system to minimize the oxidative stress. Sulfated polysaccharide conjugates isolated from the abalone, *H. discus hannai* Ino, viscera have shown promising antioxidant activities, such as scavenging of hydroxyl radical and metal-chelating activities in vitro (Zhu et al., 2008). The other important fact is that such molecules can be hydrolyzed by commercially available enzymes such as proteases (trypsin, vernase, neutrase, pepsin, and papain), and the digested products also contain similar antioxidant activities. Bioactivities of digested products can increase the range of potentials in therapy. Antioxidant and alcohol dehydrogenase (ALDH) activities were identified in water extracts using abalone-containing medicinal plant products (garlic, orange peel, honey, etc.), which have the potential to produce functional beverages (Shin et al., 2008). Results showed that the presence of phenol, flavonoids in water-soluble extracts displayed the hydroxyl radical scavenging, superoxide dismutase (SOD), ALDH activities under different combinations of plant materials with abalones. It indicates that abalone may release some bioactive agents when preparing extracts with herbal plant materials. Abalone gonad tissue has shown high antioxidant capacity (Li et al., 2009). The polysaccharide fraction has been extracted from the gonad tissue of abalone *H. discus hannai* by proteinase-hydrolysis coupled to alcohol extraction. Furthermore, gonad polysaccharide has shown higher hydroxyl, DPPH scavenging, and moderate antiperoxidation power of lipids. Li et al. (2007) reported that single homogeneous sulfated polysaccharide isolated from abalone viscera has molecular weight of 10,000–15,000 Da and 11.55% sulfate content. Various types of sugar units (rhamnose, fucose, xylose, galactose, and glucose) were identified by analyzing the chemical composition. It has been reported that there is no significant difference in the antioxidant capacity between the ethanol (80%)-extracted products of abalone (*H. discus hannai*) body and the visceral mass (Kim et al., 2006). This is an important finding that indicates that visceral mass is also a very good source for bioactive materials similar to abalone body tissues.

6.4.2 Anticoagulant activities

It is estimated that 0.7% of the population in the Western world receives anticoagulants such as heparin and warfarin in the treatment and prevention of venous thrombosis (Gustafsson et al., 2004). Molecules with anticoagulant activity are important to prevent formation of blood clots as well as to dissolve the existing blood clots. Though heparin and warfarin have shown very effective, fast action in inhibiting or delaying blood coagulation, serious side effects and drawbacks have been reported consistently. These drawbacks include drug–drug and food–drug interactions with side effects such as bleeding, skin rash, headache, and weakening of bone strength (Arbit et al., 2006). Hence, foods consisting of natural anticoagulant molecules have great potential to develop nutraceuticals with multiple benefits. Anticoagulant properties of crude extracts prepared from various abalone species have been described in recent years. An anticoagulant of glycosaminoglycan-like sulfated polysaccharide has been extracted from abalone, *H. discus hannai* Ino. Chemical analysis has shown that it consists of galactosamine, glucuronic acid, fucose, and galactose with 15.5% sulfate content. The effect of abalone anticoagulant polysaccharide as a specific inhibitor on the blood coagulation cascade at the stage of thrombin-mediated fibrin formation has been displayed (Li et al., 2011). In a rat model experiment, animals fed with freeze-dried abalone powder (5%) have shown significant increase in the prothrombin (PT) levels after 2 weeks compared to the respective control

(Kim et al., 2007). These in vivo experiment results provide strong evidence to prove that abalone-supplemented diet could be involved in the inhibition of blood coagulation.

6.4.3 Anticancer activities

According to the World Health Organization, cancer was responsible for 7.6 million deaths worldwide (around 13% of all deaths) in 2008 and is projected to continue increasing with an estimated 13.1 million deaths in 2030. Around 30% of cancer deaths could be prevented by avoiding risk factors through a healthy lifestyle and diet (http://www.who.int/cancer/en/). Currently, different strategies have been applied to prevent cancers by changing dietary factors (Donaldson, 2004). Also, there is no perfect treatment or medicine for cancer though various chemotherapies are being used. On the other hand, most of the cancer patients need long-term chemotherapy. However, these treatments can cause various negative side effects on health. Therefore, screening of new anticancer agents with minimum side effects should be implemented from potential sources. The extracts of abalone, which are composed of macromolecules such as polysaccharides, proteins, and peptides, possess antitumor activities (Cheng et al., 2008; Huang et al., 2009). Glycoprotein antineoplastic agent isolated from abalone, *H. discus hannai* has shown strong tumor inhibition in ICR mice or BALB/c mice inoculated with allogeneic sarcoma 180 or syngeneic Meth-A fibrosarcoma (Chen, 1998). Abalone extract obtained after chromatography consists of 22% of carbohydrate and 44% of protein. Furthermore, this abalone glycoprotein has enhanced the cytostatic activity of peritoneal and alveolar macrophages in vivo, indicating that it could be a potent source for activation of host-mediated antitumor activity (Uchida et al., 1987).

Antitumor effects of abalone (*H. diversicolor* Reeve) polysaccharide have been investigated on mice implanted with human nasopharyngeal cancer cells. Under the electron microscope observation, it has clearly shown that abalone polysaccharide could significantly inhibit the growth of implanted cancer cells by inducing cell apoptosis and necrosis (Wang et al., 2008). The combined effect of polysaccharide and cyclophosphamidum of abalone on tumor-bearing mice has been investigated. It has been shown that abalone polysaccharide could enhance the inhibition effects of cyclophosphamidum on HepA tumor and protection against leukocytopenia, while decreasing the weight of spleen and thymus and hemolysin content in serum (Wang et al., 1999). Very recently, breast cancer model (BALB/c-mouse derived 4T1 mammary carcinoma) has been designed to describe the anticancer efficacy of abalone visceral extract and to determine the possible clues of its mechanisms of action. It has clearly shown that oral administration of abalone visceral extract has significantly inhibited the tumor growth and metastasis in spleen and lung. The same study shows that abalone visceral extract can increase the antitumor activities and immune responses of CD8+ T cells. Moreover, certain cytokines including IFN-γ, TNF-α, and cytolytic molecules

(FasL, Gzm B, and Gzm C) have shown increased expression in CD8+ T cells while suppressing the Cox-2 expression after administration of abalone visceral extract (Lee et al., 2010). Zhu et al. (2009) have reported the antitumor effects of the proteoglycan extracted from abalone viscera in hepatic carcinoma cell H22-bearing mice. Tumor growth is inhibited significantly with abalone proteoglycan compared to control group with saline. Moreover, this study further confirmed that abalone proteoglycan is able to increase spleen weight, lymphocyte proliferation, NK cell, and phagocytotic activity of peritoneal macrophage. Additionally, it has shown enhanced level of serum TNF-α, IL-1, and IFN-γ in mice after administration of proteoglycan extract. However, structure–function relationship has not been reported for any of those bioactive molecules isolated from abalone extracts, which is required to be determined in the future to get the maximum therapeutic advantage of abalone-derived products.

6.4.4 Antimicrobial and immune-modulatory activities

Studies have shown antimicrobial and immune-modulatory effects of abalone-derived macromolecules. Several reports describe that abalone is a potential source of antiviral compounds. The canned abalone juice has protected the Swiss mice from poliovirus infections (Li, 1960).

Crude fresh abalone juice has shown antibacterial effects against *Staphylococcus aureus* (Li, 1960). A recent study demonstrated that crude hemolymph and lipophilic extract (from the digestive gland) of *Haliotis laevigata* processed inhibitory effects against herpes simplex virus type 1 (Dang et al., 2011). It further suggested that antiviral effect could be due to two different molecules specific to abalone hemolymph and digestive gland, indicating that bioactivities may vary with the type of abalone tissue or organ. Two polysaccharides (AVPI and AVPII) that are extracted from the viscera of abalone *H. discus hannai* Ino have similar monosaccharide profiling including rhamnose, glucose, xylose, mannose, galactose, and fucose but in different ratios (1.00:2.15:4.00:5.36:33.18:45.14 for AVPI whereas 1.00:1.46:1.33:4.98:16.08:12.51 for AVPII) (Wang et al., 2008). Both AVPI and AVPII have significantly stimulated lymphocyte proliferation, phagocytosis of macrophage, and natural killer cell activity in a dose-dependent manner. With in vitro experiment evidence, it convinces that abalone viscera is a potential source for immune-modulating agents.

6.4.5 Cosmeceutical activities

Cosmeceuticals are applied as topical applications to enhance the various aspects of skin care properties including transformation, regeneration, wound healing, and preventing wrinkles. Antiaging commercial cosmetic products (mask pack) have been developed from abalone. Abalone contains higher amounts of collagen, chondroitin sulfuric acid, amino acids, and minerals necessary for skin regeneration. Collagen is a fibrous protein with pharmaceutical, cosmetic, and biomedical value. Patent has been taken for an isolation process of a soluble, native collagen from the group consisting of the black-lip abalone (*Haliotis rube*), the brown-lip abalone (*Haliotis conicopora*), and the green-lip abalone (*H. laevigata*), or Roe's abalone (*Haliotis roei*) (U.S. Patent 20070179283). This abalone collagen is considered as an alternative to collagen isolated from terrestrial animals and shows potential in industrial applications.

6.4.6 Miscellaneous bioactivities

Few studies have been reported on the occurrence of antihypertensive compounds in abalone species. Crude extract (80% ethanol) of abalone *H. discus hannai* has shown promising antihypertensive (angiotensin-converting enzyme [ACE]) effect in vitro. ACE-inhibitory effects of visceral extracts were much higher than those of body extracts (Kim et al., 2006). An interesting experiment has been conducted to evaluate the effect of two abalone extracts on memory power of mice. The water and enzymatic extracts of abalone were given to mice on a daily basis for 10 days. A number of parameters such as step-down latency, escape latency in a passive avoidance, and food-hunting time in a maze have been compared. It was revealed that enzymatic extracts can improve the learning and memory more effectively than water extracts (Peng et al., 2004).

6.5 Conclusions

It is evident that consumption of abalone as food and traditional medicine has a long history in different communities. Abalones are low in fat, contain omega-3 fatty acids, are excellent protein sources, and are good sources of vitamins and minerals. Additionally, various bioactive sulfated polysaccharides, peptides, glycoproteins, chitosan, and carotenoids have been identified as the compounds that are associated with health-promoting properties including antioxidant, anticoagulant, anticancer, antimicrobial, and immune-modulatory effects. As highlighted, abalone is a source with higher nutritional value, biological active molecules supporting for multiple health benefits as common characteristics of nutraceuticals. However, nutritional value and type of bioactive properties can vary with regard to content and qualitatively depend on abalone species, season, extraction method, etc. Also, much need to be done to screen and exploit the chemical diversity of abalone with structure–function relationship related to health-promoting agents. Moreover, various bioactivities associated with isolated molecules of abalone provide strong evidence on its potential as a natural functional food. Therefore, we need to take the advantage of developing value-added healthy food products using abalone-derived nutraceuticals and biotechnological applications.

References

Ahn, S. 1974. Studies on the carotenoids in the viscera of abalone (*Haliotis discus hannai*). *J Korean Agric Chem Soc* 17:257–258.

Arbit, E., Goldberg, M., Gomez-Orellana, I., Majuru, S. 2006. Oral heparin: Status review. *Thrombois J* 4:6.

Aroma, New Zealand. Company profile and product specification. Accessed online http://aromanz.rad3web.com, May 17, 2012.

Benkendorff, K. 2010. Molluscan biological and chemical diversity: Secondary metabolites and medicinal resources produced by marine mollusks. *Biol Rev* 85:757–775.

Bing, W., Jianmin, J., Donghui, X., Shibo, X., Zhiyu, S., Yongcheng, L. 2000. Antitumor effect of abalone polysaccharide on human nasopharyngeal cancer inoculated on nude mice. *Chin Trad Herb Drugs* 8:597–599.

Brower, V. 1998. Nutraceuticals: Poised for a healthy slice of the healthcare market? *Nat Biotechnol* 16:728–731.

Chen, Q. 1998. Extraction of *Haliotis discus hannai* Ino. Polysaccharide and experimental study on its anti-tumor action. *Chin J Mod Appl Pharm* 1:8–10.

Cheng, T.T., Li, D.M., Liu, N., Zhu, B.W. 2008. Identification and activity assay of a polysaccharide from abalone harslet. *Chin J Mar Drugs* 2:9–13.

Dalian Haibao Fishery Company Ltd, China. Company profile and product specification. Accessed online http://hailongxian.com, May 17, 2012.

Dang, V.T., Benkendorfft, K., Speck, P. 2011. In vitro antiviral activity against herpes simplex virus in the abalone *Haliotis laevigata*. *J Gen Virol* 92:627–637.

Degnan, S.M., Imron, G.D.L., Degnan, B.M. 2006. Evolution in temperate and tropical seas: Disparate patterns in southern hemisphere abalone (Mollusca: Vetigastropoda: Haliotidae). *Mol Phylogenet Evol* 41:249–256.

Donaldson, M. 2004. Nutrition and cancer: A review of the evidence for an anticancer diet. *Nutr J* 3:19.

Extraction process for a pharmaceutical product, U.S. Patent Application 20070179283.

Freeman, K.A. 2001. Aquaculture and related biological attributes of abalone species in Australia—A review. Fisheries Research Report No. 128.

Geiger, D.L. 2000. Distribution and biogeography of the Haliotidae (Gastropoda: Vetigastropoda) world-wide. *Bollettino Malacologico* 35:57–120.

Gustafsson, D., Bylund, R., Antonsson, T., Nilsson, I., Nystrom, J.E., Eriksson, U. et al. 2004. A new oral anticoagulant: The 50-year challenge. *Nat Rev Drug Discov* 3:649–659.

HealthChemist, New Zealand. Company profile and product specification. Accessed online http://www.healthchemist.co.nz, May 17, 2012.

Hooker, S.H., Creese, R.G. 1995. Reproduction of Paua, *Haliotis iris* Gmelin 1791 (Mollusca, Gastropoda), in north eastern New Zealand. *Mar Freshwater Res* 46:617–622.

Huang, F.F., Yang, Y.F., Ding, G.F. 2009. Advances in the study of antitumor activities of the marine mollusks extracts. *J China Pharm* 3.

Hwang, E.K., Baek, J.M., Park, C.S. 2009. The mass cultivation of *Ecklonia stolonifera* Okamura as a summer feed for the abalone industry in Korea. *J Appl Phycol* 21:585–590.

Jha, R.K., Zi-rong, X. 2004. Biomedical compounds from marine organisms. *Mar Drugs* 2:123–146.

Jo, K.S., Park, Y.H. 1985. Studies on the organic acids composition in shellfishes 1. Nonvolatile organic acids composition of oyster, sea-mussel, baby clam, hen clam and their boiled-dried products. *Bull Korean Fish Soc* 18:227–234.

Kantha, S.S. 1989. Carotenoids of edible mollusks: A review. *J Food Biochem* 13:429–442.

Kim, H.L., Kang, S.G., Kim, I.C., Kim, S.J., Kim, D.W., Ma, S.J. et al. 2006. In vitro anti-hypertensive, antioxidant and anticoagulant activities of extracts from *Haliotis discus hannai*. *J Korean Soc Food Sci Nutr* 35:835–840.

Kim, H.L., Kim, S.J., Kim, D.W., Ma, S.J., Gao, T., Li, H. et al. 2007. The abalones, *Haliotis discus hannai*, exhibits potential anticoagulant activity in normal Sprague Dawley rats. *Korean J Food Preserv* 14:431–437.

Koh, S.M., Kim, H.S., Cho, Y.C., Kang, S.G., Kim, J.M. 2009. Preparation and physicochemical characteristics of abalone meat aged in gochujang. *J Korean Soc Food Sci Nutr* 37:773–779.

Larsen, R., Eilertsen, K.E., Elvevoll, E.O. 2011. Health benefits of marine foods and ingredients. *Biotechnol Adv* 29:508–518.

Lee, S.H. 1994. Burn remedial composition using natural materials and its production method. U.S. Patent No. 5,362,499.

Lee, C.G., Kwon, H.K., Ryu, J.H., Kang, S.J., Im, C.R., Kim, J.I. et al. 2010. Abalone visceral extract inhibit tumor growth and metastasis by modulating Cox-2 levels and CD8+ T cell activity. *BMC Complement Alternat Med* 10:60.

Lee, Y.H., Vaequier, V.D. 1995. Evolution and systematic in Haliotidae (Mollusca: Gastropoda): Inferences from DNA sequences of sperm lysin. *Mar Biol* 124:267–278.

Li, C.P. 1960. Antimicrobial effects of abalone juice. *Exp Biol Med* 103:522–524.

Li, G., Chen, S., Wang, Y., Xue, Y., Chang, Y., Li, Z. et al. 2011. A novel glycosaminoglycan-like polysaccharide from abalone *Haliotis discus hannai* Ino: Purification, structure identification and anticoagulant activity. *Int J Biol Macromol* 49:1160–1166.

Li, D.M., Tan, F.Z., Yang, J.F., Yin, H.L., Li, T., Zhu, B.W. 2007. Isolation, purification and identification of polysaccharides from abalone viscera. *Fish Sci* 9:485–488.

Li, T., Zhou, D.Y., Yang, J.F., Li, D.M., Yan, X., Bao, G.Q. et al. 2009. Antioxidant activity of polysaccharides from abalone (*Haliotis discus hannai*) Gonad. *Fish Sci* 4:179–182.

Maoka, T., Etoh, T., Akimoto, N., Yasui, H. 2011. Noval carotenoid pyropheophorbide A esters from abalone. *Tetrahedron Lett* 52:3012–3015.

Matsumoto, M., Hosokawa, M., Matsukawa, N., Hagio, M., Shinoki, A., Nishimukai, M. et al. 2010. Suppressive effect of the marine carotenoids, fucoxanthin and fucoxanthinol on triglyceride absorption in lymph duct-cannulated rats. *Eur J Nutr* 49:243–249.

Mayer, A.M.S., Rodriguez, A.D., Berlinck, R.G.S., Hamann, M.T. 2009. Marine pharmacology in 2005–6: Marine compounds with anthelmintic, antibacterial, anticoagulant, antifungal, anti-inflammatory, antimalarial, antiprotozoal, antituberculosis, and antiviral activities affecting the cardiovascular, immune and nervous systems, and other miscellaneous mechanisms of action. *Biochim Biophys Acta* 1790:283–308.

Moon, C.Y., Yoon, W.B., Hahm, Y.T., Kim, H.K., Baik, M.Y., Kim, B.Y. 2011. Optimization of processing conditions and evaluation of shelf-life of Jeonbokjang products. *Food Sci Biotechnol* 20:1419–1421.

Nettleton, J.A., Katz, R. 2005. n-3 long chain polyunsaturated fatty acids in type 2 diabetes: A review. *J Am Diet Assoc* 105:428–440.

Nkondjock, A., Ghadirian, P. 2004. Dietary carotenoids and risk of colon cancer: Case-control study. *Int J Cancer* 110:110–116.

Oh, D.Y., Talukdar, S., Bae, E.J., Imamura, T., Morinage, H., Fan, W. et al. 2010. GPR120 is an omega-3 fatty acid receptor mediating potent anti-inflammatory and insulin-sensitizing effects. *Cell* 142:687–698.

Peng, W.D., Chen, Q.L., Zhao, J.H. 2004. Effects of enzymolytic extracts of abalone on learning and memory in mice. *Acta Nutrimenta Sinica* 26:45–48.

Prashanth, K.V.H., Tharanathan, R.N. 2007. Chitin/chitosan: modifications and their unlimited application potential-an overview. *Trends Food Sci Technol* 18:117–131.

Qingdao King Industry Co, China. Company profile and product specification. Accessed online http://www.worldoftrade.com, May 17, 2012.

Rajasekaran, A., Sivagnanam, G., Xavier, R. 2008. Nutraceuticals as therapeutic agents: A review. *Res J Pharm Technol* 1:328–340.

Seifried, H.E., Anderson, D.E., Fisher, E.I., Milner, J.A. 2007. A review of the interaction among dietary antioxidants and reactive oxygen species. *J Nutr Biochem* 18:567–579.

She, Z., Hu, G., Wu, Y., Lin, Y. 2002. Study on the methanolysis of the sulphated polysaccharide hal-a from *Haliotis diverisicolor* Reeve. *Chin J Org Chem* 22:367–370.

Shin, J.H., Lee, S.J., Choi, D.J, Kang, M.J., Sung, N.J. 2008. Antioxidant and alcohol dehydrogenase activity of water extract from abalone containing medicinal plants. *Korean J Food Cookery Sci* 24:182–187.

Sun, L., Zhu, B., Li, D., Wang, L., Dong, X., Murata, Y. et al. 2010. Purification and bioactivity of a sulphated polysaccharide conjugate from viscera of abalone *Haliotis discus hannai* Ino. *Food Agric Immunol* 21:15–26.

Troell, M., Robertson-Andersson, D., Anderson, R.J., Bolton, J.J., Maneveldt, G., Halling, C. et al. 2006. Abalone farming in South Africa: An overview with perspectives on kelp resources, abalone feed, potential for on-farm seaweed production and socio-economic importance. *Aquaculture* 257:266–281.

Uchida, H., Sasaki, T., Uchida, N.A., Takasuka, N., Endo, Y., Kamiya, H. 1987.Oncostatic and immunomodulatory effects of a glycoprotein fraction from water extract of abalone, *Haliotis discus hannai*. *Cancer Immunol Immunother* 24:207–212.

Wang, B., Xu, D., Xu, S., Ling, Y., Shao, Z. 1999. Synergy and attenuation of cyclophosphamidum (CTX) activities by abalone polysaccharide. *J Chin Med Mat* 22:198–200.

Wang, J., Xu, Y., Zhao, Y., Huang, Y, Wang D., Jiang, L., Wu, J., Xu, D. 2003. Morphology and crystalline characterization of abalone shell and mimetic mineralization. *J Cryst Growth* 252:367–371.

Wang, L.S., Zhu, B.W., Sun, L.M., Li, D.M. 2008. Antitumor and immunomodulating activity in vitro of the polysaccharide from *Haliotis discus hannai*. Ino. *J Dalian Polytech Univ* 4:289–293.

Weiss, I.M., Gohring, W., Fritz, M., Mann, K. 2001. Perlustrin, a *Haliotis laevigata* (Abalone) Nacre protein, is homologous to the insulin-like growth factor binding protein N-terminal module of vertebrates. *Biochem Biophys Res Commun* 285:244–249.

Xu, D., Wang, B., Xu, S., Ling, Y., Shao, Z. 1999. Effects of abalong polysaccharide on the activity of the peritoneal macrophages and delayed-type hypersensitivity in mice bearing S180. *J Chin Med Mat* 22:88–89.

Yeung, H.C. 1983. *Handbook of Chinese Herbal Formulas*. Institute of Chinese Medicine, Rosemead, LA.

Yoneda, C., Hirayama, Y., Nakayo, M., Matsubara, Y., Ire, S., Hatae, K. et al. 1999. The occurrence of two types of collagen pro-chain in the abalone *Haliotis discus* muscle. *Eur J Biochem* 261:714–721.

Yoo, M.J., Chung, H.J. 2007. Optimal manufacturing condition and quality properties of the drinking extract of disk abalone. *Korean J Food Cult* 22:827–832.

Zenth, F., Bedouet, L., Almeida, M.J., Milet, C., Lopez, E., Giraud, M. 2001. Characterization and quantification of chitosan extracted from nacre of the abalone *Haliotis discus tuberculata* and the oyster *Pinctada maxima*. *Mar Biotechnol* 3:36–44.

Zhou, D.Y., Tong, L., Zhu, B.W., Wu, H.T., Qin, L., Tan, H. et al. 2012. Extraction of lipids from abalone (*Haliotis discus hannai Ino*) gonad by supercritical carbon dioxide and enzyme-assisted organic solvent method. *J Food Process Preserv* 36:126–132.

Zhu, L.L., Sun, L.M., Li, D.M., Zhu, B.W. 2009. The anti-tumor effects of the proteoglycan from abalone viscera against H-22 cell line in vivo. *Acta Nutrimenta Sinica* 5:478–481.

Zhu, B.W., Wang, L.S., Zhou, D.Y., Li, D.M., Sun, L.M., Yang, J.F. 2008. Antioxidant activity of sulphated polysaccharide conjugates from abalone (*Haliotis discus hannai* Ino). *Eur Food Res Technol* 227:1663–1668.

7

Marine Biopolymers in Asian Nutraceuticals

Ngo Dang Nghia and Se-Kwon Kim

Contents

7.1 Introduction

Marine biopolymers comprise a large range of macromolecules presented in plants and animals living in the sea whose native chemistry ranges from carbohydrates to proteins. They have exhibited more and more interesting properties that make them adapt to numerous applications from industries, food, and technology to advanced research in biomedicine. We can list some of the highly interesting representatives among them as alginate, fucoidan, chitin/chitosan, carrageenan, and bioactive hydrolyzed protein extract from marine animals.

In Asian countries like China, Korea, Japan, Vietnam, Thailand, and Indonesia, people are known to consume seaweed for routine meals and other animals like seahorse and edible bird's nest (EBN) for the treatment of some disorders or for health recovery. This habitude of using seaweed and animals in traditional medicine has existed for many hundreds of years or may be longer. The knowledge on the properties of these kinds of food has been accumulated only from experience for a long time. Today, with the modern facilities of scientific research, more and more structural aspects related to the functions of these foods have been elucidated. In this view, the biopolymers from brown seaweed (alginates, fucoidans) and the extracts from the two specific animals, EBN (mucoprotein) and seahorse (bioactive hydrolyzed protein), will be taken into account in this chapter.

7.2 Alginate

When we talk about alginate, it seems to be a very old topic since 1881 when the British chemist E.E.C. Stanford discovered it, and we are also familiar with the variety of its applications, such as gelation, thickening, and stabilization agents. But now, with more studies about the effects of alginate on health, it becomes a new material in nutraceuticals.

7.2.1 Structure of alginate

In chemical term, alginate is a family of polysaccharides that have similar structures. They are linear copolymers composed of 1→4 linked α-L-guluronic acid (G) and β-D-mannuronic acid (M). The two uronic acids do not distribute randomly but build up three blocks: polyguluronic (-GGGG-), polymannuronic (-MMMM-), and alternative (-MGMGMG-) blocks (Haug 1964, Haug et al. 1966). The average size of every block and their distribution in the whole chain of alginate change in a large range from species to species of brown algae, even from different parts of the same plant of alga where they are located in the cell wall and extracted. This made them possess a huge variety of structures and hence the properties (Grasdalen et al. 1979). The flexibility in the structure of alginate and so their properties is not unique but found naturally in many biopolymers like agar, chitin, and carrageenan. This feature may be responsible for the different functions of the bioorganisms for the different requirements of nature. However, this property is the greatest challenge for scientists and manufacturers in controlling the reproducibility of experiment or quality product of each batch as well as in mastering in using them in specific application related severely to structure.

7.2.2 Source of alginate

As a marine biopolymer, alginates are extracted mainly from brown algae that are distributed in many areas of the world and have the biggest individual size. The main species for commercial production are *Laminaria hyperborea*, *Laminaria japonica*, *Ascophyllum nodosum*, *Macrocystis pyrifera*, *Eclonia maxima*, *Lesonia negrescens*, and *Sargassum* (McHugh 1987). It is also found in some soil bacteria as *Azotobacter vinelandii* (Gorin et al. 1996) and some species of *Pseudomonas* (Govan et al. 1981). Alginate is the main constituent of the cell wall of these algae, and it may constitute up to 40% of the dry weight. In nature, alginate is mainly present as an insoluble calcium alginate. It made the frame flexible and strong enough for the algae in the sea water environment to live with the waves.

7.2.3 Properties of alginate

The properties of alginate are based on its structure. While the viscosity of the alginate solution is based on the molecular size, the intrinsic flexibility of alginate molecules in solution depends on the blocks in the order GG < MM < MG. The selectivity for binding of cations and gelation capacity are affected by the composition and sequence of alginate. The affinity series for various divalent cations are Pb > Cu > Cd > Ba > Sr > Ca > Co > Ni > Zn > Mn (Haug and Smidsrod 1970).

7.2.3.1 Ion-exchange and swelling properties

Alginate is an ion-exchange agent with the carboxyl groups between protons and cations as Na^+, Ca^{++}, Sr^{++}, Ba^{++}. When it comes to the body fluid that is rich in Na^+, this property is used in changing from insoluble to soluble form and releasing the contents from the gel matrix.

Alginate is insoluble in acid or multicovalent salts as calcium or strontium. When changing to sodium salt, it is soluble in water solution. Alginic acid can absorb large quantities of water and swell to a larger volume. This is a soft gel and is easily broken on strong agitation.

7.2.3.2 Immunogenicity

Alginate is produced in different grades as food, research, or ultra pure. It is shown that the mitogenic impurities are responsible for the side effects of alginate. These side effects include cytokine release and inflammatory reaction. In addition, the alginate rich in mannuronic acid activates cytokine production more than the alginate with high guluronic acid. For in vivo studies, it is necessary to choose alginates with high degree of purity and which are rich in guluronic acid to avoid the inflammatory reactions (Gombotz and Wee 1998).

7.2.3.3 Bioadhesion

The term "bioadhesion" refers to the contact or adhesion between two surfaces, with one being a biological substratum (Peppas and Buri 1985). In case of one surface being mucosal, it is called mucoadhesion. The interaction between the biopolymer and the mucosal layer is related to the charges of the molecules constituting the two surfaces. It is reported that polyanion polymers have bioadhesion better than polycation polymers or nonionic polymers (Park and Robinson 1984). Alginate, a biopolymer with many carboxyl groups, is a polyanionic mucoadhesive polymer. This property is useful in expanding the resident time of drug at the site of activity or resorption such as GI tract or the nasopharynx.

The mucoadhesive properties of a polymer depend on the flexibility of the backbone structure and its polar group. The surface of mucin, as any biological surface, is not smooth but more rough, so if the backbone chain is flexible, it is easy to contact. The more flexible the backbone of a polymer is, the more convenient is the adhesion in the mucosal layer. The interaction between adhesive materials and mucin bases on the electrostatic adsorption is through the van der Waals and hydrogen bonds. Because alginate has a carboxyl group with negative charge, it will link with the positive charge site of the mucosal surface. The hypothesis of mechanism of adhesion mentioned here is true for the free molecules of alginate. However, in practice, alginate is used with gel formation, usually the gel beads, in which the carboxyl groups are linked with the calcium ions. The number of free carboxyl groups ready for bioadhesion will reduce significantly. In addition, when forming gel, the matrix of molecules will reduce the flexibility of the backbone chain. Due to the gelation between alginate and Ca^{++} carried out in the sites of polyguluronic blocks (-GGGG-) following the egg-box model, the adhesion may be expected through available carboxyl groups in the fractions of polymannuronic (-MMMM-) or alternative units (-MGMG-) that do not take part in the gelation process.

7.2.4 Alginate benefit to health

Beside the applications of alginate as a stabilizer, thickening or gelation agent, alginate, due to low digestion, may be viewed as a dietary fiber. The role of dietary fiber is to reduce the intestinal uptake and because of that, it can lower the glycemic index in the blood and reduce the risk of cardiovascular disease. The activities of some enzymes in the GI tract like pepsin and trypsin are inhibited in the presence of alginate (Brownlee et al. 2005). This inhibition of proteases may be of benefit in reducing the glycemic index. In feeding the rats with cholesterol-free diets supplemented with sodium alginate, the weight gain and food conversion ratio (FCR) decreased when compared with the control. The presence of alginate in the small intestinal lumen decreases uptake of fat and thereby reduces the plasma cholesterol (Seal and Mathers 1996).

This result suggests that alginate might have the potential for use as a dietary supplement in hypercholesterolemia patients.

Alginate also reduces the glucose absorption rates. A 5 g supplement of sodium alginate to test meal that was given to a cohort of diabetes type II patients significantly reduced the postprandial rise in blood glucose by 31%, in serum insulin 42%, and in plasma C-peptide 35% (Torsdottir et al. 1991). When using with low levels (i.e., 1%) of alginate the indices of colonic mucosal health as maximal mucus thickness, mucus replenishment rate is improved. However, with 5% of alginate supplement, these indices reduce when compared with the control sample (Brownlee et al. 2005). The mechanism of action of alginate to the mucin is not fully understood. Alginate has effect on the colonic microflora in terms of the population of species and the quantities of short-chain fatty acid produced. Fecal levels of sulfide, ammonia, and bacterially phenolic toxins were reduced and levels of acetate and propionate production increased over the alginate consumption period (Terada et al. 1995).

In testing on rats, M-rich alginate has been shown to enhance repair of mucosal damage in GI tract while G-rich alginate did not show any effect (Del Buono et al. 2001). Although there are many studies about the effects of alginate on health, which showed that alginate can be used as a nutraceutical product, it is necessary to go further in order to find out the relation between the medical properties and the structure for controlling the quality of products. Some studies prove that the properties of alginate in relation to health benefit depend on M-rich alginate or G-rich alginate, but not yet extended further to the sophisticated structure, composition, and sequence of alginate molecules (expression by frequencies of monad, diad, triad, average length of polyM, polyG blocks).

Nowadays, alginate appears in many studies as a vehicle for the drug delivery to the target site. This application relates to the gel bead formation techniques and often combined with another polymer like chitosan, pectin, or xanthan gum. The combination of the biopolymers alters the porosity, the adhesion to the mucosal layer of the gel beads, and thereby controls the sustainable drug release (George and Abraham 2006, Yu et al. 2009). In the gastric juice with high concentration of proton, the gel beads made of calcium alginate loaded drug will exchange ions, becoming alginic acid and releasing the drug such as metronidazole to prevent recurrence of peptic ulcer disease, a phenomenon correlated with *Helicobacter pylori* (Murata et al. 2000).

7.3 Fucoidans

Fucoidans were isolated for the first time in 1913 by Killing from brown algae and later from marine invertebrates in 1948 in the study of Vasseur. This is a family of polysaccharides mainly constituted of L-fucose present with sulfate ester group occurring in the cell wall of brown algae, in egg jelly coat of sea urchins, and in the body wall of sea cucumber (Berteau and Mulloy 2003).

As well as alginate, fucoidans are extracted mainly from brown seaweeds. The extract from this algae have many activities that benefit health (Gombotz and Wee 1998, Samee et al. 2009).

The interest in fucoidans increases quickly when many bioactive effects and benefit to health from this biopolymer have been discovered. In the past few years, with the development of advanced analysis techniques, the structure is better known and many issues related to their bioactivities have been elucidated (Ale et al. 2011, Berteau and Mulloy 2003, Li et al. 2008).

7.3.1 Sources of fucoidan

Fucoidans are reported from different brown algae in orders of Fucales, Laminariales, Chordariales, Dictyotales, Ectocarpales, and Scytosiphonales. The species were studied as *Fucus vesiculosus, Fucus evanescens, Fucus distichus, Fucus serratus, Sargassum stenophyllum, Sargassum horneri, Sargassum kjellmanianum, Macrocystis pyrifera, Laminaria digitata, L. japonica, Undaria pinnatifida, Hizikia fusiforme*, etc. It seems that fucoidans are absent from green algae, red algae, golden algae, and from fresh water algae.

While many species of brown algae were investigated for fucoidans, only some species of two kinds of invertebrates were studied: sea cucumber (*Ludwigothurea grisea*) (Ribeiro et al. 1994) and sea urchin (*Lytechinus variegates, Arbacia lixula, Strongylocentrotus franciscanus, Strongylocentrotus pallidus, Strongylocentrotus droebachiensis*) (Alves et al. 1997, Mulloy et al. 1994, Viela-Silva et al. 1999, 2002).

7.3.2 Structure of fucoidan

The first time when fucoidans were isolated, the structure seemed simple with the chain constituted of fucose and sulfate groups. But more complex structure of fucoidans has been revealed through the flexibility of the carbon positions linked between focuses in the backbone of the polymer chain, of the positions, and the random distribution of functional groups. This is also the feature of polysaccharide in nature. Because of the diversity in structure, the nomenclature of fucoidans changes with the period and the source. There are many different terms as fucoidin, fucoidan, fucans, and fucansulfate. Following IUPAC, *fucoidan* refers to fucans extracted from algae, while *sulfate fucan* to polysaccharide based mainly on sulfated L-fucose with less than 10% of other monosaccharides, applied for sulfated fucans from invertebrates. Fucoidan from *F. vesiculosus* is based on L-fucose with mainly $\alpha(1\rightarrow3)$ glycosidic bonds and sulfate groups in position C-4. In *Ascophyllum nodosum*, both $\alpha(1\rightarrow3)$ and $\alpha(1\rightarrow4)$ glycosidic bonds were found in fucoidans, and there were branches (Daniel et al. 2001). In case of *Ecklonia kurome*, the structure was complicated with 3-*O*-linked with sulfate groups at C-4. In other species of algae, there are some 2-*O*-sulfation and 2-*O*-acetylation in the molecular chains (Chizhov et al. 1999). In invertebrates, the fucoidans are regular with repeating mono-, tri-, and tetrasaccharides, in which the glycosidic linkage is the same. In general, every species possesses a particular sulfated fucan. These polymers are homogeneous and unbranched. Some species may have two distinct fucans.

7.3.3 Properties of fucoidan

Fucoidans have a wide range of bioactivities, such as anticoagulant, antithrombotic, antivirus, antitumor, immunomodulatory, antioxidant, and antiinflammatory. In the scope of nutraceuticals, we limited some properties related to preventing and benefiting health, although there are some subjects that seemed to overlap between nutrition and medicine.

7.3.3.1 Anticoagulant and antithrombotic

Fucoidans have a wide variety of biological activities, of which anticoagulant and antithrombotic activities are well studied. Fucoidans from algae exhibited anticoagulant activities when tested for the activated partial thromboplastin time (APTT), thromboplastin time (TT), and antifactor Xa activity. While all fucoidans showed some TT and APTT activities, the antifactor Xa activity was not remarkable in any fucoidans (Nishimo and Nagumo 1987).

Many studies showed that the anticoagulant activity of fucoidans was related to sulfate content and position, molecular weight, and sugar composition. The higher content of sulfate groups presents stronger anticoagulant activity of fucoidans. This activity also depends on the concentration of C-2 sulfate and C-2,3 disulfate (Chevolot et al. 2001).

For binding the thrombin conveniently, it is necessary to have a conformation of fucoidans with a long enough chain. The native fucoidans with high molecular weight have strong anticoagulant activity, while the depolymerized fractions exhibited weak activity. Some studies showed that the composition of sugar in fucoidans may be related to this activity.

7.3.3.2 Antioxidant activity

Fucoidans reveal significant antioxidant activity in vitro and so has greater potential for preventing diseases related to free radicals. Fucoidan extracted from *F. vesiculosus* could protect the LLC-PK1 renal tubular epithelial cells from AAPH-induced oxidative stress. At a concentration of 100 μg/mL, fucoidan could revise cell viability from 10.84% on treatment with AAPH to 80.23% (Park and Han 2008). Fucoidans also have an inhibitory effect on lipid peroxidation, scavenging activity on superoxide anion radical and hydroxyl radical. When treated with AAPH, the activities of antioxidant enzymes such as SOD and GSH-px in the cells decreased significantly. However, if the cells were incubated with fucoidans, the effect of AAPH can be scavenged (Park and Han 2008).

The antioxidant properties of fucoidans were tested in rats being administered CCl_4-induced oxidative stress. The results showed that fucoidans could recover the levels of SOD, CAT, and GPx in the CCl_4-tested rats (Kang et al. 2008).

Because the free radicals induce cell damage, the antioxidant activity of fucoidans reduces the risk of severe diseases and becomes more and more important in nutraceuticals and medicine as well.

7.3.3.3 Reducing lipid accumulation

Fucoidans can inhibit lipid accumulation in adipocytes through the regulation of lipolysis. Lipolysis, that is, the hydrolysis of triglycerides into free fatty acids, is regulated by hormone-sensitive lipase (HSL), which is known to be a rate-limiting enzyme, and its active form is p-HSL. By increasing the protein levels of HSL and p-HSL, fucoidans reduce the lipid accumulation and inhibit the glucose uptake into adipocytes (Park et al. 2011). With these properties, fucoidans can be used for preventing or treatment of obesity.

7.3.3.4 Antitumor and antimetastatic

The fucoidan isolated from *Fucus evanescens* was examined for its potential of antitumor and antimetastatic activities in the C57B1/6 mice with transplanted Lewis lung adenocarcinoma. The results showed that a single injection dose of 25 mg/kg did not inhibit the tumor growth, but repeated injections of fucoidan three times with the dose of 10 mg/kg, inhibiting the tumor growth by 33% in comparison with the control. The number of metastases in the lungs also reduced significantly when treated with fucoidan. The combination of fucoidan and cyclophosphamide (CP) exhibited excellent antimetastatic effect (reduced 90% compared with the control) (Alekseyenko et al. 2007).

In other works, fucoidan was tested for its effect on apoptosis of human colon cancer cell. The HT-29 and HCT116 (the human colon cancer cells) and FHC (the human normal colon epithelial cell) were cultured with various concentrations of fucoidan (0–20 μg/mL). The results showed that the fucoidan reduced the viability of HT-29 and HCT116 cells by 64.9% and 36.7%, respectively. In addition, fucoidan had no effect on the viability of FHC cells. This study proved that fucoidan induces apoptosis of the human colon cancer cells and that this phenomenon is mediated via both the death receptor-mediated and mitochondria-mediated apoptotic pathway (Kim et al. 2010).

7.4 Edible bird's nest extract

In Asian countries, the EBN has been used for a long time as an excellent well-being and royal food. It is known as the Caviar of the East and has been consumed by Chinese communities all over the world and some countries in Southeast Asia. The people in this area believe that EBN can recover the health quickly after severe disease, enhance the working ability, especially for middle- and old-aged people. Although possessing many valuable properties, the number of studies about structure and properties of EBN is still low.

7.4.1 Source

EBN is the nest made from the saliva of some species of *Aerodramus* genus (formely *Collocallia*) as *Aerodramus fuciphagus* (White-nest Swiftlet) and *Aerodramus maximus* (Black-nest Swiftlet). These birds are small to medium size, distributed in Thailand, Vietnam, Indonesia, Malaysia, and the Philippines. Their preys include aerial insects such as Diptera, Homoptera, Coleoptera, Hemiptera, and Hymenoptera (Nguyen et al. 2002). The swiftlet lives in marine caves above sea level, more or less the distance from the coast where the relative humidity is high and stable all year round (**Figure 7.1**).

Figure 7.1. *The "birds nest" of swiftlet hung in the cave island in Nha Trang Bay, Vietnam.*

In the breeding season, the birds make the nest from their saliva secreted from the two sublingual glands. The nest of the White-nest Swiftlet is made entirely from saliva, while the Black-nest Swiftlet uses a large amount of feathers. Due to the adhesive property of saliva, the birds glue their nests to the cave wall. EBNs are sold mainly for the whole dried nest for the best quality product. The remainder is processed into drinks or hydrolyzed and mixed with other ingredients to make functional foods.

7.4.2 Structure

The EBN is composed mainly of glycoprotein in which, besides the protein, the carbohydrate component consists of galactose, galactosamin, glucosamine, fucose, and especially sialic acid. A small amount of minerals, mainly calcium, sulfur, magnesium, and iron, is present in EBN. The composition changes with the species and habitat. N-Acetylneuraminic acid is the best characterized sialic acid and can be isolated from EBN (Martin et al. 1977). The EBN contains sialylglycoconjugates. The structure data of O-glycans (Wieruszeski et al. 1987), glycosaminoglycans (Nakagawa et al. 2007), and N-glycans (Yagi et al. 2008) from EBN have been reported. Alkaline reductive treatment of the crude EBN led to the release of numerous monosialyl and disialyl oligosaccharides. Some of the monosialyl oligosaccharides possess the core structure Galβ(1→3)[GlcNAcβ(1→6)]GalNAcα(1→3)GalNAc-ol.

The most complex representatives of the monosialyl fraction are

$$NeuAcα(2→3)Galβ(1→3)\searrow$$
$$GalNAc\text{-}ol$$
$$Galα(1→4)Gaβ(1→4)Galβ(1→4)GlcNAcβ(1→6)\nearrow$$

and

$$NeuAcα(2→3)Galβ(1→3)\searrow$$
$$GalNAcα(1→3)GalNAc\text{-}ol$$
$$Galα(1→4)Gaβ(1→4)]Galβ(1→4)GlcNAcβ(1→6)\nearrow$$

(Wieruszeski et al. 1987).

Glycoproteins isolated from EBN were rich in proteoglycan containing nonsulfated chondroitin glycosaminoglycans (GAGs). The structure of GAGs is the chondroitin chain of $(→4GlcUAβ1→3GalNAcβ1→)_n$ with the molecular weight of 49 kDa (Nakagawa et al. 2007). N-glycans were obtained from EBN by using glycosamidase A, labeled with 2-aminopyridine. In N-glycosylation profile of EBN with eight N-glycans identified, the tri-antennary N-glycan bearing the α-2, 3-N-acetylneuraminic acid residues is displayed as a major component. EBN expressed highly branching N-glycan structures possessing α-2,3-NeuAc residues. The sialylated high-antennary N-glycans of EBN are known to contribute to the inhibition of influenza viral infection (Yagi et al. 2008).

7.4.3 Function

EBN has been used as a functional food and therapeutic herbal medicine over several centuries in China and in other Asian countries such as Vietnam. Although during this time there was not strong scientific evidence supporting the health benefits, through accumulated experiences, people in these countries have known some of the properties and bioactivities of EBN. Because of precious properties related to benefit in health and being only collected in small quantities from nature, EBN has become very expensive and appeared only in the royal official parties in the dynasty time. EBN can neutralize infection of influenza viruses in MDCK cells and inhibit the hemagglutination of human erythrocytes by influenza A virus from various hosts as human, avian, and porcine. It stimulates mitosis hormones and the growth factor for epidermal growth, resulting in repair of cells and stimulation of the immune system. EBN stimulates cell division and growth, enhances tissue growth, and regeneration. Sialic acid residues of many glycoconjugates are involved in biologically important ligand–receptor interactions, such as specific cell-to-cell, pathogen-to-cell, or drug-to-cell interactions.

Small molecule of EBN extract was more propitious for its antiviral effect than intact EBN. In case of no treatment with pancreatic enzyme (Guo et al. 2006), the EBN extract cannot inhibit the hemagglutination of influenza A virus to human type O erythrocytes. However, when treated with protease (Pancreatin F) to hydrolyze glycoprotein to glycopeptide, EBN extract had stronger hemagglutination inhibition (HAI) activities. After treatment with neuraminidase, the effect of EBN extract was reduced significantly, revealing the fact that the influenza virus binding is associated with sialic acid of EBN (Guo et al. 2006).

EBN extract strongly promoted the proliferation of human adipose-derived stem cells (hADSCs), which are the sources of stem cell therapy. The proliferation of hADSCs is promoted by the expression of IL-6 and VEGF, of which IL-6 is regulated by activation of p44/42 MAPK and NF-κB and VEGF expression is induced by EBN extract through the activation of p38 MAPK. With the potential of enhancing the stem cells, EBN extract can act as an extracellular factor that improves self-renewal via an increase in the proliferative capacity of hADSCs. The effects of EBN are limited in the normal cells and do not affect transformed cell lines (Roh et al. 2012). As mentioned in Section 7.4.2, EBN is rich in chondroitin GAG, which is one of the main components of bone. When feeding rats with EBN-hydrolyzed extract, the concentration of phosphorus and calcium in the bone becomes higher, which induces the bone strength via the maximum breaking force increases. EBN extract supplementation also increases the dermal thickness of ovariectomized rats (Matsukawa et al. 2011). Although EBN is benefit to health, there is also side effect due to its nature of protein. EBN caused anaphylaxis surpassing other well-defined food allergens as milk, egg, peanut, or crustacean seafood in children. Goh et al. (2001) showed that the fraction of 66 kDa from EBN was the major allergen.

With high price, the counterfeit EBN products have been found in the market, and in some cases, it is difficult to distinguish them. Many methods have been developed to identify the nest-shaped material from the genuine EBN, such as microscopy and physicochemical authentication. But in case the counterfeit products are made from the nest of different species or genera of birds, they are distinguished only by genetic identification (Lin et al. 2009).

7.5 Seahorse extract

Seahorses, *Hippocampus* spp., are highly unusual marine fishes due to their unique body morphology and the strange reproduction with the male becoming pregnant and incubating the young in a brood pouch. At least 46 species of seahorses are currently reported, distributed in tropical and temperate shallow coastal habitats including sea grass bed, coral reefs, mangroves, and estuaries (Koldewey and Keith 2010). In Asian countries, such as China and Vietnam, seahorses are used as traditional medicine, while in many countries, the live seahorses are traded as aquarium fishes. There are nearly 80 countries that take part in the market of seahorses in the world, and in Asia more than 45 tons of dried seahorses were sold in 1995, and now the production may have been considerably greater (Koldewey and Keith 2010). Vietnam is among the larger exporters of seahorses with at least 5 tons/year including seven species, of which species *Hippocampus kuda* dominated (Lourie et al. 1999). Because of the increasing demand in the world, the aquaculture of seahorses has developed quickly in China, Australia, Vietnam, Thailand, Indonesia, etc.

When extracting seahorse, *Hippocampus kuda*, with water, methanol, and ethanol, the total phenolic obtained reveals the potential of free radical and reactive oxygen species (ROS) scavenging activities. Among three extracts, the seahorse extract with methanol exhibits the highest antioxidant activity in linoleic acid system, effective reducing power, DPPH radical scavenging, hydroxyl radical scavenging, superoxide radical scavenging, alkyl radical scavenging, inhibitory intracellular ROS, and inhibited MPO activity (Qian et al. 2008). Seahorse protein hydrolysates obtained when hydrolyzing the seahorse by six proteases (alcalase, neutrase, papain, pepsin, pronase, and trypsin) exhibit different alkaline phosphate activities, in which the pronase E hydrolysate possessed the highest ALP activity, a phenotype of marker of osteoblast and chondrocyte differentiation. The amino acid of this peptide was determined to be LEDPKDKDDWDNWK, composed of 14 amino acids, and its molecular weight is 1821 Da. The effect of this peptide on differentiation and inflammation in osteoblastic MG-63 and chondrocytic SW-1353 cells demonstrates the benefit of seahorse in the amelioration of arthritis (Ryu et al. 2010).

Another peptide was found by Ryu et al. (2010) from *Hippocampus kuda* by purifying the hydrolysate of seahorse through chromatographic methods, combining fast protein liquid chromatography (FPLC) and repeated reversed-phase high-performance liquid chromatography (RP-HPLC). This peptide comprises of 15 amino acids following the sequence of LEDPFDKDDWDNWKS. The experiment shows that this peptide inhibits protein and collagen release, blocks TPA-induced collagenases, and inhibits LPS-induced NO production in chondrocyte-like SW-1353 cells and osteoblast-like MG-63 cells. The two peptides found show that seahorses can be used in antiarthritis therapeutics. From seahorses, other important substances have been found as new phthalate derivatives inhibiting cathepsin B (Li et al. 2008), 1-(5-bromo-2-hydroxy-4-methoxyphenyl)ethanone suppressing proinfammatory responses (Himaya et al. 2011).

The discoveries of the antioxidants, peptides, and other bioactive substances elucidate the effect of traditional medicine in Asian countries for hundreds of years and provide the new potential application of seahorse in nutraceuticals.

7.6 Conclusions

With the consuming habitude of traditional foods, especially the specific ones mentioned earlier, people in East and Southeast Asia have excellent opportunities to protect the body from risk of disorders or severe diseases. Only in some seaweed and marine animals in which the precious biopolymers are presented, the potential of keeping well and avoiding the risk of diseases by food based on biomaterials originated from marine animals is remarkably considered.

References

Ale, M.T., J.D. Mikkelsen, A.S. Meyer. 2011. Important determinants for fucoidan bioactivity: A critical review of structure-function relation and extraction methods for fucose-containing sulfated polysaccharides from brown seaweed. *Mar. Drugs* 9: 2106–2130.

Alekseyenko, T.V., S.Y. Zhanayeva, A.A. Venediktova, T.N. Zvyagintseva, T.A. Kuznetsova, N.N. Besednova, T.A. Korolenko. 2007. Antitumor and antimetastatic activity of fucoidan, a sulfated polysaccharide isolated from the Okhotsk Sea *Fucus evanescens* brown alga. *Bull. Exp. Biol. Med.* 143: 730–732.

Alves, A.P., B. Mulloy, J.A. Diniz, P.A.S. Mourao. 1997. Sulfated polysaccharides from the egg jelly layer are species-specific inducers of acrosomal reaction sperms of sea urchins. *J. Biol. Chem.* 272: 6965–6971.

Berteau, O., B. Mulloy. 2003. Sulfated fucans, fresh perspectives: Structures, functions, and biological properties of sulfated funcans and an overview of enzymes active towards this class of polysaccharides. *Glycobiology* 13: 29R–40R.

Brownlee, I.A., A. Allen, J.P. Pearson, P.W. Dettmar, M.E. Havler, M.R. Atherton, E. Onsøyen. 2005. Alginate as a source of dietary fiber. *Crit. Rev. Food Sci.* 45: 497–510.

Chevolot, L., B. Mulloy, J. Jacqueline. 2001. A disaccharides repeat unit is the structure in fucoidans from two species of brown algae. *Carbohydr. Res.* 330: 529–535.

Chizhov, A.O., A. Dell, H.R. Morris, S.M. Haslam, R.A. McDowell, A.S. Shashkov, N.E. Nifant'ev, E.A. Khatuntseva, A.I. Usov. 1999. A study of fucoidan from the brown seaweed *Chorda filum. Carbohydr. Res.* 320: 108–119.

Daniel, R., O. Berteau, L. Chevolot, A. Varenne, P. Garei, N. Goasdoue. 2001. Regioselective desulfation of sulfated L-fucopyranoside by a new sulfoesterase from the marine mollusk *Pecten maximus*: Application to the structural study of algal fucoidan (*Ascophyllum nodosum*). *Eur. J. Biochem.* 268: 5617–5628.

Del Buono, R., E.M. Dunne, P.W. Dettmar, I.G. Joliffe, M. Pignatelli. 2001. Sodium alginate decreases gastric damage *in vivo. J. Pathol.* 193: 4A.

George. M., T.E. Abraham. 2006. Polyionic hydrocolloids for the intestinal delivery of protein drugs: Alginate and chitosan—A review. *J. Controlled. Release* 114: 1–14.

Goh, D.L.M., K.Y. Chua, F.T. Chew, T.K. Seow, K.L. Ou, F.C. Yi, B.W. Lee. 2001. Immunochemical characterization of edible bird's nest allergens. *J. Allergy Clin. Immunol.* 107: 1082–1088.

Gombotz, W.R., S.F. Wee. 1998. Protein release from alginate matrices. *Adv. Drug Deliv. Rev.* 31: 267–285.

Gorin, P.A.J., J.F.T. Spencer. 1996. Exocellular alginic acid from *Azotobacter vinelandii. Can. J. Chem.* 44: 993–998.

Govan, J.R.W., J.A.M Fyte, T.R. Jarman. 1981. Isolation of alginate-producing mutant of *Pseudomonas fluorescens, Pseudomonas putida* and *Pseudomonas mendocina. J. Gen. Microbiol.* 125: 217–220.

Grasdalen, H., B. Larsen, O. Smidsrød. 1979. A PMR study of the composition and sequence of uronate residues in alginate. *Carbohydr. Res.* 68: 23–31.

Guo, C.T., T. Takahashi, W. Bukawa, N. Takahashi, H. Yagi, K. Kato, K.I.P.J. Hidari, D. Miyamoto, T. Suzuki, Y. Suzuki. 2006. Edible bird's nest extract inhibits influenza virus infection. *Antiviral Res.* 70: 140–146.

Haug, A. 1964. Composition and properties of alginates. *Rep. Norw. Inst. Seaweed Res.* 30: 25–45.

Haug, A., B. Larsen, O. Smidsrød. 1966. A study of the constitution of alginic acid by partial hydrolysis. *Acta Chem. Scand.* 20: 183–190.

Haug, A., O. Smidsrød. 1970. Selectivity of some anionic polymers for divalent metal ions. *Acta Chem. Scand.* 24: 843–854.

Himaya, S.W.A., B.M. Ryu, Z.J. Qian, Y. Li, S.K. Kim. 2011. 1-(5-bromo-2-hydroxy-4-methoxy-phenyl)ethanol [SE1] suppresses pro-inflammatory responses by blocking NF-κB and MAPK signaling pathways in activated microglia. *Eur. J. Pharmacol.* 670: 608–616.

Kang, K.S., I.D. Kim, R.H. Kwon, J.Y. Lee, J.S. Kang, B.J. Ha. 2008. The effects of fucoidan extracts on CCl_4-induced liver injury. *Arch. Pharm. Res.* 31: 622–627.

Kim, E.J., S.Y. Park, J.Y. Lee, J.H.Y. Park. 2010. Fucoidan present in brown algae induces apoptosis of human colon cancer cells. *Biomed. Cent. Gastroenterol.* 10: 96–107.

Koldewey, H.J., M.S. Keith. 2010. A global review of seahorse aquaculture. *Aquaculture* 302: 131–152.

Li, B., F. Lu, X. Wei, R. Zhao. 2008. Fucoidan: Structure and bioactivity. *Molecules* 13: 1671–1695.

Lin, J.R., H. Zhou, X.P. Lai, Y. Hou, X.M. Xian, J.N. Chen, P.X. Wang, L. Zhou, Y. Dong. 2009. Genetic identification of edible bird's nest based on mitochondrial DNA sequences. *Food Res. Int.* 42: 1053–1061.

Lourie, S.A., J.C. Pritchard, S.P. Casey, S.K. Truong, J.H. Heather, C.J.V. Amanda. 1999. The taxonomy of Vietnam's exploited seahorses (family Syngnathidae). *Biol. J. Linn. Soc.* 66: 231–256.

Martin, J.E., S.W. Tanenbaum, M. Flashner. 1977. A facile procedure for the isolation of N-acetylneuramic acid. *Carbohydr. Res.* 56: 423–425.

Matsukawa, N., M. Matsumoto, W. Bukawa, H. Chiji, K. Nakayama, H. Hara, T. Tsukahara. 2011. Improvement of bone strength and dermal thickness due to dietary bird's nest extract in ovariectomized rats. *Biosci. Biotechnol. Biochem.* 75: 590–592.

McHugh, D.J. 1987. Production, the properties and uses of alginates. In: *Production and Utilization of Products from Commercial Seaweeds*, FAO Fisheries Technical Paper No 288, ed. McHugh, pp. 58–115. Rome, Italy: FAO.

Mulloy, B., A.C. Ribeiro, A.P. Alves, R.P. Vieira, P.A.S. Mourao. 1994. Sulfated funcans from echinoderms have a regular tetrasaccharide repeating unit defined by specific patterns of sulfation at the O-2 and O-4 positions. *J. Biol. Chem.* 269: 22113–22123.

Murata, Y., N. Sasaki, E. Miyamoto, S. Kawashima. 2000. Use of floating alginate gel beads for stomach-specific drug delivery. *Eur. J. Pharm. Biopharm.* 50: 221–226.

Nakagawa, H., Y. Hama, T. Sumi, S. Li, K. Maskos, K. Kalayanamitra, S. Mizumoto, K. Sugahara, Y.T. Li. 2007. Occurrence of a nonsulfated chondroitin proteoglycan in the dries saliva of *Collocalia swiftlet* (edible bird's-nest). *Glycobiology* 17: 157–164.

Nguyen, Q.P., V.Q. Yen, J.F. Voisin. 2002. *The White-nest Swiftlet and Black-nest Swiftlet.* Paris, France: Société Nouvelle des Éditions Boubée.

Nishimo, T., T. Nagumo. 1987. Sugar constituents and blood-anticoagulant activities of fucose-containing sulfated polysaccharides in nine brown seaweed species. *Nippon Nogeikagaku Kaishi* 61: 361–363.

Park M.K., U. Jung, C. Roh. 2011. Fucoidan from marine brown algae inhibits lipid accumulation. *Mar. Drugs* 9: 1359–1367.

Park, M.J., J.S. Han. 2008. Fucoidan protect LLC-PK1 cells against AAPH-induced damage. *J. Food Sci. Nutr.* 13: 259–265.

Park, K., J.R. Robinson. 1984. Bioadhesive polymers as platforms for oral-controlled drug delivery: Method to study adhesive interaction. *Int. J. Pharm.* 19: 107–127.

Peppas, N.A., P.A. Buri. 1985. Surface interfacial and molecular aspects of polymer bioadhesion on soft tissues. *J. Control. Rel.* 2: 257–275.

Qian, Z.J., B.M. Ryu, M.M. Kim, S.K. Kim. 2008. Free radical and reactive oxygen species scavenging activities of the extracts from seahorse, Hippocampus kuda Bleeler. *Biotechnol. Bioprocess Eng.* 13: 705–715.

Ribeiro, A.C., R.P. Vieira, P.A.S. Mourao, B. Moulloy. 1994. A sulfated α-L-fucan from sea cucumber. *Carbohydr. Res.* 255: 225–240.

Roh, K.B., J. Lee, Y.S. Kim, J. Park, J.H. Kim, J. Lee, D. Park. 2012. Mechanism of edible bird's nest extract-induced proliferation of human adipose-derived stem cells. *Evid-Based Complement. Alternat. Med.* 2012: 11.

Ryu, B.M., Z.J. Qian, S.K. Kim. 2010. Purification of a peptide from seahorse, that inhibits TPA-induced MMP, iNOS and COX-2 expression through MAPK and NF-kB activation, and induces human osteoblastic and chondrocytic differentiation. *Chem-Biol. Interact.* 184: 413–422.

Samee, H., Z.X. Li, H. Lin, J. Khalid, Y.C. Guo. 2009. Anti-allergic effects of ethanol extracts from brown seaweeds. *J. Zhejiang Univ. Sci. B* 10: 47–153.

Seal, C.J., J.C. Mathers. 1996. Comparative gastrointestinal responses to guar gum and seaweed polysaccharide (sodium alginate) in rats. *Br. J. Nutr.* 85: 317–324.

Terada, A., H. Hara, T. Mitsuoka. 1995. Effect of dietary alginate on the faecal microbiota and faecal metabolic activity in human. *Microb. Ecol. Health D* 8: 259–266.

Torsdottir, I., M. Alpsten, G. Holm, A.S. Sandberg, J. Tolli. 1991. A small dose of soluble alginate-fiber affects postprandial glycemia and gastric emptying in humans with diabetes. *J. Nutr.* 121: 795–799.

Viela-Silva, A.C., A.P. Alves, A.P. Valente. 1999. Structure of the sulfated α-L-fucan from the egg jelly coat of the sea urchin *Strongylocentrotus franciscanus*: Patterns of preferential 2-O- and 4-O-sulfation determine sperm cell recognition. *Glycobiology* 9: 927–933.

Viela-Silva, A.C., M.O. Castro, A.P. Valente, C.H. Biermann, P.A.S. Mourao. 2002. Sulfated fucans from the egg jellies of the closely related sea urchins *Strongylocentrotus droebachiensis* and *Strongylocentrotus pallidus* ensure species-specific fertilization. *J. Biol. Chem.* 277: 379–387.

Wieruszeski, J.M., J.C. Michalski, J. Montreuil, G. Strecker, J. Peter-Katalinic, H. Egge, H. van Halbeek, J.H.G.M. Mutsaers, J.F.G. Vliegenthart. 1987. Structure of the monosialyl oligosaccharides derived from salivary gland mucin glycoprotein of the Chinese swiftlet (Genus Collocalia). *J. Biol. Chem.* 262: 6650–6657.

Yagi, H., N. Yasukawa, S.Y. Yu, C.T. Guo, N. Takahashi, T. Takahashi, W. Bukawa, T. Suzuki, K.H. Khoo, Y. Suzuki, K. Kato. 2008. The expression of sialylated high-antennary N-glycans in edible bird's nest. *Carbohydr. Res.* 343: 1373–1377.

Yu, C.Y., B.C. Yin, W. Zhang, S.X. Cheng, X.Z. Zhang, R.X. Zhou. 2009. Composite microparticle drug delivery systems based on chitosan, alginate and pectin with improved pH-sensitive drug release property. *Colloids Surf. B* 68: 245–249.

8

Marine Natural Antihypertensive Peptides from *Styela clava* Having Multifunctions of ACE Inhibition and NO Production in Endothelial Cells

Seok-Chun Ko, Se-Kwon Kim, and You-Jin Jeon

Contents

8.1 Introduction

Bioactive peptides can be obtained from a variety of marine bioresource proteins (Byun and Kim, 2001). Improved nutritional and potent bioactive peptides have been shown to be generated by the enzymatic hydrolysis of proteins (Ondetti, 1997). Bioactive peptides may function as potential physiological modulators in the process of metabolism during the intestinal digestion of the diet (Kato and Suzuki, 1975). Bioactive peptides are liberated depending on their structure, composition, and amino acid sequence (Lee et al., 2009). Some of these bioactive peptides have been identified to possess nutraceutical potentials that are beneficial for the promotion of human health. Recently, the possible roles for food-derived bioactive peptides in reducing the risk of cardiovascular diseases (CVDs) have been explored (Erdmann et al., 2008). These peptides evidenced a variety of bioactivities, including antioxidative (Liu et al., 2010), antimicrobial (Kim et al., 2001a), and antihypertensive effects (Vercruysse et al., 2008). Recently, many antihypertensive peptides have been isolated from various marine bioresource proteins, such as rotifer (Lee et al., 2009), hard clam (Tsai et al., 2008), tuna frame (Lee et al., 2010a), sea cucumber (Zhao et al., 2007), and skate skin (Lee et al., 2011).

Among various marine bioresources, ascidians are known to provide an abundance of bioactive materials with valuable pharmaceutical and biomedical potential. According to the results of previous studies, bioactive substances present in ascidians, and various other constituents have been demonstrated to possess antioxidant (Jo et al., 2010), antimicrobial (Kim et al., 2007), and anticancer (Kim et al., 2006) properties. Among the ascidians, *Styela clava* belongs to Class Ascidiacea, and is commonly known as a sea squirt (**Figure 8.1**).

S. clava has also been aquacultured in Korea, as Koreans use it for food. The flesh tissue of *S. clava* contains about 65% protein (**Table 8.1**), and it could be a good candidate as nutraceutical. Several studies on *S. clava* have pointed out a variety of biological benefits, including antioxidant and anticancer effects (Kim et al., 2006). However, the earlier studies employed solvent extraction methods, and there remains a paucity of data on the bioactivities of *S. clava* protein.

This chapter focuses to attempt an enzymatic hydrolysis technique to prepare enzymatic hydrolysates, such as smaller peptides, from *S. clava* protein, and to characterize the purified peptide with regard to antihypertensive effects.

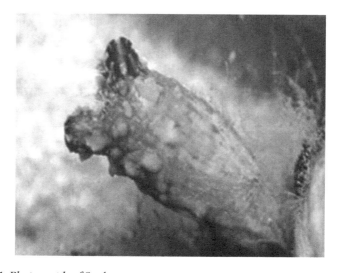

Figure 8.1 Photograph of S. clava.

Table 8.1 Chemical Compositions of *S. clava*

Composition	Content (%)		
	Whole	**Flesh**	**Tunic**
Moisture	9.34 ± 0.21	1.84 ± 0.18	1.78 ± 0.37
Ash	10.77 ± 0.33	7.05 ± 0.32	3.57 ± 0.25
Protein	33.12 ± 0.29	67.80 ± 0.22	31.51 ± 0.21
Carbohydrate	42.52 ± 0.41	16.77 ± 0.07	60.38 ± 0.21
Lipid	4.25 ± 0.43	6.54 ± 0.21	2.76 ± 0.11

Source: Ko, S.C. et al., *Biochemistry*, 47, 34, 2012b.

8.2 Hypertensive mechanisms

Recently, metabolic syndrome has become a major problem, and poses a major worldwide threat to human health. It is a cluster of metabolic abnormalities that has been associated with CVD, risk factors of diabetes, abdominal obesity, high blood pressure, and high cholesterol (Alberti et al., 2005, 2006). Hypertension is one of the major risk factors relevant to the development of CVDs including arteriosclerosis, stroke, and myocardial infarction (Je et al., 2005; Lee et al., 2010a). It is triggered by environmental influences including salt intake, obesity, insulin resistance, stress, smoking, and lack of exercise (Bohr et al., 1991). Among the processes related to hypertension, angiotensin I-converting enzyme (ACE, EC 3.4.15.1) plays an important role in the physiological regulation of blood pressure in the renin–angiotensin system (RAS) and kallikrein kinnin system (KKS). In RAS, ACE acts as an exopeptidase that cleaves His–Leu from the C-terminal of angiotensin I (decapeptide: Asp–Arg–Val–Tyr–Ile–His–Pro–Phe–His–Leu), and produces the potent vasoconstrictor angiotensin II (octapeptide: Asp–Arg–Val–Tyr–Ile–His–Pro–Phe), while in the KKS, ACE inactivates the antihypertensive vasodilator bradykinin, and increases the blood pressure (Segura-Campos et al., 2011). Therefore, inhibition of ACE activity is considered to be an important therapeutic approach for controlling hypertension. Several synthetic ACE inhibitors have been developed, including captopril, enalapril, fosinopril, ramipril, and zofenopril, all of which are currently extensively used in the treatment of essential hypertension and heart failure in humans (Je et al., 2005; Ondetti, 1977). However, these synthetic ACE inhibitors have certain side effects, including cough, taste disturbances, and skin rashes (Lee et al., 2011; Thurman and Schrier, 2003). Therefore, the development of ACE inhibitors from natural products has become a major area of research.

The vascular endothelium plays an important role in controlling vascular tone via secretion of both relaxant and contractile factors (Dimo et al., 2007). The endothelium, a single layer of the vascular wall, regulates vascular tone via the production of vasoactive factors (Asselbergs et al., 2005). Endothelial cells synthesize nitric oxide (NO) through endothelial NO synthase (eNOS) by oxidation of the amino acid L-arginine. The endothelium generates potent vasodilators such as endothelium-derived hyperpolarizing factor (Beny and Brunet, 1988) and endothelium-derived relaxing factor (EDRF) (Moncada and Vane, 1978). The EDRF has been identified as NO (Kim et al., 2011). Endothelial dysfunction, which manifests as reduced bioactive NO levels, is one of the most common pathologic changes occurring in a variety of CVDs (Christopher et al., 2008). NO is produced continuously by eNOS in the healthy endothelium in certain amounts in response to shear and pulsatile stretch of the vascular wall (Asselbergs et al., 2005).

8.2.1 Preparation of antihypertensive peptide from *Styela clava*

Antihypertensive peptides are the resulting products of the hydrolysis of proteins, and this can be achieved by enzymes (either by digestive enzymes or enzymes derived from microorganisms and plants), alkali or acid, and microbial fermentation (Je et al., 2005; Kim et al., 2001b). Enzymatic hydrolysis is one of the major methods for obtaining antihypertensive peptides from protein sources using appropriate proteolytic enzymes. The physicochemical conditions of the reaction media, such as temperature and pH of the protein solution, must then be adjusted in order to be optimized for the activity of the enzyme used (**Figure 8.2**). Proteolytic enzymes from microbes, plants, and animals can be used for the hydrolysis process of marine proteins to develop antihypertensive peptides (Simpson et al., 1998). Furthermore, Protamex, Alcalase, Neutrase, Flavourzyme, Kojizyme, α-chymotrypsin, pepsin, papain, and trypsin have been used for the hydrolysis of *S. clava* under optimal conditions of pH and temperature (Ko et al., 2012b).

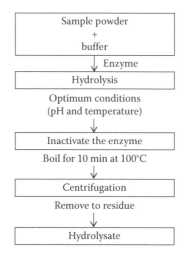

Figure 8.2 *Schematic diagram of the enzymatic hydrolysis process.*

8.2.2 Purification and identification of antihypertensive peptide from *Styela clava*

The antihypertensive peptide from *S. clava* was hydrolyzed using various commercial enzymes. Nine proteases (Protamex, Kojizyme, Neutrase, Flavourzyme, Alcalase, pepsin, trypsin, α-chymotrypsin, and papain) were used, and their respective enzymatic hydrolysates and an aqueous extract were screened to evaluate their potential ACE inhibitory activity. Among the various hydrolysates, Protamex hydrolysate possessed the highest ACE inhibitory activity (**Table 8.2**; IC_{50} = 1.023 mg/mL).

Also, the Protamex hydrolysate of flesh tissue showed relatively higher ACE inhibitory activity (IC_{50} = 0.455 ± 0.011 mg/mL) compared with the Protamex hydrolysate of tunic tissue (IC_{50} = 2.060 ± 0.007 mg/mL) (**Table 8.3**). Compared to what has been observed in previous reports, the ACE inhibitory activities of enzymatic hydrolysates were more effective than those of the organic solvents and aqueous extracts from *S. clava* (Lee et al., 2010b). This indicates that ACE inhibition was not the only action mode of Protamex hydrolysate from *S. clava* flesh tissue, but other mechanisms could also be implicated in its antihypertensive effects.

Therefore, the effect of Protamex hydrolysate from *S. clava* flesh tissue on vasorelaxation was characterized using rat thoracic aorta prepared with or without endothelium to assess any improvements in endothelium-dependent vasorelaxation (**Figure 8.3**).

The vasorelaxation of Protamex hydrolysate from *S. clava* flesh tissue was increased in a concentration-dependent manner (0.187–3 mg/mL) on an intact

Table 8.2 ACE Inhibitory Activity of Enzymatic Hydrolysates and Aqueous Extract from *S. clava*

Enzyme	IC_{50} Value (mg/mL)[a]
Kojizyme	2.481 ± 0.032
Flavourzyme	2.343 ± 0.022
Neutrase	2.234 ± 0.019
Alcalase	1.781 ± 0.018
Protamex	1.023 ± 0.047
Pepsin	2.147 ± 0.051
Trypsin	2.427 ± 0.033
α-Chymotrypsin	2.263 ± 0.028
Papain	2.282 ± 0.042
Aqueous	6.887 ± 0.031

Source: Ko, S.C. et al., *Biochemistry*, 47, 34, 2012b.

[a] The concentration of an inhibitor required to inhibit 50% of the ACE activity. The values of IC_{50} were determined by triplicate individual experiments.

Table 8.3 ACE Inhibitory Activity of Protamex Hydrolysate from Flesh and Tunic Tissues of *S. clava*

Tissue	IC$_{50}$ Value (mg/mL)[a]
Whole	1.023 ± 0.003
Flesh	0.455 ± 0.011
Tunic	2.060 ± 0.007

Source: Ko, S.C. et al., *Biochemistry*, 47, 34, 2012b.

[a] The concentration of an inhibitor required to inhibit 50% of the ACE activity. The values of IC$_{50}$ were determined by triplicate individual experiments.

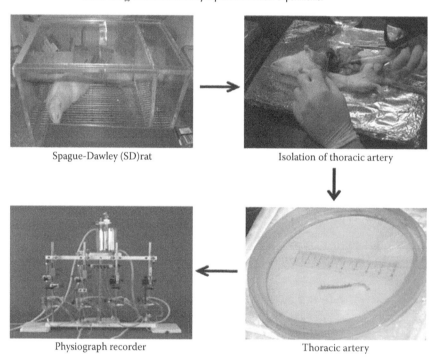

Spague-Dawley (SD)rat

Isolation of thoracic artery

Physiograph recorder

Thoracic artery

Figure 8.3 Photograph of determination of vasorelaxation effect.

endothelium. However, endothelial denudation abolished vasorelaxation completely. To further evaluate the mechanism of the vasorelaxation response in Protamex hydrolysate from *S. clava* flesh tissue in endothelium-intact aorta rings, they are pre-incubated with L-NAME (100 μM), a NO synthesis inhibitor. The pretreatment of aorta rings with L-NAME markedly attenuated the vasorelaxation effect of Protamex hydrolysate from *S. clava* flesh tissue (**Figure 8.4**). These results demonstrated that Protamex hydrolysate from *S. clava* flesh tissue could mediate nitric oxide, thereby exerting an endothelium-dependent vasorelaxation activity. ACE catalyzes the bradykinin inactivation that regulates different biological processes including vascular endothelial NO production (Rosa et al., 2010).

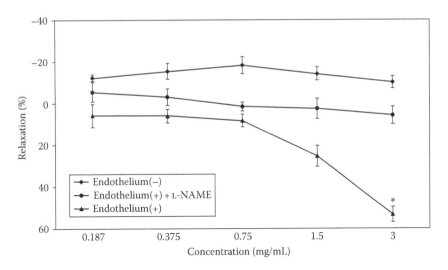

Figure 8.4 *Concentration-dependent vasorelaxation of Protamex hydrolysate from S. clava flesh tissue in aortic segments with and without endothelium. The data are expressed as the means ± SEM. Statistical evaluations were conducted to compare the endothelial (–). *p < 0.05. (From Ko, S.C. et al., Food Chem., 134,1141, 2012a.)*

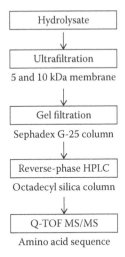

Figure 8.5 *Schematic diagram of the purification and identification of the antihypertensive peptide.*

The antihypertensive peptide of Protamex hydrolysate from *S. clava* flesh tissue was fractionated using ultrafiltration, sephadex G-25, and octadecyl silica (ODS) column (Ko et al., 2012a,b; **Figure 8.5**).

After a three step purification process, the purified antihypertensive peptide was identified as being a five amino acid residue of Ala–His–Ile–Ile–Ile, with a molecular weight of 565.3 Da (**Figure 8.6**). An ACE inhibitory

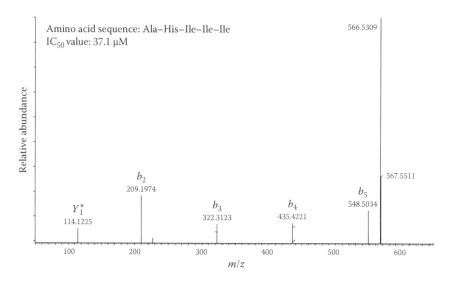

Figure 8.6 labels within image:
Amino acid sequence: Ala–His–Ile–Ile–Ile
IC$_{50}$ value: 37.1 μM

566.5309
567.5511
b_2 209.1974
b_5 548.5034
b_3 322.3123
b_4 435.4221
Y_1^* 114.1225
Relative abundance
m/z
100 200 300 400 500 600

Figure 8.6 *Identification of molecular mass and amino acid sequence of the antihypertensive peptide from Protamex hydrolysate from* S. clava *flesh tissue. (From Ko, S.C. et al., Food Chem., 134, 1141,2012a.)*

activity of the antihypertensive peptide exhibited an IC$_{50}$ value of 37.1 μM, and vasorelaxation was enhanced.

8.3 Effects of the antihypertensive peptide on NO production and eNOS phosphorylation level in human endothelial cells

The direct effects of antihypertensive peptide on the endothelium were investigated in human endothelial cells. In order to observe the effect of antihypertensive peptide on NO production, human endothelial cells (EA. hy 926 cells) were incubated in a medium, and the levels of released NO were measured at varying concentrations of the antihypertensive peptide. The levels of NO released in the incubation media of endothelial cells were increased by treatment with the antihypertensive peptide at a concentration of 0.1 mM. On the basis of Western blot analysis, the treatments with the antihypertensive peptide markedly enhanced eNOS phosphorylation (ser 1177) in endothelial cells (Ko et al., 2012a). These results demonstrate that the antihypertensive peptide could mediate NO production, thereby exerting an endothelium-dependent vasorelaxation activity.

8.4 Antihypertensive effect of the antihypertensive peptide on spontaneously hypertensive rats

The antihypertensive effect of the antihypertensive peptide was evaluated by measuring the change in systolic blood pressure (SBP) at 1, 3, 6, 12, and 24 h after the oral administration of the peptide (100 mg/kg body weight; **Figure 8.7**).

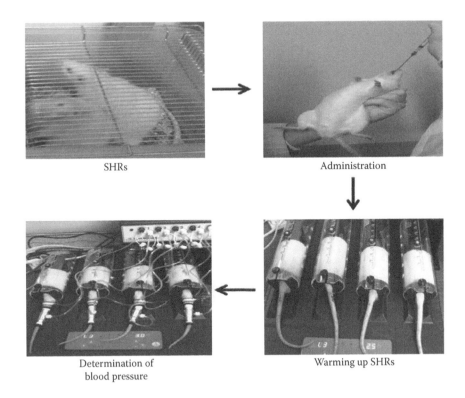

SHRs

Administration

Determination of
blood pressure

Warming up SHRs

Figure 8.7 *Photograph of determination of blood pressure by tail-cuff method.*

Amlodipine (30 mg/kg body weight) was employed as a positive control, and the control group was injected with an identical volume of saline. In acute oral administration experiments, minimum SBP changes were observed in the spontaneously hypertensive rat (SHR) model control group. The maximum SBP reduction levels of these treatment groups were in descending order: amlodipine group (53.67 mmHg) and the antihypertensive peptide group (46.83 mmHg). The antihypertensive effects of these treatments were maintained for 12 h. The SBP began to recover 12 h after treatment (the antihypertensive peptide) and returned to initial levels 24 h after administration (**Figure 8.8**).

8.5 Conclusions

Marine natural antihypertensive peptide was purified from the enzymatic hydrolysate of *S. clava* flesh tissue, and exhibited multiple antihypertensive effects. Through consecutive chromatographic methods, the antihypertensive peptide was found to exhibit potent ACE inhibitory activity, with an IC_{50} value of 37.1 μM. Moreover, in human endothelial cells, NO synthesis was found to be increased and eNOS phosphorylation was upregulated when the cells were cultured with the antihypertensive peptide. Additionally, the antihypertensive effect in SHR

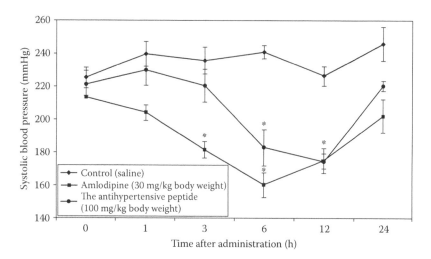

Figure 8.8 *Change of systolic blood pressure (SBP) of SHRs after the oral administration of test group. (♦) Control (Saline); (■) Positive control (Amlodipine, 30 mg/kg body weight); (●) The antihypertensive peptide (100 mg/kg body weight). The data are expressed as the means ± SEM. Statistical evaluation was carried out and compared with the control group. *p < 0.01. (From Ko, S.C. et al., Food Chem., 134,1141, 2012a.)*

also showed that the oral administration of the antihypertensive peptide could reduce SBP significantly. The results of this study indicate that the marine natural antihypertensive peptide from *S. clava* flesh tissue could be employed as a functional food ingredient with potential therapeutic benefits in the prevention and treatment of hypertension and other associated diseases.

References

Alberti, K. G., Zimmet, P., and Shaw. J. (2005). IDF Epidemiology Task Force Consensus Group. The metabolic syndrome new worldwide definition. *Lancet, 366,* 1059–1062.

Alberti, K. G., Zimmet, P., and Shaw, J. (2006). Metabolic syndrome—A new world-wide definition. A consensus statement from the International Diabetes Federation. *Diabetic Medicine, 23,* 469–480.

Asselbergs, F. W., Harst, P., Jessurum, G. A. J., Tio, R. A., and van Gilst, W. H. (2005). Clinical impact of vasomotor function assessment and the role of ACE-inhibitors and statins. *Vascular Pharmacology, 42,* 125–140.

Beny, J. L. and Brunet, P. C. (1988). Electrophysiological and mechanical effects of substance P and acetylcholine on rabbit aorta. *Journal of Physiology, 398,* 277–289.

Bohr, D. F., Dominiczak, A. F., and Webb, R. C. (1991). Pathophysiology of the vasculature in hypertension. *Hypertension, 18*(Suppl 111), S69–S75.

Byun, H. G. and Kim, S. K. (2001). Purification and characterization of angiotensin I converting enzyme (ACE) inhibitory peptides from Alaska pollack (*Theragra chalcogramma*) skin. *Process Biochemistry, 36,* 1155–1162.

Christopher, T. A., Lopez, B. L., Stillwagon, J. C., Gao, F., Gao, E., Ma, X. L., Ohlstein, E. H., and Yue, T. L. (2008). Idoxifene causes endothelium-dependent, nitric oxide-mediated vasorelaxation in male rats. *European Journal of Pharmacology, 446,* 139–143.

Dimo, T., Bopda Mtopi, O.-S., Nguelefack, T. B., Kamtchouing, P., Zapfack, L., Asongalem, E. A., and Dongo, E. (2007). Vasorelaxant effects of *Brillantaisia nitens* Lindau (Acanthaceae) extracts on isolated rat vascular smooth muscle. *Journal of Ethnopharmacology, 111,* 104–109.

Erdmann, K., Cheung, B. W. Y., and Schroder, H. (2008). The possible roles of food–derived bioactive peptides in reducing the risk of cardiovascular disease. *Journal of Nutritional Biochemistry, 19,* 643–654.

Je, J. Y., Park, J. Y., Jung, W. K., Park, P. J., and Kim, S. K. (2005). Isolation of angiotensin I converting enzyme (ACE) inhibitor from fermented oyster sauce, *Crassostrea gigas. Food Chemistry, 90,* 809–814.

Jo, J. E., Kim, K. H., Yoon, M. H., Kim, N. Y., Lee, S. C., and Yook, H. S. (2010). Quality characteristics and antioxidant activity research of *Halocynthia roretzi* and *Halocynthia aurantium. Journal of Korean Food Science and Nutrition, 39,* 1481–1486.

Kato, H. and Suzuki, T. (1975). Bradykinin-potentiating peptides from the venom of *Agkistrodon halys blomhoffii*: Their amino acid sequence and the inhibitory activity on angiotensin-I converting enzyme from rabbit lung. *Life Sciences, 16,* 810–811.

Kim, S. K., Choi, K. S., and Son, S. M. (2007). Isolation and purification of natural antimicrobial peptides from Munggae, *Halocynthia roretzi. Food Engineering Progress, 11,* 54–59.

Kim, J. B., Iwamuro, S., Knoop, F. C., and Conlon, J. M. (2001a). Antimicrobial peptides from the skin of the Japanese mountain brown frog, *Rana ornativentris. Journal of Peptide Research, 58,* 349–356.

Kim, S. K., Kim, Y. T., Byun, H. G., Park, P. J., and Ito, H. (2001b). Purification and characterization of antioxidative peptides from bovine skin. *Journal of Biochemistry. Molecular Biology, 34,* 214–219.

Kim, J. J., Kim, S. J., Kim, S. H., Park, H. R., and Lee, S. C. (2006). Antioxidant and anticancer activities of extracts from *Styela clava* according to the processing methods and solvents. *Journal of Korean Food Science and Nutrition, 35,* 278–283.

Kim, H. Y., Oh, H., Li, X., Cho, K. W., Kang, D. G., and Lee, H. S. (2011). Ethanol extract of seeds of *Oenothera odorata* induces vasorelaxation via endothelium-dependent NO-cGMP signaling through activation of Akt-eNOS-sGC pathway. *Journal of Ethnopharmacology, 133,* 315–323.

Ko, S. C., Kim, D. G., Han, C. H., Lee, Y. J., Lee, J. K., Byun, H. G., and Jeon, Y. J. (2012a). Nitric oxide-mediated vasorelaxation effects of anti-angiotensin I-converting enzyme (ACE) peptide from *Styela clava* flesh tissue and its anti-hypertensive effect in spontaneously hypertensive rats. *Food Chemistry, 134,* 1141–1145.

Ko, S. C., Lee, J. K., Byun, H. G., Lee, S. C., and Jeon, Y. J. (2012b). Purification and characterization of angiotensin I-converting enzyme (ACE) inhibitory peptide from enzymatic hydrolysates of *Styela clava* flesh tissue. *Process Biochemistry, 47,* 34–40.

Lee, J. K., Hong, S., Jeon, J. K., Kim, S. K., and Byun, H. G. (2009). Purification and characterization of angiotensin I converting enzyme inhibitory peptides from the rotifer, *Brachionus rotundiformis. Bioresource Technology, 100,* 5255–5259.

Lee, J. K., Jeon, J. K., and Byun, H. G. (2011). Effect of angiotensin I converting enzyme inhibitory peptide purified from skate skin hydrolysate. *Food Chemistry, 125,* 495–499.

Lee, S. H., Qian, Z. J., and Kim, S. K. (2010a). A novel angiotensin I converting enzyme inhibitory peptide from tuna frame protein hydrolysate and its antihypertensive effect in spontaneously hypertensive rats. *Food Chemistry, 118,* 96–102.

Lee, D. W., You, D. H., Yang, E. K., Jang, I. C., Bae, M. S., Jeon, Y. J., Kim, S. J., and Lee, S. C. (2010b). Antioxidant and ACE inhibitory activities of *Styela clava* according to harvesting time. *Journal of Korean Food Science and Nutrition, 39,* 331–336.

Liu, R., Wang, M., Duan, J. A., Guo, J. M., and Tang, Y. P. (2010). Purification and identification of three novel antioxidant peptides from *Cornu Bubali* (water buffalo horn). *Peptides, 31,* 786–793.

Moncada, S. and Vane, J. R. (1978). Pharmacology and endogenous roles of prostaglandin endoperoxides, thromboxane A2 and prostacyclin. *Pharmacology Reviews, 30,* 293–331.

Ondetti, M. A. (1977). Design of specific inhibitors of angiotensin converting enzyme: New class of orally active antihypertensive agents. *Science, 196,* 441–444.

Rosa, A. P. B., Montoya, A. B., Martínez-Cuevas, P., Hernández-Ledesma, B., León-Galván, M. F., León-Rodríguez, A. D., and González, B. (2010). Tryptic amaranth glutelin digests induce endothelial nitric oxide production through inhibition of ACE: Antihypertensive role of amaranth peptides. *Nitric Oxide, 23,* 106–111.

Segura-Campos, M. R., Chel-Guerrero, L. A., and Betancur-Ancona, D. A. (2011). Purification of angiotensin I-converting enzyme inhibitory peptides from a cowpea (*Vigna unguiculata*) enzymatic hydrolysate. *Process Biochemistry, 46,* 864–872.

Simpson, B. K., Nayeri, G., Yaylayan, V., and Ashie, I. N. A. (1998). Enzymatic hydrolysis of shrimp meat. *Food Chemistry, 61,* 131–138.

Thurman, J. M. and Schrier, R. W. (2003). Comparative effects of angiotensin-converting enzyme inhibitors and angiotensin receptor blockers on blood pressure and the kidney. *The American Journal of Medicine, 114,* 588–598.

Tsai, J. S., Chen, J. L., and Pan, B. S. (2008). ACE-inhibitory peptides identified from the muscle protein hydrolysate of hard clam (*Meretrix lusoria*). *Process Biochemistry, 43,* 743–747.

Vercruysse, L., Smagghe, G., Matsui, T., and Camp, J. V. (2008). Purification and identification of an angiotensin I converting enzyme (ACE) inhibitory peptide from the gastrointestinal hydrolysate of the cotton leafworm, *Spodoptera littoralis. Process Biochemistry, 43,* 900–904.

Zhao, Y., Li, B., Liu, Z., Dong, S., Zhao, X., and Zeng, M. (2007). Antihypertensive effect and purification of an ACE inhibitory peptide from sea cucumber gelatin hydrolysate. *Process Biochemistry, 42,* 1586–1591.

9

Beneficial Effects of Marine Natural Products on Autoimmune Diseases

Mi Eun Kim, Jun Sik Lee, Se-Kwon Kim, and Won-Kyo Jung

Contents

9.1 Introduction

Autoimmune diseases arise from an inappropriate immune response of the body against substances and tissues normally present in the body. In the same way, the immune system mistakes some parts of the body as pathogens and attacks its own cells. The breakdown of the mechanism assuring recognition of nonself and self by the immune system is a characteristic feature of autoimmune diseases. The immune system particularly recognizes and eliminates foreign agents, thereby protecting the host against infection. During maturation of the immune system, immune cells that react against self-tissues are eliminated providing an immune system that is "tolerant" to self. Historically, reactivity or autoimmunity of the immune system to self-antigens was thought of as an aberrant response. A combination of genetic predisposition and environmental factors contributes to the development of an autoimmune disease. Furthermore, regulating and understanding the mechanisms that lead to deregulation of the immune response resulting in autoimmune diseases is necessary to develop better therapies to treat and possibly even prevent these diseases. This chapter will discuss the effects of marine natural products on the regulation and bioactivity of autoimmune disease and their regulation mechanisms.

9.2 Mechanism and regulation of autoimmune disease

9.2.1 Effects of genetics on autoimmune disease

The development of an autoimmune disease depends on the combination of genetic and environmental factors. Most autoimmune diseases are thought to be polygenic, involving more than one gene. The idea that individuals are genetically predisposed to develop an autoimmune disease arose from clinical reports that patients often describe a family history of autoimmune diseases. For example, human lymphocyte antigen or HLA haplotype is the best available predictor of developing autoimmune diseases. HLA haplotype, or the major histocompatibility complex (MHC) in mice, is anticipated to increase autoimmune disease by enhancing antigen presentation in the periphery resulting in increased T-cell activation. Genes outside of the MHC also contribute to the risk of developing autoimmune disease. Many studies of type I diabetes mellitus and lupus or their animal models, have revealed a number of non-MHC genes that contribute to susceptibility. Common susceptibility loci have been found for a number of different autoimmune diseases, including myocarditis and diabetes, suggesting that shared genes are involved in the pathogenesis of autoimmune disease. Recent evidence suggests that many of the genes conferring susceptibility control immunoregulatory factors. Studies on the prevalence of autoimmune disease in monozygotic and dizygotic twins indicate that environmental factors are necessary for the development of the disease. If an autoimmune disease is due entirely to genes, then its concordance rate in identical monozygotic twins should be 100% and its concordance in nonidentical dizygotic twins 50%. However, if an autoimmune disease is due to environmental factors, the concordance rate should be similar in both monozygotic and dizygotic twins. The occurrence of autoimmune diseases in genetically identical, monozygotic twins was found to have a concordance rate in the range of 10%–50% in different studies when compared to 2%–40% in dizygotic twins. The low disease concordance in monozygotic twins indicates that environmental agents are important in the development of autoimmune diseases.

9.2.2 Effects of environment on autoimmune disease

Fundamentally, all autoimmune diseases occur as a consequence of impaired immune function that results from interaction of genetic and environmental factors. Despite important progress, much remains to be learned about these factors and their interactions. External environmental factors, such as hormones, drugs, diet, toxins, and/or infections are important in determining whether an individual will develop autoimmune disease. Environmental agents are able to amplify autoimmunity in genetically susceptible individuals and to break tolerance in genetically resistant individuals, thereby increasing the risk of developing autoimmune disease. The possible role of exposure to various marine natural products in autoimmune disease has been explored, primarily through laboratory and animal studies.

9.2.3 Effects of toxins and drugs on autoimmune disease

For a long time, toxins like heavy metals or drugs intended for therapy have been associated with disease syndromes resembling autoimmune diseases. For example, drugs like procainamide and hydralazine can induce auto-antibodies and lupus-like disorders in patients. Penicillamine has been associated with myasthenia gravis, and a-methyldopa is known to cause a form of hemolytic anemia. However, in all cases of drug-induced autoimmune diseases described thus far, the disease disappears when the drug is removed. Various heavy metals, such as mercury, silver, or gold, can induce an auto-antibody response to cell nuclear antigens in susceptible strains of mice. By yet unknown mechanisms, mercurial compounds have been shown to exacerbate autoimmune disease in experimental animal models. Recently, administration of mercuric chloride to susceptible strains of mice was found to increase auto-antibodies and cell-mediated autoimmunity in a collagen-induced model of arthritis. These findings suggest that environmental factors like the microbial component of adjuvant in the collagen-induced model and mercury exposure can act synergistically to promote autoimmune disease.

9.2.4 Effects of infections on autoimmune disease

Recent studies show that infectious agents have variable effects on autoimmune disease, with protection from autoimmune disease being a more frequent response to infection than previously thought. Bacterial and viral infections were some of the first agents associated with autoimmune diseases more than a century ago. However, most of the clinical evidence linking autoimmune diseases with preceding infections is only circumstantial. For example, diabetes has been associated with coxsackievirus and cytomegalovirus infections, multiple sclerosis with Epstein–Barr virus and measles virus infections, rheumatoid arthritis with mycobacteria and Epstein–Barr infections, and myocarditis with coxsackievirus and cytomegalovirus infections, to name a few. Since infections generally occur well before the onset of signs and symptoms of auto-immune disease, linking a specific causative agent to a particular autoimmune disease is difficult. The most direct evidence that infectious agents can induce autoimmune disease is the development of disease in experimental animals following inoculation with self-antigens in combination with adjuvant containing uninfectious microbial antigens. The fact that multiple, diverse types of microorganisms are associated with a single autoimmune disease suggests that infectious agents induce autoimmune disease through common mechanisms.

9.2.5 Cell-mediated damage in autoimmune disease

Cell-mediated immunity is an immune response that does not involve antibodies but rather involves the activation of macrophage, antigen-specific cytotoxic T cells, natural killer cells, and the release of various cytokines in response to an antigen. Damage induced by cells of the immune system plays a major pathogenic role in many autoimmune diseases. The predominant infiltrating cells include phagocytic macrophages, neutrophils, self-reactive CD4[+] T helper cells, and self-reactive CD8[+] cytolytic T cells, with smaller numbers of natural killer cells, mast cells, and dendritic cells. Immune cells damage tissues directly by killing cells or indirectly by releasing cytotoxic cytokines, prostaglandins, reactive nitrogen or oxygen intermediates. Tissue macrophages and monocytes can act as antigen-presenting cells to initiate an autoimmune response, or as effector cells once an immune response has been initiated. Macrophages act as killer cells through antibody-dependent cell-mediated cytotoxicity and by secreting cytokines, such as tumor necrosis factor (TNF) or interleukin (IL)-1, which act as protein signals between cells. Macrophages and neutrophils damage tissues (and microorganisms) by releasing highly cytotoxic proteins like nitric oxide and hydrogen peroxide. Cytokines and other mediators released by macrophages recruit other inflammatory cells, like neutrophils and T cells, to the site of inflammation. CD4[+] T cells have been classified as T helper 1 (Th1) or T helper 2 (Th2) cells, depending on the release of the cytokines interferon-γ (IFN-γ) or IL-4, respectively (**Figure 9.1**). IFN-γ is a proinflammatory cytokine associated with many organ-specific autoimmune diseases like type I diabetes (**Figure 9.2**) and thyroiditis, while IL-4 activates B cells to produce antibodies and is associated with auto-antibody/immune complex-mediated autoimmune diseases like lupus and arthritis. Suppressor or regulatory T-cell populations, including activated CD25[+]CD4[+] regulatory T cells, exist in peripheral tissues and are important in controlling inflammation and autoimmune responses by killing autoreactive cells. These regulatory cells also secrete anti-inflammatory cytokines like IL-10 and transforming growth factor (TGF)-β that further inhibit Th1 immune responses, thereby reducing inflammation and autoimmune disease. If regulation of self-reactive T cells and auto-antibody production by regulatory T-cell populations is disrupted by environmental agents like infections or toxins, then chronic autoimmune disease may result.

9.3 Effects of marine natural products on autoimmune disease

Marine algae are being consumed in Asia since early times, which are rich in minerals, vitamins, dietary fibers, polysaccharides, proteins, and various functional polyphenols (Jung et al. 2009). Moreover, marine natural products are considered to be rich sources of anti-inflammatory and antioxidant compounds. For this

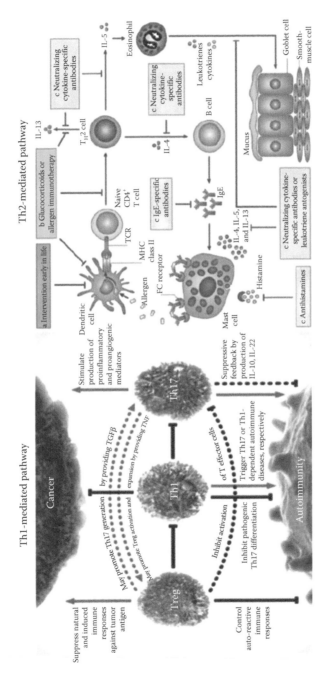

Figure 9.1 *Th1- and Th2-mediated immune response pathways. (Chen, X and J.J. Oppenheim. www.SABiosciences.com/support_literature.php,2009; Hawrylowicz C.M., and A.O'Garra. Nat Rev Immunol, 5(4), 271, 2005.)*

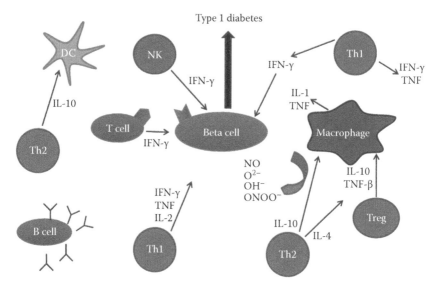

Figure 9.2 *Diabetes: A form of diabetes mellitus that results from autoimmune destruction of insulin-producing beta cells of the pancreas.*

reason, many types of marine natural products have been investigated to identify new and effective antioxidant compounds, as well as to elucidate the regulation mechanisms of anti-inflammation, cell proliferation, antidiabetic, and anticancer effects (Yuan and Walsh 2006; Kim et al. 2008; Maeda et al. 2009; Kang et al. 2010a; Schumacher et al. 2011). Recently, many countries are more closely looking at marine natural products as potent targets for bioactive substances, because they have suggested some possible valuable uses and applications in the fields of nutraceutics and pharmaceutics (**Table 9.1**; Schumacher et al. 2011).

9.3.1 Effects of the extracts from *Laurencia undulate* and *Ecklonia cava* on the regulation of asthma

In 2008, Kim et al. published the effects of *Ecklonia cava* (EC) ethanolic extracts on airway hyperresponsiveness (AHR) and inflammation in a murine asthma mole (Kim et al. 2008). This report demonstrated whether pretreatment with *E. cava* induces a significant inhibition of asthmatic reactions in a mouse asthma model and suggested that *E. cava* extracts treatment resulted in significant reductions of matrix metalloproteinase-9 (MMP-9) and suppressor of cytokine signaling-3 (SOCS-3) expression and a reduction in the increased eosinophil peroxidase activity (**Figure 9.3a** and **b**). This report indicated that *E. cava* extracts reduce airway inflammation and hyperresponsiveness via the inhibition of SOCS-3 protein expression in a murine model of ovalbumin (OVA)-induced asthma. Therefore, these results indicated that *E. cava* extracts may prove to be a useful therapeutic approach to the treatment of allergic airway diseases. Interestingly, Jung et al. described the antiasthmatic effect of marine red alga (*Laurencia undulate* [LU]) polyphenolic extracts in a murine model of asthma (Jung et al. 2009).

Table 9.1 The Physiological Role of Marine Natural Products on Autoimmune Diseases

Target Disease	Compound/Organism	Country	References
Diabetes	Methanol extract (*Ecklonia cava* attenuates)	Korea	Kang et al. (2010b)
	Methanol extract (*Ecklonia stolonifera*)	Japan	Iwai et al. (2008)
	Phloroglucinol derivative (*Eisenia bicyclis*)	Japan	Okada et al. (2004)
	Chlorella vulgaris	Malaysia	Aizzat et al. (2010)
Obesity	Ethanol extract (*Undaria pinnatifida*)	Japan	Okada et al. (2011)
	Water-soluble extract (*Petalonia binghamiae*)	Korea	Kang et al. (2010b)
	Fucoxanthin (*Undaria pinnatifida*)	Japan	Maeda et al. (2009)
	Xanthigen (fucoxanthin + pomegranate seed oil)	Russia	Abidov et al. (2010)
Asthma	Fatty acid	Finland	Lumia et al. (2011)
	Cyclotheonamide E4 and E5 (*Ircinia*)	Japan	Murakami et al. (2002)
	Polyphenolic extract (*Laureneia undulata*)	Korea	Jung et al. (2009)
	Lipid extract (*Perna canaliculus*)	Russia	Emelyanov et al. (2002)

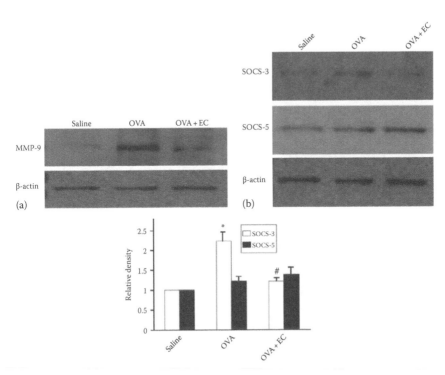

Figure 9.3 *Effects of E. cava extracts on the expression of MMP-9 (a) and SOCS-3, SOCS-5 (b) expression in lung tissues of OVA-sensitized mice and OVA-challenged mice. (From Kim, S.K. et al., Biomed. Pharmacother., 62(5), 289, 2008.)*

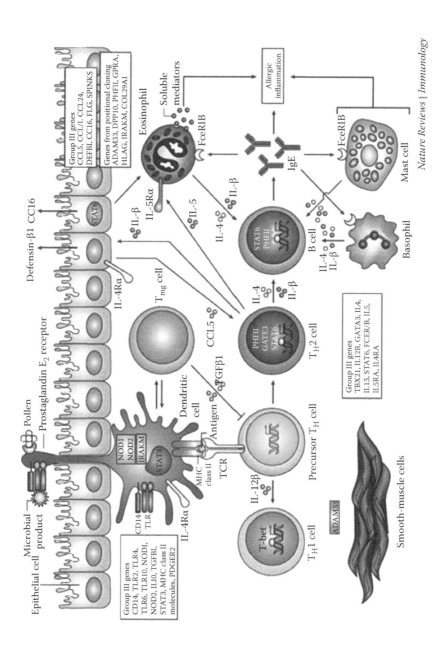

Figure 9.4 *Asthma: The common chronic inflammatory disease of the airways characterized by variable and recurring symptoms, reversible airflow obstruction, and bronchospasm. (From Vercelli, D. Nat Rev Immunol., 8, 169, 2008.)*

Nature Reviews | Immunology

Allergic diseases are common chronic conditions in children, and so far primary prevention strategies have not been very successful. Among various allergic diseases, asthma is a complex inflammatory disease of the lung characterized by variable airflow obstruction, AHR, and airway inflammation (**Figure 9.4**). The inflammatory response in the asthmatic lung is characterized by infiltration of the airway wall by mast cells, lymphocytes, and eosinophils (Elias et al. 2003), and is associated with the increased expression of several inflammatory proteins, including cytokines, adhesion molecules, and enzymes in the airways (**Figure 9.5**). The Th2-type cytokines, such as IL-4, IL-5, and IL-13, produced by activated CD4[+] T cells play a central role in the pathogenesis of asthma by controlling the key process of immunoglobulin E (IgE) production, growth of mast cells, and the differentiation and activation of mast cells and eosinophils (Jung et al. 2009). The red alga genus *Laurencia* is known to produce a wide array of natural products exhibiting some kind of biological activities, using the natural bioactive materials of polyphenols, polysaccharides, and other halogenated secondary metabolites. They have been shown to process antioxidant (Li et al. 2009), antibacterial (Bansemir et al. 2004), and anticancer activities (Liang et al. 2007) against marine bacteria as in the previous reports. Jung et al. (Jung et al. 2009) investigated how *L.undulate* made its effect on the development of OVA-induced eosinophilia in murine model of asthma. It has been suggested that eosinophils contribute to several of the clinical features of allergic asthma, including tissue damage and AHR.

The effect of *L.undulate* extracts on cellular changes in bronchoalveolar lavage (BAL) fluid and pathological changes in lung tissues are shown in **Figure 9.5a** and **b**.

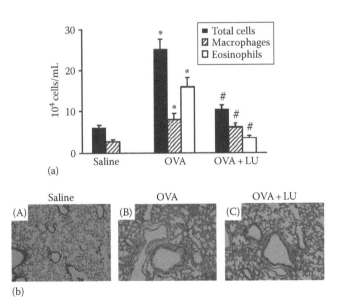

(a)

(b)

Figure 9.5 *Effects of* L. undulata *extracts (LU) on the recruitment of inflammatory cells into BAL in OVA-induced allergic asthmatic mice (a) and pathological changes (b). (From Jung, W.K. et al., Food Chem. Toxicol., 47(2), 293, 2009.)*

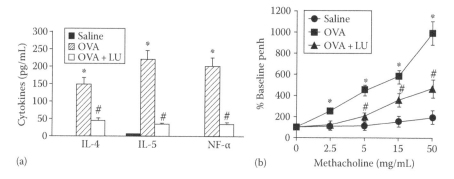

Figure 9.6 *Effects of* L. undulata *extracts (LU) on the cytokines levels of IL-4, IL-5, TNF-α (a), and airway responsiveness in OVA-sensitized mice (b). (From Jung, W.K. et al., Food Chem. Toxicol., 47(2), 293, 2009.)*

As shown in **Figure 9.5a**, the number of eosinophils in BAL fluids was increased about 16-fold, compared to those in the control group at 2 days after the OVA challenge. Moreover, the number of eosinophils observed in BAL fluids in the *L.undulate*-treated group of mice decreased to the level of 0.22-fold of OVA-challenged group. Mice treated with *L.undulate* showed marked reductions in the infiltration of inflammatory cells in the peribronchiolar and perivascular regions. Administration of *L.undulate* resulted in a marked improvement of luminal narrowing in the airway. These results suggested that *L.undulate* is highly capable of inhibiting the development of allergic status induced by OVA in mice (**Figure 9.6**). Jung et al. (Jung et al. 2009) determined whether *L.undulate* was related to inflammatory cytokine production. Initiation of allergic response appears to occur with the presentation of the allergen by antigen-presenting cells to CD4⁺ T cells. Bronchoalveolar lavage fluid (BALF) supernatants were collected, and the products of IL-4 and IL-5 were analyzed by ELISA (**Figure 9.6a**). The administration of *L.undulate* reduced the concentrations of IL-4 and IL-5 by 70% and 85%, respectively, and led to an 83.5% reduction in TNF-α secretion. Also, OVA-induced asthma is characterized by AHR and inflammation of the airways. As shown in **Figure 9.6b**, OVA-sensitized and OVA-challenged mice treated with *L.undulate* showed a dose–response curve of percent Penh that shifted to the right compared with that of untreated mice. Although it still remains to identify the chemical properties and structures of the polyphenolic components derived from the red alga, these results support that *L.undulate* could be used as adjuvant therapeutic materials for patients with bronchial asthma. In addition, the results can motivate research on other type I allergic reactions, such as allergic rhinitis, atopic dermatitis, and anaphylaxis.

9.3.2 Effects of marine natural products on the regulation of diabetes

Since 2004, many researchers investigated the effects of materials derived from marine natural products on diabetes, including phloroglucinol derivatives

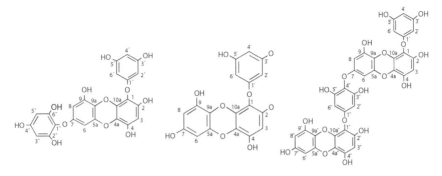

Figure 9.7 *A new phloroglucinol derivatives. (From Okada, Y. et al., J. Nat. Prod., 67(1), 103, 2004.)*

(Okada et al. 2004), polyphenols from brown alga *Ecklonia stolonifera* (Iwai 2008), fucoxanthin (Maeda et al. 2009), *Chlorella vulgaris* (Aizzat et al. 2010), water-soluble extract of *Petalonia binghamiae* (Kang et al. 2010b), Xanthigen (Abidov et al. 2010), and *E. cava* (Kang et al. 2010a). In 2004, Okada et al. described that a new phloroglucinol derivative from the brown alga *Eisenia bicyclis* has the potential for the effective treatment of diabetic complications (**Figure 9.7**). Obesity and diabetes are the major obstacles to efforts aimed at improving human health and quality of life. Type 1 diabetes mellitus is characterized by loss of insulin-producing β-cells of the islets of Langerhans in the pancreas leading to insulin deficiency. This type of diabetes can be further classified as immune-mediated or idiopathic. The majority of type 1 diabetes is of the immune-mediated nature, where β-cell loss is a T-cell mediated autoimmune attack. Most affected people are otherwise healthy and of a healthy weight when onset occurs. Sensitivity and responsiveness to insulin are usually normal, especially in the early stages. Type 1 diabetes can affect children or adults, but is traditionally termed "juvenile diabetes" because it represents a majority of the diabetes cases in children. As reported in the study by Maeda et al. (2009), fucoxanthin has antidiabetic effects on diet-induced obesity conditions in a murine model (**Figure 9.8**). As shown in **Figure 9.8** and **Table 9.2**, the administration of fucoxanthin from Wakame (*Undaria pinnatifida*) diet additionally promotes the recovery of blood glucose uptake by muscle cells by regulating GLUT4 mRNA expression. Kang et al. (2010b) demonstrated that a water-soluble extract of *P. binghamiae* inhibited the expression of adipogenic regulators in 3T3-L1 preadipocytes, and reduced adiposity and weight gain in rats fed a high-fat diet. The effect of water-soluble extract of *P. binghamiae* (PBEE) on cell viability and cytotoxicity of 3T3L-1 cells was evaluated by the MTT and LDH assays. PBEE at a concentration of 500 mg/mL did not affect the viability (2.43% ± 3.30% compared to control) as well as the cytotoxicity (−1.59% ± 0.95% compared to control) of the 3T3-L1 cells, as determined by MTT and LDH assays. Moreover, this report investigated whether PBEE inhibits MDI-induced differentiation of 3T3-L1 preadipocytes. On day 0, PBEE was added to the MDI differentiation medium (which contains IBMX, dexamethasone, and insulin); on day 8, the adipocytes were stained using Oil Red O. The Oil Red O

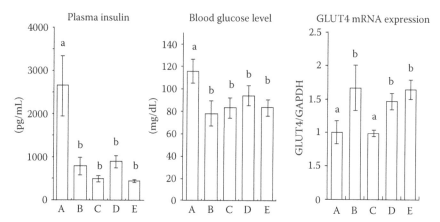

Figure 9.8 *Plasma insulin and blood glucose levels in C57BL/6J mice and glucose transporter 4 (GLUT4) mRNA expression in the muscle tissue of C57BL/6J mice. (A) High fat (NF), (B) high fat control (NFC), (C) HF-NFC, (D) HF-fucoxanthin-rich Wakame lipids (WL) 1, and (E) HF-WL2. Values are presented as the mean ± SE (n = 6). Differences were considered significant at p < 0.05. (From Maeda, H. et al., Mol. Med. Rep., 2(6), 897, 2009.)*

staining results demonstrated that PBEE treatment at 8, 40, and 200μg/mL inhibited 3T3-L1 adipocyte differentiation in a dose-dependent manner. The positive-control cells, which were treated with 100μmol/L genistein, exhibited dramatic inhibition of lipid accumulation (**Figure 9.9a** and **b**). Kang et al. (2010b) also determined whether PBEE inhibits adipocyte differentiation by negatively regulating the expression of key transcriptional regulators and examined the expression levels of C/EBPα, C/EBPβ, and PPARγ during adipocyte differentiation in the presence and absence of PBEE. Expression levels of all the three factors were reduced in PBEE-treated cells, as was the expression level of aP2 (**Figure 9.9c** and **d**). However, PBEE did not affect the expression levels of phospho-ERK or SREBP1c (**Figure 9.9e** and **f**). These results suggested that PBEE may

Table 9.2 Plasma Lipid Parameters and Leptin Levels of C57BL/6J Mice Fed with the Experimental Diets

	HF$_C$	NF$_C$	HF-NF$_C$	HF-WL1	HF-WL2
Total cholesterol (mg/dL)	168 ± 5	127 ± 8	127 ± 4	192 ± 10	185 ± 7
HDL cholesterol (mg/dL)	65 ± 2	67 ± 5	67 ± 3	71 ± 3	66 ± 2
LDL cholesterol (mg/dL)	13 ± 1.8	7 ± 0.4	7 ± 0.4	9 ± 0.9	7 ± 0.4
Triacylglycerol (mg/dL)	56 ± 9	84 ± 15	72 ± 13	82 ± 19	62 ± 20
Phospholipid (mg/dL)	278 ± 12	251 ± 15	239 ± 8	289 ± 6	266 ± 12
FFA (μEq/L)	1424 ± 116	1807 ± 199	1686 ± 187	1599 ± 67	1346 ± 167
Leptin (ng/dL)	13.0 ± 2.8	7.6 ± 2.4	6.0 ± 0.7	3.2 ± 2.1	1.5 ± 1.0

Source: Maeda, H. et al., *Mol. Med. Rep.*, 2(6), 897, 2009.
Values are presented as the mean ±SE (n = 6). Differences were considered significant at p < 0.05. HDL, high-density lipoprotein; LDL, low-density lipoprotein; FFA, free fatty acids.

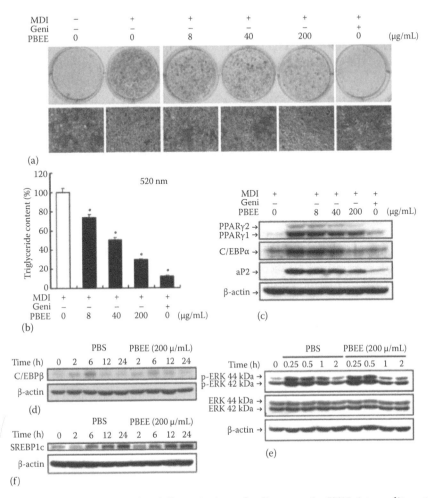

Figure 9.9 *PBEE inhibits the differentiation of adipocytes in 3T3L-1 preadipocytes. Cells were cultured in MDI differentiation medium in the presence or absence of PBEE (Geni: genistein 100 μmol/L). (a) Differentiated adipocytes were stained with Oil Red O on Day 8 (after 4 days of PBEE treatment). (b) Lipid accumulation was assessed by the quantification of OD520 as described in Materials and Methods. Results are shown as means±S.D. (n=3;P<.05 compared with no PBEE). (c) Western blot analysis of PPARγ, C/EBPα and aP2 expression. Proteins were prepared from 3T3-L1 cells on Day 6. (d–f) Western blot analysis of C/EBPβ (d) phospho-ERK (e) and SREBP1c (f) expression in post-confluent differentiated 3T3-L1 cells. Proteins were harvested at the indicated times. (From Kang, S.I. et al., J. Nutr. Biochem., 21(12), 1251, 2010b.)*

protect against high-fat diet-induced obesity by inhibiting adipocyte differentiation and glucose uptake in mature adipocytes. Brown alga *E. cava* attenuates type I diabetes by activating AMPK and Akt signaling pathway, as reported by Kang et al. (2010b). This report indicated that the polyphenol-rich extract of *E. cava* has a potent radical scavenging activity, and its administration showed significant improvement from all the diabetic indications in experimental animals of the present report. The extract of *E. cava* activates both AMPK/ACC and PI3K/Akt signaling in C_2C_{12} skeletal muscle cells, which is consistent with the in vivo results. Although the therapeutic potential of the extract of *E. cava* is demonstrated in type 1 diabetes mellitus model in the present report, the detailed molecular mechanism of its actions needs to be further investigated in the future.

9.4 Conclusions

Although a number of reports have widely investigated the effects of marine natural products and their derivatives, a few reports on autoimmune disease regulation adjuvants have recently been published. Marine natural products can be promising candidates as cytokine regulation and supporting material for autoimmune disease therapeutic applications, because of their antioxidant, antiasthmatic, antiobesity, and antidiabetic activities. However, much more work is required to accomplish the regulation mechanisms of autoimmune disease by marine natural products.

Acknowledgments

This chapter was supported by the Technology Development Program for Fisheries, Ministry for Food, Agriculture, Forestry and Fisheries, and also by the Basic Science Research Program through the National Research Foundation of Korea (NRF) funded by the Ministry of Education, Science and Technology (2010–2003111).

References

Abidov, M., Z. Ramazanov et al. (2010). The effects of Xanthigen in the weight management of obese premenopausal women with non-alcoholic fatty liver disease and normal liver fat. *Diabetes, Obesity & Metabolism* **12**(1): 72–81.

Aizzat, O., S. W. Yap et al. (2010). Modulation of oxidative stress by *Chlorella vulgaris* in streptozotocin (STZ) induced diabetic Sprague-Dawley rats. *Advances in Medical Sciences* **55**(2): 281–288.

Bansemir, A., N. Just et al. (2004). Extracts and sesquiterpene derivatives from the red alga Laurencia chondrioides with antibacterial activity against fish and human pathogenic bacteria. *Chemistry and Biodiversity* **1**(3): 463–467.

Chen, X., and J. J. Oppenheim (2009). Regulatory T Cells, Th17 Cells, and TLRs: Crucial Roles in Inflammation, Autoimmunity, and Cancer. www.SABiosiences.com/support_literature.php

Emelyanov, A., G. Fedoseev, O. Krasnoschekova, A. Abulimity, T. Trendeleva, P. J. Barnes (2002). Treatment of asthma with lipid extract of New Zealand green-lipped mussel: a randomised clinical trial. *Eur Respir J* **20**(3): 596–600.

Elias, J. A., C. G. Lee et al. (2003). New insights into the pathogenesis of asthma. *Journal of Clinical Investigation* **111**(3): 291–297.

Hawrylowicz, C. M., and A. O'Garra (2005). Potential role of interleukin-10-secreting regulatory T cells in allergy and asthma. *Nat Rev Immunol* **5**(4): 271–283.

Iwai, K. (2008). Antidiabetic and antioxidant effects of polyphenols in brown alga *Ecklonia stolonifera* in genetically diabetic KK-A(y) mice. *Plant Foods for Human Nutrition* **63**(4): 163–169.

Jung, W. K., I. Choi et al. (2009). Anti-asthmatic effect of marine red alga (*Laurencia undulata*) polyphenolic extracts in a murine model of asthma. *Food and Chemical Toxicology: An International Journal Published for the British Industrial Biological Research Association* **47**(2): 293–297.

Kang, C., Y. B. Jin et al. (2010a). Brown alga *Ecklonia cava* attenuates type 1 diabetes by activating AMPK and Akt signaling pathways. *Food and Chemical Toxicology: An International Journal Published for the British Industrial Biological Research Association* **48**(2): 509–516.

Kang, S. I., M. H. Kim et al. (2010b). A water-soluble extract of *Petalonia binghamiae* inhibits the expression of adipogenic regulators in 3T3-L1 preadipocytes and reduces adiposity and weight gain in rats fed a high-fat diet. *Journal of Nutritional Biochemistry* **21**(12): 1251–1257.

Kim, S. K., D. Y. Lee et al. (2008). Effects of *Ecklonia cava* ethanolic extracts on airway hyperresponsiveness and inflammation in a murine asthma model: Role of suppressor of cytokine signaling. *Biomedicine and Pharmacotherapy* **62**(5): 289–296.

Li, Y. X., Y. Li et al. (2009). In vitro antioxidant activity of 5-HMF isolated from marine red alga *Laurencia undulata* in free-radical-mediated oxidative systems. *Journal of Microbiology and Biotechnology* **19**(11): 1319–1327.

Liang, H., J. He et al. (2007). Effect of ethanol extract of alga *Laurencia* supplementation on DNA oxidation and alkylation damage in mice. *Asia Pacific Journal of Clinical Nutrition* **16** (Suppl 1): 164–168.

Lumia M., P. Luukkainen, H. Tapanainen, M. Kaila, M. Erkkola, L. Uusitalo, S. Niinistö, M. G. Kenward, J. Ilonen, O. Simell, M. Knip, R. Veijola, S. M. Virtanen (2011). Dietary fatty acid composition during pregnancy and the risk of asthma in the offspring **22**(8): 827–835.

Maeda, H., M. Hosokawa et al. (2009). Anti-obesity and anti-diabetic effects of fucoxanthin on diet-induced obesity conditions in a murine model. *Molecular Medicine Reports* **2**(6): 897–902.

Murakami Y., M. Takei, K. Shindo, C. Kitazume, J. Tanaka, T. Higa, H. Fukamachi (2002). Cyclotheonamide E4 and E5, new potent tryptase inhibitors from an Ircinia species of sponge. *J Nat Prod* **65**(3): 259–261.

Okada, Y., A. Ishimaru et al. (2004). A new phloroglucinol derivative from the brown alga *Eisenia bicyclis*: Potential for the effective treatment of diabetic complications. *Journal of Natural Products* **67**(1): 103–105.

Okada T., Y. Mizuno, S. Sibayama, M. Hosokawa, K. Miyashita (2011). Antiobesity effects of Undaria lipid capsules prepared with scallop phospholipids. *J Food* Sci **76**(1): H2–6.

Schumacher, M., M. Kelkel et al. (2011). A survey of marine natural compounds and their derivatives with anti-cancer activity reported in 2010. *Molecules* **16**(7): 5629–5646.

Vercelli, D. (2008) Discovering susceptibility genes for asthma and allergy. *Nat Rev Immunol* **8**: 169–182.

Yuan, Y. V. and N. A. Walsh (2006). Antioxidant and antiproliferative activities of extracts from a variety of edible seaweeds. *Food and Chemical Toxicology: An International Journal Published for the British Industrial Biological Research Association* **44**(7): 1144–1150.

Extraction of Nutraceuticals from Shrimp By-products

Trang Si Trung and Willem Frans Stevens

Contents

10.1 Introduction

Nutraceuticals—foods with medical-health benefits beyond their basic nutritional function—are of increasing interest for the prevention and treatment of disease. Marine products are good sources of nutraceuticals with antimicrobial, antioxidant, immunoenhancing, antitumor, and anti-inflammatory activities (**Table 10.1**). The majority of the nutraceuticals have multiple therapeutic benefits. The nutraceuticals chitin, chitosan, glucosamine, carotenoids (mainly astaxanthin), and carotenoprotein have been extracted from shrimp, crab shells, and squid pens (Shahidi, 2005). Collagen, gelatin, peptides, and unsaturated essential fatty acids can be extracted from fish

Table 10.1 Common Nutraceuticals Extracted from Marine Resources

Nutraceutical Compounds from Marine Resources	Properties	References
Chitin, chitosan, glucosamine	Antimicrobial, anti-inflammatory, antioxidant, antitumor, antiulcer, immunoenhancing, weight and cholesterol reduction	Shahidi (2005), Rasmussen and Morrissey (2008), Kim et al. (2008)
Carotenoids, collagen, gelatin, peptides, carotenoproteins, hemoproteins	Antioxidant, immunoenhancing, antitumor	Shahidi (2005), Kim et al. (2008), Trung (2010), Riccioni et al. (2011)
Omega-3 fatty acids	Prevention of cardiovascular disease, macular degeneration, inflammatory disorders	Shahidi (2005, 2008)
Chondroitin sulfate	Anti-inflammatory, osteoarthritis treatment	Shahidi (2005), Ronca et al. (1998)
Minerals (calcium)	Strengthening of teeth and bones, nerve function, and many enzymatic reactions that require calcium as a cofactor	Shahidi (2005, 2008), Jung et al. (2008)
Speciality chemicals	Various functional properties	Shahidi (2005), Lordan et al. (2011)

waste (Kim et al., 2008). Hemoprotein was recovered from blood waters of fish processing (Trung, 2010). Organic minerals can be obtained from fish bone and shrimp shells. Chondroitin was produced as a shark by-product (Garnjanagoonchorn et al., 2007). Fishery by-products are cheap. It can be concluded that fishery by-products are abundant and attractive sources for extraction of nutraceuticals.

10.2 Shrimp by-products as sources for the production of chitin, chitosan, and carotenoprotein

The major cultured shrimp species in Vietnam and South East Asia are black tiger shrimp (*Penaeus monodon*) and white shrimp (*Penaeus vannamei*). In 2007, shrimp production in Vietnam was approximately 350,000 metric tons, comprised of 270,000 tons of black tiger shrimp and 80,000 tons of white shrimp (Merican, 2008). The heads and shells account for 30%–40% of the raw material and are considered in the shrimp meat production industry as waste. It is estimated that in Vietnam alone, the volume of shrimp by-products will reach 200,000 tons (Trung and Phuong, 2012). In the past, this waste/by-product caused environmental problems as it was not utilized for the production of value-added products. Nowadays, drying this material is considered as a valuable addition to the feed. However, it is equally important as a resource for extraction of valuable components: chitin, protein, carotenoids, and minerals (mainly calcium carbonate) (**Table 10.2**). Currently, the emphasis is on the

Table 10.2 Chemical Composition of Shrimp By-products in Vietnam

Shrimp Species	Black Tiger Shrimp		White Shrimp	
	Heads	Shells	Heads	Shells
Chitin (%)	11.5–12.4	28.8–30.9	10.8–11.7	27.7–29.8
Protein (%)	52.2–54.4	22.6–25.5	52.8–55.6	22.8–26.2
Minerals (%)	21.3–23.6	29.3–32.5	21.4–22.5	28.9–31.8
Carotenoids (mg/kg)	163.6–178.9	88.3–99.6	144.3–159.3	79.8–83.6

Values based on dry basis.

extraction of chitin and chitosan. The composition and properties of these by-products depend on the shrimp species and season. The average chitin content is approximately 28% and 11% for shells and heads, respectively, with no significant variation between the two shrimp species. However, protein content is higher in shrimp heads. Numerous studies focus on the utilization of shrimp by-products as raw materials for the production of nutraceuticals: chitin, chitosan, protein, and carotenoids (Bough et al., 1978; Benjakul and Sophanodora, 1993; Stevens, 1996; Chakrabarti, 2002; Holanda and Netto, 2006; Timme et al., 2009). A mixture of protein and carotenoids from shrimp waste, carotenoprotein is widely used in food and aquaculture feed (Cremades et al., 2003). Carotenoprotein contains a small but very valuable amount of carotenoids (mainly astaxanthin). This compound has many applications in food and feed industry, especially in aquaculture feed for salmon (Meyers 1994; Auerswald and Gäde, 2008).

10.2.1 Extraction of chitin from shrimp by-product

Chitin in shrimp shells forms a highly ordered but complex structure with protein. The chitin is surrounded by proteins to form chitin–protein fibrils with various diameters. The chitin–protein fibrils build up sheets of successive planes of horizontal and parallel fibrils with directions changing from one layer to the other by continuous rotation. Chitin was first isolated by a French chemist Henri Braconnot (1811). Chitin can be extracted from crustacean shells, fungi, bacteria, and insects (Hirano, 1996; Stevens, 1996). Currently, chitin is primarily produced from shrimp industry by-products by chemical treatment. This comprises two major steps: deproteinization and demineralization (**Figure 10.1**). The deproteinization process usually implies a treatment with dilute NaOH (No and Meyers, 1997; Roberts, 1998). Recently, chitin production from shrimp by-product has been carried out by both biocatalytic and chemical procedures, and the chemical treatment method has been modified in order to obtain chitin of high quality through environment-friendly procedures (Rao et al., 2000; Stevens, 2001; Sini et al., 2007). Chitin can be extracted from all species of shrimp: black tiger shrimp, white shrimp, deep-water sea shrimp (*Pandalus borealis*), and squilla (*S. empusa*) (Stevens, 2001; Rao et al., 2007; Trung and Phuong, 2012).

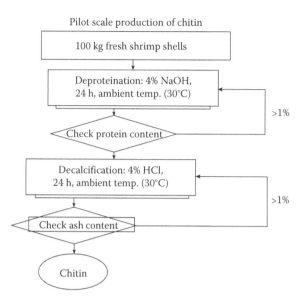

Pilot scale production of chitin

Figure 10.1 *Chitin production by chemical treatment. (From Stevens, W.F., Production of chitin and chitosan: Refinement and sustainability of chemical and biological processing. Chitin and Chitosan in Life Science, Paper read at Proceedings 8th International Conference on Chitin and Chitosan and 4th Asian Pacific Chitin and Chitosan Symposium, Yamaguchi, Japan, 2001.)*

10.2.1.1 Extraction of chitin from shrimp by-product by chemical method

Shrimp by-products are commonly treated with dilute sodium hydroxide at high temperature to dissolve protein in order to remove protein from the shrimp shells (Roberts, 1998). This alkaline treatment at high temperature is an efficient procedure to remove protein from the shells, but it also causes depolymerization and deacetylation. Conditions of 5 N NaOH and temperatures in the range of 100°C or above are usually applied in industries where large masses of by-products have to be treated in a short time. On small-scale and semi-large-scale laboratories, degradation of chitin can largely be prevented by application of mild conditions. Depolymerization can be reduced significantly by deproteinization at lower temperatures. The protein content of final chitin can be decreased to lower than 1% (w/w) when the shrimp shell is treated with 1 M NaOH for 20 h at 30°C (Stevens, 2001). Protein content can be further reduced by treatment in 0.2 M NaOH for 20 h at 50°C–70°C.

Demineralization is usually accomplished by treatment with dilute hydrochloric acid at room temperature to dissolve the minerals (mostly calcium carbonate) (No and Meyers, 1997; Roberts, 1998). In the present research, deproteinized samples were demineralized by soaking them in 4% HCl solution for 24 h with solid/liquid ratio of 1/5 (w/v) at room temperature, and then washed to obtain the chitin.

10.2.1.2 Recovery of chitin from shrimp by-product using biological methods

Chemical deproteinization and demineralization treatments for chitin production lead to a low viscosity of the obtained chitosan (Hall, 1996). Besides, these processes lead to degradation of protein and generate a large amount of chemical waste that significantly affects the environment (Holanda and Netto, 2006). To overcome these problems, deproteinization and demineralization of shrimp waste using protease enzymes or microbes by lactic acid fermentation have been reported by several authors (Rao et al., 2000; Stevens, 2001; Chakrabarti, 2002; Xu et al., 2008; Valdez-Pena et al., 2010). The biological treatment results in deproteinization and demineralization but does not cause the depolymerization of the chitin chain as long as the microorganism does not produce significant amounts of the enzyme chitinase. Besides, it also contributes to the reduction of chemical use in chitin production and proves to be a more environment-friendly process. A disadvantage is that complete removal of protein cannot be obtained (Beaney et al., 2005; Holanda and Netto, 2006) and that an extra treatment with small amounts of alkali is required to obtain a clean product. The treatment can also be improved significantly by a combined treatment with protease and *Lactobacillus*. This treatment can reach 90% deproteinization (Rao et al., 2000). Chakrabarti (2002) used protease enzymes (papain, pepsin, and trypsin) combined with lime to remove protein from shrimp shell waste and to produce carotenoprotein.

A new process is presented here for the treatment of shrimp waste. The new process combines biological deproteinization with chemical purification for chitin production. The ground shrimp waste was first deproteinized by commercial protease (Flavourzyme or Alcalase) with an enzyme/waste ratio of 0.2% (v/w) for 8 h. The partially deproteinized waste was further treated with dilute NaOH at a concentration of 2% (w/v) for 12 h with a solid/liquid ratio of 1/5 (w/v) at room temperature in order to remove the remaining protein. The waste was then demineralized by soaking in 4% HCl solution for 12 h with solid/liquid ratio of 1/5 (w/v) at room temperature to get chitin. The quality of the chitin obtained is presented in **Table 10.3**. Chitin produced by this treatment has a pinkish white color with low residual protein and ash in amounts less than 1% as required, and has high degree of acetylation (**Table 10.3**). This proves that this chitin is of good quality and can be used for the production of chitosan and glucosamine.

10.2.2 Chitosan preparation by chemical deacetylation

Chitosan (**Figure 10.1**) has a free amine group on a hexose sugar ring and is chemically much more active than chitin in which the amine group is acetylated. Chitosan is prepared from chitin by removing this N-acetyl group by treatment with strong alkali solution. Deacetylation is usually accomplished in a very strong sodium or potassium hydroxide solution (40%–50%) at

Table 10.3 Quality of Chitin Extracted from Shrimp By-product by Combined Process

Characteristics	Value
Color	Pinkish white
Ash (%)	0.98 ± 0.2
Protein (%)	0.99 ± 0.1
Degree of deacetylation (%)	6.1 ± 0.1

temperatures of 70°C–140°C (No and Meyers, 1997). Optimal conditions have been identified in order to obtain chitosan with good solubility and high molecular weight. The deacetylation process is affected by many factors, such as time, temperature, alkali concentration, origin and way of preparation of the chitin, chitin particle size, atmosphere, and the ratio of chitin to alkali solution.

Higher deacetylation rates can be obtained by using multi-alkaline treatment. Repetitions of the deacetylation protocol produces chitosan with higher degrees of deacetylation (Benjakul and Sophanodora, 1993). The deacetylation efficiency of shrimp chitin slows down when 60%–70% degree of deacetylation is reached. However, after washing the material with water, followed by a new deacetylation treatment, the degree of deacetylation of chitosan can reach more than 95% DD (Stevens, 2001).

In this study, shrimp chitin was further treated with 50% NaOH at 65°C for 24 h for the preparation of chitosan (Trung et al., 2006). The chitosan was characterized, and the results are presented in **Table 10.4**. The chitosan obtained from this treatment is of good quality, and the protein and ash content is less than 1%, as required (Rao et al., 2007). The degree of deacetylation reached 83%, and the chitosan obtained has a high viscosity of nearly 1500 cPs. The high viscosity of chitosan is due to the light treatment conditions of enzyme

Table 10.4 Quality of Chitosan Extracted from Shrimp By-product

Characteristics	Values
Color[a]	White
Ash[a] (%)	0.95 ± 0.2
Protein[a] (%)	0.93 ± 0.1
Bulk density[a] (g/mL)	0.51 ± 0.01
Water binding capacity[a] (%)	484 ± 30
Viscosity (cps)	1440 ± 70
Degree of deacetylation (%)	83 ± 0.5
Solubility[a] (%)	99.2 ± 0.2
Turbidity (NTU)	18.3 ± 0.8

[a] Based on dry basis.

combined with dilute NaOH. This chitosan has high solubility and its solutions have low turbidity. These characteristics indicate that the chitosan is of good quality and can be used in food and agriculture.

In order to get fully deacetylated chitosan with high molecular weight, heterogeneous deacetylation by means of Freeze-Pump out-Thaw (FPT) cycles have been developed by Lamarque et al. (2005). Chitin was deacetylated in a multistep process by means of FPT cycles in the presence of 50% (w/v) NaOH, at temperatures ranging from 80°C to 110°C. It was reported that FPT cycles can improve the reaction effectiveness by opening the crystalline structure of chitin and lead to better deacetylation efficiency; the final chitosan product has higher deacetylation and molecular weight.

10.2.3 Chitosan by biocatalytic deacetylation

The fungal enzyme chitin deacetylase, engaged in the deacetylation of chitin during the assembly of fungal cell wall, might be a candidate to achieve enzymatical deacetylation in a bioreactor. Chitin deacetylase was first identified and partially purified from the fungus *Mucor rouxii* and used to deacetylate chitin very partially (1%–2% of acetyl groups) by Araki and Ito (1975). Since then, this enzyme has been found to be present in other fungi (Gao et al., 1995; Tsigos and Bouriotis, 1995). Tsigos et al. (1994) reported that only deacetylase from *Colletotrichum lindemuthianum* appears to liberate a small amount of acetic acid if acting on natural chitin. The other deacetylases have only a minor activity on chitin but can act on partially deacetylated chitin (degree of deacetylation is about 60%). In none of these cases a significant rise in the degree of deacetylation could be demonstrated using natural chitin as substrate. Win and Stevens (2001) speculated that most acetyl groups of natural chitin cannot be attacked by the enzyme because they are hidden within the highly crystalline structure of the chitin. In natural chitin, only a very limited number of acetyl groups, present at the periphery of the crystal particle, would be accessible for the enzyme. They prepared a decrystallized chitin by dissolution in formic acid and precipitation to obtain a very fine chitin powder. Using powdered chitin, it was possible to obtain nearly 100% enzymatic deacetylation. The method is however quite complicated and not yet suitable for large-scale application.

10.2.4 Purification of chitosan extracted from shrimp by-product

Chitosan for application in biomedical and cosmetic products must have high purity. The purification method involves dissolution of chitosan in acetic acid and regeneration of solid chitosan by precipitation using sodium hydroxide (Hirano, 1996). It has been shown that various recovery methods and conditions have effect on the yield, solubility, molecular weight, and creep compliance of purified chitosan (Chen and Lui, 2002). They reported that the molecular weight of purified chitosan decreased with the increase of alkali concentration. Chitosans with various degrees of deacetylation and different recovery media produce purified hydrogels with different structures and tactile properties. In order to purify chitosan by regeneration, chitosan 2% (w/v) solutions were prepared in diluted aqueous solutions of various organic acids. After shaking at 25°C ± 2°C at 150 rpm for 10 h, the chitosan solution was filtered through a textile cloth to remove insoluble components. The dissolved chitosan in the filtrate was precipitated by adjusting the pH by adding 1 M NaOH. The chitosan was collected on a filter, then washed and dried. This chitosan was named "purified chitosan." Purified chitosans (**Table 10.5**) have been obtained using 1% formic acid, 1% acetic acid, 1% propionic acid, 1% lactic acid, 5% citric acid, 1% ascorbic acid, and 0.25% hydrochloric acid, followed by precipitation using alkali. During the regeneration process, the pH for complete precipitation of chitosan from the chitosan solutions varied from 9.5 to 10, except for chitosan purified from citric acid solution. The pH for complete precipitation was significantly lower, from 4.2 to 4.5. Chitosans after purification are white in color except chitosan purified from ascorbic acid that is light yellow in color. Yields of these treatments are 85%, 83%, 84%, 82%, 79%, 64%, 79% for formic, acetic, propionic, lactic, citric, ascorbic and hydrochloric acid treatments, respectively. The low pH precipitation of chitosan in citric acid solution may be due to citric acid being a relatively strong, multivalent acid. When chitosan in citric acid solution becomes neutralized, the trivalent acid at lower pH (3–4) is strong enough to keep itself dissociated, and the chitosan is protonated. Complex formation between the polyanion and the cationic

Table 10.5 Characteristics[a] of Original and Purified Chitosans

Chitosans	Ash (%)	N (%)	DD (%)	Bulk Density (g/mL)	Turbidity[b] (NTU)	M_w[c] (10^{-6} Da)	Viscosity[d] (cP)
CO	1.19 ± 0.1[e]	7.74 ± 0.25	86.4 ± 0.5	0.55 ± 0.01	29.3 ± 3.5[f]	1.12 ± 0.02[e]	2240 ± 50[g]
CRF	0.28 ± 0.02[h]	7.76 ± 0.22	87.0 ± 0.8	0.33 ± 0.01[h]	13.0 ± 2.6[h]	1.16 ± 0.02[e]	2260 ± 60[g]
CRA	0.27 ± 0.04[h]	7.68 ± 0.05	86.3 ± 0.5	0.47± 0.01[g]	14.7 ± 0.6[e,h]	1.11 ± 0.03[e]	2220 ± 40[g]
CRP	0.27 ± 0.04[h]	7.79 ± 0.25	87.0 ± 0.7	0.50 ± 0.01[i]	12.3 ± 1.5[h]	1.14 ± 0.04[e]	2280 ± 70[g]
CRL	0.30 ± 0.04[h]	7.88 ± 0.27	86.9 ± 0.7	0.31 ± 0.02[h]	13.0 ± 1.0[h]	1.09 ± 0.04[e]	1480 ± 80[f]
CRC	0.30 ± 0.09[h]	7.91 ± 0.10	85.7 ± 0.4	0.32 ± 0.01[h]	13.6 ± 2.1[h]	1.05 ± 0.02[e]	720 ± 60[h]
CRAs	0.33 ± 0.06[h]	7.88 ± 0.13	87.1 ± 1.1	0.58 ± 0.01[i]	16.6 ± 0.6[e]	0.88 ± 0.07[h]	670 ± 70[h]
CRH	0.33 ± 0.07[h]	7.95 ± 0.02	87.1 ± 1.0	0.44 ± 0.02[f]	13.3 ± 2.5[h]	1.10 ± 0.02[e]	1290 ± 60[e]

CO, original chitosan used as starting material; CRF, CRA, CRP, CRL, CRC, CRAs, CRH, chitosan purified by formic, acetic, propionic, lactic, citric, ascorbic, and hydrochloric acid; N, nitrogen content; DD, degree of deacetylation; NTU, nephelometric turbidity unit.

[a] Mean ± standard deviation of triplicate determinations.
[b] Turbidity was measured with 1% chitosan in 1% acetic acid.
[c] Average molecular weight.
[d] Viscosity was measured with 1% chitosan in 1% acetic acid.
[e–i] Means with different superscripts within a column indicate significant differences ($p < 0.05$).

chitosan occurs, and the complex precipitates. With the other acids, chitosan precipitates only when it becomes deprotonated. The acid plays no active role in precipitating earlier. At lower pH, the acid is insufficiently dissociated to provide the complexing base, and due to monovalency, no complex is formed. All purified chitosans had a very low ash content of less than 0.3%. Therefore, the purification process can further reduce 80% of the remaining ash of the chitosans at the start of the procedure. There was no significant difference in ash content removal among different acid treatments. It is conceivable that some minerals have not been removed during the extraction of the original chitosan from shrimp by-products, because they were enclosed in the interior of the solid material and protected against HCl used for demineralization. In the regeneration procedure, the chitosan is completely dissolved, the minerals will be freed and react with acid solvents. The ash content of all purified chitosans was significantly lower than that of the original chitosan. The turbidity values of purified chitosans were also significantly lower ($p < 0.05$) than those of the original chitosan. This may be attributed to the removal of impurities after regeneration. The values for the degree of deacetylation remained the same. Molecular weight of most purification chitosan preparations from these solvents was not changed except for ascorbic acid treatment. The depolymerization of some polysaccharides by ascorbic acid was reported by Herp et al. (1967). The treatments with lactic acid, hydrochloric acid, and ascorbic acid resulted in a significant decrease in viscosity compared to that of other acid treatments and original chitosan. The purification chitosan from ascorbic acid had the lowest viscosity. This may be due to its lower molecular weight.

10.2.4.1 X-Ray diffraction pattern of chitosans

The x-ray diffraction patterns of the original and the purified chitosans are presented in **Figure 10.2**. The diffractograms of most purified chitosans showed only one major peak at approximately 20° (2θ), whereas the original chitosan showed two peaks at approximately 10° and 20° (2θ). The peak at 10° (2θ) of chitosan after purification had disappeared. Ogawa (1991) reported that the hydrated crystal peak at 10° (2θ) present in deacetylated chitosan is not observed when the polymer is regenerated by NaOH from acetic acid solution. From x-ray diffraction patterns, it can be concluded that purified chitosans have a lower crystallinity than the original chitosan. The reduction in crystallinity of purified chitosans might be due to the decrease or rearrangement of hydrogen bonding among chitosan chains after regeneration. IR spectra of all purified chitosans were identical to that of original chitosan. It shows that the original and purified chitosans have the same functional groups.

The aforementioned results demonstrate that critical physicochemical properties of chitosan can be changed after regeneration from some solvents. Therefore, the functional properties may be expected to differ as well.

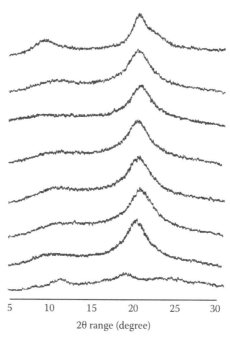

2θ range (degree)

Figure 10.2 *X-ray diffraction spectra of purified chitosan preparations, from top to bottom: original chitosan, purified chitosans from propionic, ascorbic, lactic, acetic, hydrochloric, formic and citric acid treatments.*

10.2.4.2 Dye-binding capacity

Dye-binding capacity (DBC) of original chitosan and purified chitosan preparations have been compared. The results are presented in **Table 10.6**. A marked increase in DBC was observed in purified chitosans compared to that of original chitosan, ranging from 78.6% to 85.5% (as % adsorption), compared to DBC of original chitosan of 46.7%. Significant differences in DBC were also obtained among some acid treatments. Chitosan purified from citric acid and formic acid had highest DBC, and regeneration from hydrochloric acid had the lowest DBC among the acids tested. The high DBC of purified chitosan might be due to its loose structure and more purity compared to original chitosan. The reduction in ash content of purified chitosan can also play a role in DBC. No et al. (2000) reported that high ash content contributes to low DBC of chitosan.

10.2.4.3 Water-binding capacity

Water-binding capacity (WBC) of original and purified chitosan preparations was measured with the results shown in **Table 10.5**. WBC of original and purified chitosan varied from acid to acid treatments, ranging from 464% to 628%. Among acid treatments, acetic acid and propionic acid resulted in purified chitosan with a WBC not significantly changed from that of the original chitosan. WBC increased significantly for chitosan purified from ascorbic acid,

Table 10.6 Functional Properties of Purified Chitosan Preparations

Chitosans[a]	DBC (%)	WBC[b] (%)	FBC[b] (%)
CO	46.7 ± 1.3[c]	464 ± 29[c]	257 ± 26[d]
CRF	85.6 ± 0.8[e]	628 ± 12[d]	226 ± 31[d]
CRA	82.7 ± 1.5[d,e]	476 ± 27[c]	160 ± 25[c,f]
CRP	84.0 ± 1.3[d,e]	479 ± 25[c]	139 ± 14[c]
CRL	84.5 ± 1.6[e]	620 ± 28[d]	242 ± 33[d]
CRC	90.3 ± 1.7[g]	617 ± 34[d]	166 ± 13[c,f]
CRAs	81.7 ± 1.4[d]	530 ± 22[f]	143 ± 21[c,f]
CRH	78.6 ± 1.2[f]	536 ± 20[f]	183 ± 15[f]

[a] The chitosan samples were ground to small particle size (<150 μm) prior to analysis.
[b] Represents the weight increase due to water/fat absorption compared with original weight.
[c–g] Means with different superscripts within a column indicate significant differences.

hydrochloric acid, citric acid, lactic acid, and formic acid. The treatment with citric acid, formic acid, and lactic acid gave the highest WBC. So, the WBC differed depending on the solvent used for the regeneration. According to Knorr (1983), the differences in WBC of chitosan may be due to the differences in crystallinity.

10.2.4.4 Fat-binding capacity

Fat-binding capacity (FBC) differed considerably with original chitosan and purified chitosan from some acid treatments, ranging from 13.9 to 25.7 (% w/w). Purified chitosans from formic acid and lactic acid had the highest FBC, almost similar to that of the original chitosan. FBC of purified chitosans from ascorbic acid and propionic acid was lowest among treatments. The original chitosan had a higher FBC than any of the purified chitosans. The range of FBC found was somewhat lower than that reported by No et al. (2000). All purified chitosans showed higher purity in terms of ash content and turbidity. The reduction of crystallinity in purified chitosans was concluded from the XRD patterns (**Figure 10.2**). The degree of deacetylation did not change. Chitosan samples purified from ascorbic acid treatment showed significant change in the degree of polymerization, but not with other acid treatments. The purified chitosans have the same functional groups as in the original chitosan (**Figure 10.3**). The functional properties such as DBC, WBC, FBC of purified chitosans differ from that of original chitosan depending on the acid solvent used. Therefore, this study shows that it is possible to purify or physically modify chitosan by dissolving it in various acid solvents and regenerating using sodium hydroxide.

10.2.5 Recovery of carotenoprotein from shrimp by-products

Several studies have been focused on the recovery of protein and carotenoid complexes (carotenoprotein) from shrimp waste for animal food (Synowiecki

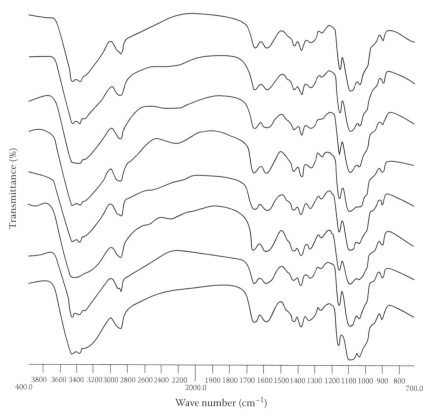

Figure 10.3 *FTIR spectra of purified chitosan preparations. Top to bottom: original chitosan, purified chitosans from propionic, ascorbic, lactic, acetic, hydrochloric, formic and citric acid treatments.*

and Al-Khateeb, 2000; Chakrabarti, 2002; Holanda and Netto, 2006). For feed applications, protein and carotenoids have been obtained as a complex mixture carotenoprotein. Biological methods have been applied in order to recover carotenoprotein from shrimp waste. Several enzymes such as papain, pepsin, Alcalase, and flavourzyme have been tested for protein hydrolysis from shrimp waste for food and feed application (Synowiecki and Al-Khateeb, 2000; Chakrabarti, 2002).

Carotenoprotein from shrimp waste is a complex between protein and carotenoid pigment. In crustaceans, the most abundant carotenoid is astaxanthin (Armenta and Guerrero-Legarreta, 2009). Astaxanthin has valuable functional properties, such as antioxidant and immunostimulant that lead to important applications in food, feed, and health industry (Shahidi, 2005; Riccioni et al., 2011; Vilchez et al., 2011). Recently, Trung and Phuong (2012) reported a new process for extraction of lipid–mineral-rich carotenoprotein from shrimp heads. Ground shrimp heads (10 kg) were treated with Alcalase (Novozyme, Denmark) with the following conditions: Enzyme/waste ratio of 0.2% v/w, pH 8, 55°C, treatment duration of 8 h. Lipid–mineral-rich carotenoprotein was

Table 10.7 Characteristics of the Lipid-rich
Carotenoprotein

Characteristics	Value
Color	Reddish
Smell	Shrimp powder
Minerals content (%)	12.7 ± 1.6
Protein content (%)	33.8 ± 2.3
Lipid content (%)	31.1 ± 1.4
Chitin content (%)	6.4 ± 1.2
Carotenoid content (mg/kg)	886 ± 14

recovered from the hydrolysis solution by isoelectric precipitation at pH 4–4.5 by adding chitosan at a concentration of 100 ppm as a coagulant and flocculent. The quality of the obtained carotenoprotein was determined and presented in **Table 10.7**. The protein in the lipid–mineral-rich carotenoprotein has good amino acid composition with high amount of glutamic acid (**Table 10.8**).

Table 10.8 Amino Acid Composition in the Lipid-rich
Carotenoprotein

Amino Acids	$(mg/kg \times 10^{-3})$
Alanine	16.2
Glycine	16.6
Serine	5.4
Proline	9.5
Valine	8.0
Threonine	0
Trans-4 hydroxy-L-proline	0
Leucine–isoleucine	11.5
Methionine	12
Phenylalanine	3.3
Arginine	0
Aspartic acid	19.3
Glutamic acid	30.5
Tryptophan	0.3
Cysteine	0.3
Lysine	10.1
Histidine	1.2
Tyrosine	7.4
Cystine	0

A previous study showed that shrimp waste has a high protein content with good amino acid balance (Ibrahim et al., 1999). Additionally, carotenoid–protein complexes are more stable than the carotenoids alone (Chakrabarti, 2002). Lipid–mineral-rich carotenoprotein is a promising product for use in food, and especially in aquafeed.

10.3 Conclusions

Nutraceutical and functional products, such as chitin, chitosan, and carotenoprotein can be extracted from shrimp by-products by suitable processes. The better utilization of fishery waste can lead to environmental protection and improve the efficiency of fishery resource.

References

Araki, Y. and E. Ito. 1975. A pathway of chitosan formation in *Mucor rouxii*. Enzymatic deacetylation of chitin. *European Journal of Biochemistry* 55 (1):71–78.

Armenta, R. E. and I. Guerrero-Legarreta. 2009. Stability studies on astaxanthin extracted from fermented shrimp byproducts. *Journal of Agricultural and Food Chemistry* 57:6095–6100.

Auerswald, G. and L. Gäde. 2008. Simultaneous extraction of chitin and astaxanthin from waste of lobsters *Jasus lalandii* and use of astaxanthin as an aquacultural feed additive. *African Journal of Marine Science* 30:35–44.

Beaney, P., J. Lizardi-Mendoza, and M. Healy. 2005. Comparison of chitins produced by chemical and bioprocessing methods. *Journal of Chemical Technology and Biotechnology* 80:145–150.

Benjakul, S. and P. Sophanodora. 1993. Chitosan production from carapace and shell of black tiger shrimp (*Penaeus monodon*). *ASEAN Food Journal* 34:145–148.

Bough, W. A., W. L. Salter, A. C. M. Wu, and B. E. Perkins. 1978. Influence of manufacturing variables on the characteristics and effectiveness of chitosan products. I. Chemical composition, viscosity, and molecular weight distribution of chitosan products. *Journal of Biotechnology and Bioengineering* 20:1931–1966.

Braconnot, H. 1811. Sur la nature des champignons. *Annales des Chimie (Paris)* 79:265–304.

Chakrabarti, R. 2002. Carotenoprotein from tropical brown shrimp shell waste by enzymatic process. *Food Biotechnology* 16:81–90.

Chen, R. H. and C. S. Lui. 2002. Effect of recovery methods and conditions on the yield, solubility, molecular weight, and creep compliance of regenerated chitosan. *Journal of Applied Polymer Science* 84:193–202.

Cremades, O., J. Parrado, M. Alvarez-Ossorio, M. Jover, L. Collantes de Terán, J. F. Gutierrez, and J. Bautista. 2003. Isolation and characterization of carotenoproteins from crayfish (*Procambarus clarkii*). *Food Chemistry* 82:559–566.

Gao, X. D., T. Katsumoto, and K. Onodera. 1995. Purification and characterization of chitin deacetylase from *Absidia coerulea*. *Journal of Biochemistry* 117:257–263.

Garnjanagoonchorn, W., L. Wongekalak, and A. Engkagul. 2007. Determination of chondroitin sulfate from different sources of cartilage. *Chemical Engineering and Processing* 46:465–471.

Hall, G. M. 1996. Biological approaches to chitin recovery. Paper read at *Proceedings of 2nd Asia Pacific Symposium*, Bangkok, Thailand, pp. 26–33.

Herp, A., T. Rickards, G. Matsumura, L. B. Jakosalem, and W. Pigman. 1967. Depolymerization of some polysaccharides and synthetic polymers by L-ascorbic acid. *Carbohydrate Research* 4:63–71.

Hirano, S. 1996. Chitin biotechnology application. *Biotechnology Annual Review* 2:237–258.

Holanda, H. D. D. and F. M. Netto. 2006. Recovery of components from shrimp (*Xiphopenaeus kroyeri*) processing waste by enzymatic hydrolysis. *Journal of Food Science* 71:298–303.

Ibrahim, H. M., M. F. Salama, and H. A. El-Banna. 1999. Shrimp's waste: Chemical composition, nutritional value and utilization. *Nahrung* 43:418–423.

Jung, W-K., F. Shahidi, and S-K. Kim. 2008. Calcium from fish bone and other marine resources. In *Marine Nutraceuticals and Functional Foods*, C. J. Barrow and F. Shahidi (eds). Boca Raton, FL: Taylor & Francis Group, pp. 419–429.

Kim, S-K., E. Mendis, and F. Shahidi. 2008. Marine fisheries by-products as potential nutraceuticals: An overview. In *Marine Nutraceuticals and Functional Foods*, C. J. Barrow and F. Shahidi (eds). Boca Raton, FL: Taylor & Francis Group, pp. 1–22.

Knorr, D. 1983. Dye binding properties of chitin and chitosan. *Journal of Food Science* 48:36–41.

Lamarque, G., M. Cretenet, C. Viton, and A. Domard. 2005. New route of deacetylation of alpha and beta-chitins by means of freeze-pump out-thaw cycles. *Biomacromolecules* 6:1380–1388.

Lordan, S., R. P. Ross, and C. Stanton. 2011. Marine bioactives as functional food ingredients: Potential to reduce the incidence of chronic diseases. *Marine Drugs* 9(6):1056–1100.

Merican, Z. 2008. SPF shrimp hatchery: A good start integration in Vietnam. *Shrimp News International* 4:1–10.

Meyers, S. P. 1994. Developments in world aquaculture, feed formulations, and role of carotenoids. *Pure Applied Chemistry* 66:1069–1076.

No, H. K., K. S. Lee, and S. P. Meyers. 2000. Correlation between physical chemical characteristics and binding capacity of chitosan products. *Journal of Food Science* 65:1134–1137.

No, H. K. and S. P. Meyers. 1997. Preparation of chitin and chitosan. In *Chitin Handbook*, R. A. A. Muzzarelli and M. G. Peter (eds). Ancona, Italy: Atec Edizioni.

Ogawa, K. 1991. Effect of heating an aqueous suspension of chitosan on the crystallinity and polymorphs. *Agricultural Biology and Chemistry* 55:2375–2397.

Rao, M. S., J. Munoz, and W. F. Stevens. 2000. Critical factors in chitin production by fermentation of shrimp biowaste. *Applied Microbiology and Biotechnology* 54:808–813.

Rao, M. S., K. A. Nyein, T. S. Trung, and W. F. Stevens. 2007. Optimum parameters for production of chitin and chitosan from Squilla (*S. empusa*). *Journal of Applied Polymer Science* 103:3694–3700.

Rasmussen, R. S. and M. T. Morrissey. 2008. Chitin and chitosan. In *Marine Nutraceuticals and Functional Foods*, C. J. Barrow and F. Shahidi (eds). Boca Raton, FL: Taylor & Francis Group, pp. 155–182.

Riccioni, G., N. D'Orazio, S. Franceschelli, and L. Speranza. 2011. Marine carotenoids and cardiovascular risk markers. *Marine Drugs* 9 (7):1166–1175.

Roberts, G. A. F. 1998. Chitosan production routes and their role in determining the structure and properties of the product. In *Advances in Chitin Science*, Vol. II, A. Domard, G. A. F. Roberts, and K. M. Varum (eds). Paris, France: Jacques Andre, pp. 23–31.

Ronca, F., L. Palmieri, P. Panicucci, and G. Ronca. 1998. Anti-inflammatory activity of chondroitin sulfate. *Osteoarthritis Cartilage* 6 (Suppl A):14–21.

Shahidi, F. 2005. Nutraceuticals from seafood and seafood by-products. In *Asian Functional Foods*, J. Shi, C.-T. Ho, and F. Shahidi (eds). Boca Raton, FL: Taylor & Francis Group, pp. 267–288.

Shahidi, F.. 2008. Omega-3 oils: Sources, applications, and health effects. In *Marine Nutraceuticals and Functional Foods*, C. J. Barrow and F. Shahidi (eds). Boca Raton, FL: Taylor & Francis Group, pp. 23–62.

Sini, T. K., S. Santhosh, and P. T. Mathew. 2007. Study on the production of chitin and chitosan from shrimp shell by using *Bacillus subtilis* fermentation. *Carbohydrate Research* 342:2423–2429.

Stevens, W. F. 1996. Chitosan: A key compound in biology and bioprocess technology. Paper read at *Proceedings of 2nd Asia Pacific Symposium*, Bangkok, Thailand, pp. 13–21.

Stevens, W. F. 2001. Production of chitin and chitosan: Refinement and sustainability of chemical and biological processing. Chitin and chitosan in life science. Paper read at *Proceedings 8th International Conference on Chitin and Chitosan and 4th Asian Pacific Chitin and Chitosan Symposium*, Yamaguchi, Japan, pp. 293–300.

Synowiecki, J. and N. A. A. Q. Al-Khateeb. 2000. The recovery of protein hydrolysate during enzymatic isolation of chitin from shrimp *Crangon crangon* processing discards. *Food Chemistry* 68:147–152.

Timme, E., D. Walwyn, and A. Bailey. 2009. Characterisation of the carotenoprotein found in carapace shells of *Jasus lalandii*. *Comparative Biochemistry and Physiology, Part B* 153:39–42.

Trung, T. S. 2010. The innovative utilization of fishery by-products in Vietnam. Paper read at *FFTC-KU Joint Seminar on Improved Utilization of Fishery By-products as Potential Nutraceuticals and Functional Foods*, Bangkok, Thailand, pp. 76–84.

Trung, T. S. and P. T. D. Phuong. 2012. Bioactive compounds from by-products of shrimp processing industry in Vietnam. *Journal of Food and Drug Analysis* 20(Suppl. 1):194–197.

Trung, T. S., W. W. Thein-Han, N. T. Qui, C. H. Ng, and W. F. Stevens. 2006. Functional characteristics of shrimp chitosan and its membranes as affected by the degree of deacetylation. *Bioresource Technology* 97:659–663.

Tsigos, I. and V. Bouriotis. 1995. Purification and characterization of chitin deacetylase from *Colletotrichum lindemuthianum*. *Journal of Biological Chemistry* 270:26286–26291.

Tsigos, I., A. Martinou, K. M. Varum, and V. Bouriotis. 1994. Enzymatic deacetylation of chitinous substrates employing chitin deacetylases. In *Proceedings of the 6th international conference on Chitin and Chitosan*: In *Chitin World*, Z. Karnichi, A. Wojtasz-Pajak, M. Brzeski, and P. Bykowski (eds), Poland, pp. 98–107.

Valdez-Pena, A. U., J. D. Espinoza-Perez, G. C. Sandoval-Fabian, N. Balagurusamy, A. Hernandez-Rivera, I. M. De-la-Garza-Rodriguez, and J. C. Contreras-Esquivel. 2010. Screening of industrial enzymes for deproteinization of shrimp head for chitin recovery. *Food Science and Biotechnology* 19:553–557.

Vilchez, C., E. Forjan, M. Cuaresma, F. Bedmar, I. Garbayo, and J. M. Vega. 2011. Marine carotenoids: Biological functions and commercial applications. *Marine Drugs* 9 (3):319–333.

Win, N. N. and W. F. Stevens. 2001. Shrimp chitin as substrate for fungal chitin deacetylase. *Applied Microbiology and Biotechnology* 57:334–341.

Xu, Y., C. Gallert, and J. Winter. 2008. Chitin purification from shrimp wastes by microbial deproteination and decalcification. *Applied Microbiology and Biotechnology* 79:687–697.

Fucoidan
A Potential Ingredient of Marine Nutraceuticals

Se-Kwon Kim, Thanh-Sang Vo, and Dai-Hung Ngo

Contents

11.1 Introduction

Chronic diseases are increasing globally and are a serious threat to health and longevity in developing countries (Rachel, 2008). Meanwhile, the food habits have been considered to be important factors associated with chronic illness. Consumption of junk food has increased manifold, which has led to a number of diseases related to nutritional deficiencies (Manisha et al., 2010). Recently, consumers are understandably more interested in the potential benefits of nutritional support for disease control or prevention (Hardy, 2000).

Thus, nutraceuticals are known to play an important role in reducing health risks and improving health quality. Substantially, nutraceuticals is a term formed by combining the words "nutrition" and "pharmaceutical" in 1989 by Stephen Defelice, founder and chairman of the foundation for innovation in medicine. Herein, nutraceuticals have been considered to be foods or food products that reportedly provide health and medical benefits, including prevention and treatment of diseases (Brower, 1998). According to Kalra et al. (2003), there is a difference between the functional foods and nutraceuticals. When food is being cooked or prepared using "scientific intelligence" with or without the knowledge of how or why it is being used, the food is called "functional food." Thus, functional food provides the body with the required amount of vitamins, fats, proteins, carbohydrates, etc., needed for its healthy survival. When functional food aids in the prevention and/or treatment of disease(s) and/or disorder(s) other than anemia, it is called a nutraceutical.

During the last few decades, the increasing health consciousness has been one of the most important stimulating factors for rapid global growth of the nutraceutical and functional food industry (Hasler, 2000; Saikat et al., 2007). In this sense, natural products have been identified to be rich sources of nutraceutical agents. The capacity of some plant-derived foods to reduce the risk of chronic diseases has been associated, at least in part, to the occurrence of secondary metabolites that have been shown to exert a wide range of biological activities. In general, these metabolites have low potency as bioactive compounds when compared to pharmaceutical drugs, but since they are ingested regularly and in significant amounts as part of the diet, they might have a noticeable long-term physiological effect (Espín et al., 2007). Notably, these products are preferred due to their medicinal synergy, safety, economical status, and fewer side effects than many drugs routinely prescribed for the treatment of certain symptoms (Raskin et al., 2002). Herein, a number of nutraceuticals have been derived from natural sources such as lipids, vitamins, proteins, glycosides, phenolic compounds, etc. (Bernal et al., 2011; Rajasekaran et al., 2008; Vincenzo et al., 2009).

Recently, a great deal of interest has been developed by consumers toward novel bioactive compounds as ingredients in nutraceuticals from marine natural resources. It is well-known that the world's oceans, covering more than 70% of the earth's surface, provide a diverse living environment for invertebrates (Vignesh et al., 2011). Marine organisms have evolved biochemical and physiological mechanisms that include the production of bioactive compounds for reproduction, communication, and protection against predation, infection, and competition (Halvorson, 1998). Thus, marine environment has been a rich source of both biological and chemical diversity. During the last decades, numerous novel compounds have been isolated from marine organisms, and many of these compounds have the potential for the industrial development of pharmaceuticals and nutritional supplements due to their interesting biological activities (Blunden, 2001; Blunt et al., 2006; Mayer et al., 2011). Among them, fucoidans derived from marine algae have been found to be potential nutraceuticals due to their antioxidant, anti-inflammatory, antiallergic, antitumor, antiobesity, anticoagulant, antiviral, antihepatopathy, antiuropathy, and antirenalpathy effects (Bo et al., 2008). These special properties of fucoidans have supported them to be applied to nutraceuticals for the enhancement of human health. Accordingly, this chapter focuses on potential nutraceuticals of fucoidans derived from marine algae and presents a brief overview of their nutraceutical activities with health benefits.

11.2 Fucoidans: sources, molecular structure, and physiological properties

11.2.1 Sources

Fucoidans are a complex series of sulfated polysaccharides found widely in the cell walls of brown seaweed. In recent years, different brown algae were analyzed for their content of fucoidans, including *Pelvetia canaliculata* (Descamps et al., 2006), *Fucus vesiculosus* (Béress et al., 1993; Obluchinskaya and Minina, 2004), *Sargassum stenophyllum* (Duarte et al., 2001), *Ascophyllum nodosum* (Medcalf and Larsen, 1977), *Cladosiphon okamuranus* (Sakai et al., 2003), *Dictyota menstrualis* (Albuquerque et al., 2004), *Fucus evanescens* (Kuznetsova et al., 2003), *Fucus serratus* (Bilan et al., 2006), *Fucus distichus* (Bilan et al., 2004), *Kjellmaniella crassifolia* (Sakai et al., 2002), *Hizikia fusiforme* (Li et al., 2006), *Analipus japonicus* (Bilan et al., 2007), and *Chorda filum* (Chizhov et al., 1999). The low-molecular-weight fractions of algal fucoidans (less than 30 kDa) obtained by depolymerization have been shown to exhibit some heparin-like properties, with less side effects (Karim et al., 2011). Besides, fucoidans can also be isolated from marine invertebrates such as sea cucumbers (*Ludwigothurea grisea*) or sea urchins (*Lytechinus variegatus, Arbacia lixula, Strongylocentrotus purpuratus, Strongylocentrotus franciscanus, Strongylocentrotus pallidus*, and *Strongylocentrotus droebachiensis*) (Alves et al., 1997, 1998; Mulloy et al., 1994; Ribeiro et al., 1994; Vilela-Silva et al., 1999, 2002). These fucoidans are simpler than fucoidans found in marine algae (Karim et al., 2011).

11.2.2 Molecular structure

Since Kylin first isolated fucoidan in 1913, the structures of fucoidans from different brown seaweeds have been investigated. Fucoidans are mainly composed of fucose and sulfate. Besides, they also contain other monosaccharides (mannose, galactose, glucose, xylose, etc.) and uronic acids, and even acetyl groups and protein (Bo et al., 2008). In general, invertebrate-derived fucoidans have a linear backbone of sulfated monosaccharides, whereas the fucoidans of algae are branched in various ways (Karim et al., 2011). Fucoidans can differ in structure among species and can vary even within the same species. The fucoidans of most algae consist mainly of sulfated L-fucose with a fucose content of about 34%–44% (Kloareg et al., 1986). However, some fucoidans have minor fucose components and majority of other monosaccharides like galactose (Xue et al., 2001) or uronic acids (Mabeau et al., 1990; Nishino et al., 1994). It is observed that the sulfation may occur at positions 2, 3, and 4, and the monosaccharides are associated via α-1,2, α-1,3, or α-1,4 glycosidic bonds (Holtkamp et al., 2009).

11.2.3 Physiological properties

The role of most fucoidans in marine organisms is not investigated well. For algae, some studies have shown a correlation between fucoidan content and the depth at which they grow. The content of fucoidan is more at the intertidal zone and less at the zone under the low water line. This difference was suggested to be due to the conservation against dehydration (Black, 1954; Black et al., 1952). Moreover, fucoidans are as well supposed to enhance cell wall stability and involved in the morphogenesis of algae embryos (Bisgrove and Kropf, 2001; Mabeau et al., 1990). In addition, the role of fucoidans is also known to be involved in the fertilization of sea urchin and in the maintenance of the body wall's integrity of sea cucumber (Berteau and Mulloy, 2003).

11.3 Nutraceutical properties of fucoidans

11.3.1 Anticoagulant activity

Coagulation is a complex process related to the formation of clots to end bleeding at an injured site. It is an important part of hemostasis, the cessation of blood loss from a damaged vessel, wherein a damaged blood vessel wall is covered by a platelet and fibrin-containing clot to stop bleeding and begin repair of the damaged vessel. Disorders in blood coagulation can lead to an increased risk of bleeding (hemorrhage) or clotting (thrombosis) (David et al., 2009). These illnesses have increased over the last decades and no useful new substances have been discovered to remediate them. So far, heparin, a highly sulfated polysaccharide present in mammalian tissues, has been used as an anticoagulant drug for more than 50 years (Lindahl, 2000). However, the clinical use of heparin has been known to cause several side effects such as excessive bleeding, thrombocytopenia, mild transaminase elevation, and hyperkalemia (Tolwani and Wille, 2009). Thus, it is necessary to find alternative drugs for heparin with safe and efficient anticoagulant properties. Notably, marine algae-derived fucoidans have been determined to be effective against blood coagulation. Anticoagulant activity is one of the best studied biological activities of fucoidans. Many studies have proposed fucoidan as an alternative agent to the anticoagulant heparin (Mourao, 2004; Mourao and Pereira, 1999). So far, the anticoagulant activity of fucoidan was reported by Springer et al. (1957). It was shown that a certain fraction of fucoidan from *F. vesiculosus* possessed powerful anticoagulant activity that qualified fucoidan to belong to the group of heparinoids. Moreover, Nishino and Nagumo (1987) have examined the anticoagulant activities of fucoidans isolated from nine brown seaweed species, including activated partial thromboplastin time (APTT), thromboplastin time (TT), and antifactor Xa activity in comparison with the values of heparin (167 units/mg). It was found that fucoidan from *Ecklonia kurome* and *Hijikia fusiforme* exhibited the highest activity with respect to APTT (38 units/mg) and TT (35 units/mg) for *E. kurome* and APTT (25 units/mg) and TT (22 units/mg) for *H. fusiforme*. Presently, several different fucoidan preparations from various algal species, including *F. vesiculosus* (Nishino et al., 1994), *Laminaria brasiliensis* (Mourao and Pereira, 1999), *E. kurome* (Nishino et al., 1999), *A. nodosum* (Millet et al., 1999), and *P. canaliculata* (Colliec et al., 1994) have been reported for their anticoagulant activities. Recently, anticoagulant activities of fucose-containing sulfated polysaccharide isolated from brown seaweed *E. cava*, including APTT, TT, and PT have been reported by Athukorala et al. (2006). The anticoagulant effect of this compound was observed to be similar with heparin. Further, fucose-containing sulfated polysaccharide from *E. cava* has been shown to inhibit the activities of coagulation factors via interaction with antithrombin III in both the extrinsic and common coagulation pathways (Jung et al., 2007). The structures of fucoidans vary from their algal source species to species and must give rise to variation in the degree of most biological activities, including anticoagulation action (Boisson-Vidal et al., 2000; Chevolot et al., 1999; Pereira et al., 1999). Many studies showed that the anticoagulant activity of fucoidan might have some relation with the sulfate content and position, molecular weight, and sugar composition. In particular, Nishino and colleagues have revealed that higher content of fucose and sulfate groups presents higher anticoagulant activity in native fucoidans from *E. kurome* (Nishino and Nagumo, 1991, 1992; Nishino et al., 1989). Moreover, the position of sulfate groups on sugar residues is also very important for the anticoagulant activity of fucoidan. It was identified that the concentrations of C-2 sulfate and C-2,3 disulfate of fucoidans are related to the anticoagulant activity (Chevolot et al., 1999, 2001; Yoon et al., 2007). Duarte et al. (2001) determined that the anticoagulant effect of fucoidans was mainly related to the fucose-sulfated chains, especially the disulfated fucosyl units. Silva et al. (2005) reported that 3-*O*-sulfation at C-3 of 4-α-L-fucose-1→ units was responsible for the anticoagulant properties of fucoidan from *Padina gymnospora*. On the other hand, fucoidans with higher molecular weight such as 27 and 58 kDa showed stronger anticoagulant effect than the fucoidans with lower molecular weight (~10 kDa) (Nishino et al., 1991). The native fucoidan (MW 320,000) from *Lessonia vadosa* showed good anticoagulant activity, whereas the radical depolymerized fraction (MW 32,000) presented weak anticoagulant activity (Chandía and Matsuhiro, 2008). Similarly, Pomin and colleagues also confirmed the relationship between the molecular weight of fucoidans and their anticoagulant activity. Selective cleavage to reduce molecular size of the fucoidan dramatically reduced its effect on thrombin inactivation mediated by heparin cofactor II (Pomin et al., 2005). Accordingly, fucoidans have been suggested to be the potential biological materials of nutraceuticals in the treatment of blood coagulant disorders.

11.3.2 Antiviral activity

Sulfated polysaccharides have been known to be capable of inhibiting the replication of enveloped viruses including herpes simplex virus (HSV), human immunodeficiency virus (HIV), human cytomegalovirus, dengue virus, and respiratory syncytial virus (Jiao et al., 2011). Their inhibition exhibits low cytotoxicity compared with other antiviral drugs currently used in clinical medicine. Notably, fucoidans can effectively prevent the penetration of viruses into cells on account of the modification of properties of cellular surface, although direct interaction of polysaccharides with viral surface proteins or viral enzymes is also possible (Usov and Bilan, 2009). The blockade of fucoidans on HSV has been most popular. Feldman et al. (1999) isolated fucoidan fractions (Ee, Ec, and Ea) from *Leathesia difformis* and determined their selective antiviral abilities against HSV-1 and HSV-2. Fucoidan Ea was shown to be the most active agent, with IC_{50} value in the range 0.5–1.9 μg/mL. Continually, fucoidans were found in different marine brown macroalgae due to their anti-HSV properties, including *Adenocystis utricularis*, *S. horneri*, *Cystoseira indica*, *Stoechospermum marginatum*, and *S. tenerrimum* (Adhikari et al., 2006; Mandal et al., 2007; Ponce et al., 2003; Preeprame et al., 2001; Sinha et al., 2010). Noticeably, *Undaria pinnatifitida*, the most commonly eaten brown seaweed in Japan, contains sulfated polyanions and other components with appreciable anti-HSV effect. Galactofucan, the major component of an aqueous extract of *U. pinnatifida*, was evaluated for antiviral activity against 32 clinical strains of HSV, including 12 ACV-resistant strains (four HSV-1 and eight HSV-2) and 20 ACV-susceptible strains (10 HSV-1 and 10 HSV-2). The median IC_{50} of galactofucan for the 14 strains of HSV-1 and 18 strains of HSV-2 was 32 and 0.5 μg/mL, respectively. It was indicated that galactofucan is significantly more active against the clinical strains of HSV-2 than HSV-1. The mode of action of the galactofucan was shown to be through the inhibition of viral binding and entry into the host cell (Thompson and Dragar, 2004). In addition, a fucoidan from the sporophyll of *U. pinnatifida* was examined for its antiviral activity. The IC_{50} values for HSV-1 and HSV-2 were 2.5 and 2.6, respectively, under conditions in which the fucoidan was added at the same time of viral infection (Lee et al., 2004). In the in vivo conditions, ingestion of fucoidan from *U. pinnatifida* was associated with increased healing rates in patients with active infections (Cooper et al., 2002). Moreover, oral administration of the fucoidan from *U. pinnatifida* could protect mice from infection with HSV-1 as judged from the survival rate and lesion scores (Hayashi et al., 2008b). Substantially, natural killer and cytotoxic T lymphocytes activity in HSV-1-infected mice was enhanced by oral administration of the fucoidan. The production of neutralizing antibodies in the mice inoculated with HSV-1 was significantly promoted during the oral administration of the fucoidan for 3 weeks. According to these results, fucoidan from *U. pinnatifida* was suggested as a topical microbicide for the prevention of transmission of HSV through direct inhibition of viral replication and stimulation of both innate and adaptive immune defense functions. On the other hand, fucoidans have been found to exhibit anti-HIV activity with different mechanisms of action. According to Queiroz et al. (2008), the fucoidan from *Dictyota mertensii*, *Lobophora variegata*, *Spatoglossum schroederi*, and *F. vesiculosus* were reported to inhibit HIV reverse transcriptase (RT). They have indicated that the galactofucan fraction from *L. variegate*, which is rich in galactose, fucose, and glucose with a lower sulfate content, had a marked inhibitory effect on RT enzyme, with 94% inhibition for synthetic polynucleotides at a concentration of 1.0 µg/mL. Moreover, fucan A from *S. schroederi* and *D. mertensii*, which contains mainly fucose with a lower sulfate level, showed a high inhibitory effect on RT enzyme at 1.0 mg/mL, with 99.03% and 99.3% inhibition, respectively. Meanwhile, fucan B from *S. schroederi*, which contains galactose, fucose, and high sulfate level, showed a lower inhibitory activity (53.9%) at the same concentration. In another approach, the authors purified a fucan fraction from *F. vesiculosus*, a homofucan containing only sulfated fucose with high sulfate content, which exhibited high inhibitory activity of HIV on RT enzyme. This fraction inhibited 98.1% of the reaction with poly(rA)-oligo(dT) at a concentration of 0.5 mg/mL (Queiroz et al., 2008). In a recent study, Trinchero et al. (2009) have shown that galactofucan fractions from the brown algae *A. utricularis* exhibited anti-HIV-1 activity in vitro. Among five fractions, EA1-20 and EC2-20 had a strong inhibitory effect on HIV-1 replication with low IC_{50} values (0.6 and 0.9 µg/mL, respectively). Additionally, EA1-20 and EC2-20 displayed this capacity against wild type and drug-resistant HIV-1 strains. For active fractions, it was also shown that the inhibitory effect was not due to an inactivating effect on the viral particles but rather to a blockade of early events of viral replication. Based on these results, seaweed-derived fucoidans are regarded as good candidates for further studies on prevention of HIV-1 infection.

11.3.3 Anti-inflammatory activity

Inflammation is a critically important aspect of host responses to various stimuli including physical damage, ultraviolet irradiation, microbial invasion, and immune reactions. It is associated with a large range of mediators that initiate the inflammatory response, recruit and activate other cells to the site of inflammation. However, excessive or prolonged inflammation can prove harmful, contributing to the pathogenesis of a variety of diseases, including chronic asthma, rheumatoid arthritis, multiple sclerosis, inflammatory bowel disease, psoriasis, and cancer (Vo et al., 2012). Meanwhile, fucoidans derived from marine algae have been demonstrated to inhibit inflammatory response in many recent studies. According to Kang et al. (2011), the inhibitory effect of sulfated polysaccharide containing fucose from *E. cava* on inflammatory response has been investigated in lipopolysaccharide (LPS)-stimulated RAW 264.7 cells. It was found that this algal fucoidan dose-dependently inhibited nitric oxide (NO) and prostaglandin E_2 (PGE_2) production by suppressing the expression of NO synthase (iNOS) and cyclooxygenase (COX)-2 at the protein levels. Also, fucoidans were found to exhibit suppressive effect on neuroinflammatory response in LPS-induced microglia cells (Cui et al., 2010; Park et al., 2011a). Cui et al. (2010) showed that fucoidan from *Laminaria japonica* exhibited inhibitory effect on NO production and expression. In the next study, Park et al. (2011a) have also reported that the treatment of fucoidan from *F. vesiculosus* significantly inhibited excessive production of NO and PGE_2 accompanied by suppressing the expression of iNOS and COX-2. Moreover, fucoidan treatment caused decrease in the production and expression of monocyte chemoattractant protein-1 (MCP-1) and proinflammatory cytokines, including interleukin-1β (IL-1β) and tumor necrosis factor (TNF)-α. Notably, fucoidan exhibited anti-inflammatory properties by suppression of nuclear factor-kappa B (NF-κB) activation and down-regulation of extracellular signal-regulated kinase (ERK), c-Jun N-terminal kinase (JNK), p38 mitogen-activated protein kinase (MAPK), and AKT pathways (Cui et al., 2010; Park et al., 2011a). In an in vivo experiment, potential inhibitory mechanisms of fucoidan on rat myocardial ischemia-reperfusion (I/R) model have been evaluated by Li et al. (2011). The administration of fucoidan resulted in reduction of myocardial infarct size, serum levels of TNF-α and IL-6, and the activity of myeloperoxidase. Furthermore, fucoidan down-regulated the expression of high-mobility group box 1, phosphor-IκB-α, and NF-κB. Besides, the infiltration of polymorphonuclear leukocytes and histopathological damages in myocardium were decreased in fucoidan-treated groups. These findings revealed that the administration of fucoidan could regulate the inflammation response via high-mobility group box 1 and NF-κB inactivation in I/R-induced myocardial damage. In another sense, it has been known that connective tissue destruction during inflammatory diseases, such as chronic wound, chronic leg ulcers, or rheumatoid arthritis, is the result of continuous supply of inflammatory cells and exacerbated production of inflammatory cytokines and matrix proteinases (Senni et al., 2006). Herein, fucoidan from *A. nodosum* is known to be a potent modulator of connective tissue proteolysis (Senni et al., 2006). Thus, fucoidan was suggested to be used for treating some inflammatory pathologies in which uncontrolled extracellular matrix degradation takes place. So far, the selectin family, which expressed on endothelial cells, leukocytes, and platelets, has been evidenced to contribute to the interactions of leukocytes and platelets at the side of vascular injury. Such interactions enhance inflammatory reactions during the arterial response to injury (Ley, 2003). Especially, the interaction of selectin with its ligand is effectively inhibited by fucoidans (Bachelet et al., 2009; Chauvet et al., 1999; Preobrazhenskaya et al., 1997; Semenov et al., 1998), thus reducing inflammation process at its earlier stages.

11.3.4 Antiallergic activity

Allergy is a disorder of the immune system due to an exaggerated reaction of the immune system to harmless environmental substances, such as animal dander, house dust mites, foods, pollen, insects, and chemical agents. The initial event responsible for the development of allergic diseases is the generation of allergen-specific CD^{4+} type 2 helper (Th2) cells. Once generated, effector Th2 cells produce IL-4, IL-5, IL-9, and IL-13, which cause the production of allergen-specific IgE by B cells. Subsequently, allergic reactions are induced upon binding of allergen to IgE, which is tethered to the high-affinity IgE receptor on the surface of mast cells and basophils (Vo et al., 2012). Hence, IgE and Th2 cytokines are considered as the potential targets for antiallergic therapeutics. Interestingly, algal fucoidans have been found to suppress IgE and Th2 cytokine production as shown in recent studies. According to Maruyama et al. (2005), fucoidan obtained from *Undaria pinnatifida* is able to augment Th1 cell response in normal BALB/c mice, which contributes to the inhibition of Th2 cell response. Indeed, the production of Th2 cytokines including IL-4 and IL-13 in bronchoalveolar lavage fluid was suppressed when fucoidan was injected intraperitoneally. Moreover, anti-ovalbumin (OVA) immunoglobulin E (IgE) and IgE levels in serum determined after challenge with aerosolized OVA at the end of the experiment were reduced in the fucoidan-treated mice (Maruyama et al., 2005). Likewise, Yanase et al. (2009) determined that the OVA-induced increase of plasma IgE was significantly suppressed when fucoidan was intraperitoneal. Further, the production of IL-4 in response to OVA in spleen cells isolated from OVA-sensitized mice treated with fucoidan was lower than that from mice treated without fucoidan. Specially, the flow-cytometric analysis and ELISpot assay revealed that the administration of fucoidan suppressed a number of IgE-expressing and IgE-secreting B cells, respectively. In an in vitro experiment, Oomizu et al. (2006) have confirmed that fucoidan inhibited the production of IgE and Cε germline transcription in murine B cells induced by IL-4 and anti-CD40 antibodies. Yet, the inhibitory activity of fucoidan has not been observed if B cells were prestimulated with IL-4 and anti-CD40 antibody before the administration of fucoidan. Thus, it suggested that fucoidan may not prevent a further increase of IgE in patients who have already developed allergic diseases and high levels of serum IgE. However, Iwamoto et al. (2011) have recently determined that fucoidan effectively reduced IgE production in both peripheral blood mononuclear cells from atopic dermatitis patients and healthy donors. These findings indicated that fucoidan suppresses IgE production by inhibiting immunoglobulin class-switching to IgE in human B cells, even after the onset of atopic dermatitis.

11.3.5 Antioxidant activity

The oxidants such as superoxide anion, hydrogen peroxide, hydroxyl radicals, and singlet oxygen are well-known to cause various chronic diseases (Waris and Ahsan, 2006). Antioxidants might have a positive effect on human health as they can protect the human body against damage by reactive oxygen species (ROS), which attack macromolecules such as membrane lipids, proteins, and DNA, lead to many health disorders such as cancer, diabetes mellitus, neurodegenerative, and inflammatory diseases with severe tissue injuries (Ngo et al., 2011). Recently, there is a considerable interest in the food industry as well as pharmaceutical industry for the development of antioxidants from natural sources as safe alternatives of many synthetic commercial anti-oxidants. Among them, fucoidans derived from marine algae have been found as great antioxidants through their scavenging effect on biologically harmful oxidants. Fucoidan from the edible seaweed *F. vesiculosus* was shown to prevent the formation of superoxide radicals (IC_{50} 58 μg/mL), hydroxyl radicals (IC_{50} 157 μg/mL), and lipid peroxidation (IC_{50} 1250 μg/mL) (Micheline et al., 2007). Moreover, fucoidan fractions F-A and F-B from *L. japonica* exhibit excellent scavenging capacities on superoxide radical and hypochlorous acid, except the highly sulfated fraction L-B. Especially, low-molecular-weight fractions L-A and L-B possess great inhibitory effects on low-density lipoprotein (LDL) oxidation induced by Cu^{2+} (Zhao et al., 2005). The superoxide radical scavenging ability of fucoidan obtained from *L. japonica* has also been confirmed by Wang et al. (2008). In the same regard, Zhao et al. (2011) have determined that fucoidan F-C from *L. japonica* with low molecular weight of 2000–8000 and sulfate content 24.3% had much strong protective effect on both hydrophilic radical AAPH and lipophilic radical AMVN-induced LDL oxidation. Further, the highly sulfated fucoidan fraction L-B with molecular weight 20,000 effectively suppressed the oxidant of LDL induced by AMVN. In an in vivo experiment, fucoidan from *L. japonica* was observed to be able to prevent the increase of lipid peroxide in serum, liver, and spleen of diabetic mice obviously (Li et al., 2002). Collectively, these results clearly indicate the beneficial effects of algal fucoidans as antioxidants which have great potential for preventing the free radical-mediated diseases.

11.3.6 Antiobesity activity

Obesity is a chronic metabolic disorder caused by an imbalance between energy intake and expenditure. Obesity increases the likelihood of various diseases, particularly heart disease, type 2 diabetes, obstructive sleep apnea, certain types of cancer, and osteoarthritis (Kopelman, 2000; Spiegelman and Flier, 2001). In general, obesity is associated with the extent of adipocyte differentiation, intracellular lipid accumulation, and lipolysis (Park et al., 2011b). Recently, there has been increasing interest in potential biological activity against obesity of fucoidan. Herein, Kim et al. (2009) evaluated the protective effect of fucoidan from brown algae in 3T3-L1 adipocyte differentiation. They have suggested that fucoidan could be used for inhibiting fat accumulation, which is mediated by suppressing gene expression of fatty acid binding proteins, acetyl CoA carboxylase, and peroxisome proliferation-activated receptor γ. Moreover, Park et al. (2011b) have clearly investigated the inhibitory effects of fucoidan from *Focus vesiculosus* on lipid accumulation through the regulation of lipolysis in 3T3-L1 adipocytes. It was observed that the expressed protein levels of total hormone-sensitive lipase (HSL) and its activated form, phosphorylated HSL (p-HSL), were significantly increased at the concentration of 200 μg/mL fucoidan. Further, insulin-induced 2-deoxy-D-[³H] glucose uptake was decreased up to 51% in fucoidan-treated cells as compared to control. Evidently, the increase of HSL and p-HSL expression and decrease of glucose uptake into adipocytes lead to the stimulation of lipolysis, thus contributing to the reduction of lipid accumulation.

Fucoidans are a kind of active materials which can enhance the negative charges of cell surface so as to effect the aggradation of cholesterol in blood, thus decreasing the cholesterol in serum. Fucoidan of *L. japonica* significantly reduced total cholesterol, triglyceride, and LDL-C, remarkedly increased HDL-C in serum of mice with hypercholesterolemia and rats with hyperlipidemia, and efficiently prevented the formation of experimental hypercholesterolemia in mice (Li et al., 1999a, 2001). Low molecular weight sulfated fucan (average Mw = 8000 Da) prepared from *L. japonica* also attenuated blood lipids of hyperlipidemic rats (Li et al., 1999b). Interestingly, fucoidan can remarkably reduce the contents of cholesterol and triglyceride in serum of patients with hyperlipidemia, without the side effect of damaging liver and kidney (Wang and Bi, 1994).

11.3.7 Antitumor activity

Since fucoidans do not exert cytotoxic activity, their antitumor activity is mainly accounted for by inhibiting the proliferation of tumor cells, stimulating the apoptosis of tumor cells, altering the adhesive behavior of cells, and enhancing various immune responses. Indeed, the antiproliferative activity of oversulfated fucoidan from commercially cultured *C. okamuranus* TOKIDA in U937 cells has been observed (Teruya et al., 2007). The inhibition of cell proliferation was caused by induction of apoptosis via caspase-3 and caspase-7 activation-dependent pathways. Moreover, fucoidan from the brown seaweed *C. okamuranus* significantly inhibited the growth of peripheral blood mononuclear cells of adult T-cell leukemia patients and human T-cell leukemia virus type 1-infected T-cell lines but not that of normal peripheral blood mononuclear cells. Fucoidan induced apoptosis of HTLV-1-infected T-cell lines mediated through down-regulation of cellular inhibitor of apoptosis protein-2 and surviving. In vivo use of this fucoidan resulted in partial inhibition of growth of tumors of an HTLV-1-infected T-cell line transplanted subcutaneously in severe combined immune-deficient mice (Haneji et al., 2005). Additionally, fucoidan activates a caspase-independent apoptotic pathway in MCF-7 cancer cells through activation of ROS-mediated MAP kinases and regulation of the Bcl-2 family protein-mediated mitochondrial pathway (Zhang et al., 2011). Likewise, the Miyeokgui fucoidan showed antitumor activity against PC-3 (prostate cancer), HeLa (cervical cancer), A549 (alveolar carcinoma), and HepG2 (hepatocellular carcinoma) cells, in a similar pattern to that of commercial fucoidan (Synytsya et al., 2010). Also, fucoidan from *F. vesiculosus* induces apoptosis of human HS-Sultan cells accompanied by activation of caspase-3 and down-regulation of ERK pathways (Aisa et al., 2005). Fucoidans from *L. saccharina*, *L. digitata*, *F. serratus*, *F. distichus*, and *F. vesiculosus* strongly blocked MDA-MB-231 breast carcinoma cell adhesion to platelets, an effect which might have critical implications in tumor metastasis

(Cumashi et al., 2007). According to Liu et al. (2005), fucoidan inhibits the adhesion of MDA-MB-231 cells to fibronectin by blocking the protein's heparin- and cell-binding domains, modulating the reorganization of the integrin alpha5 subunit, down-regulating the expression of vinculin.

Another mechanism of the antitumor activity of fucoidans is due to the antiangiogenic effect, since fucoidans suppress the intensive formation of vessels and so reduce the active supply of blood to tumor tissues (Koyanagi et al., 2003; Soeda et al., 1997). Besides, fucoidans inhibit tumor growth and metastatic process by the enhancement of immune responses. Fucoidan increases the quantity of macrophages (Song et al., 2000), and mediates tumor destruction through type 1 Th1 cell and NK cell responses (Maruyama et al., 2006). Fucoidan activates lymphocytes and macrophages mediated by the production of free radicals (NO and H_2O_2) and cytokines (TNF-α and IL-6), thus contributing to their effectiveness in the immunoprevention of tumor (Choi et al., 2005). Overall, finding of antitumor properties of brown algal fucoidans could elevate the value of brown seaweeds as functional ingredients in nutraceuticals or pharmaceuticals.

11.3.8 Other biological activities of fucoidan

Fucoidan derived from *C. okamuranus* tokida has been known to be a safe compound for gastric protection (Shibata et al., 2000). It was shown to inhibit the growth of stomach cancer cells but did not show any effects on normal cells (Kawamoto et al., 2006). Fucoidan prevented concanavalin A-induced liver injury by mediating the endogenous IL-10 production and inhibition of proinflammatory cytokine in mice (Saito et al., 2006). Moreover, hepatic fibrosis induced by CCl_4 was also attenuated by injection of fucoidan (Hayashi et al., 2008a). On the other hand, fucoidan administration was able to maintain the integrity of erythrocyte membrane and decrease the damage to erythrocytes in hyperoxaluria (Veena et al., 2007b). Fucoidan treatment can prevent the increased excretion of calcium oxalate monohydrate crystals in urine along with crystal deposition in renal tissues (Veena et al., 2007a). The oral intubation of fucoidan significantly reduced the elevated urinary protein excretion and plasma creatinine due to the induction of Heymann nephritis. This indicated that fucoidan has a renoprotective effect on active Heymann nephritis and is a promising therapeutic agent for nephritis (Zhang et al., 2005).

11.4 Conclusions

During the recent past, marine sources have received much attention of the nutraceutical industries since they are valuable sources of chemically diverse compounds with numerous biological activities and health-benefit effects. Recent studies have provided evidence that fucoidans derived from marine algae play a vital role in human health and nutrition due to their various nutraceutical properties. Thus, the extensive studies of algal fucoidans will discover novel biological properties that can contribute to the development of nutraceutical industries in the near future.

Acknowledgment

This study was supported by a grant from Marine Bioprocess Research Center of the Marine Bio 21 Project funded by the Ministry of Land, Transport, and Maritime, Republic of Korea.

References

Adhikari, U., Mateu, C. G., Chattopadhyay, K., Pujol, C. A., Damonte, E. B., and Ray, B. (2006). Structure and antiviral activity of sulfated fucans from *Stoechospermum marginatum*. *Phytochemistry*, *67*, 2474–2482.

Aisa, Y., Miyakawa, Y., Nakazato, T., Shibata, H., Saito, K., Ikeda, Y., and Kizaki, M. (2005). Fucoidan induces apoptosis of human HS-Sultan cells accompanied by activation of caspase-3 and down-regulation of ERK pathways. *American Journal of Hematology*, *78*, 7–14.

Albuquerque, I. R. L., Queiroz, K. C. S., Alves, L. G., Santos, E. A., Leite, E. L., and Rocha H. A. O. (2004). Heterofucans from *Dictyota menstrualis* have anticoagulant activity. *Brazilian Journal of Medical and Biological Research*, *37*, 167–171.

Alves, A. P., Mulloy, B., Diniz, J. A., and Mourao, P. A. S. (1997). Sulfated polysaccharides from the egg jelly layer are species-specific inducers of acrosomal reaction in sperms of sea urchins. *Journal of Biological Chemistry*, *272*, 6965–6971.

Alves, A. P., Mulloy, B., Moy, G. W., Vacquier, V. D., and Mourao, P. A. S. (1998). Females of the sea urchin *Strongylocentrotus purpuratus* differ in the structure of their egg jelly sulphated fucans. *Glycobiology*, *8*, 939–946.

Athukorala, Y., Jung, W. K., Vasanthan, T., and Jeon, Y. J. (2006). An anticoagulative polysaccharide from an enzymatic hydrolysate of *Ecklonia cava*. *Carbohydrate Polymers*, *66*, 184–191.

Bachelet, L., Bertholon, I., Lavigne, D., Vassy, R., Jandrot-Perrus, M., Chaubet, F., and Letourneur, D. (2009). Affinity of low molecular weight fucoidan for P-selectin triggers its binding to activated human platelets. *Biochimica et Biophysica Acta*, *1790*, 141–146.

Béress, A., Wassermann, O., Tahhan, S., Bruhn, T., Béress, L., Kraiselburd, E. N., Gonzalez, L. V., de Motta, G. E., and Chavez, P. I. (1993). A new procedure for the isolation of anti-HIV compounds (polysaccharides and polyphenols) from the marine alga *Fucus vesiculosus*. *Journal of Natural Products*, *56*, 478–488.

Bernal, J., Mendiola, J. A., Ibáñez, E., and Cifuentes, A. (2011). Advanced analysis of nutraceuticals. *Journal of Pharmaceutical and Biomedical Analysis*, *55*, 758–774.

Berteau, O. and Mulloy, B. (2003). Sulfated fucans, fresh perspectives: Structures, functions, biological properties of sulfated fucans and overview of enzymes active towards this class of polysaccharide. *Glycobiology*, *13*, 29R–40R.

Bilan, M. I., Grachev, A. A., Shashkov, A. S., Nifantiev, N. E., and Usov, A. I. (2006). Structure of a fucoidan from the brown seaweed *Fucus serratus* L. *Carbohydrate Research*, *341*, 238–245.

Bilan, M. I., Grachev, A. A., Ustuzhanina, N. E., Shashkov, A. S., Nifantiev, N. E., and Usov, A. I. (2004). A highly regular fraction of a fucoidan from the brown seaweed *Fucus distichus* L. *Carbohydrate Research*, *339*, 511–517.

Bilan, M. I., Zakharova, A. N., Grachev, A. A., Shashkov, A. S., Nifantiev, N. E., and Usov, A. I. (2007). Polysaccharides of algae: 60. Fucoidan from the pacific brown alga *Analipus japonicus* (Harv.) winne (Ectocarpales, Scytosiphonaceae). *Russian Journal of Bioorganic Chemistry, 33*, 38–46.

Bisgrove, S. R. and Kropf, D. L. (2001). Cell wall deposition during morphogenesis in fucoid algae. *Planta, 212*, 648–658.

Black, W. A. P. (1954). The seasonal variation in the combined L-fucose content of the common British Laminariaceae and Fucaceae. *Journal of the Science of Food and Agriculture, 5*, 445–448.

Black, W. A. P., Dewar, E. T., and Woodward, F. N. (1952). Manufacture of algal chemicals. IV. Laboratory-scale isolation of fucoidin from brown marine algae. *Journal of the Science of Food and Agriculture, 3*, 122–129.

Blunden, G. (2001). Biologically active compounds from marine organisms. *Phytotherapy Research, 15*, 89–94.

Blunt, J. W., Copp, B. R., Munro, M. H. G., Northcote, P. T., and Prinsep, M. R. (2006). Marine natural products. *Natural Product Reports, 23*, 26–78.

Bo, L., Fei, L., Xinjun, W., and Ruixiang, Z. (2008). Fucoidan: Structure and bioactivity. *Molecules, 13*, 1671–1695.

Boisson-Vidal, C., Chaubet, F., Chevolot, L., Sinquin, C., Theveniaux, J., Millet, J., Sternberg, C., Mulloy, B., and Fischer, A. M. (2000). Relationship between antithrombotic activities of fucans and their structure. *Drug Development Research, 51*, 216–224.

Brower, V. (1998). Nutraceuticals: Poised for a healthy slice of the healthcare market? *Nature Biotechnology, 16*, 728–731.

Chandía, N. P. and Matsuhiro, B. (2008). Characterization of a fucoidan from *Lessonia vadosa* (Phaeophyta) and its anticoagulant and elicitor properties. *International Journal of Biological Macromolecules, 42*, 235–240.

Chauvet, P., Bienvenu, J. G., Théorêt, J. F., Latour, J. G., and Merhi, Y. (1999). Inhibition of platelet-neutrophil interactions by fucoidan reduces adhesion and vasoconstriction after acute arterial injury by angioplasty in pigs. *Journal of Cardiovascular Pharmacology, 34*, 597–603.

Chevolot, L., Foucault, A., Chaubet, F., Kervarec, N., Sinquin, C., Fisher, A. M., and Boisson-Vidal, C. (1999). Further data on the structure of brown seaweed fucans: Relationships with anticoagulant activity. *Carbohydrate Research, 319*, 154–165.

Chevolot, L., Mulloy, B., and Racqueline, J. (2001). A disaccharide repeat unit is the structure in fucoidans from two species of brown algae. *Carbohydrate Research, 330*, 529–535.

Chizhov, A. O., Dell, A., and Morris, H. R. (1999). A study of fucoidan from the brown seaweed *Chorda filum. Carbohydrate Research, 320*, 108–119.

Choi, E. M., Kim, A. J., Kim, Y., and Hwang, J. K. (2005). Immunomodulating activity of arabinogalactan and fucoidan *in vitro. Journal of Medicinal Food, 8*, 446–453.

Colliec, S., Boisson-Vidal, C., and Jozefonvicz, J. (1994). A low molecular weight fucoidan fraction from the brown seaweed *Pelvetia caniculata. Phytochemistry, 35*, 697–700.

Cooper, R., Dragar, C., Elliot, K., Fitton, J. H., Godwin, J., and Thompson, K. (2002). GFS, a preparation of Tasmanian *Undaria pinnatifida* is associated with healing and inhibition of reactivation of Herpes. *BMC Complementary and Alternative Medicine, 2*, 11.

Cui, Y. Q., Zhang, L. J., Zhang, T., Luo, D. Z., Jia, Y. J., Guo, Z. X., Zhang, Q. B., Wang, X., and Wang, X. M. (2010). Inhibitory effect of fucoidan on nitric oxide production in lipopolysaccharide-activated primary microglia. *Clinical and Experimental Pharmacology and Physiology, 37*, 422–428.

Cumashi, A., Ushakova, N. A., Preobrazhenskaya, M. E., D'Incecco, A., Piccoli, A., Totani, L., Tinari, N. et al. (2007). A comparative study of the anti-inflammatory, anticoagulant, antiangiogenic, and antiadhesive activities of nine different fucoidans from brown seaweeds. *Glycobiology, 17*, 541–552.

David, L., Nigel, K., Michael, M., and Denise, O. (2009). *Practical Hemostasis and Thrombosis*, Wiley-Blackwell, New York, pp. 1–5.

Descamps, V., Colin, S., Lahaye, M., Jam, M., Richard, C., Potin, P., Barbeyron, T., Yvin, J. C., and Kloareg, B. (2006). Isolation and culture of a marine bacterium degrading the sulfated fucans from marine brown algae. *Marine Biotechnology, 8,* 27–39.

Duarate, M., Cardoso, M., and Noseda, M. (2001a). Structural studies on fucoidans from the brown seaweed *Sargassum stenophyllum. Carbohydrate Research, 333,* 281–293.

Duarte, M. E. R., Cardoso, M. A., Noseda, M. D., and Cerezo, A. S. (2001b). Structural studies on fucoidans from the brown seaweed *Sargassum stenophyllum. Carbohydrate Research, 333,* 281–293.

Espín, J. C., García-Conesa, M. T., and Tomás-Barberán, F. A. (2007). Nutraceuticals: Facts and fiction, *Phytochemistry, 68,* 2986–3008.

Feldman, S. C., Reynaldi, S., Stortz, C. A., Cerezo, A. S., and Damont, E. B. (1999). Antiviral properties of fucoidan fractions from *Leathesia difformis. Phytomedicine, 6,* 335–340.

Halvorson, H. O. (1998). Aquaculture, marine sciences and oceanography: A confluence connection. *New England's Journal of Higher Education and Economic Development, 13,* 28–42.

Haneji, K., Matsuda, T., Tomita, M., Kawakami, H., Ohshiro, K., Uchihara, J. N., Masuda, M. et al. (2005). Fucoidan extracted from *Cladosiphon okamuranus* Tokida induces apoptosis of human T-cell leukemia virus type 1-infected T-cell lines and primary adult T-cell leukemia cells. *Nutrition and Cancer, 52,* 189–201.

Hardy, G. (2000). Nutraceuticals and functional foods: Introduction and meaning. *Nutrition, 16,* 688–689.

Hasler, C. M. (2000). The changing face of functional food. *Journal of American College of Nutrition, 19,* 499S–506S.

Hayashi, S., Itoh, A., Isoda, K., Kondoh, M., Kawase, M., and Yagi, K. (2008a). Fucoidan partly prevents CCl$_4$-induced liver fibrosis. *European Journal of Pharmacology, 580,* 380–384.

Hayashi, K., Nakano, T., Hashimoto, M., Kanekiyo, K., and Hayashi, T. (2008b). Defensive effects of a fucoidan from brown alga *Undaria pinnatifida* against herpes simplex virus infection. *International Immunopharmacology, 8,* 109–116.

Holtkamp, A. D., Kelly, S., Ulber, R., and Lang, S. (2009). Fucoidans and fucoidanases-focus on techniques for molecular structure elucidation and modification of marine polysaccharides. *Applied Microbiology and Biotechnology, 82,* 1–11.

Iwamoto, K., Hiragun, T., Takahagi, S., Yanase, Y., Morioke, S., Mihara, S., Kameyoshi, Y., and Hide, M. (2011). Fucoidan suppresses IgE production in peripheral blood mononuclear cells from patients with atopic dermatitis. *Archives of Dermatological Research, 303,* 425–431.

Jiao, G., Yu, G., Zhang, J., and Ewart, H. S. (2011). Chemical structures and bioactivities of sulfated polysaccharides from marine algae, *Marine Drugs, 9,* 196–223.

Jung, W. K., Athukorala, Y., Lee, Y. J., Cha, S. H., Lee, C. H., Vasanthan, T., Choi, K. S., Yoo, S. H., Kim, S. K., and Jeon, Y. J. (2007). Sulfated polysaccharide purified from *Ecklonia cava* accelerates antithrombin III-mediated plasma proteinase inhibition. *Journal of Applied Phycology, 19,* 425–430.

Kalra, E. K. (2003). Nutraceutical—Definition and introduction. *AAPS PharmSci, 5,* Article 25 (http://www.pharmsci.org).

Kang, S. M., Kim, K. N., Lee, S. H., Ahn, G., Cha, S. H., Kim, A. D., Yang, X. D., Kang, M. C., and Jeon, Y. J. (2011). Anti-inflammatory activity of polysaccharide purified from AMG-assistant extract of *Ecklonia cava* in LPS-stimulated RAW264.7 macrophages. *Carbohydrate Polymers, 85,* 80–85.

Karim, S., Jessica, P., Farida, G., Christine, D., Corinne, S., Jacqueline, R., Gaston, G., Anne-Marie, F., Dominique, H., and Sylvia, C. J. (2011). Marine polysaccharides: A source of bioactive molecules for cell therapy and tissue engineering. *Marine Drugs, 9,* 1664–1681.

Kawamoto, H., Miki, Y., Kimura, T., Tanaka, K., Nakagawa, T., Kawamukai, M., and Matsuda, H. (2006). Effects of fucoidan from Mozuku on human stomach cell lines. *Food Science and Technology Research, 12,* 218–222.

Kim, M. J., Chang, U. J., and Lee, J. S. (2009). Inhibitory effects of fucoidan in 3T3-L1 adipocyte differentiation. *Marine Biotechnology*, 5, 557–562.

Kloareg, B., Demarty, M., and Mabeau, S. (1986). Polyanionic characteristics of purified sulfated homofucans from brown algae. *International Journal of Biological Macromolecules*, 8, 380–386.

Kopelman, P. G. (2000). Obesity as a medical problem. *Nature*, 404, 635–643.

Koyanagi, S., Tanigawa, N., Nakagawa, H., Soeda, S., and Shimeno, H. (2003). Oversulfation of fucoidan enhances its anti-angiogenic and antitumor activities. *Biochemical Pharmacology*, 65, 173–179.

Kuznetsova, T. A., Besednova, N. N., Mamaev, A. N., Momot, A. P., Shevchenko, N. M., and Zvyagintseva, T. N. (2003). Anticoagulant activity of fucoidan from brown algae *Fucus evanescens* of the Okhotsk Sea. *Bulletin of Experimental Biology and Medicine*, 136, 471–473.

Lee, J. B., Hayashi, K., Hashimoto, M., Nakano, T., and Hayashi, T. (2004). Novel antiviral fucoidan from sporophyll of *Undaria pinnatifida* (Mekabu). *Chemical and Pharmaceutical Bulletin*, 52, 1091–1094.

Ley, K. (2003). The role of selectins in inflammation and disease. *Trends in Molecular Medicine*, 9, 263–268.

Li, C., Gao, Y., Xing, Y., Zhu, H., Shen, J., and Tian, J. (2011). Fucoidan, a sulfated polysaccharide from brown algae, against myocardial ischemia–reperfusion injury in rats via regulating the inflammation response. *Food and Chemical Toxicology*, 49, 2090–2095.

Li, B., Wei, X. J., Sun, J. L., and Xu, S.Y. (2006). Structural investigation of a fucoidan containing a fucose-free core from the brown seaweed, *Hizikia fusiforme. Carbohydrate Research*, 341, 1135–1146.

Li, D. Y., Xu, Z., Huang, L. M., Wang, H. B., and Zhang, S. H. (2001). Effect of fucoidan of *L. japonica* on rats with hyperlipidaemia. *Food Science*, 22, 92–95.

Li, D. Y., Xu, Z., and Zhang, S. H. (1999a). Prevention and cure of fucoidan of *L. japonica* on mice with hypercholesterolemia. *Food Science*, 20, 45–46.

Li, D. Y., Xu, R. Y., Zhou, W. Z., Sheng, X. B., Yang, A. Y., and Cheng, J. L. (2002). Effects of fucoidan extracted from brown seaweed on lipid peroxidation in mice. *Acta Nutrimenta Sinica*, 24, 389–392.

Li, Z. J., Xue, C. H., and Lin, H. (1999b). The hypolipidemic effects and antioxidative activity of sulfated fucan on the experimental hyperlipidemia in rats. *Acta Nutrimenta Sinica*, 21, 280–283.

Lindahl, U. (2000). "Heparin" — From anticoagulant drug into the new biology. *Glycoconjugate Journal*, 17, 597–605.

Liu, J. M., Bignon, J., Haroun-Bouhedja, F., Bittoun, P., Vassy, J., Fermandjian, S., Wdzieczak-Bakala, J., and Boisson-Vidal, C. (2005). Inhibitory effect of fucoidan on the adhesion of adenocarcinoma cells to fibronectin. *Anticancer Research*, 25, 2129–2133.

Mabeau, S., Kloareg, B., and Joseleau, J. P. (1990). Fractionation and analysis of fucans from brown algae. *Phytochemistry*, 29, 2441–2445.

Mandal, P., Mateu, C. G., Chattopadhyay, K., Pujol, C. A., Damonte, E. B., and Ray, B. (2007). Structural features and antiviral activity of sulphated fucans from the brown seaweed *Cystoseira indica. Antiviral Chemistry and Chemotherapy*, 18, 153–162.

Manisha, P., Rohit, K. V., and Shubhini, A. S. (2010). Nutraceuticals: New era of medicine and health. *Asian Journal of Pharmaceutical and Clinical Research*, 3, 11–15.

Maruyama, H., Tamauchi, H., Hashimoto, M., and Nakano, T. (2005). Suppression of Th2 immune responses by Mekabu fucoidan from *Undaria pinnatifida* sporophylls. *International Archives of Allergy and Immunology*, 137, 289–294.

Maruyama, H., Tamauchib, H., Iizuka, M., and Nakano, T. (2006). The role of NK cells in antitumor activity of dietary fucoidan from *Undaria pinnatifida* sporophylls (Mekabu). *Planta Medica*, 72, 1415–1417.

Mayer, A. M., Rodriguez, A. D., Berlinck, R., and Fusetani, N. (2011). Marine pharmacology in 2007–8: Marine compounds with antibacterial, anticoagulant, antifungal, anti-inflammatory, antimalarial, antiprotozoal, antituberculosis and antiviral activities; affecting the immune and nervous system, and other miscellaneous mechanisms of action. *Comparative Biochemistry and Physiology—Part C: Toxicology and Pharmacology, 153*, 191–222.

Medcalf, D. G. and Larsen, B. (1977) Fucose-containing polysaccharides in the brown algae *Ascophyllum nodosum* and *Fucus vesiculosus. Carbohydrate Research, 59*, 531–537.

Micheline, R. S., Cybelle, M., Celina, G. D., Fernando, F. S., Hugo, O. R., and Edda, L. (2007). Antioxidant activities of sulfated polysaccharides from brown and red seaweeds. *Journal of Applied Phycology, 19*, 153–160.

Millet, J., Jouault, S. C., Mauray, S., Theveniaux, J., Sternberg, C., Boisson, V. C., and Fischer, A. M. (1999). Antithrombotic and anticoagulant activities of a low molecular weight fucoidan by the subcutaneous route. *Journal of Thrombosis and Haemostasis, 81*, 391–395.

Mourao, P. A. S. (2004). Use of sulfated fucans as anticoagulant and antithrombotic agents: Future perspectives. *Current Pharmaceutical Design, 10*, 967–981.

Mourao, P. A. S. and Pereira, M. S. (1999). Searching for alternatives to heparin: Sulfated fucans from marine invertebrates. *Trends in Cardiovascular Medicine, 9*, 225–232.

Mulloy, B., Ribeiro, A. C., Alves, A. P., Vieira, R. P., and Mourao, P. A. S. (1994). Sulfated fucans from echinoderms have a regular tetrasaccharide repeating unit defined by specific patterns of sulfation at the 0–2 and 0–4 positions. *Journal of Biological Chemistry, 269*, 22113–22123.

Ngo, D. H., Wijesekara, I., Vo, T. S., Ta, Q. V., and Kim, S. K. (2011). Marine food-derived functional ingredients as potential antioxidants in the food industry: An overview. *Food Research International, 44*, 523–529.

Nishino, T., Aizu, Y., and Nagumo, T. (1991). The influence of sulfate content and molecular weight of a fucan sulfate from the brown seaweed *Ecklonia kurome* on its antithrombin activity. *Thrombosis Research, 64*, 723–731.

Nishino, T., Fukuda, A., Nagumo, T., Fujihara, M., and Kaji, E. (1999). Inhibition of the generation of thrombin and factor Xa by a fucoidan from the brown seaweed *Ecklonia kurome. Thrombosis Research, 96*, 37–49.

Nishino, T. and Nagumo, T. (1987). Sugar constituents and blood-anticoagulant activities of fucose-containing sulfated polysaccharides in nine brown seaweed species. *Nippon Nogeikagaku Kaishi, 61*, 361–363.

Nishino, T. and Nagumo, T. (1991). The sulfate-content dependence of the anticoagulant activity of a fucan sulfate from the brown seaweed *Ecklonia kurome. Carbohydrate Research, 214*, 193–197.

Nishino, T. and Nagumo, T. (1992). Anticoagulant and antithrombin activities of oversulfated fucans. *Carbohydrate Research, 229*, 355–362.

Nishino, T., Nishioka, C., Ura, H., and Nagumo, T. (1994). Isolation and partial characterization of a novel amino sugar-containing fucan sulfate from commercial *Fucus vesiculosus* fucoidan. *Carbohydrate Research, 255*, 213–224.

Nishino, T., Yokoyama, G., Dobashi, K., Fujihara, M., and Nagumo, T. (1989). Isolation, purification, and characterization of fucose-containing sulfated polysaccharides from the brown seaweed *Ecklonia kurome* and their blood-anticoagulant activities. *Carbohydrate Research, 186*, 119–129.

Obluchinskaya, E. D. and Minina, S. A. (2004). Development of extraction technology and characterization of extract from wrack algae grist. *Pharmaceutical Chemistry, 38*, 323–326.

Oomizu, S., Yanase, Y., Suzuki, H., Kameyoshi, Y., and Hide, M. (2006). Fucoidan prevents Cε germline transcription and NF-κB p52 translocation for IgE production in B cells. *Biochemical and Biophysical Research Communications, 350*, 501–507.

Park, H. Y., Han, M. H, Park, C, Jin, C. Y., Kim, G. Y., Choi, I. W., Kim, N. D., Nam, T. J., Kwon, T. K., & Choi, Y. H. (2011a). Anti-inflammatory effects of fucoidan through inhibition of NF-κB, MAPK and Akt activation in lipopolysaccharide-induced BV2 microglia cells. *Food and Chemical Toxicology, 49*, 1745–1752.

Park, M. K., Jung, U., and Roh, C. (2011b). Fucoidan from marine brown algae inhibits lipid accumulation. *Marine Drugs, 9*, 1359–1367.

Pereira, M. S., Mulloy, B., and Mourao, P. A. S. (1999). Structure and anticoagulant activity of sulfated fucans. Comparison between the regular, repetitive, and linear fucans from echinoderms with the more heterogeneous and branched polymers from brown algae. *Journal of Biological Chemistry, 274*, 7656–7667.

Pomin, V. H., Pereira, M. S., Valente, A. P., Tollefsen, D. M., Pavao, M. S. G., and Mourao, P. A. S. (2005). Selective cleavage and anticoagulant activity of a sulfated fucan: Stereospecific removal of a 2-sulfate ester from the polysaccharide by mild acid hydrolysis, preparation of oligosaccharides, and heparin cofactor II-dependent anticoagulant activity. *Glycobiology, 15*, 369–381.

Ponce, N. M., Pujol, C. A., Damonte, E. B., Flores, M. L., and Stortz, C. A. (2003). Fucoidans from the brown seaweed *Adenocystis utricularis*: Extraction methods, antiviral activity and structural studies. *Carbohydrate Research, 338*, 153–165.

Preeprame, S., Hayashi, K., Lee, J. B., Sankawa, U., and Hayashi, T. (2001). A novel antivirally active fucan sulfate derived from an edible brown alga, *Sargassum horneri*. *Chemical and Pharmaceutical Bulletin, 49*, 484–485.

Preobrazhenskaya, M. E., Berman, A. E., Mikhailov, V. I., Ushakova, N. A., Mazurov, A. V., Semenov, A. V., Usov, A. I., Nifant'ev, N. E., and Bovin, N. V. (1997). Fucoidan inhibits leukocyte recruitment in a model peritoneal inflammation in rat and blocks interaction of P-selectin with its carbohydrate ligand. *Biochemistry and Molecular Biology International, 43*, 443–451.

Queiroz, K. C. S., Medeiros, V. P., Queiroz, L. S., Abreu, L. R. D., Rocha, H. A. O., Ferreira, C. V., Juca, M. B., Aoyama, H., and Leite, E. L. (2008). Inhibition of reverse transcriptase activity of HIV by polysaccharides of brown algae. *Biomedicine and Pharmacotherapy, 62*, 303–307.

Rachel, N. (2008). Chronic diseases in developing countries health and economic burdens. *Annals of the New York Academy of Sciences, 1136*, 70–79.

Rajasekaran, A., Sivagnanam, G., and Xavier, R. (2008). Nutraceuticals as therapeutic agents: A review. *Research Journal of Pharmacy and Technology, 1*, 328–340.

Raskin, I., Ribnicky, D. M., Komarnytsky, S., Ilic, N., Poulev, A., Borisjuk, N., Brinker, A. et al. (2002). Plants and human health in the twenty-first century. *Trends in Biotechnology, 20*, 522–531.

Ribeiro, A. C., Vieira, R. P., Mourao, P. A. S., and Mulloy, B. (1994). A sulfated alpha-L-fucan from sea cucumber. *Carbohydrate Research, 255*, 225–240.

Saikat, K. B., James, E. T., and Surya, N. A. (2007). Prospects for growth in global nutraceutical and functional food markets: A Canadian perspective. *Australian Journal of Basic and Applied Sciences, 1*, 637–649.

Saito, A., Yoneda, M., Yokohama, S., Okada, M., Haneda, M., and Nakamura, K. (2006). Fucoidan prevents concanavalin A-induced liver injury through induction of endogenous IL-10 in mice. *Hepatology Research, 35*, 190–198.

Sakai, T., Ishizuka, K., Shimanaka, K., Ikai, K., and Kato, I. (2003). Structures of oligosaccharides derived from *Cladosiphon okamuranus* fucoidan by digestion with marine bacterial enzymes. *Marine Biotechnology, 5*, 536–544.

Sakai, T., Kimura, H., and Kato, I. (2002). A marine strain of Flavobacteriaceae utilizes brown seaweed fucoidan. *Marine Biotechnology, 4*, 399–405.

Semenov, A. V., Mazurov, A. V., Preobrazhenskaia, M. E., Ushakova, N. A., Mikhaïlov, V. I., Berman, A. E., Usov, A. I., Nifant'ev, N. E., and Bovin, N. V. (1998). Sulfated polysaccharides as inhibitors of receptor activity of P-selectin and P-selectin-dependent inflammation. *Voprosy Meditsinskoi Khimii, 44*, 135–144.

Senni, K., Gueniche, F., Bertaud, A. F., Tchen, S. I., Fioretti, F., Jouault, S. C., Durand, P., Guezennec, J., Godeau, G., and Letourneur, D. (2006). Fucoidan a sulfated polysaccharide from brown algae is a potent modulator of connective tissue proteolysis. *Archives of Biochemistry and Biophysics, 445*, 56–64.

Shibata, H., Kimura-Takagi, I., Nagaoka, M., Hashimoto, S., Aiyama, R., Iha, M., Ueyama, S., and Yokokura, T. (2000). Properties of fucoidan from *Cladosiphon okamuranus* tokida in gastric mucosal protection. *Biofactors, 11*, 235–245.

Silva, T. M. A., Alves, L. G., Queiroz, K. C. S., Santos, M. G. L., Marques, C. T., Chavante, S. F., Rocha, H. A. O., and Leite, E. L. (2005). Partial characterization and anticoagulant activity of a heterofucan from the brown seaweed *Padina gymnospora*. *Brazilian Journal of Medical and Biological Research*, *38*, 523–533.

Sinha, S., Astani, A., Ghosh, T., Schnitzler, P., and Ray, B. (2010). Polysaccharides from *Sargassum tenerrimum*: Structural features, chemical modification and anti-viral activity. *Phytochemistry*, *71*, 235–242.

Soeda, S., Shibata, Y., and Shimeno, H. (1997). Inhibitory effect of oversulfated fucoidan on tube formation by human vascular endothelial cells. *Biological Pharmaceutical Bulletin*, *20*, 1131–1135.

Song, J. Q., Xu, Y. T., and Zhang, H. K. (2000). Immunomodulation action of sulfate polysaccharide of *Laminaria japonica* on peritoneal macrophages of mice. *Chinese Journal of Microbiology and Immunology*, *16*, 70.

Spiegelman, B. M. and Flier, J. S. (2001). Obesity and the regulation of energy balance. *Cell*, *104*, 531–543.

Springer, G. F., Wurzel, H. A., McNeal, G. M., Ansell, N. J., and Doughty, M. F. (1957). Isolation of anticoagulant fractions from crude fucoidin. *Proceedings of the Society for Experiment Biology and Medicine*, *94*, 404–409.

Synytsya, A., Kim, W. J., Kim, S. M., Pohl, R., Synytsya, A., Kvasnicka, F., Copíková, J., and Park, Y. I. (2010). Structure and antitumour activity of fucoidan isolated from sporophyll of Korean brown seaweed *Undaria pinnatifida*. *Carbohydrate Polymers*, *81*, 41–48.

Teruya, T., Konishi, T., Uechi, S., Tamaki, H., and Tako, M. (2007). Anti-proliferative activity of oversulfated fucoidan from commercially cultured *Cladosiphon okamuranus* TOKIDA in U937 cells. *International Journal of Biological Macromolecules*, *41*, 221–226.

Thompson, K. D. and Dragar, C. (2004). Antiviral activity of *Undaria pinnatifida* against herpes simplex virus. *Phytotherapy Research*, *18*, 551–555.

Tolwani, A. J. and Wille, K. M. (2009). Anticoagulation and continuous renal replacement therapy. *Seminars in Dialysis*, *22*, 141–145.

Trinchero, J., Ponce, N. M. A., Cordoba, O. L., Flores, M. L., Pampuro, S., Stortz, C. A., Salomon, H., and Turk, G. (2009). Antiretroviral activity of fucoidans extracted from the brown seaweed *Adenocystis utricularis*. *Phytotherapy Research*, *23*, 707–712.

Usov, A. I. and Bilan, M. I. (2009). Fucoidans—Sulfated polysaccharides of brown algae. *Russian Chemical Reviews*, *78*, 785–799.

Veena, C. K., Josephine, A., Preetha, S. P., and Varalakshmi, P. (2007a). Beneficial role of sulfated polysaccharides from edible seaweed *Fucus vesiculosus* in experimental hyperoxaluria. *Food Chemistry*, *100*, 1552–1559.

Veena, C. K., Josephine, A., Preetha, S. P., and Varalakshmi, P. (2007b). Effect of sulphated polysaccharides on erythrocyte changes due to oxidative and nitrosative stress in experimental hyperoxaluria. *Human and Experimental Toxicology*, *26*, 923–932.

Vignesh, S., Raja, A., and Arthur James, R. (2011). Marine drugs: Implication and future studies. *International Journal of Pharmacology*, *7*, 22–30.

Vilela-Silva, A. C. E. S., Alves, A. P., Valente, A. P., Vacquier, V. D., and Mourao, P. A. S. (1999). Structure of the sulfated alpha-L-fucan from the egg jelly coat of the sea urchin Strongylocentrotus franciscanus: Patterns of preferential 2-O- and 4-O-sulfation determine sperm cell recognition. *Glycobiology*, *9*, 927–933.

Vilela-Silva, A. C., Castro, M. O., Valente, A. P., Biermann, C. H., and Mourao, P. A. S. (2002). Sulfated fucans from the egg jellies of the closely related sea urchins *Strongylocentrotus droebachiensis* and *Strongylocentrotus pallidus* ensure species-specific fertilization. *Journal of Biological Chemistry*, *277*, 379–387.

Vincenzo, L., Paul, A. K., Vito, L., and Angela, C. (2009). Globe artichoke: A functional food and source of nutraceutical ingredients. *Journal of Functional Foods*, *1*, 131–144.

Vo, T. S., Ngo, D. H., and Kim, S. K. (2012). Potential targets for anti-inflammatory and anti-allergic activities of marine algae: An overview. *Inflammation and Allergy—Drug Targets*, *11*, 90–101.

Wang, S. Z. and Bi, A. F. (1994). Clinic observation of fucoidan on patients with hyperlipidae-mia. *Medical Journal of Qilu*, 173–174.

Wang, J., Zhang, Q., Zhang, Z., and Li, Z. (2008). Antioxidant activity of sulfated polysaccharide fractions extracted from *Laminaria japonica. International Journal of Biological Macromolecules*, *42*, 127–132.

Waris, G. and Ahsan, H. (2006). Reactive oxygen species: Role in the development of cancer and various chronic conditions. *Journal of Carcinogenesis*, *5*, 14.

Xue, C. H., Fang, Y., Lin, H., Chen, L., Li, Z. J., Deng, D., and Lu, C. X. (2001). Chemical characters and antioxidative properties of sulfated polysaccharides from *Laminaria japonica. Journal of Applied Phycology*, *13*, 67–70.

Yanase, Y., Hiragun, T., Uchida, K., Ishii, K., Oomizu, S., Suzuki, H., Mihara, S. et al. (2009). Peritoneal injection of fucoidan suppresses the increase of plasma IgE induced by OVA-sensitization. *Biochemical and Biophysical Research Communications*, *387*, 435–439.

Yoon, S. J., Pyun, Y. R., Hwang, J. K., and Mourão, P. A. S. (2007). A sulfated fucan from the brown alga *Laminaria cichorioides* has mainly heparin cofactor II-dependent antico-agulant activity. *Carbohydrate Research*, *342*, 2326–2330.

Zhang, Q. B., Li, N., Zhao, T. T., Qi, H. M., Xu, Z. H., and Li, Z. E. (2005). Fucoidan inhibits the development of proteinuria in active *Heymann nephritis. Phytotherapy Research*, *19*, 50–53.

Zhang, Z., Teruya, K., Eto, H., and Shirahata, S. (2011). Fucoidan extract induces apopto-sis in mcf-7 cells via a mechanism involving the ROS-dependent JNK activation and mitochondria-mediated pathways. *PLoS One*, *6*, e27441.

Zhao, X, Wang, J. F., and Xue, C. H. (2011). The inhibitory effects of fucoidans from *Laminaria japonica* on oxidation of human low-density lipoproteins. *Advanced Materials Research*, *236–238*, 2067–2071.

Zhao, X., Xue, C. H., Cai, Y. P., Wang, D. F., and Fang, Y. (2005). The study of antioxidant activities of fucoidan from *Laminaria japonica. High Technology Letters*, *11*, 91–94.

Chitosan for Body Weight Management

Current Issues and Future Directions

Soon Kong Yong and Tin Wui Wong

Contents

12.1 Introduction to chitosan

12.1.1 Sources and physicochemical properties

Chitin was first isolated from mushroom by Braconnot in 1811 (Roberts 2008). Rouget was first to discover chitosan but it was not until Hoppe-Seyler who first named this material as chitosan. The structure of chitosan was only resolved in 1950 where it was described as a polymer consisting of repeating units of glucosamine (Baldrick 2010). Chitosan is a type of

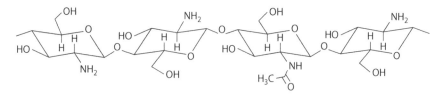

Figure 12.1 Chemical structure of chitosan.

polysaccharide resembling cellulose (**Figure 12.1**). To date, most chitosans are derived from chitin-rich shells of prawn, crab, and lobster due to the abundance of waste from seafood processing industries. Other sources of chitosan include krill shells, squid pen, and fungi mycelium. The pathway for extracting chitin and its subsequent conversion to chitosan is summarized in **Figure 12.2**.

Overall, the production process of chitosan involves alkaline (NaOH or KOH) and acidic (HCl) treatment of raw materials. The reaction with dilute NaOH aims to remove protein from chitin (Synowiecki and Al-Khateeb 2000). The reaction of chitin with dilute HCl dissolves and eliminates $CaCO_3$. The acid treatment may not be necessary when squid pen and fungi mycelium are used as starting materials. The main processing step in chitosan production is deacetylation, where acetyl moiety from C_2 amide is converted into amine group. Due to the requirement of a large quantity

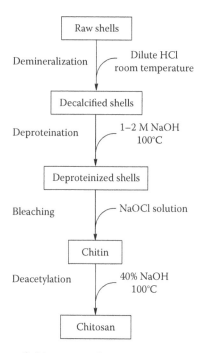

Figure 12.2 Process flow of chitosan production.

Table 12.1 Acid Solubility of Chitosan

Solvent	Solvent Concentration (mol/L)	Chitosan Solubility (g/L)	Reference
Formic acid	0.1	1.5	Li et al. (2006)
Acetic acid	1	50	Hamdine et al. (2005)
Hydrochloric acid	0.25	50	
Lactic acid	0.50	50	
Malic acid	0.50	50	
Citric acid	0.50	50	

of NaOH in this process, the cost of chitosan is mainly dependent on the market price of NaOH. Other means of producing chitosan includes enzymatic and microbial pathways (Rao and Stevens 2005). In comparison to chemical processing method, these techniques are not practical and receive limited applications due to low conversion efficiency and product yield (Zhao et al. 2010).

Chitosan is mostly white or beige in color, depending on the extent of decoloration during the production of chitosan (Youn et al. 2007). The solubility of chitosan is a function of its chemical structure, namely content of amine groups and molecular size of chitosan (Hein et al. 2008). Generally, chitosan is not soluble in organic solvent. It is however soluble in organic acidic solution. The acid solubility of chitosan is promoted by a high degree of polymer deacetylation and low solution pH (Wong 2009). Chitosan begins to coagulate in an aqueous milieu of pH 6.5 where it can precipitate to form insoluble particles due to loss of protonation of amine moiety (Pillai et al. 2009). The solubility of chitosan is summarized in **Table 12.1**.

The gelation of chitosan involves formation of intermolecular crosslinks via hydrogen, ionic, and/or covalent bonds. Gelation of chitosan may also be achieved with only nonpolar dispersion forces, using a balanced mixture of alcohol and acetic acid (Montembault et al. 2005). Similar to other polysaccharides, chitosan has the ability to hold a high quantity of moisture via interaction of its polar amine and hydroxyl functional groups with water, forming swollen chitosan gels of viscous liquid (Park and Kim 2010).

12.1.2 Biological properties

Chitosan has been reported to exhibit a vast range of bioactivities such as anti-inflammatory (Cho et al. 2011; Xia et al. 2011), antiulcer (Ito et al. 2000), wound healing (Kim et al. 2007; Koide 1998), and anticarcinogenic activities (Koide 1998; Xia et al. 2011). In addition, recent studies indicate that chitosan is a potential remedy for hyperlipidemia (Xia et al. 2011; Zhang et al. 2011) due to its fat-binding and cholesterol-lowering properties. Chitosan carries antioxidative characteristics where it scavenges hydroxyl and superoxide radicals, donates proton (Park et al. 2003; Sun et al. 2003; Xia et al. 2011), and chelates Cu(II) or Fe(II) ions, which are responsible to produce free radicals (Feng et al. 2008). Chitosan possesses antimicrobial activity against fungi, bacteria, yeast, and viruses (Koide 1998; Rabea et al. 2003; Xia et al. 2011). Its antimicrobial activity is ascribed by surface interaction between amine with lipopolysaccharide of microbe cell membrane (Helander et al. 2001) as well as deoxyribonucleic acid which lead to cell membrane leakage and genetic material condensation (Sudarshan et al. 1992). The antimicrobial and antioxidative properties of chitosan suggest that it is appropriate for use in food preservation (Friedman and Juneja 2010; Xia et al. 2011), treatment of oxidative damage caused by Cd(II) poisoning (Li et al. 2011) and plant disease (Rabea et al. 2003; Xia et al. 2011), as well as shelf-life prolongation of postharvest crops (Meng et al. 2010). Biomedically, chitosan has been recognized as a good scaffold for wound healing owing to its excellent hemostatic, antimicrobial, and wound-healing activities (Kim et al. 2007; Koide 1998; Xia et al. 2011). With reference to current healthcare market, chitosan is mainly applied as a dietary supplement to manage body weight, blood cholesterol, and osteoarthritis (Baldrick 2010; Xia et al. 2011). Other popular usages of chitosan include drug delivery (Wong 2009; Wong and Harjoh in press; Wong and Sumiran 2008; Xia et al. 2011; Zakaria and Wong 2009), orthopedic, and tissue engineering (Khor and Lim 2003; Xia et al. 2011).

12.1.3 Toxicity

In general, chitosan is considered as nontoxic to higher organisms such as animals and humans. The oral LD_{50} values of chitosan for rat and mice are >1,500 and 16,000 mg/kg, respectively, while oligochitosan has an oral LD_{50} of >10,000 mg/kg in mice (Baldrick 2010). The oral LD_{50} value for human is deduced to be 1/12 of that of mice (1.33 g/kg) (Prajapati 2009). However, intravenously administered chitosan at a dosage of 50 mg/kg can be lethal probably due to blood cell aggregation (Kean and Thanou 2010). This is inferred from the fact that dogs injected with 200 mg/kg of chitosan die from severe hemorrhagic pneumonia after 8 days (Minami et al. 1996).

Studies related to carcinogenicity and reproduction toxicology of chitosan is limited, especially on human subject. Fortunately, there are little chitosan-associated complications in human. No complications were reported by Gades and Stern (2003) and Zahorska-Markiewicz et al. (2002) when male and female subjects were given high daily chitosan dosages (4.5 g) for 12 and 90 days, respectively. This is further supported by a study using 3 g chitosan/day for 168 days where none of the subjects express toxicity-related complications (Mhurchu et al. 2004). At an oral LD_{50} value of 1.33 g/kg, the subjects with an average adult body weight of 60 kg are considered to have consumed a remarkably low chitosan dosage.

At high oral doses of 6.75 g daily, short-term human trials up to 12 weeks indicate no clinically significant symptoms including allergic responses except for a low incidence of mild to transitory nausea and constipation (Baldrick 2010). Oral absorption and systemic exposure of chitosan do not pose significant risks. It is not likely that the chitosan will be retained and accumulated in body due to conversion to glucosamine derivatives which are either excreted or used in amino sugar pool.

12.2 Body weight management of chitosan

Obesity is characterized by a body mass index of 30 kg/m² and above (DeWald et al. 2006). It is linked with illnesses such as cardiovascular disease, stroke, hypertension, pulmonary hypertension, insulin resistance, type 2 diabetes mellitus, musculoskeletal disorder, gall bladder disorder, hyperuricemia and gout, osteoarthritis, hyperlipidemia, and cancer (DeWald et al. 2006; Lois and Kumar 2008). Obesity represents one of the medical disorders which involves complex interaction between genetic, environment, and psychosocial factors. Common strategy of weight management is by achieving negative energy balance, which can be reached through increasing energy expenditure or decreasing energy intake into the body, namely, via physical activity, reduced calorie/food intake, and pharmacotherapy (Biesemeier and Cummings 2008; DeWald et al. 2006; Witkamp 2011). This can be aided by behavior therapy as the foundation of body weight management program and surgical treatments (Biesemeier and Cummings 2008; DeWald et al. 2006). Notably, a degree of body weight reduction by as little as 5%–10% can reduce the risks of cardiovascular diseases and associated comorbidities (DeWald et al. 2006).

Pulverized chitosan powder produced from the exoskeleton of crustaceans is initially employed to remediate crude oil contamination (Abe et al. 1974). Based on this idea, chitosan is used to absorb edible oil. Its high absorbency to electrically neutral oil molecules implies possible lipid adsorption in human's gastrointestinal tract, thereby inhibiting fat adsorption and promoting body weight loss. Trials on animals have verified the fat-binding effect of chitosan. The uses of chitosan in pharmaceutical products have been documented in several patents worldwide. These patents have claimed numerous properties, such as lipid binding (Furda 1980) and cholesterol removal from food products (Sundfeld et al. 1994), and binding of negatively charged lipid to reduce its gastrointestinal uptake (Deuchi et al. 1995a).

Principally, the management of body weight by means of chitosan can be effected via fat binding and excretion. Other possible mechanisms include immobilization of bile salts and suppression of appetite as indicated by the recent research findings. The effects of chitosan on body weight reduction, if any, are commonly attributed to interplay outcomes of fat and bile salt binding with appetite suppression. The management of body weight by chitosan is expected to be rarely dependent on a single mode of mechanism.

12.2.1 Binding of fats

The fat-binding mechanism of chitosan is mostly explained by the ionic interaction between negatively charged lipid molecules with positively charged chitosan (Wydro et al. 2007; Xia et al. 2011). Fatty acids consist of a large hydrocarbon moiety with a polar and negatively charged head. Highly deacetylated chitosan, on the one hand, possesses an extensive number of amine groups. These amine groups undergo protonation in acidic milieu to give positively charged $-NH_3^+$ moieties, thereby favoring its binding to fat.

The interaction of chitosan with fat may not solely rely on electrostatic attraction between the said molecules. Using Langmuir monolayer model, it is found that electrostatic, hydrophobic, van der Waals, as well as hydrogen bonds participate in the binding process of chitosan with cholesterols and fatty acids characterized by different degrees of saturation and chain length (Wydro et al. 2007). Hydrophobic interaction between chitosan and fat, in another study by Deuchi et al. (1994), is nonetheless deemed to be negligible. Other bonding types, namely, electrostatic attraction can be of a greater significance as dissociation of chitosan from fat takes place at pH 7 as a result of loss of amine protonation (Mun et al. 2006).

The interaction of chitosan and fat in acidic gastric medium may give rise to the formation of micelles or complexes which are not absorbable (Czechowska-Biskup et al. 2005; Muzzarelli 1996; Xia et al. 2011). The nonabsorbable matrix is then precipitated in the intestinal media of pH 6 to 6.5 following chitosan chains losing their charge and undergoing aggregation. This entraps the fat in matrix, hinders the digestive action of lipase, and promotes the fat passage through intestinal lumen into feces and excretion. There are three main lipases in human: lingual, gastric, and pancreatic (Wilde and Chu 2011). The fat entrapment in the matrix through chitosan aggregation in intestinal media is deemed critical as pancreatic lipase is responsible for the majority of fat hydrolysis, whereas lingual and gastric lipases hydrolyze up to 30% of dietary fats only.

Apart from fat binding, it has been suggested that chitosan itself can act as an alternative substrate to which lipase binds to, leading to partial inhibition of fat digestion and possible body weight loss (Muzzarelli 1996). Using chitosan as a fat binder, the loss of body weight is lately found to be not ascribed by fecally excreted fat and water (Bondiolotti et al. 2011). The increased fat excretion (0.44 g/day) in animals treated with chitosan does not account for the 28 g difference in body weight over 63 days, nor does the higher fecal water content. The excretion of fat and water in the feces may not be the body weight reduction index of chitosan. The binder property of chitosan has brought about a large amount of fat and glucose to colon for use as fuel by bacteria in a chitosan concentration-dependent manner. The body weight reduction effect of chitosan is likely associated with "bacterial" energy wasting (Bondiolotti et al. 2011).

12.2.2 Immobilization of bile salts

Ingestion of chitosan has been reported to lead to bile acids excreted in free forms into feces without absorption, in addition to fat particles (Muzzarelli 1996; Xia et al. 2011). This is attributed to the fact that chitosan can bind to bile salts via dispersion forces or polar interaction between its cationic $-NH_3^+$ moiety and anionic $-COO^-$ group of bile salts (Aranaz et al. 2009). The binding of bile salts by chitosan interrupts the micellization process of fat, bile salt-dependent activation process of lipase, digestion of fat under the influence of lipase, and fat absorption in lacteal (Kim and Rajapakse 2005; Koide 1998; Muzzarelli et al. 2006). This in turn increases fat as well as bile acids excretion, and leads to possible body weight loss.

12.2.3 Suppression of appetite

The body weight reduction effect of chitosan can be mediated via appetite suppression. Chitosan is relatively difficult to be digested in the gastrointestinal tract. In acidic gastric milieu, it undergoes protonation of amine moiety, hydration, and swelling to form a viscous gel. This swollen gel possesses a high water-holding capacity and may induce satiety and satiation (Rasmussen and Morrissey 2007). Its highly viscous nature can too act to decrease food intake due to slow gastric emptying and sustained effect of fullness (Wanders et al. 2011).

The fat-binding characteristics of chitosan can be another mode of reducing appetite and prolonging the feelings of hunger (Wilde and Chu 2011). Inhibiting or slowing down the fat digestion allows the fat particles to travel along toward the distal end of ileum. These fats are sensed by the cell linings of the intestinal tract and in response secrete hormones and peptides which slow down digestion and send signals to the central nervous system to reduce the appetite.

12.2.4 Genetic complications

A distinct body weight reduction is expressed by ob/ob mice fed on chitosan oligosaccharides (Yun 2010). Proteome analysis of mouse plasma before and after chitosan treatment suggests that the expression of many genes is altered significantly with respect to body weight loss. Downregulation of obesity-related genes is found following 12% decrease in body weight gain of mice which receive 200 mg chitosan of molecular weight ranges between 3000 and 5000 Da over 28 days.

12.2.5 In vitro trials

Chitosan has been known to have the capacity to bind to both fat and bile acids (Xia et al. 2011; Zhou et al. 2006). However, the outcomes on the physicochemical influences of chitosan on its fat and bile acid binding properties are contradictory. In a digestion model of which 3% to 10% of oil droplets are added to 0.1% chitosan solution at pH 3 to simulate consumption of an oil-containing meal after chitosan ingestion, the amount of chitosan adsorbed onto the oil droplets decreases with decreasing molecular weight and with increasing degree of deacetylation (Helgason et al. 2009). In another study by Zhou et al. (2006), the fat-binding capacity of chitosan is 1077–1239 g oil/kg polymer sample, and no correlation is found between the degree of deacetylation, swelling capacity, and solution viscosity of chitosan with its fat-binding property. Long molecular chains can translate to fat binding by chitosan via electrostatic interaction as well as physical entanglement by viscous polysaccharides (Xia et al. 2011). Contrary to findings reported by Helgason et al. (2009), chitosan in a low-molecular-weight range between 25,000 and 400,000 Da exhibits a high fat-binding ability where 1 g of chitosan can bind up to 20 g of fat (Czechowska-Biskup et al. 2005). Depending on the molecular weight, it is envisaged that an overly long and rigid polymer chain cannot form oil-trapping micelles unlike shorter chains of a higher mobility. Nonetheless, when a chain becomes excessively short, it probably tends to form individual solubilized molecule rather than oil-trapping matrix.

The bile salt binding capacity of chitosan is reported to be greater with an increase in molecular weight and is not dependent on the degree of deacetylation of polymer (Xia et al. 2011; Zhou et al. 2006). Though the solution viscosity of chitosan is a function of its molecular weight, the bile acid binding property of chitosan is found to be independent of the effect of solution viscosity (Zhou et al. 2006).

12.2.6 Animal trials

Animal trials involving Wistar rats, ob/ob mice, C57BL/6J mice, and ICR mice with high fat diet have principally exhibited positive fat binding, fat excretion, and body weight loss outcomes (Koide 1998; Xia et al. 2011; Yun 2010). The fat-binding ability of chitosan increases with a reduction in its particle size due to a rise in specific surface area and porous structure for adsorption (Xia et al. 2011). It increases with an increase in polymer molecular weight and degree of deacetylation. Inferring from plasma cholesterol studies where glucosamine has not been effective, it is suggested that some polymerization is required for chitosan to exert its fat-binding activity (Kim and Rajapakse 2005; Muzzarelli 1996). Chitosan is known to form a highly viscous solution in the stomach and precipitate in the small intestine (Xia et al. 2011). It however does not modulate fat digestion via increasing viscosity of intestinal content, unlike dietary fibers.

12.2.7 Clinical trials

The effects of chitosan on human body weight reduction are controversial (**Table 12.2**). The chitosan exhibits a positive influence on body weight reduction in short term (Biesemeier and Cummings 2008). Nonetheless, this effect has yet to be labeled as clinically significant. The fat-binding capacity of chitosan is found to vary from 0.02 to 0.72 g of fat/g of chitosan (Gades and Stern 2005). Limitations and inconsistencies of research are possibly attributed to the use of non-obese subjects in some studies (Bokura and Kobayashi 2003; Guerciolini et al. 2001; Tsujikawa et al. 2003); use of mixture of chitosan, guar's meal, ascorbic acid, and other micronutrients instead of pure chitosan (Colombo and Sciutto 1996; Veneroni et al. 1996); undocumented chitosan type, timing of administration, and diet pattern (Bokura and Kobayashi 2003; Gades and Stern 2003; Guerciolini et al. 2001; Ho et al. 2001; Kaats et al. 2006; Pittler et al. 1999; Tsujikawa et al. 2003; Zahorska-Markiewicz et al. 2002).

12.3 Complications of chitosan

12.3.1 Gastrointestinal disorders

Almost all reported complications of chitosan are related to gastrointestinal disorders such as indigestion, bloating, and abdominal pain, with constipation being the most common side effect (Egras et al. 2011; Koide 1998). The gastrointestinal disorders are prevalent at high daily chitosan dosages (6.75 g) (Tapola et al. 2008). Chitosan undergoes a limited digestion in human gastrointestinal tract (Zhang and Neau 2002). Its ability to transport fat to colon in the form of micelles or complexes enables fat fermentation in large intestines, causing bloating and abdominal pain (Rasmussen and Morrissey 2007). Other possible causes of gastrointestinal disorders are increased fecal mass and deprivation of colonic mucus by chitosan. The cytotoxic effect of chitosan on intestinal flora may induce gastrointestinal complications. On prolonged ingestion, it is suggested that chitosan may alter the normal flora of the intestinal tract and result in growth of pathogens resistant to antimicrobial activity (Koide 1998). However, further investigation is needed to ascertain these effects.

12.3.2 Interaction with lipid-soluble vitamins and minerals

Chitosan is able to bind to the fat in human gastrointestinal tract. It is similarly possible to bind to the fat-soluble nutrients such as vitamin A, D, E, and K

Table 12.2 Outcomes of Clinical Trials on Chitosan with Respect to Its Effects on Body Weight and Fecal Excretion

Test Subject	Chitosan Characteristics	Design	Chitosan Dosage/Daily Frequency/Duration	Timing	Diet	Results	Conclusion	Reference
15 (13 women, 2 men)	Not reported	R, PC, DB	Four 250 mg/twice/28 days	Not reported	Unrestricted but documented	Baseline body weight (71.8 kg) increased to 72.4 kg	Chitosan did not significantly decrease body weight but significantly decreased vitamin K absorption	Pittler et al. (1999)
68 (31 women, 37 men)	Not reported	R, PC, DB	Four 250 mg/thrice/84 days	Not reported	Unrestricted and not documented	Fat percentage decreased (Women: −0.95% ± 2.43%; Men: −0.07% ± 2.19%)	Chitosan did not significantly decrease body weight	Ho et al. (2001)
12 (5 women, 7 men)	Not reported	R, 2PC, OL	Two 445 mg/thrice/14 days	Not reported	Standardized; 2500 kcal/day (83 g fat/day)	Mean fecal fat excretion against baseline: 0.27 ± 1.02 g/day	Chitosan did not significantly increase fecal fat excretion	Guerciolini et al. (2001)
50 women	Not reported	R, PC, DB	Two 750 mg/thrice/90 days	Before meals	Standardized; 1000 kcal/day	Body weight loss was higher in the chitosan-supplemented group (15.9 kg) than in the placebo group (10.9 kg)	Chitosan significantly decreased body weight	Zahorska-Markiewicz et al. (2002)
15 men	Not reported	OL	Two 450 mg/five times/12 days	30 min before meal	Standardized; 3084 ± 23 kcal/day (133 ± 23 g fat/day)	Mean fecal fat excretion increased from 6.1 ± 1.2 g/day to 7.2 ± 1.8 g/day	Chitosan marginally but significantly increased fecal fat excretion by 1.1 ± 1.8 g/day	Gades and Stern (2003)

Subjects	Chitosan characteristics	Study design	Dosage	Timing	Diet	Results	Conclusion	Reference
44 women	Viscosity 160 mPa s; 89.5% degree of deacetylation	R, PC, DB	Three 199.1 ± 6.7 mg/twice/56 days	After meal	Unrestricted but documented	Body weight loss was unchanged in the chitosan-supplemented group (56.7 ± 10.5 kg)	Chitosan did not decrease body weight	Bokura and Kobayashi (2003)
11 (4 women, 7 men)	Not reported	OL	1.05 g/once/56 days	Not reported	Unrestricted and undocumented; vitamin C supplemented	Fecal fat excretion increased from about 60 to 80 mg/g of chitosan	Chitosan significantly increased fecal fat excretion	Tsujikawa et al. (2003)
250 (205 women, 45 men)	β chitosan; 75.5% degree of deacetylation; 130 kDa	R, PC, DB	Four 250 mg/thrice/168 days	Before meals	Standardized low fat diet	Mean body weight loss by 0.4 kg with chitosan treatment	Chitosan decreased body weight but it is considered as clinically insignificant	Mhurchu et al. (2004)
24 (12 women, 12 men)	Not reported	OL	Two 250 mg/five times/12 days	30 min before meal	Standardized; >2400 kcal and 75 g fat/day (men); >1800 kcal and 50 g fat/day (women)	Mean fecal fat excretion increased with chitosan by 1.8 ± 2.4 g/day in men, but did not increase in women (0.0 ± 1.4 g/day)	Chitosan marginally but significantly increased fecal fat excretion in men but not in women	Gades and Stern (2005)
134 (111 women, 23 men)	Not reported	R, PC, DB	Six 500 mg/not reported/60 days	Not reported	Unrestricted but documented	Loss of body weight: Chitosan vs. control (−2.8 lb vs. +0.8 lb); Chitosan vs. placebo (−2.8 lb vs. −0.6 lb) Loss of fat mass: Chitosan vs. control (−2.6 lb vs. +0.1 lb); Chitosan vs. placebo (−2.6 lb vs. +0.6 lb)	Chitosan significantly decreased body weight and fat mass	Kaats et al. (2006)

R, randomized; PC, placebo controlled; DB, Double blind; 2PC, two-period crossover; OL, open label.

(Koide 1998). A reduction in fat-soluble vitamin absorption has been observed in rats fed with chitosan and vitamin C (Deuchi et al. 1995b). Such complication can nevertheless be circumvented via coadministration of chitosan with vitamin K. A mixed effect of chitosan on absorption profiles of fat-soluble vitamins in human subjects is observed. At high chitosan dosages, serum vitamin A, E, 25-hydroxyvitamin D, α- and β-carotene in chitosan-fed subjects cannot be statistically different from the placebo (Tapola et al. 2008).

The chelating ability of chitosan in animals has translated to depletion of iron and calcium which can have a serious implication on bone calcium and hemoglobin iron contents and their physiological functions (Xia et al. 2011). Combination effects of vitamin and mineral losses as well as reduced energy intake following fat binding and indigestion can result in growth retardation (Koide 1998; Muzzarelli 1996).

12.3.3 Allergic reaction to shellfish-derived chitosan

Almost all chitosan available in the market are derived from shellfish, namely prawn and crab. One in 50 American adults is allergic to shellfish (Sicherer et al. 2004). Antigens in the flesh of shellfish trigger the release of immunoglobulin E antibodies (IgE), which in turn cause allergic reaction. The latest review suggests that chitin or chitosan is not allergenic (Muzzarelli 2010). Two pilot studies on shrimp-derived glucosamine supplement (Gray et al. 2004) and chitosan bandage (Waibel et al. 2011) have found no allergic reaction on human subjects. However, both authors suggest that a similar test shall be run on a larger population as commercial chitosan may contain protein which is not completely removed during its production.

12.4 Chitosan modification

From animal and clinical trials, it is noted that the body weight reduction property of chitosan requires further improvement in order to achieve its clinical significance. The chitosan has been subjected to chemical modification and formulation with additives or into different dosage forms. Using 3T3-L1 adipocyte cell line, sulfated (Karadeniz et al. 2011), phosphorylated (Kong et al. 2010), carboxymethylated (Kong et al. 2011), and amine methylated (Ozhan Aytekin et al. 2012) chitosan/chitin/glucosamine derivatives are found to be able to suppress adipocyte differentiation and adipogenesis through upregulation of adenosine monophosphate-activated protein kinase pathway (**Figure 12.3**). O-carboxymethyl chitosan and N-[(2-hydroxy-3-N,N-dimethylhexadecyl ammonium) propyl] chitosan chloride can elevate hepatic lipase and lecithin cholesterol acyltransferase to aggravate the metabolism of fat (Liu et al. 2011). The conversion of chitosan into salts of oleic, linoleic, palmitic, stearic, and linolenic acids renders it able to bind to the fat in the gastrointestinal tract via hydrophobic interaction (Muzzarelli 1996).

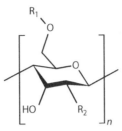

	R_1	R_2	n
SGlc	SO_3^-	NH_2	1
PGlc	PO_3^{2-}	NH_2	1
TMC	H	$N(CH_3)_3$	>1
CM-chitin	CH_2COO^-	$NHCOOCH_3$	>1
N,O-sulfated	SO_3^-	$NHSO_3^-$	>1
chitosan			

Figure 12.3 *Chemical structures of sulfated glucosamine (SGlc), phosphorylated glucosamine (PGlc), N-trimethyl chitosan (TMC), carboxymethyl chitin (CM-chitin), and N,O-sulfated chitosan.*

The chitosan fatty acid salts are not hydrolyzable in the stomach. They grow in size through binding to fat as they travel through the gastrointestinal tract and promote fat excretion.

Chitosan, in combination with ascorbic acid, is frequently reported to exert a higher level of fat binding and indigestion activity in Sprague–Dawley rats (Deuchi et al. 1995a; Kanauchi et al. 1995), guinea pigs (Jun et al. 2010), and human (Woodgate and Conquer 2003). The modes of mechanism of ascorbic acid include (Deuchi et al. 1995b; Kanauchi et al. 1995; Muzzarelli 1985; Muzzarelli et al. 1984):

1. Reducing the viscosity of chitosan in stomach and improving its mixing and binding with fat. The viscosity lowering effect of ascorbic acid may be ascribed to Schiff base reaction of dehydroascorbic acid with chitosan.
2. Enhancing the flexibility of chitosan gel to minimize brittleness-induced leakage of trapped fat.
3. Protecting chitosan from degradation before its passage to the distal end of the gastrointestinal tract by undergoing oxidation and depriving oxygen available for chitosan in stomach. This could probably aid to maintain the fat-binder action of chitosan in the gastrointestinal tract.

Formulating chitosan in the form of ionotropically cross-linked nanoparticles in the size range between 500 and 1000 nm has likewise improved fat binding and provided a greater decrease in body weight of rats than the unmodified chitosan (Zhang et al. 2011)

12.5 Future of chitosan as dietary nutraceuticals

Oral chitosan as dietary nutraceuticals has been approved in Japan, Italy, and Finland (Muzzarelli 1996). The U.S. Food and Drug Administration only approves the application of chitosan as hemostatic dressing (Wedmore et al. 2006) and not for dietary application due to limited scientific evidence on its body weight reduction effectiveness. The European Food Safety Authority considers chitosan to have no significant effect in reducing body weight (Agostoni et al. 2011). The physicochemical and biological relationship of chitosan on body weight reduction has yet to be fully established. The unconvincing scientific evidence of chitosan in fighting obesity may have caused limited approval and recognition. For the purpose of commercialization and consumer application, it is imperative to conduct structured animal and clinical trials on chitosan formulation and preferably chitosan or chitosan derivative alone through the standpoint of pharmaceutics and polymer science. The best grades of chitosan products shall be identified on the basis of scientific fundamentals and further tested for potential adverse reactions with reference to dosage regimen design.

References

Abe, K., Tomita, S., Matsuda, Y., Terajima, K., and Kanayama, T. 1974. Treatment of oil spills using basic polymer oil-ball forming agents, Patent JP49089687A.

Agostoni, C., Bresson, J.-L., Fairweather-Tait, S., Flynn, A., Golly, I., Korhonen, H., Lagiou, P., Lovik, M., Marchelli, R., Martin, A., Moseley, B., Neuhauser-Berthold, M., Przyrembel, H., Salminen, S., Sanz, Y., Strain, S., Strobel, S., Tetens, I., Tome, D., van, L.H., Verhagen, H., Heinonen, M., de, G.K., Harrold, J., Hansen, M., Kristensen, M., and Sjodin, A. 2011. Scientific opinion on the substantiation of health claims related to chitosan and reduction in body weight (ID 679, 1499), maintenance of normal blood LDL-cholesterol concentrations (ID 4663), reduction of intestinal transit time (ID 4664) and reduction of inflammation (ID 1985) pursuant to Article 13(1) of Regulation (EC) No 1924/2006. *EFSA Journal* 9, 6: 2214, 1–21.

Aranaz, I., Mengibar, M., Harris, R., Panos, I., Miralles, B., Acosta, N., Galed, G., and Heras, A. 2009. Functional characterization of chitin and chitosan. *Current Chemical Biology* 3: 203–230.

Baldrick, P. 2010. The safety of chitosan as a pharmaceutical excipient. *Regulatory Toxicology and Pharmacology* 56 (3): 290–299.

Biesemeier, C.K. and Cummings, S.M. 2008. Ethic opinion: Weight loss products and medications. *Journal of the American Dietetic Association* 108: 2109–2113.

Bokura, H. and Kobayashi, S. 2003. Chitosan decreases total cholesterol in women: A randomized, double-blind, placebo-controlled trial. *European Journal of Clinical Nutrition* 57 (5): 721–725.

Bondiolotti, G., Cornelli, U., Strabbioli, R.S., Frega, N.G., Cornelli, M., and Bareggi, S.R. 2011. Effect of a polyglucosamine on the body weight of male rats: Mechanisms of action. *Food Chemistry* 124: 978–982.

Cho, Y.-S., Lee, S.-H., Kim, S.-K., Ahn, C.-B., and Je, J.-Y. 2011. Aminoethyl-chitosan inhibits LPS-induced inflammatory mediators, iNOS and COX-2 expression in RAW264.7 mouse macrophages. *Process Biochemistry* 46 (2): 465–470.

Colombo, P. and Sciutto, A.M. 1996. Nutritional aspects of chitosan employment in hypocaloric diet. *Acta Toxicologica et Therapeutica* 17 (4): 287–302.

Czechowska-Biskup, R., Rokita, B., Ulanski, P., and Rosiak, J.M. 2005. Radiation-induced and sonochemical degradation of chitosan as a way to increase its fat-binding capacity. *Nuclear Instruments and Methods in Physics Research Section B: Beam Interactions with Materials and Atoms* 236 (1–4): 383–390.

Deuchi, K., Kanauchi, O., Imasato, Y., and Kobayashi, E. 1994. Decreasing effect of chitosan on the apparent fat digestibility by rats fed on a high-fat diet. *Biosciences Biotechnology, and Biochemistry* 58 (9): 1613–1616.

Deuchi, K., Kanauchi, O., Imasato, Y., and Kobayashi, E. 1995a. Effect of the viscosity or deacetylation degree of chitosan on fecal fat excreted from rats fed on a high-fat diet. *Bioscience, Biotechnology, and Biochemistry* 59 (5): 781–785.

Deuchi, K., Kanauchi, O., Shizukuishi, M., and Kobayashi, E. 1995b. Continuous and massive intake of chitosan affects mineral and fat-soluble vitamin status in rats fed on a high-fat diet. *Bioscience, Biotechnology, and Biochemistry* 59 (7): 1211–1216.

DeWald, T., Khaodhiar, L., Donahue, M.P., and Blackburn, G. 2006. Pharmacological and surgical treatments for obesity. *American Heart Journal* 151 (3): 604–624.

Egras, A.M., Hamilton, W.R., Lenz, T.L., and Monaghan, M.S. 2011. An evidence-based review of fat modifying supplemental weight loss products. *Journal of Obesity* 2011, 297315: 1–8.

Feng, T., Du, Y., Li, J., Hu, Y., and Kennedy, J.F. 2008. Enhancement of antioxidant activity of chitosan by irradiation. *Carbohydrate Polymers* 73 (1): 126–132.

Friedman, M. and Juneja, V.K. 2010. Review of antimicrobial and antioxidative activities of chitosans in food. *Journal of Food Protection* 73 (9): 1737–1761.

Furda, I. 1980. Nonabsorbable lipid binder, Google Patents.

Gades, M.D. and Stern, J.S. 2003. Chitosan supplementation and fecal fat excretion in men. *Obesity* 11, 5: 683–688.

Gades, M.D. and Stern, J.S. 2005. Chitosan supplementation and fat absorption in men and women. *Journal of the American Dietetic Association* 105 (1): 72–77.

Gray, H.C., Hutcheson, P.S., and Slavin, R.G. 2004. Is glucosamine safe in patients with seafood allergy? *Journal of Allergy and Clinical Immunology* 114 (2): 459–460.

Guerciolini, R., Radu-Radulescu, L., Boldrin, M., Dallas, J., and Moore, R. 2001. Comparative evaluation of fecal fat excretion induced by orlistat and chitosan. *Obesity* 9 (6): 364–367.

Hamdine, M., Heuzey, M.-C., and Bégin, A. 2005. Effect of organic and inorganic acids on concentrated chitosan solutions and gels. *International Journal of Biological Macromolecules* 37 (3): 134–142.

Hein, S., Ng, C.H., Stevens, W.F., and Wang, K. 2008. Selection of a practical assay for the determination of the entire range of acetyl content in chitin and chitosan: UV spectrophotometry with phosphoric acid as solvent. *Journal of Biomedical Materials Research—Part B Applied Biomaterials* 86 (2): 558–568.

Helander, I.M., Nurmiaho-Lassila, E.L., Ahvenainen, R., Rhoades, J., and Roller, S. 2001. Chitosan disrupts the barrier properties of the outer membrane of Gram-negative bacteria. *International Journal of Food Microbiology* 71 (2–3): 235–244.

Helgason, T., Gislason, J., McClements, D.J., Kristbergsson, K., and Weiss, J. 2009. Influence of molecular character of chitosan on the adsorption of chitosan to oil droplet interfaces in an in vitro digestion model. *Food Hydrocolloids* 23 (8): 2243–2253.

Ho, S., Tai, E., Eng, P., Tan, C., and Fok, A. 2001. In the absence of dietary surveillance, chitosan does not reduce plasma lipids or obesity in hypercholesterolaemic obese Asian subjects. *Singapore Medical Journal* 42 (1): 6–10.

Ito, M., Ban, A., and Ishihara, M. 2000. Anti-ulcer effects of chitin and chitosan, healthy foods, in rats. *The Japanese Journal of Pharmacology* 82 (3): 218–225.

Jun, S.C., Jung, E.Y., Kang, D.H., Kim, J.M., Chang, U.J., and Suh, H.J. 2010. Vitamin C increases the fecal fat excretion by chitosan in guinea-pigs, thereby reducing body weight gain. *Phytotherapy Research* 24 (8): 1234–1241.

Kaats, G.R., Michalek, J.E., and Preuss, H.G. 2006. Evaluating efficacy of a chitosan product using a double-blinded, placebo-controlled protocol. *Journal of the American College of Nutrition* 25 (5): 389.

Kanauchi, O., Deuchi, K., Imasato, Y., Shizukuishi, M., and Kobayashi, E. 1995. Mechanism for the inhibition of fat digestion by chitosan and for the synergistic effect of ascorbate. *Bioscience, Biotechnology, and Biochemistry* 59 (5): 786–790.

Karadeniz, F., Karagozlu, M.Z., Pyun, S.-Y., and Kim, S.-K. 2011. Sulfation of chitosan oligomers enhances their anti-adipogenic effect in 3T3-L1 adipocytes. *Carbohydrate Polymers* 86 (2): 666–671.

Kean, T. and Thanou, M. 2010. Biodegradation, biodistribution and toxicity of chitosan. *Advanced Drug Delivery Reviews* 62 (1): 3–11.

Khor, E. and Lim, L.Y. 2003. Implantable applications of chitin and chitosan. *Biomaterials* 24 (13): 2339–2349.

Kim, S.-K. and Rajapakse, N. 2005. Enzymatic production and biological activities of chitosan oligosaccharides (COS): A review. *Carbohydrate Polymers* 62 (4): 357–368.

Kim, I.-Y., Seo, S.-J., Moon, H.-S., Yoo, M.-K., Park, I.-Y., Kim, B.-C., and Cho, C.-S. 2007. Chitosan and its derivatives for tissue engineering applications. *Biotechnology Advances* 26 (1): 1–21.

Koide, S.S. 1998. Chitin-chitosan: Properties, benefits and risks. *Nutrition Research* 18 (6): 1091–1101.

Kong, C.-S., Kim, J.-A., Bak, S.-S., Byun, H.-G., and Kim, S.-K. 2011. Anti-obesity effect of carboxymethyl chitin by AMPK and aquaporin-7 pathways in 3T3-L1 adipocytes. *The Journal of Nutritional Biochemistry* 22 (3): 276–281.

Kong, C.-S., Kim, J.-A., Eom, T.-K., and Kim, S.-K. 2010. Phosphorylated glucosamine inhibits adipogenesis in 3T3-L1 adipocytes. *The Journal of Nutritional Biochemistry* 21 (5): 438–443.

Li, Q.X., Song, B.Z., Yang, Z.Q., and Fan, H.L. 2006. Electrolytic conductivity behaviors and solution conformations of chitosan in different acid solutions. *Carbohydrate Polymers* 63 (2): 272–282.

Li, R., Zhou, Y., Wang, L., and Ren, G. 2011. Low-molecular-weight-chitosan ameliorates cadmium-induced toxicity in the freshwater crab, *Sinopotamon yangtsekiense*. *Ecotoxicology and Environmental Safety* 74 (5): 1164–1170.

Liu, X., Yang, F., Song, T., Zeng, A., Wang, Q., Sun, Z., and Shen, J. 2011. Effects of chitosan, O-carboxymethyl chitosan and N-[(2-hydroxy-3-N,N-dimethylhexadecyl ammonium) propyl] chitosan chloride on lipid metabolism enzymes and low-density-lipoprotein receptor in a murine diet-induced obesity. *Carbohydrate Polymers* 85: 334–340.

Lois, K. and Kumar, S. 2008. Pharmacotherapy of obesity. *Therapy* 5 (2): 223–235.

Meng, X., Yang, L., Kennedy, J.F., and Tian, S. 2010. Effects of chitosan and oligochitosan on growth of two fungal pathogens and physiological properties in pear fruit. *Carbohydrate Polymers* 81 (1): 70–75.

Mhurchu, C.N., Poppitt, S.D., McGill, A.T., Leahy, F.E., Bennett, D.A., Lin, R.B., Ormrod, D., Ward, L., Strik, C., and Rodgers, A. 2004. The effect of the dietary supplement, chitosan, on body weight: A randomised controlled trial in 250 overweight and obese adults. *International Journal of Obesity* 28 (9): 1149–1156.

Minami, S., Oh-oka, M., Okamoto, Y., Miyatake, K., Matsuhashi, A., Shigemasa, Y., and Fukumoto, Y. 1996. Chitosan-inducing hemorrhagic pneumonia in dogs. *Carbohydrate Polymers* 29 (3): 241–246.

Montembault, A., Viton, C., and Domard, A. 2005. Physico-chemical studies of the gelation of chitosan in a hydroalcoholic medium. *Biomaterials* 26 (8): 933–943.

Mun, S., Decker, E., Park, Y., Weiss, J., and McClements, D. 2006. Influence of interfacial composition on in vitro digestibility of emulsified lipids: potential mechanism for chitosan's ability to inhibit fat digestion. *Food Biophysics* 1 (1): 21–29.

Muzzarelli, R.A.A. 1985. Removal of uranium from solutions and brines by a derivative of chitosan and ascorbic acid. *Carbohydrate Polymers* 5 (2): 85–89.

Muzzarelli, R.A.A. 1996. Chitosan-based dietary foods. *Carbohydrate Polymers* 29 (4): 309–316.

Muzzarelli, R. 2010. Chitins and chitosans as immunoadjuvants and non-allergenic drug carriers. *Marine Drugs* 8 (2): 292–312.

Muzzarelli, R.A.A., Orlandini, F., Pacetti, D., Boselli, E., Frega, N.G., Tosi, G., and Muzzarelli, C. 2006. Chitosan taurocholate capacity to bind lipids and to undergo enzymatic hydrolysis: An in vitro model. *Carbohydrate Polymers* 66 (3): 363–371.

Muzzarelli, R.A.A., Tanfani, F., and Emanuelli, M. 1984. Chelating derivatives of chitosan obtained by reaction with ascorbic acid. *Carbohydrate Polymers* 4 (2): 137–151.

Ozhan Aytekin, A., Morimura, S., and Kida, K. 2012. Physiological activities of chitosan and N-trimethyl chitosan chloride in U937 and 3T3-L1 cells. *Polymers for Advanced Technologies* 23 (2): 228–235.

Park, P.J., Je, J.Y., and Kim, S.K. 2003. Free radical scavenging activity of chitooligosaccharides by electron spin resonance spectrometry. *Journal of Agricultural and Food Chemistry* 51 (16): 4624–4627.

Park, J. and Kim, D. 2010. Effect of polymer solution concentration on the swelling and mechanical properties of glycol chitosan superporous hydrogels. *Journal of Applied Polymer Science* 115 (6): 3434–3441.

Pillai, C.K.S., Paul, W., and Sharma, C.P. 2009. Chitin and chitosan polymers: Chemistry, solubility and fiber formation. *Progress in Polymer Science* 34 (7): 641–678.

Pittler, M.H., Abbot, N.C., Harkness, E.F., and Ernst, E. 1999. Randomized, double-blind trial of chitosan for body weight reduction. *European Journal of Clinical Nutrition* 53 (5): 379–381.

Prajapati, B.G. 2009. Chitosan a marine medical polymer and its lipid lowering capacity. *The Internet Journal of Health* 9: 1–7.

Rabea, E.I., Badawy, M.E.T., Stevens, C.V., Smagghe, G., and Steurbaut, W. 2003. Chitosan as antimicrobial agent: Applications and mode of action. *Biomacromolecules* 4 (6): 1457–1465.

Rao, M.S. and Stevens, W.F. 2005. Chitin production by *Lactobacillus* fermentation of shrimp biowaste in a drum reactor and its chemical conversion to chitosan. *Journal of Chemical Technology and Biotechnology* 80 (9): 1080–1087.

Rasmussen, R. and Morrissey, M. 2007, Chitin and chitosan, in *Marine Nutraceuticals and Functional Foods*, Barrow, C., Shahidi, F. (Eds), CRC Press, Boca Raton, FL, pp. 155–182.

Roberts, G.A.F. 2008. Thirty years of progress in chitin and chitosan. *Progress on Chemistry and Application of Chitin and Its Derivatives* 13: 7–15.

Sicherer, S.H., Muñoz-Furlong, A., and Sampson, H.A. 2004. Prevalence of seafood allergy in the United States determined by a random telephone survey. *Journal of Allergy and Clinical Immunology* 114 (1): 159–165.

Sudarshan, N.R., Hoover, D.G., and Knorr, D. 1992. Antibacterial action of chitosan. *Food Biotechnology* 6 (3): 257–272.

Sun, T., Xie, W., and Xu, P. 2003. Antioxidant activity of graft chitosan derivatives. *Macromolecular Bioscience* 3 (6): 320–323.

Sundfeld, E., Krochta, J.M., and Richardson, T. 1994. Aqueous process to remove cholesterol from food products, Google Patents.

Synowiecki, J. and Al-Khateeb, N.A.A.Q. 2000. The recovery of protein hydrolysate during enzymatic isolation of chitin from shrimp *Crangon crangon* processing discards. *Food Chemistry* 68 (2): 147–152.

Tapola, N.S., Lyyra, M.L., Kolehmainen, R.M., Sarkkinen, E.S., and Schauss, A.G. 2008. Safety aspects and cholesterol-lowering efficacy of chitosan tablets. *Journal of the American College of Nutrition* 27 (1): 22–30.

Tsujikawa, T., Kanauchi, O., Andoh, A., Saotome, T., Sasaki, M., Fujiyama, Y., and Bamba, T. 2003. Supplement of a chitosan and ascorbic acid mixture for Crohn's disease: A pilot study. *Nutrition* 19 (2): 137–139.

Veneroni, G., Veneroni, F., Contos, S., Tripodi, S., De, B.M., Guarino, C., and Marletta, M. 1996. Effect of a new chitosan dietary integrator and hypocaloric diet on hyperlipidemia and overweight in obese patients. *Acta Toxicologica et Therapeutica* 17 (4): 53–70.

Waibel, K.H.M.C.U.S.A., Haney, B., Moore, M., Whisman, B., and Gomez, R. 2011. Safety of chitosan bandages in shellfish allergic patients. *Military Medicine* 176 (10): 1153–1156.

Wanders, A.J., van den Borne, J.J.G.C., de Graaf, C., Hulshof, T., Jonathan, M.C., Kristensen, M., Mars, M., Schols, H.A., and Feskens, E.J.M. 2011. Effects of dietary fibre on subjective appetite, energy intake and body weight: A systematic review of randomized controlled trials. *Obesity Reviews* 12 (9): 724–739.

Wedmore, I., McManus, J.G., Pusateri, A.E., and Holcomb, J.B. 2006. A special report on the chitosan-based hemostatic dressing: Experience in current combat operations. *The Journal of Trauma* 60 (3): 655–658. DOI: 10.1097/01.ta.0000199392.91772.44.

Wilde, P.J. and Chu, B.S. 2011. Interfacial and colloidal aspects of lipid digestion. *Advances in Colloid and Interface Science* 165: 14–22.

Witkamp, R. 2011. Current and future drug targets in weight management. *Pharmaceutical Research* 28 (8): 1792–1818.

Wong, T.W. 2009. Chitosan and its use in design of insulin delivery system. *Recent Patents on Drug Delivery and Formulation* 3: 8–25.

Wong, T.W. and Harjoh, N. In Press. Sustained-release alginate-chitosan pellets prepared by melt pelletization technique. *Drug Development and Industrial Pharmacy*, DOI: 10.3109/03639045.2011.653364.

Wong, T.W. and Sumiran, N. 2008. Drug release property of chitosan-pectinate beads and its changes under the influence of microwave. *European Journal of Pharmaceutics and Biopharmaceutics* 69: 176–188.

Woodgate, D.E. and Conquer, J.A. 2003. Effects of a stimulant-free dietary supplement on body weight and fat loss in obese adults: A six-week exploratory study. *Current Therapeutic Research* 64 (4): 248–262.

Wydro, P., Krajewska, B., and Hąc-Wydro, K. 2007. Chitosan as a lipid binder: A Langmuir monolayer study of chitosan–lipid interactions. *Biomacromolecules* 8 (8): 2611–2617.

Xia, W., Liu, P., Zhang, J., and Chen, J. 2011. Biological activities of chitosan and chitooligosaccharides. *Food Hydrocolloids* 25: 170–179.

Youn, D.K., No, H.K., and Prinyawiwatkul, W. 2007. Physical characteristics of decolorized chitosan as affected by sun drying during chitosan preparation. *Carbohydrate Polymers* 69 (4): 707–712.

Yun, J.W. 2010. Possible anti-obesity therapeutics from nature—A review. *Phytochemistry* 71: 1625–1641.

Zahorska-Markiewicz, B., Krotkiewski, M., Olszanecka-Glinianowicz, M., and Zurakowski, A. 2002. Effect of chitosan in complex management of obesity. *Polski merkuriusz lekarski: organ Polskiego Towarzystwa Lekarskiego* 13 (74): 129.

Zakaria, Z. and Wong, T.W. 2009. Chitosan spheroids with microwave modulated drug release. *Progress in Electromagnetics Research* 99: 355–382.

Zhang, H. and Neau, S.H. 2002. In vitro degradation of chitosan by bacterial enzymes from rat cecal and colonic contents. *Biomaterials* 23 (13): 2761–2766.

Zhang, H.-L., Tao, Y., Guo, J., Hu, Y.-M., and Su, Z.-Q. 2011. Hypolipidemic effects of chitosan nanoparticles in hyperlipidemia rats induced by high fat diet. *International Immunopharmacology* 11 (4): 457–461.

Zhao, Y., Park, R.-D., and Muzzarelli, R.A.A. 2010. Chitin deacetylases: Properties and applications. *Marine Drugs* 8 (1): 24–46.

Zhou, K., Xia, W., Zhang, C., and Yu, L. 2006. In vitro binding of bile acids and triglycerides by selected chitosan preparations and their physico-chemical properties. *LWT—Food Science and Technology* 39 (10): 1087–1092.

13

Prospects of Indonesian Uncultivated Macroalgae for Anticancer Nutraceuticals

Hari Eko Irianto and Ariyanti Suhita Dewi

Contents

13.1 Introduction

Within the last decade, the Indonesian GDP has raised 4.6 times, according to International Monetary Fund (IMF). The improvement in the country's welfare has been followed by the change of living standards that leads to the shift of healthcare needs. Nowadays, awareness on maintaining well-being is raised, resulting in the consumption of natural products as supplement rather than chemically synthesized drugs by increasing number of people. Nutraceuticals have gained considerable interests due to their presumed safety, in addition to potential nutritional and therapeutic effects. The terminology of "nutraceuticals" refers to substances that are intended to *prevent* diseases. Nutraceuticals may be interchangeable with "functional foods" when they are intended to *prevent* or *treat* disease(s) or disorder(s) other than anemia (Rajasekaran et al., 2008). Lordan et al. (2011) added that functional foods are designed to complement basic nutritional value from daily diet and to improve the general conditions of the body against risks of illness and diseases. In current applications, they are consumed individually or used as fortificants in food.

Although not as much as in developed countries, cancer in Indonesia has shown an increasing trend during the last decade. It is estimated that 170–190 new cases of cancer are found for every 100,000 people annually, with cervical cancer being the most common (Tjindarbumi and Mangunkusumo, 2002). In relation with cancer-preventive agents, it is important to pay attention on natural substances with abilities to lower cancer risks, such as antioxidant and anti-inflammatory agents.

Reactive oxygen species (ROS) such as hydroxyl, superoxide, and peroxyl radicals are formed in human cells by endogenous factors and exogenously result in extensive oxidative damage that in turn leads to geriatric degenerative conditions, cancer, and a wide range of other human diseases (Chandini et al., 2008). Meanwhile, chronic inflammation is associated with a high cancer risk due to its ability to induce deleterious gene mutation and posttranslational modifications of key cancer-related proteins. It is also related to immune-suppression, which is another risk factor of cancer (Rajasekaran et al., 2008). Thus, bioactive compounds with antioxidant and anti-inflammatory activities can be used as anticancer nutraceuticals.

Indonesian waters store enormous untapped resources for the development of natural drugs and nutraceuticals due to its biodiversity. Among these resources, macroalgae are possibly the most abundant biota in Indonesian coastlines. Macroalgae are classified based on the nature of their chlorophyll, cell wall chemistry, and the presence or absence of flagella. However, macroalgae are commonly characterized by their pigments, other than chlorophyll; thus they are usually classified into three algal divisions, i.e., brown algae (Phaeophyceae), green algae (Chlorophyceae), and red algae (Rhodophyceae) (Lordan et al., 2011).

Seaweed farming in Indonesia has contributed to 19,000 MT since the early 1990s with *Eucheuma* as the most dominant species (Crawford, 2002) and area as much as 21,100 ha (Firdausy and Tisdell, 1992). Seaweed production of Indonesia in 2009 has reached 2,791,688 MT (Anonymous, 2010) bringing Indonesia as the largest seaweed-producing country in the world. Six other genera of seaweed with economic importance are *Gracilaria* and *Gelidium* for the production of agar; *Laminaria* and *Sargassum* for the production of algin; and *Hypnea* species, along with *Eucheuma*, as the main source of carrageenan (Firdausy and Tisdell, 1992). Additionally, Indonesia is also one of the largest producers of *Kappaphycus* for raw materials in carrageenan industries, according to Bixler and Porse (2011). The socioeconomic study of seaweed culture indicated an internal rate return as much as 47%. With net benefit cost ratio of 7.9%, the potential of seaweed farming in Indonesia is considered as high (Firdausy and Tisdell, 1992). Since 1980s, this particular farming has played an important role in coastal area development in Indonesia as a complementary to other fishery products.

Other tropical marine macroalgae in Indonesia have not been developed mainly due to lack of demand, despite their massive supply. Exploration on the potential of these macroalgae as anticancer agents has been widely reported. Macroalgae contain polysaccharides and dietary fibers that compose its nutritional value, in addition to proteins, lipids, and minerals. They also contain micronutrients such as vitamins, polyphenols (phlorotannin), and carotenoids (Burtin, 2003). The polyphenols and carotenoids are responsible for their free radical-scavenging properties. Recently, other valuable compounds in macroalgae have been investigated for their capability to prevent, if not inhibit, cancer progression. Some of these compounds are isolated from uncultivated macroalgae wildly found in Indonesian waters, such as *Turbinaria* sp., *Ulva* sp., and *Laurencia* sp. that respectively represent brown, green, and red algae. This chapter reviews valuable compounds from those macroalgae with potential activities as anticancer nutraceuticals.

13.2 Anticancer agents from uncultivated macroalgae

13.2.1 Brown algae (*Turbinaria* sp.)

The brown color of this alga is due to the presence of xanthophyll pigments and fucoxanthin that masks other pigments, such as chlorophylls and other xanthophylls. Food reserves in brown algae are typically complex polysaccharides and higher alcohols. The principal carbohydrate reserve is laminarin. The cell walls are made of cellulose and alginic acid (Gamal et al., 2010; Januar et al., 2011; Wikanta et al., 2010).

Turbinaria is usually found in tropical marine waters and classified as Phaeophyceae in the order of Fucales and family Sargassaceae. *Turbinaria* is characterized as having radially branched, monopodially developed axes, bearing leaves, or "laterals" that have the form of firm, more or less peltate or turbinate structures. The habit of most species is a stiff, compact, cone-like aspect. Typically, these "leaf-vesicles" are stipitate, turbinate, or pyramidal, and crowned with a peltate lamina. In all of the currently known species of *Turbinaria*, the lateral is terminated by a more or less triangular or rectangular, flat or concave outer face, with possibly one or two edges of teeth, depending on the species. The proximal part of the leaf is terete in most species, but is ridged or winged in some species. Leaves in most species of *Turbinaria* have an air vesicle embedded within the peltate enlargement. Thalli are usually attached by a well-developed system of spreading hapterous branches emanating from the main axes. Short, densely branched receptacles arise on the upper sides of the stalks of the leaves (Wynne, 2002). There are four species of *Turbinaria* of Indonesian origin, namely *Turbinaria ornata*, *Turbinaria conoides*, *Turbinaria decurrens*, and *Turbinaria murayana* (IPTEK, 2012).

Sulfated polysaccharides (SPSs) from brown algae containing 20%–60% of fructose are known as fucoidans, while those with less than 10% of fructose are referred to as sulfated fucans (**Figure 13.1**). Structures of fucoidans from marine algae have alternating α-(1→3) and α-(1→4) glycosidic bond on their oligosaccharide chain with sulfate group at C-4. Variations of structures include sulfated group at C-3 and glycosidic bond at alpha (1→2)

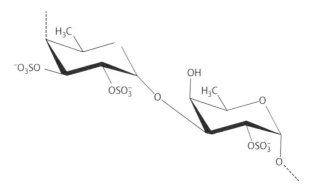

Figure 13.1 *The repeating dimeric units of fucoidan. (From Wijesekara, I. et al., Carbohydr. Polym., 84, 14, 2011.)*

(Jaswir and Monsur, 2011). Fucoidans are composed of substantial percentages of L-fucose and sulfate ester group, in addition to monosaccharides, uronic acids, acetyl groups, and protein. Although ubiquitously available in brown algae, the structures of fucoidans vary among species. Extraction methods and different parts of the plant also affect their chemical composition (Li et al., 2008).

Research conducted by Eluvakkal et al. (2010) showed that *T. ornata* produces the highest yield of fucoidans (**Figure 13.2**). However, fucoidans from *T. decurrens* contain higher levels of carbohydrate, fucose, and sulfates, compared to *T. ornata* and *T. conoides*. Fucoidans from *Turbinaria* sp. are more complex than those isolated from other brown algae. For example, a fucoidan from *T. conoides* contains 33%–34% terminals, 27%–28% linked, and 21%–22% branched in the (1→3)-linked main chain (Jiao et al., 2011).

Fucoidans and fucans have been investigated for their numerous biological activities. Studies on the aqueous extract of *T. ornata* revealed that the extract exhibited maximum scavenging activity against 2,2-diphenyl-1-picrylhydrazyl (DPPH) and nitrous oxide (NO) at 500 and 125 μg/mL, respectively (Ananthi et al., 2010, 2011). Chattopadhyay et al. (2010) compared the capacity of fucoidan to inhibit ROS in comparison with other polysaccharides, i.e., alginate and laminaran, and found that fucoidan has the highest antioxidant activity in DPPH and ferric-reducing ability of plasma (FRAP) assays. Similar study confirmed that SPS from *T. ornate* was high in total antioxidant of 47.49 mg GAE/g and a high percentage of 81.21% in DPPH radical-scavenging activity (Arivuselvan et al., 2011). Studies on the structure–activity relationships of fucans suggested that the ratio of sulfate content/fucose affected their free radical and hydroxyl radical scavenging activities (Jiao et al., 2011).

Anti-inflammatory assay of water-soluble crude polysaccharide of *T. ornata* in mice showed that oral administration of polysaccharide reduced the paw

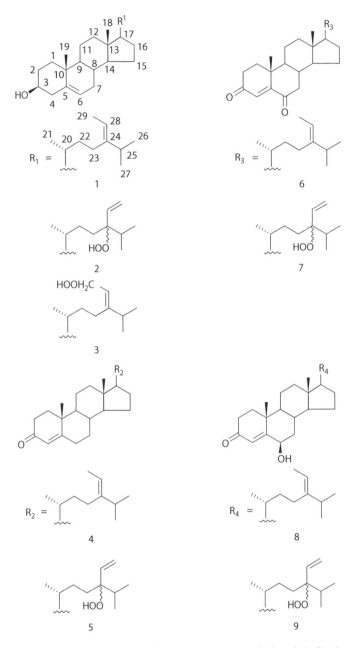

Figure 13.2 *Oxygenated fucosterols from* T. ornata. *Metabolite 4–9 displayed cytotoxicity toward various cancer cell lines. (From Sheu, J.H. et al., J. Nat. Prod., 62(2), 224, 1999.)*

edema considerably compared to carrageenan-induced rats. Additionally, polysaccharide intakes also inhibited vascular permeability in mice. Both activities are dose-dependent (Ananthi et al., 2010, 2011).

Early studies on the anticancer activity of SPSs from *Turbinaria* sp. found that a sulfated fucan like polysaccharide from *T. ornata* was cytotoxic against murine melanoma and colon cancer cells (Asari et al., 1989). A fucoidan from the same source exhibited antiproliferative effect against human non-small cell bronchopulmonary carcinoma line (NSCLC-N6) in vitro. It is suggested that they inhibit the cell growth in G1-phase by triggering terminal differentiation of cancerous cells (Deslandes et al., 2000). SPSs from *T. conoides* demonstrated that they inhibited proliferation of colon carcinoma cell line (COLO 320DM) cells at a concentration of 1 mg/mL by 40% in 24 h. The seaweed extract also exhibited anti-invasive activity in a dose-dependent manner. Additionally, SPSs inhibited vasculogenesis in chick embryo, which indicates its potential for antimetastatic drug development (Delma et al., 2008).

Although fucoidan possesses various therapeutic effects, its application is hampered by its high molecular weight and temporal variation. Since the bioactivities of fucoidan are mostly affected by its structure and sulfation content, the modification of its structure by means of hydrolysis is a logical solution to obtain applicable fucoidan-based products. Enzymatic hydrolysis is considered to be ideal for this process due to its ability to maintain the presence of sulfate content, compared to chemical hydrolysis. Study case on fucoidan from *Undaria pinnatifida* showed that its anticancer activity is considerably increased after mild depolymerization without desulfation. The enhancement of antioxidant activities of fucoidan was also found to be size-dependent (Morya et al., 2012).

Bromophenols and phenols from brown algae, also known as phlorotannins, play an important role against herbivory in tropical marine species. They are acetate–malonate derived phloroglucinol polymers and are found in concentrations as high as 25%–40% dry weight. Research on the cytotoxicity of phenols from *Turbinaria turbinate* showed that they exhibited activities against various cancer cell lines, i.e., normal canine kidney (MDCK), human laryngeal carcinoma (Hep-2), human cervical adenocarcinoma (HeLa), and human nasopharyngeal carcinoma (KB) at IC_{50} within 20–45 µg/mL (Moo-Puc et al., 2009). Methanolic extract from *T. conoides* was also found to be active in DPPH assay in dose-dependent pattern due to its high phenol content (Chandini et al., 2008; Devi et al., 2011).

Studies on the cytotoxicity of *T. decurrens* extracts collected from Binuangen, province of Banten, Indonesia showed that its crude extract is toxic against HeLa cancer cell lines with LC_{50} value of 29.9 µg/mL, whereas its *n*-hexane extract exhibited even more potent cytotoxicity with LC_{50} value of 15.1 µg/mL. Moreover, crude, methanolic, ethyl acetic, and *n*-hexane extracts of *T. decurrens* enhanced proliferation of lymphocytes (Fajarningsih et al., 2008).

Sheu et al. (1997, 1999) also studied the cytotoxic activities of extracts from *T. conoides* and *T. ornata* against murine lymphocytic leukemia (P-388), human lung adenocarcinoma epithelial (A549), and human colon cancer (HT29) cell lines at concentrations less than 5 µg/mL. It was revealed that sterols were responsible for this activity. However, the mechanism of their anticancer activity has not been fully understood. In relation to *T. ornata*, a cytotoxic new secosqualene carboxylic acid, namely turbinaric acid, was also reported from this species (Asari et al., 1989).

13.2.2 Green algae (*Ulva* sp.)

Green algae are photosynthetic eukaryotes bearing double membrane-bound plastids containing chlorophyll *a* and *b*, accessory pigments found in embryophytes (beta-carotene and xanthophylls) and a unique stellate structure linking nine pairs of microtubules in the flagellar base. Starch is stored inside the plastid, and cell walls when present are usually composed of cellulose (Lewis and McCourt, 2004). *Ulva* or sea lettuce is a rich source of carbohydrates, protein, vitamins, amino acids, trace elements, dietary fibers, and pigments, while containing low levels of lipids (El Baky et al., 2009).

Species of the genus *Ulva* Linnaeus (Ulvophyceae, Ulvales) occur in all aquatic habitats from freshwaters through brackish to fully saline environments. *Ulva* species have either a monostromatic tubular or distromatic, foliose thallus that can vary in length. It is reported that the genus includes more than 100 species (Loughnane et al., 2002). The diversity of *Ulva* in Indonesia includes *Ulva lactuca*, *Ulva fasciata*, *Ulva pertusa*, and *Ulva reticulata* (IPTEK, 2012).

The minor polysaccharide in *Ulva* consists of SPSs, known as ulvan. Similar to fucoidans, ulvans are not digested by human; hence they are considered as dietary fibers (Burtin, 2003). Ulvan represents about 8%–29% of the *Ulva* dry weight (El Baky et al., 2009). It is composed of different repeating chemical sequences mostly based on disaccharides made of rhamnose, glucuronic acid, iduronic acid, xylose, and sulfate (the mainly repeating disaccharide units are ([β-d-GlcA-(1→4)-α-L-Rhap3S]) (A) and (-α-L-IdopA-(1→4)-α-L-Rhap3S]) (B)) (**Figure 13.3**).

Ulvan from *U. pertusa* showed a significant scavenging activity against superoxide and hydroxyl radicals at a concentration less than 20 µg/mL (Qi et al., 2005). It is further reported that the antioxidant activity of ulvan is related to its sulfate content, in which higher sulfate content showed more effectiveness against hydroxyl radical, compared to natural ulvan (Qi et al., 2005; Zhang et al., 2010). Further modifications of ulvan via acetylation and benzoylation also contributed to improvement of its antioxidant activity (Qi et al., 2006).

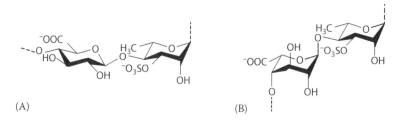

(A)   (B)

Figure 13.3 *The main repeating disaccharide units of ulvan.* (A) [→4)-b-D-Glcp-
(1→4)-a-L-Rhap3S-(1→]n. (B) [→4)-a-L-Idop-(1→4)-a-L-Rhap3S-(1→]n *(From
Jiao, G. et al., Mar. Drugs, 9, 196, 2011.)*

Moreover, the scavenging effects of ulvan from *U. pertusa* against hydroxyl
radicals were concentration-dependent, and did not exhibit bad effects at
high concentration as those from *Laminaria japonica* and *Bryopsis pulmosa*
(Zhang et al., 2010). Interestingly, ulvan from *U. lactuca* showed a moderate
antioxidant activity in DPPH and 2,2′-azino-bis(3-ethylbenzothiazoline-6-sul-
fonic acid) (ABTS) radical assays (El Baky et al., 2009).

Although having moderate scavenging properties, ulvan from *U. lactuca* has
been proven to inhibit cell proliferation of breast adenocarcinoma (MCF-7) and
hepatocellular carcinoma (HepG2) cell lines with an IC_{50} within 0.5–10 μg/mL
(El Baky et al., 2009). Further study on the structure relationship of ulvan in
relation with its anticancer activity toward cancerous colonic epithelial cell
lines suggested the influence of sulfate content and molecular weight. The
reduction of uronic acid also enhanced the ratio of sulfate residues accessible
to surface cell receptors (Kaeffer et al., 1999).

As mentioned previously, the genus of *Ulva* is a rich source of chlorophylls
and other pigments that are responsible for its green color. Many studies
have reported the relation of these pigments with their scavenging activities.
Studies conducted by El Baky et al. (2008) showed that the crude organic
extracts of *U. lactuca* exhibited significant antioxidant activities due to the
presence of carotenoids, phenolics, and chlorophylls. Moreover, Meenakshi
and Gnanambigai (2009) reported that phenolics have greater radical-scavenging activities than flavonoids in the methanolic extract of *U. lactuca*.

Only a few reports have been recorded on the anti-inflammatory and cytotoxic
activities from the extract of *Ulva*. Kim et al. (2009) studied the anti-inflammatory effects of *U. pertusa* extract on the production of NO and prostaglandin E2
(PGE2) in murine macrophage RAW 264.7 cells. It was found that the dichloromethane extract of *U. pertusa* inhibited NO production and PGE2 within 100
and 50 μg/mL, respectively, in a dose-dependent manner. On the other hand,
the methanolic extract of *U. reticulata* only showed antiphlogistic effect, but
not antiproliferative effect against acute inflammation in carrageenan-induced
paw edema in mice (Hong et al., 2011).

Preliminary studies on the extract of *U. fasciata* demonstrated its cytotoxicity toward brine shrimp lethality test (BSLT) and HeLa cancer cell line. It is reported that the crude organic extract of *U. fasciata* displayed moderate cytotoxicity in BSLT assay (Ayesha et al., 2010; Selvin and Lipton, 2004). On the other hand, its ethyl acetic extract was active against HeLa cancer cell line with LC_{50} value of 19 µg/mL (Wikanta et al., 2010). Water-soluble fraction of methanolic extract from *U. lactuca* was also found to inhibit 50% of the human leukemia cell line (U937) in growth at a concentration of 140 µg/mL (Lee et al., 2004).

13.2.3 Red algae (*Laurencia* sp.)

The genus *Laurencia* was established by J. V. Lamouroux (1813) with the original recognition of eight species. *Laurencia obtuse* (Hudson) J. V. Lamouroux is the type species (Senties et al., 2010). *L. obstusa* is described as "globose tufts of brittle, cartilaginous, narrow, cylindrical, reddish brown to yellowish red fronds, 150 mm long, from small discoid base. Axis simple, branches patent, often opposite, spirally arranged, shorter toward apex giving irregularly pyramidal outline. Ultimate ramuli very short, truncate, with special refringent inclusions in cortical cells" (Braune and Guiry, 2011). Macroalgae from the genus *Laurencia* are one of the most investigated among all marine genera because they are extremely widespread all over the world in any latitude and they contain prolific secondary metabolite content (Abdel-Mageed et al., 2010). *Laurencia obtusa*, *Laurencia poitei*, *Laurencia nidifica*, *Laurencia intricate*, and *Laurencia elata* are commonly distributed along Indonesia coastal (IPTEK, 2012).

Similar to their counterparts, red algae (Rhodophyceae) also contain cell wall polysaccharides, commonly known as galactans, that consist of carrageenans and agarans. The structural feature of galactans is composed of a linear backbone built up of alternating 3-linked β-D-galactopyranose and 4-linked α-galactopyranose residues. The β-galactose residues always belong to D-series, whereas the α-galactose are D in carrageenans and L in agarans (Usov, 1998) (**Figure 13.4**).

The use of a polysaccharide comprising galactans and fucans as an anti-inflammatory agent was reported by Thibodeau et al. (2009). It was found that the combination of fucans (MW between 0.1 and 100 kDa) from brown algae and galactans (MW > 100 kDa) from red algae between 2.5/1 (w/w) and 40/1 (w/w) inhibited the release of one or more interleukin-8 (IL-8), PGE2, and vascular endothelial growth factor (VEGF) by a cell activated during an inflammatory process. Additionally, polysaccharide from *Laurencia microcladia* was also reported to exhibit moderate antimitotic activity on *Lytechinus variegatus* larvae and angiogenesis activity in chick embryo assay (Maraschin et al., 2000).

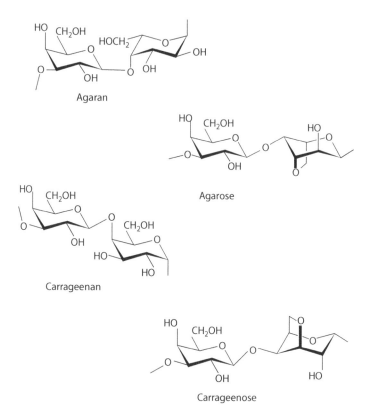

Agaran

Agarose

Carrageenan

Carrageenose

Figure 13.4 *The repeating disaccharide units of galactans. (From Usov, A., Food Hydrocoll. 12(3), 301, 1998.)*

The secondary metabolites of *Laurencia* (Ceramiales, Rhodomelaceae) are also known to exhibit various biological activities. Over 300 compounds, such as sesquiterpenes, diterpenes, triterpenes, C_{15} acetogenins, acetylenes, and chamigrenes, have been isolated from this macroalgae all over the world since the 1960s (Demirel et al., 2011; Dias and Urban, 2011). The widespread and prolific production of secondary metabolites observed in *Laurencia* spp. may be interpreted as an ecological adaptive response (Pereira et al., 2003). The stereochemicals of sesquiterpenes from this genus of red algae may even be used as a distinctive trait that reflects phylogeny (Guella and Pietra, 1997).

Anggadiredja et al. (1997) reported that the n-hexane extract of *L. obtusa* from Seribu Islands, Indonesia, demonstrated higher antioxidant properties than that of methanol and diethyl ether. It also showed better activity than the extracts from *Sargassum polycystum* (brown algae). Likewise, the chloroform extract of *L. obtusa* was found to exhibit antioxidant properties in DPPH and ATBS assays, whereas its methanolic extract and essential oil were not (Demirel et al., 2011; Nahas et al., 2007). The dichloromethane extract of *Laurencia papillosa* also demonstrated the same occurrence, compared to its ethanolic extract (Shanab, 2007). Therefore, it is suggested that its radical-scavenging activity is not correlated with the phenolic content (Demirel et al., 2011). Rather, from the study conducted by He et al. (2005) and Liang et al. (2007), the supplementation of terpene from *L. triscticha* (LTE) improves antioxidation and decreases DNA damage in mice, and is considered safe to be taken orally. In a similar case, LTE also showed preventive effect against oxidant injury after alcohol exposure, which presumably associated with increased activity of antioxidant enzymes, reduced lipid peroxidation and increased heme-oxygenase (HO-1) in mice liver cells (Liang et al., 2009).

Several studies have reported the anti-inflammatory activity of secondary metabolites from *Laurencia* sp. *Laurencia okamurae* was found to be a potent inhibitor for a series of proinflammatory mediators, such as NO, PGE2, interlekuin-6 (IL-6), and tumor necrosis factor (TNF-α) at a concentration of 100 µg/mL. This activity was found correlated with its phlorotannin content (Yang et al., 2010). Meanwhile, the administration of 1 mg/kg of neorogioltriol, a tricyclic brominated diterpenes from *Laurencia glandulifera*, significantly reduced carrageenan-induced rat edema. In addition, in an in vitro assay, the addition of 62.5 µM of neorogioltriol decreased the luciferase activity in LPS-stimulated RAW 264.7 cells. It also demonstrated inhibition toward the production of NO and the expression of cyclooxygenase-2 (COX-2) (Chatter et al., 2011).

Moreover, numerous studies have been conducted on the cytotoxic activity of *Laurencia* extracts. Preliminary study on the cytotoxicity of *Laurencia brandenii* extract revealed that it showed a LD_{50} value of 93 µg/mL in brine shrimp lethality test (BSLT). It is presumed that fatty acids were responsible for this biological activity (Manilal et al., 2009). Another preliminary study on the cytotoxicity of extracts of *Laurencia catarinensis*, *Laurencia dendroidea*, and *Laurencia translucida* toward human uterine sarcoma (MES-SA) cell line showed that their n-hexane extracts exhibited better activity compared to their methanolic extracts (Stein et al., 2011).

Among some of the *Laurencia* sp., notable bioactive compounds are terpenes and acetogenins. Polyether triterpenes from *Laurencia viridis* displayed anti-tumor activity against lymphoid neoplasm (P-388), human lung carcinoma (A-549), human colon carcinoma (HT-29), and human melanoma (MEL-28) cell lines. This activity was found to relate with the arrangement of flexible chains in the polyether structure (Fernandez et al., 1998). Further isolation of these compounds showed even higher activity in a panel of cancer cell lines, i.e., breast cancer, human T-cell acute leukemia (Jurkat), human multiple myeloma (MM144), human cervical carcinoma (HeLa), and human Ewing's sarcoma (CADO-ES1). These bioactive compounds were presumably inducing apoptosis in all of the cancer cell lines tested (Pacheco et al., 2011). Isolation of triterpenes with a squalene carbon skeleton from the same species

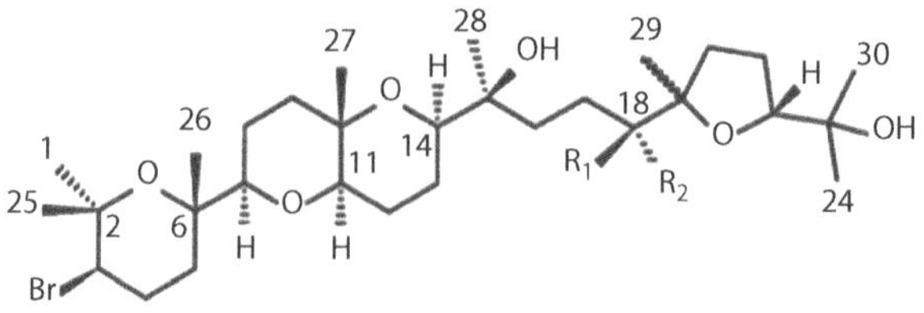

1 Thyrsiferol; $R_1 = OH$, $R_2 = H$, C–29 = αCH_3
2 Venustatriol; $R_1 = H$, $R_2 = OH$, C–29 = βCH_3

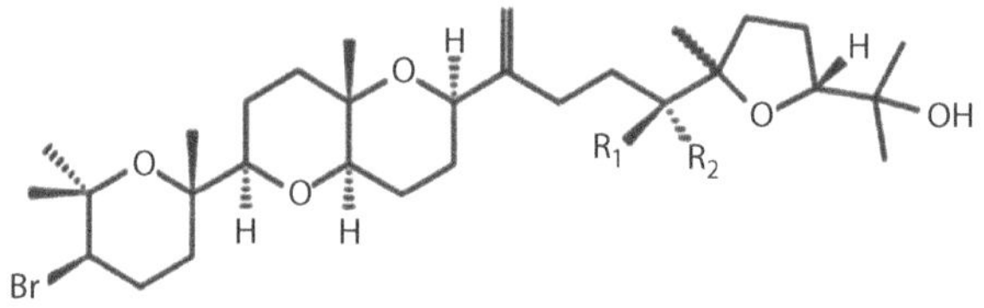

3 Dehydrothyrsiferol; $R_1 = OH$, $R_2 = H$, C–29 = αCH_3
4 Dehydrovenustatriol; $R_1 = H$, $R_2 = OH$, C–29 = βCH_3

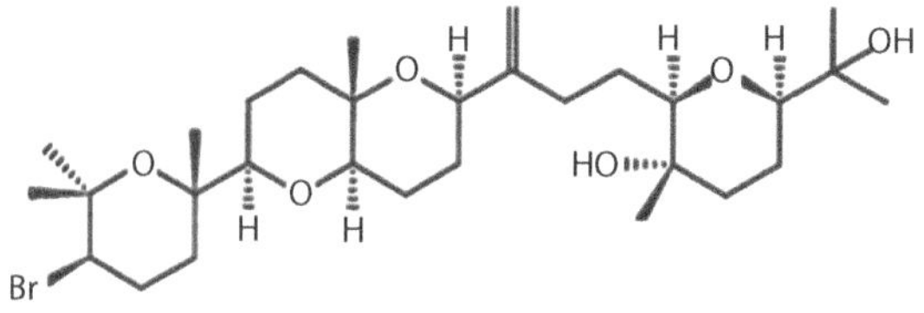

5 Isodehydrothyrsiferol

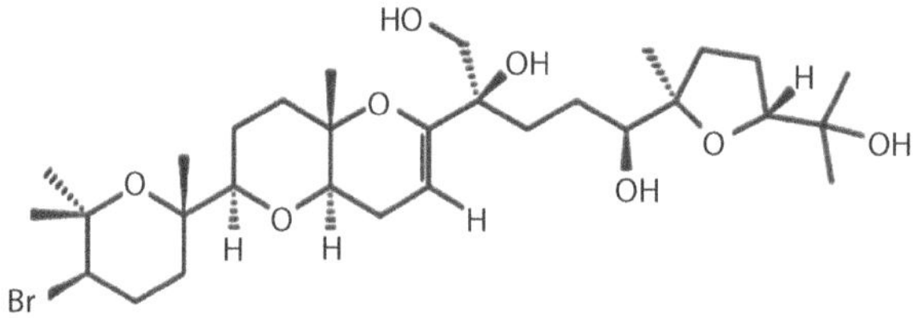

6 Thyrsenol B

Figure 13.5 *Several cytotoxic polyethers from* L. viridis. *(From Manriquez, C.P. et al., Tetrahedron, 57, 3117, 2001.)*

also demonstrated similar activity within the concentration of 1–10 µg/mL (Manriquez et al., 2001) (**Figure 13.5**).

On the contrary, cuparene sesquiterpenes from *L. microcladia* and *L. obtusa* (Kladi et al., 2005, 2006) as well as laurefurenynes, acetogenins from *Laurencia* sp. showed moderate cytotoxicity toward lung and leukemia cell lines, respectively (Abdel-Mageed et al., 2010) (**Figure 13.6**).

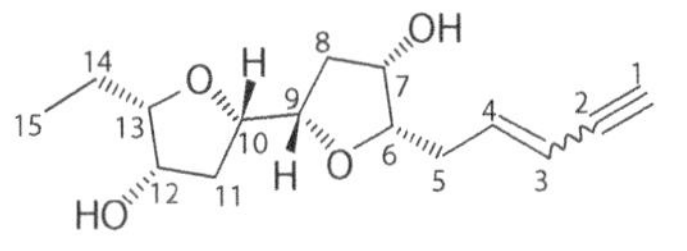

1 $\Delta_{3,4}$ *cis*
2 $\Delta_{3,4}$ *trans*

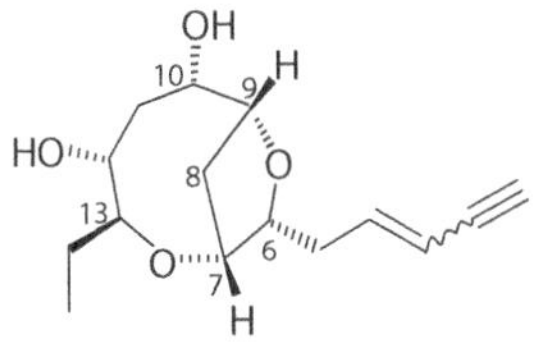

3 $\Delta_{3,4}$ *cis*
4 $\Delta_{3,4}$ *trans*

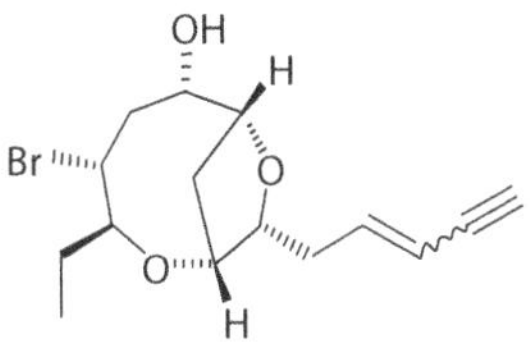

5 $\Delta_{3,4}$ *cis*
6 $\Delta_{3,4}$ *trans*

Figure 13.6 *Cytotoxic acetogenins from* Laurencia *sp.: Laurefurenyns A–F (1–6). (From Abdel Mageed, W.M. et al., Tetrahedron, 66(15), 2855, 2010.)*

13.3 Prospects for nutraceuticals

Production of anticancer nutraceutical products has a prospective future seen from some aspects, particularly in terms of health benefit, consumer preference, and raw material availability.

Recent developments in macroalgae natural products research, as previously discussed, have significantly contributed to the diversity of natural compounds with particular interest on anticancer. Cancer is a scary disease; therefore the invention of anticancer products will be beneficial for patients with cancer, or those who are willing to prevent it. Due to its low availability, drugs for cancer cost dearly; hence, those that are available at an affordable price are needed.

The idea of natural supplements is not novel to Indonesian, since herbal remedies known locally as *jamu* have been part of the local culture to treat ailments. However, most developed nutraceuticals are composed of plants from terrestrial origin, such as rhizomes, and only a few are made from marine biota. Since herbal remedies and functional foods for cancer, specifically, are scarce, marine natural products may provide alternatives to the existing compounds with better efficacy and limited side effects (**Figure 13.7**).

The success of nutraceutical development should be guaranteed with the continuous and consistent supply of raw materials with respect to quality and quantity. Uncultivated macroalgae such as *Laurencia* sp., *Turbinaria* sp., and *Ulva* sp. growing abundantly and wildly in Indonesia waters (Subaryono, 2011) indicate the availability of raw material for producing nutraceutical products. However,

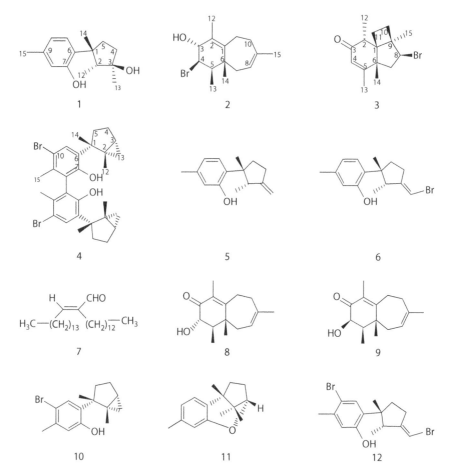

Figure 13.7 *Cuparene sesquiterpenes from* L. obtusa *(1–3, 5–9) and* L. microcladia *(4, 7, 10–12). (From Kladi, M. et al., Tetrahedron, 62(1), 182, 2006.)*

because the growing areas are spread out along the coastal regions, the collection of harvested macroalgae will be a serious problem if not being managed properly. Good handling practices for the harvested macroalgae should be introduced to those involving in supplying raw material for nutraceutical production. The development of nutraceutical industry can be started by introducing some findings from exploration and research to either "jamu" or pharmaceutical industries. The industries can probably directly produce nutraceutical products, if the production technologies are already available. However, the industries may have to conduct further research to develop the products prior to introduction to the market.

References

Abdel Mageed, WM, Ebel, R, Valeriote, F, Jaspars, M. Laurefurenynes A-F, New cyclic ether acetogenins from a marine algae, *Laurencia* sp. *Tetrahedron* 2010; 66(15): 2855–2862.

Ananthi, S, Gayathri, V, Chandronitha, C, Lakshmisundaram, R, Vasanthi, HR. Free radical scavenging and anti-inflammatory potential of a marine brown alga *Turbinaria ornata* (Turner) J. Agardh. *Indian Journal of Geo-Marine Sciences* 2011; 40(5): 664–670.

Ananthi, S, Raghavendran, HRB, Sunil, AG, Gayathri, V, Ramakhrisnan, G, Vasanthi, HR. In vitro antioxidant and in vivo anti-inflammatory potential of crude polysaccharide from *Turbinaria ornata* (marine brown alga). *Food and Chemical Toxicology* 2010; 48(1): 187–192.

Anggadiredja, J, Andyani, R, Hayati, M. Antioxidant activity of *Sargassum polycystum* (Phaeophyta) and *Laurencia obtusa* (Rhodophyta) from Seribu Islands. *Journal of Applied Phycology* 1997: 9(5): 477–479.

Anonymous. 2010. Indonesia aquaculture statistics. Directorate General of Aquaculture. Jakarta.

Arivuselvan, N, Radhiga, M, Anantharaman, P. In vitro antioxidant and anticoagulant activities of sulphated polysaccharides from brown seaweed (*Turbinaria ornata*) (Turner) J. Agardh. *Asian Journal of Pharmaceutical and Biological Research* 2011; 1(3): 232–239.

Asari, F, Kusumi, T, Kakisawa, H. Turbinaric acid, a cytotoxicsecosqualene carboxylic acid from the brown algae *Turbinaria ornata*. *Journal of Natural Products* 1989; 52(5): 1167–1169.

Ayesha, H, Sultana, V, Ara, J, Ehteshamul-haque, S. In vitro cytotoxicity of seaweeds from Karachi coast on brine shrimp. *Pakistan Journal of Botany* 2010; 42(5): 3555–3560.

Bixler, HJ, Porse, H. A decade of change in the seaweed hydrocolloids industry. *Journal of Applied Phycology* 2011; 23(3): 321–335.

Braune, W, Guiry, MD. Seaweeds: A color guide to common benthic green, brown, and red algae of the world's ocean. Koeltz Scientific Books, Königstein, Germany, 2011.

Burtin, P. Nutritional value of seaweeds. *Electronic Journal of Environmental, Agricultural and Food Chemistry* 2003; 2(4): 498–503.

Chandini, SK, Ganesan, P, Bhaskar, N. In vitro antioxidant activities of three selected brown seaweeds of India. *Food Chemistry* 2008; 107(2): 707–713.

Chatter, R, Othman, RB, Rabhi, S, Kladi, M, Tarhouni, S, Vagias, C, Roussis, V, Guizani-Tabbane, L, Kharrat, R. *In vivo* and in vitro anti-inflammatory activity of neorogioltriol, a new diterpene extracted from the red algae *Laurencia glandulifera*. *Marine Drugs* 2011; 9(7): 1293–1306.

Chattopadhyay, N, Ghosh, T, Sinha, S, Chattopadhyay, K, Karmakar, P, Ray, B. Polysaccharides from *Turbinaria conoides*: Structural features and antioxidant capacity. *Food Chemistry* 2010;118(3): 823–829.

Crawford, BB. *Seaweed Farming: An Alternative Livelihood for Small-Scale Fishers?* Coastal Resources Center, University of Rhode Island, Kingston, RI, 2002.

Delma, C, Ramalingam, K, Pandian, V, Baskar, A, Savarimuthu, I, Thangavelu, B, Somasundaram, S. Antagonistic effects of sulphated polysaccharides from *Turbinaria conoides* (J. Agardh) on tumor cell migration and angiogenesis. *Cancer Prevention Research* 2008; 1(7 Suppl.): A4.

Demirel, Z, Yilmaz-Koz, FF, Karabay-Yasavoglu, NU, Ozdemir, G, Sukatar, A. Antimicrobial and antioxidant activities of solvent extracts and the essential oil composition of *Laurencia obtusa* and *Laurencia obtusa* var. *pyramidata*. *Romanian Biotechnological Letters* 2011; 16(1): 5927–5936.

Deslandes, E, Pondaven, P, Auperin, T et al. Preliminary study of the in vitro antiproliferative effect of a hydroethanolic extract from the subtropical seaweed *Turbinaria ornata* (Turner J. Agardh) on a human non-small-cell broncho pulmonary carcinoma line (NSCLC-N6). *Journal of Applied Phycology* 2000: 257–262.

Devi, GK, Manivannan, K, Thirumaran, G, Rajathi, FAA, Anantharaman, P. In vitro antioxidant activities of selected seaweeds from Southeast coast of India. *Asian Pacific Journal of Tropical Medicine* 2011: 205–211.

Dias, DA, Urban, S. Phytochemical studies of the southern Australian marine alga, *Laurencia elata*. *Phytochemistry* 2011; 72(16): 2081–2089.

El Baky, HHA, El Baz, FK, El Baroty, GS. Evaluation of marine algae *Ulva lactuca* as a source of natural preservative ingredient. *Electronic Journal of Environmental, Agricultural and Food Chemistry* 2008; 7(11): 3353–3367.

El Baky, HHA, El Baz, FK, El Baroty, GS. Potential biological properties of sulphated polysaccharides extracted from the marine algae *Ulva lactuca* L. *Academic Journal of Cancer Research* 2009; 2(1): 1–11.

Eluvakkal, T, Sivakumar, SR, Arunkumar A. Fucoidan in some brown seaweeds found along the coast Gulf of Mannar. *International Journal of Botany* 2010; 6(2): 176–181.

Fajarningsih, ND, Nursid, M, Wikanta, T, Marraskuranto, E. Bioactivity of *Turbinaria decurrens* extracts as antitumor (HeLa and T47D) and their effects on proliferation of lymphocytes. *Journal of Marine and Fisheries Product Processing and Biotechnology* (Indonesian version) 2008; 3(1): 21–27.

Fernández, JJ, Souto, ML, Norte, M. Evaluation of the cytotoxic activity of polyethers isolated from *Laurencia. Bioorganic and Medicinal Chemistry* 1998; 6(12): 2237–2243.

Firdausy, C, Tisdell, C. Seafarming as part of Indonesia's economic development strategy: Seaweed and giant clam culture as cases. Giant clams in the sustainable development of the South Pacific: Socioeconomic issues in mariculture and conservation. *ACIAR* 1992: 80–100.

Gamal, AAE. Biological importance of marine algae. *Saudi Pharmaceutical Journal* 2010; 18(1): 1–25.

Guella, G, Pietra, F. Stereochemical features of sesquiterpene metabolites as a distinctive trait of red seaweeds in the genus. *Science* 1997; 38(47): 8261–8264.

He, J, Liang, H, Shi, D. Effects of *Laurencia* extract on antioxidant activities in mice. *Chinese Journal of Public Health* 2005; 21(9): 1082–1083.

Hong, DD, Hien, HM, Thi, H, Anh, L. Studies on the analgesic and anti-inflammatory activities of *Sargassum swartzii* (Turner) C. Agardh (Phaeophyta) and *Ulva reticulata* Forsskal (Chlorophyta) in experiment animal models. *Journal of Biotechnology* 2011; 10(12): 2308–2313.

International Monetary Fund. World Economic Outlook Report: Indonesia GDP current prices. www.imf.org. Accessed on January 30, 2012.

IPTEK Information Center. Diversity of Indonesian seaweeds. www.iptek.net.id. Accessed on January 23, 2012.

Januar, HI, Wikanta, T. Correlation between fucoxanthin contents in Turbinaria sp. and sea water nutrients at Binuangeun and Krakal coasts. *Squalen* (Indonesian version) 2011; 6(1): 18–25.

Jaswir, I, Monsur, HA. Anti-inflammatory compounds of macro algae origin: A review. *Journal of Medicinal Plants Research* 2011; 5(33): 7146–7154.

Jiao, G, Yu, G, Zhang, J, Ewart, HS. Chemical structures and bioactivities of sulphated polysaccharides from marine algae. *Marine Drugs* 2011; 9: 196–223.

Kaeffer, B, Bernard, C, Lahaye, M, Blottiere, HM, Cherbut, C. Biological properties of ulvan, a new source of green seaweed sulfated polysaccharides, on cultured normal and cancerous colonic epithelial cells. *Planta Medica* 1999; 65: 527–531.

Kim, JY, Kim, DS, Yang, EJ, Yoon, WJ, Baik, JS, Lee, WJ, Lee, NH, Hyun, CG. Green algae *Ulva pertusa* inhibit nitric oxide and prostaglandin-E2 formation in murine macrophage RAW 264.7 cells. *Journal of Applied Biological Chemistry* 2009; 52(1): 38–40.

Kladi, M, Vagias, C, Furnari, G, Moreau, D, Roussakis, C, Roussis, V. Cytotoxic cuparene sesquiterpenes from *Laurencia microcladia. Tetrahedron Letters* 2005; 46(34): 5723–5726.

Kladi, M, Xenaki, H, Vagias, C, Papazafiri, P, Roussis, V. New cytotoxic sesquiterpenes from the red algae *Laurencia obtusa* and *Laurencia microcladia. Tetrahedron* 2006; 62(1): 182–189.

Lee, DG, Hyun, JW, Kang, KA, Lee, JO, Lee, SH, Ha, BJ, Ha, JM, Lee, EY, Lee, JH. *Ulva lactuca*: A potential seaweed for tumor treatment and immune stimulation. *Biotechnology and Bioprocess Engineering* 2004; 9(3): 236–238.

Lewis, LA, McCourt, RM. Green algae and the origin of land plants. *American Journal of Botany* 2004; 91(10): 1535–1556.

Li, B, Lu, F, Wei, X, Zhao, R. Fucoidan: Structure and bioactivity. *Molecules* 2008; 13(8): 1671–1695.

Liang, H, He, J, Ma, AG, Zhang, PH, Bi, SL. Effect of ethanol extract of alga *Laurencia* supplementation on DNA oxidation and alkylation damage in mice. *Asia Pacific Journal of Clinical Nutrition* 2007; 16(38): 164–168.

Liang, H, Pang, D, He, J, Ma, A. The preventive effect of *Laurencia* terpenoids extract on antioxidant system after alcohol exposure in rats. *Acta Nutrimenta Sinica* 2009.

Lordan, S, Ross, RP, Stanton, C. Marine bioactives as functional food ingredients: Potential to reduce the incidence of chronic diseases. *Marine Drugs* 2011: 1056–1100.

Loughnane, CJ, McIvor, LM, Rindi, F, Stengel, DB, Guiry, MD. Morphology, *rbc*L phylogeny and distribution of distromatic *Ulva* (Ulvaphyceae, Chlorophyta) in Ireland and southern Britain. *Phycologia* 2002; 47(4): 416–429.

Manilal, A, Sujith, S, Kiran, GS, Selvin, J, Shakir, C. Cytotoxic potentials of red alga, *Laurencia brandenii* collected from the Indian Coast. *Science* 2009; 3(2): 90–94.

Manriquez, CP, Souto, ML, Gavin, JA, Norte, M, Fernandez, JJ. Several new squalene-derived triterpenes from *Laurencia*. *Tetrahedron* 2001; 57: 3117–3123.

Maraschin, M, Goncalves, C, Passos, R, Dias, PF, Valle, RMR, Fontana, JD, Pessatti, ML. Isolation, chemical characterization and biological activities of cell wall polysaccharides of *Laurencia microcladia*. *Notas Tecnicas da Facimar* 2000; 4: 37–41.

Meenakshi, S, Gnanambigai, DM. Total flavonoid and in vitro antioxidant activity of two seaweeds of Rameshwaram coast. *Global Journal of Pharmacology* 2009; 3(2): 59–62.

Moo-Puc, R, Robledo, D., Freile-Pelegrin, Y. In vitro cytotoxic and antiproliferative activities of marine macroalgae from Yucatan, Mexico. *Ciencias Marinas* 2009; 35(4): 345–358.

Morya, VK, Kim, J, Kim, EK. Algal fucoidan: Structural and size-dependent bioactivities and their perspectives. *Applied Microbial Biotechnology* 2012; 93: 71–82.

Nahas, R, Abatis, D, Anagnostopoulou, MA, Kefalas, P, Vagias, C, Roussis, V. Radical-scavenging activity of Aegean Sea marine algae. *Food Chemistry* 2007; 102: 577–581.

Pacheco, FC, Villa-Pulgarin, JA, Mollinedo, F, Martin, MN, Fernandez, JJ, Daranas, AH. New polyether triterpenoids from *Laurencia viridis* and their biological evaluation. *Marine Drugs* 2011; 9(11): 2220–2235.

Pereira, RC, Da Gama, BAP, Teixeira, VL, Yoneshigue-Valentin, Y. Ecological roles of natural products of the Brazilian red seaweed *Laurencia obtusa*. *Brazilian Journal of Biology* 2003; 63(4): 665–672.

Qi, H, Zhang, Q, Zhao, T, Chen, R. Antioxidant activity of different sulfate content derivatives of polysaccharide extracted from *Ulva pertusa* (Chlorophyta) *in vitro*. *International Journal of Biological Macromolecules* 2005; 37: 195–199.

Qi, H, Zhang, Q, Zhao, T, Hu, R. In vitro antioxidant activity of acetylated and benzoylated derivatives of polysaccharide extracted from *Ulva pertusa* (Chlorophyta). *Bioorganic and Medicinal Chemistry Letters* 2006; 16: 2441–2445.

Rajasekaran, A, Sivagnanam, G, Xavier, R. Nutraceuticals as therapeutic agents: A review. *Research Journal of Pharmacology and Technology* 2008; 1(4): 328–340.

Selvin, J, Lipton, AP. Biopotentials of *Ulva fasciata* and *Hypnea musciformis* collected from the peninsular coast of India. *Science* 2004; 12(1): 1–6.

Sentíes, A, Areces, A, Díaz-Larrea, J, Fujii, MT. First records of *Laurencia caduciramulosa* and *Laurencia minuscula* (Ceramiales, Rhodophyta) from the Cuban archipelago. *Botanica Marina* 2010; 53(5): 433–438.

Shanab, SMM. Antioxidant and antibiotic activities in some seaweeds (Egyptian isolates). *International Journal of Agriculture and Biology* 2007; 9(2): 220–225

Sheu, JH, Wang, GH, Sung, J, Chiu, YH, Duh, CY. Cytotoxic sterols from the formosan brown alga *Turbinaria ornata*. *Planta Medica* 1997; 63: 571–572.

Sheu, JH, Wang, GH, Sung, PJ, Duh, CY. New cytotoxic oxygenated fucosterols from the brown alga *Turbinaria conoides*. *Journal of Natural Products* 1999; 62(2): 224–227.

Stein, EM, Andreguetti, DX, Rocha, CS et al. Search for cytotoxic agents in multiple *Laurencia* complex seaweed species (Ceramiales, Rhodophyta) harvested from the Atlantic Ocean with emphasis on the Brazilian State of Espírito Santo. *Revista brasileira de farmacognosia* 2011; 21(2): 239–243.

Subaryono. Distribution of brown seaweed producing alginate in Indonesia and the potential utilization. *Squalen* (Indonesia version) 2011; 6(2): 55–62.

Thibodeau, A, Lavoie, A, Dionne, P, Moigne, JY. Polysaccharide compositions comprising fucans and galactans and their use to reduce extravasation and inflammation. U.S. Patent Application Publication, 2009.

Tjindarbumi, D, Mangunkusumo, R. Cancer in Indonesia, present and future. *Japanese Journal of Clinical Oncology* 2002; 32(Suppl. 1): S17–S21.

Usov, A. Structural analysis of red seaweed galactans of agar and carrageenan groups. *Food Hydrocolloids* 1998; 12(3): 301–308.

Wijesekara, I, Pangestuti, R, Kim, S. Biological activities and potential health benefits of sulfated polysaccharides derived from marine algae. *Carbohydrate Polymers* 2011; 84: 14–21.

Wikanta, T, Prabukusuma, A, Ratih, D, Januar, HI. Bioactivities of *U. fasciata* crude acetone extract, fractions and its sub-fractions towards HeLa tumor cell line. *Journal of Marine and Fisheries Product Processing and Biotechnology* (Indonesian version) 2010; 5(1): 1–9.

Wynne, MJ. *Turbinaria foliosa* sp. nov. (Fucales, Phaeophyceae) from the Sultanate of Oman, with a census of currently recognized species in the genus *Turbinaria. Phycological Research* 2002; 50: 283–293.

Yang, EJ, Moon, JY, Kim, MJ, Kim, DS, Kim, CS, Lee, WJ, Lee, NH, Hyun, CG. Inhibitory effect of Jeju endemic seaweeds on the production of pro-inflammatory mediators in mouse macrophage cell line RAW 264.7. *Journal of Zheijang University-Science B (Biomedicine and Biotechnology)* 2010; 11(5): 315–322.

Zhang, Z, Wang, F, Wang, X et al. Extraction of the polysaccharides from five algae and their potential antioxidant activity in vitro. *Carbohydrate Polymers* 2010; 82: 118–121.

Active Ingredients from Marine Microorganisms for Modern Nutraceuticals

Se-Kwon Kim and Pradeep Dewapriya

Contents

14.1 Introduction

Natural products have been used as a prominent source to prevent and cure various kinds of diseases for centuries. Earlier, diet was composed of these biologically active natural products with the aim of earning additional health benefits rather than basic nutritional requirements. The improvement of applications of these biologically active natural metabolites in food products leads to the term "nutraceutical" by merging two main sectors, "nutrition" and "pharmaceutical," which have direct effect on human health. Nutraceuticals became popular among health-conscious community and therefore, global nutraceutical market is growing day by day. The classical definition of nutraceutical is "a food or part of a food that provides medical or health benefits, including the prevention and/or treatment of a disease" is no longer applicable for modern nutraceuticals. In response to growing consumer interest on products that have potential to improve wellness and to combat with novel diseases, a number of nonfood-derived ingredients that are recognized as safe are used in nutraceuticals. Changes in modern life style and eating patterns have created nutrient deficiencies or excesses and have steepened the need of alternatives to maintain health (Ahmad et al. 2011). This was the main reason behind the development of modern nutraceuticals. The National Center for Complementary and Alternative Medicine (NCCAM) in the United States has categorized these types of nutraceuticals under the domain of biologically based treatments for cure or prevention of chronic diseases. Further, a national survey conducted in America found that over 17% of adult population had used dietary supplements other than conventional vitamins and minerals (Whitman 2001). Many kinds of sources have been identified for biologically active metabolites that can be used in modern nutraceutical, and one of the best examples is microorganism-derived biologically active ingredients. In addition to the basic purposes of microbial metabolites in food such as enhancing flavor, color, and texture of food, some ingredients possess remarkable health-promoting abilities. The beneficial action of these ingredients ranges from supply of essential nutrients to maintaining a healthy body by protecting from several chronic diseases.

Many microbial metabolites are currently used as functional ingredients in food and pharmaceutical products (**Table 14.1**). Single cell protein, amino acids, vitamins, colors, organic acids, and many more food-grade ingredients are currently produced using microorganisms as an economically viable method. Recombinant DNA technology has opened up the possibility of identifying, isolating, and separating valuable genes of microbes that are difficult to culture and express in suitable hosts to produce metabolites in large quantities. While the conventional ingredients are produced by microbes, many novel metabolites with advanced functional and health benefits have been identified from microorganisms. These types of metabolites have been considered as lead compounds for the development of modern nutraceuticals. In the continuous search of biologically active novel ingredients to protect from a wide variety of diseases

Table 14.1 Examples of Microorganism-Derived Metabolites Which Are Currently Used in Food Supplements and Their Health Claims

Product	Microorganism	Health Claim/s
Probiotics	Lactic acid bacteria and Bifidobacteria	Improve gut microflora
Single cell protein	Candida, Saccharomyces, and Torulopsis	Source of vitamin B
B-Carotene	Microalgae	Source of vitamin A, antioxidant, and enhances immune system
Ankaflavin	Monascus sp.	Anti-inflammation
Astaxanthin	Escherichia coli	Antioxidant, prevention of inflammatory and neurodegenerative diseases
Vitamin C	Yeast	Antioxidant
Citric acid	Aspergillus niger	Prevention of kidney stone
Glucosamine	A. niger	Prevention of osteoarthritis
Amino acid	Corynebacterium glutamicum and E. coli	Essential amino acid
Enzymes	Bacteria, yeast, and mould	Facilitates digestion and absorbance of food

and health problems, marine-derived microbial metabolites have been proved for their therapeutic potentials. The diversity of marine microbes is substantially high and they exhibit diverse physiological adaptations for marine environment (Waters et al. 2010). Thus, marine microbes would be an ideal source to explore novel biologically active metabolites that have potential to be used as modern nutraceutical. Here, the focus has been given to the chemical potential of the marine microorganism-derived active ingredients under the light of developing modern nutraceuticals.

14.2 Marine microbes

Historically, marine science has focused on either large charismatic creatures or economically important food species in the marine environment. Till the late ninetieth century, the invisible creatures of the oceans, marine microorganisms, were a negligible part even though they make up more than 90% of the marine biomass. These tiny microscopic organisms are the frontline of marine food chain and act as living lung for the planet by producing more than half of the world's oxygen. All microscopic organisms in salt water are referred as marine microorganisms. They are basically an ecological group. Current estimations show that global oceanic density is composed of 3.6×10^{29} bacterial cells, 1.3×10^{28} archaeal cells, and 4×10^{30} viruses. Since the biodiversity of marine microbial communities is incredibly high and some conventional culturing techniques are not applicable for marine microorganisms, phylogeny and functions of marine microbes have remained largely unexplored (Webster and Hill 2007). Marine microbes inhabit all kinds of different environments of the ocean such as polar ice, hydrothermal vent, deep sea, coral reef, mangroves, etc. Recent molecular approaches on the analysis of marine metagenomes have revealed a large number of phylogenetic lines.

With the identification of some constrains associated with marine animal– and plant–derived biologically active metabolites in drug development process, marine microbes have received growing attention to isolate active pharmaceutical ingredients in a sustainable manner. This tremendously diverse marine microbial community produces compounds with unique structural properties. These compounds possess broad spectrum of pharmaceutical properties such as antimicrobial, antituberculosis, antiviral, antiparasitic, anthelmintic, antimalarial, antiprotozoal, anticoagulant, antiplatelet, anti-inflammatory, antidiabetic, and antitumor effects (Imoff et al. 2011). Dozens of research articles are being published every year to reveal the potential of marine microbial metabolites in pharmaceutical applications. Waters et al. (2010) have highlighted that more than half of the molecules currently in the marine drug development pipeline are highly likely to be produced by bacteria (**Table 14.2**). Many food-grade metabolites with promising pharmaceutical properties have been reported from marine microbes and there is a potential to develop these active ingredients as modern nutraceuticals.

Table 14.2 Marine Natural Products in Drug Development Pipeline Which Have Evidence for Microbial Production

Clinical Status	Trademark	Compound	Biological Activity	Microorganism
Phase III	Soblidotin	Peptide	Anticancer	Bacteria
Phase II	Plinabulin	Diketopiperazine	Anticancer	Fungi
	Isokahalide	Depsipeptide	Anticancer	Bacteria
	Tasidotin	Peptide	Anticancer	Bacteria
Phase I	Bryostatin 1	Polyketide	Anticancer	Bacteria
	Marizomib	Beta-lactone-gamma-lactam	Anticancer	Bacteria

Source: Waters, A.L. et al., *Curr. Opin. Biotechnol.*, 21, 780, 2010.

14.3 Active ingredients for nutraceuticals

Although marine microbes produce wide variety of compounds with promising pharmacological potentials, all cannot be applied for the development of modern nutraceuticals. As the final objective of the compound is to apply it in ingestible product, the compounds are required to comply with the regulatory requirements. Generally, nutraceutical cannot be easily categorized into food or drug and often fall between the two. In most of the countries, nutraceuticals are monitored and controlled with the same scrutiny as dietary supplements. Since modern nutraceutical products are getting closer to medicinal form, some drug regulatory policies are also applicable (Gulati and Ottaway 2006). Thus, metabolites that have been identified as safe or derived from food-grade microorganisms would be ideal for the development of improved version of nutraceuticals.

14.3.1 Polyunsaturated fatty acids

Straight hydrocarbon chains with a terminal carboxyl group are generally known as fatty acids. The physiological properties of fatty acids largely depend on the degree of unsaturation and the position of first unsaturation relative to the end position. The ability to introduce double bonds beyond carbon 9 and 10 in hydrocarbon chain of fatty acid is generally low in mammals. Hence, omega-6 linoleic acid and omega-3 α-linolenic acid are considered as essential fatty acids and these types of fatty acids have to be supplemented. While the primary sources of essential fatty acids are plants and seafoods, some animals and microbes also synthesize these fatty acids. Recent studies have shown that these essential fatty acids, especially long-chain omega-3 fatty acids, possess ability to prevent and cure some chronic diseases. Due to well-documented healing properties, long-chain polyunsaturated fatty acids (PUFA) has earned considerably high market share as a dietary supplement or nutraceutical (Rodríguez et al. 2010). Further, in 2004 U.S. Food and Drug Administration has declared a "qualified health claim" status to omega-3 fatty acids eicosapentaenoic acid (EPA) and docosahexaenoic acid (DHA) (FDA 2004). Therefore, both EPA and DHA have been extensively used as nutraceuticals to prevent some chronic diseases. Among various kinds of PUFA sources, marine sources such as herring, mackerel, sardine, and salmon are considered as good sources of omega-3 fatty acids.

Recently, because of increased awareness of the health benefits of EPA and DHA, demands for fish and fish oil have increased worldwide. This increasing demand threatens the global fisheries. Likewise, seafood-derived omega-3 fatty acids raise some quality problems such as heavy metal contamination and unpleasant smell and taste. Thus, there is a growing interest for alternative and sustainable sources for the production of omega-3 fatty acids. Plant-based production of PUFA was hampered by the main fact that more genes must be engineered to produce PUFA in commercial quantities (Ward and Singh 2005). In the search of alternative source for PUFA production, marine microorganisms are under the spotlight since they are the primary producers of PUFA in marine environment.

Numerous marine organisms including microalgae, bacteria, yeast, and filamentous fungi produce long-chain PUFA. Among them, oleaginous organisms (microorganisms that produce more than 25% lipid on a dry cell weight basis) can be tailored to produce high amounts of omega-3 fatty acids, particularly DHA and EPA. *Thraustochytrids*, belong to the group of marine protists has gained much attention due to their ability to produce PUFA in high quantity, especially DHA. A recent comprehensive article on *Thraustochytrids* describes various positive characteristics of these organisms to develop as alternative producer of omega-3 fatty acids (Gupta et al. 2012). In addition, marine microalgae have been identified as a rich source of long-chain PUFA, where it accounts for 10%–20% of cell weight in some species. Interestingly, the presence of PUFA in marine microalgae is higher compared to that of freshwater microalgal species as they produce higher amounts of PUFA to survive in marine environments. *Nannochloropsis* spp., *Porphyridium cruentum*, *Phaeodactylum tricornutum*, and *Chaetoceros calcitrans* have been identified as the best EPA-producing microalgae, while *Isochrysis galbana*, *Crypthecodinium* spp., and *Schizotrichium* spp. have been reported as promising sources of DHA (Guedes et al. 2011a). Moreover, marine bacteria, particularly those found in the gut flora of fish in deep, low-temperature waters, produce PUFA. Several studies have proved that these bacterial species can grow in large-scale fermentation process to produce omega-3 fatty acids in sustainable manner. Recently, advanced technical developments have been made to facilitate rapid screening and isolation of EPA-producing marine bacteria from mixed cultures (Ryan et al. 2010).

14.3.2 Biologically active carbohydrates of microbial origin

As most abundant naturally occurring compounds, polysaccharides are present in all types of animals, plants, and microorganisms. Generally, polysaccharides are used as thickening, emulsifying, and stabilizing agents in food industry. However, owing to various types of pharmaceutical properties, polysaccharides have been used over the years as a therapeutic agent, and dietary fibers are a well-known example. A wide variety of biological activities have been documented for polysaccharides including, anticoagulant, antiinflammatory, antiviral, and antitumor activities (Paulsen 2002). By the end of 1940, few useful findings were made to describe the importance of polysaccharides produced by microorganisms. One of the greatest finding was xanthan gum, the most widely used microbial polysaccharides derived from the bacterial coat of *Xanthomonas campestris*. This finding boosts the interest on microbial polysaccharides research and dozens of microbial polysaccharides are currently produced in commercial quantities for various industrial purposes. Microbial polysaccharides are generally divided into three groups: cell wall polysaccharides, intercellular polysaccharides, and exocellular polysaccharides. The exocellular polysaccharides are commercially valuable products of microorganisms since they are constantly diffused into the cell culture medium and easy to isolate from the culture media, free from protein and cell debris. It is widely accepted that these microbial extracellular polysaccharides (EPS) offer many advantages over plant polysaccharides due to their novel functions and constant chemical and physical properties (Poli et al. 2010).

In extreme environmental conditions of oceans, marine microorganisms secrete structurally and functionally diverse polysaccharides. Currently, these marine microbial polysaccharides have received a great interest with their potentials in fields such as pharmaceuticals, adhesives, and textiles. Marine filamentous fungal species *Keissleriella* sp., *Penicillium* sp., and *Epicoccum* sp. are well-known antioxidative EPS producers. Recent advanced researches have proved that the antioxidative activity in the marine EPS is comparatively higher and there is a clear potential to develop these as natural antioxidants (Laurienzo 2010).

Glucosamine, one of the most abundant monosaccharide, is produced from naturally present structural polysaccharide chitin. It is a well-established marine nutraceutical that is consumed as a dietary supplement for the prevention or the treatment of osteoarthritis. Alternative sources for producing glucosamine on commercial scale, instead of shellfish-derived glucosamine, are currently in experimental stage to cater the demand for vegetarian glucosamine. Glucosamine is part of the structure of the polysaccharides chitosan and chitin, which compose the exoskeletons of crustaceans and other arthropods, cell walls in fungi, and many higher organisms. Although the *Aspergillus* sp. can be used to produce glucosamine in a similar manner as is produced from shrimp shells, it is a more expensive process. Several novel techniques have been applied to produce glucosamine from recombinant *Escherichia coli* with pure glucosamine yield up to 17 g/L. Wild-type bacteria were studied to produce glucosamine as an alternative but could only reach 264 mg/L of glucosamine (Sitanggang et al. 2009). Due to several adaptations to hard marine environment, marine-derived novel microbial species have been reported for higher yield of both glucosamine and chitosan compared to that of terrestrial counterpart and have a potential to be used as an alternative source for the production of vegetarian glucosamine (Logesh et al. 2012).

14.3.3 Microbial pigments

The basic purpose of pigments is changing or enhancing the color of a material. Since color plays an important role in food and food supplements, various kinds of natural as well as synthetic colorants are used. Pigments from natural sources have been obtained since long time ago, and currently their interest has markedly increased due to the side effects associated with those of synthetic origin. Carotenoids, flavonoids (anthocyanins), and some tetrapyrroles (chlorophylls, phycobiliproteins) are most common naturally occurring pigments. From the simplest organisms like cyanobacteria, microalgae, and fungi, to plants and animals are used as natural sources of pigments. Further, some of these natural pigments possess promising health-promoting abilities, including prevention of chronic diseases. Thus, bioactive natural pigments gain much attention as modern nutraceuticals (Mortensen 2006).

Among the sources of natural pigments, microbial pigments offer several advantages over others. A number of carotenoids such as β-carotene, astaxanthin, canthaxanthin, lutein, and violaxanthin, which bear promising pharmaceutical potentials, have been identified from marine microalgal species (Plaza et al. 2009). Health-promoting abilities of these carotenoids have been comprehensively studied. β-Carotene enhances the immunity via facilitating the monocyte function to increase the number of surface molecules expressed while protecting cellular damage from oxidative stress. Based on this antioxidative and immune modulatory power of carotenoids, they can extend their activity against several disease conditions, including cancer, cardiovascular disease, rheumatoid arthritis, and several neurodegenerative diseases (Abe et al. 2005). Green unicellular marine microalga *Dunaliella salina* has been identified as one of the most promising source of β-carotene with 14% of dry weight of β-carotene at appropriate culture conditions. With the development of optimum culture and extraction conditions, *D. salina* has been commercially cultured as a source of β-carotene (Guedes et al. 2011). Other than marine microalgae, marine bacteria and fungi are well-known producers of biologically active pigments. Prodiginines (red pigment), carotenes, violacein (violet pigment), and quinones (colored compounds with an aromatic ring) are commercially important natural pigments of marine bacterial origin. Owing to potent health-promoting ability, these pigments provide promising avenues for the development of modern nutraceuticals (Soliev et al. 2011).

14.3.4 Protein and bioactive peptides from marine microbes

Changing life style continually demands alternative sources to fulfill basic nutritional requirements. Natural sources with higher protein contents and high biological value are being searched as ideal candidates for supplementations. In the nutritional point of view, all the high protein sources cannot be used to develop supplements without considering their biological value, which is the amount or percentage of protein that the body is able to absorb. Likewise, there are several drawbacks of dairy and soy protein, which are used as the most common protein supplements. Besides having high biological value, proteins of animal origin have faced a growing challenge in modern society, a vegetarian diet. In this regard, marine microorganisms–derived proteins gained much attention as single-cell proteins that have high biological value.

The protein contents of some microalgal species including *Chlorella, Spirulina, Scenedesmus, Dunaliella, Micractinium, Oscillatoria, Chlamydomonas*, and *Euglena* have accounted for more than 50% of the dry weight while proving that these microalgae are promising protein sources. The biological value is considerably high in some microalgae species such as *Spirulina* sp. and *Chlorella* sp., where it is 77.6% and 71.6%, respectively. Moreover, the proteins are rich in essential amino acids, specifically lysine. Due to high lysine content, microalgae proteins can be effectively used as a supplement and formulation for other nutraceuticals (Becker 2007).

Protein-based therapeutic is another leading section in biotechnology since proteins are highly specific, sensitive molecules that have fewer side effects. In the development of therapeutic proteins, marine algae offer the potential to produce high yields of recombinant proteins more rapidly and at much lower cost than traditional cell culture. In recent years, light has been shed on *Chlamydomonas reinhardtii*, a single-celled eukaryotic marine microalga for the production of recombinant therapeutic proteins. This microalga offers several advantages over other systems employed today. The ability to fold correctly and assemble complex proteins facilitates the production of proteins with high biological value. Moreover, these microalgae can be grown in complete containment while reducing any risk of environmental contaminations (Mayfield et al. 2007).

Some peptides play a characteristic role in metabolic regulation and modulation, after releasing from the parent protein. These types of peptides are referred as bioactive peptide and have clear potential to be used as nutraceuticals and functional food ingredients for health promotion and disease risk reduction. Further, bioactive peptides are 2–20 amino acids long protein fragments; they generally remain latent within the parent protein molecule. Once released by hydrolysis, the active peptides exert hormone-like effects on physiological complications of human body beyond their nutritional value (Himaya et al. 2012). Hence, these types of peptides possess wide range of therapeutic activities, including antihypertensive, antioxidant, anticancer, antimicrobial, immunomodulatory, and cholesterol-lowering effects. The peptides are produced by the proteolytic action of microorganisms on proteins or in vitro enzymatic hydrolysis of proteins with enzymes from microbial or gastrointestinal origin. Corresponding to the specific amino acid composition, microalgal proteins have gained attention as a promising source to isolate bioactive peptides with pharmaceutical applications. The characteristic structural and compositional properties of the marine algae–derived peptides exhibit unique therapeutic actions (Kim and Kang 2011). Hydrophobic amino acid residues with aromatic or branched side chains at the C-terminal position and branched amino acids at the N-terminal position in isolated peptides enhance the functional values (Agyei and Danquah 2011). Marine microalgae *Chlorella vulgaris*, *Spirulina platensis*, *Navicula incerta*, and *Pavlova lutheri* have been proved as potential algal species to produce biologically active peptides with significant therapeutic potentials (Ryu et al. 2012).

14.3.5 Marine probiotics

The gastrointestinal tract harbors an array of microbial species. The most prominent microbial species, including *Bacteroides*, *Prevotella*, *Eubacterium*, *Clostridium*, *Bifidobacterium*, *Lactobacillus*, *Staphylococcus*, *Enterococcus*, *Streptococcus*, *Enterobacter*, and *Escherichia* serve as beneficial as well as harmful strains. Thus reducing the amount of potentially harmful or pathogenic species and promoting the growth of favorable species, which have beneficial effects on host health, are vital for well-being. A live microorganism that is beneficial for gut microflora and restricts the growth of decay or disease-causing bacteria is referred as a probiotic strain. Prebiotics stands for a selectively fermented ingredient that allows specific changes, both in the composition and/or activity in the gastrointestinal microflora that confers benefits upon host well-being and health (Reddy et al. 2011).

Modulating the activities of gut microflora with food and food supplements is not a novel concept. However, the applications of probiotics and prebiotics were boosted with modern nutraceutical and functional food market. Strains from the genera *Lactobacillus* and *Bifidobacterium* are the bacteria predominantly used as probiotics. Fermented dairy products especially yoghurts, drinks, and capsules with freeze-dried bacteria are the most popular vehicles for delivering these organisms to the gastrointestinal tract (Gibson 2007). Recent studies reveal that marine-derived *Lactobacillus* species have exerted several advantages over conventional terrestrial species. Though only few strains of lactobacillus are known from marine environment, their unique characteristics at in vivo experiments have proved that there is clear potential to use them as probiotics with advanced health benefits. Among the identified lactic acid bacterial species of marine origin, majority have been isolated from Pacific Ocean region of Japan and strains belonging to genus *Carnobacterium*, *Marinilactobacillus*, and *Halolactobacillus* are common (Kathiresan and Thiruneelakandan 2008).

14.4 Conclusion and prospects of marine microbial nutraceuticals

Undoubtedly, marine environment offers unique and diverse bioactive compounds with an array of health-promoting abilities. Even today, several marine-derived biologically active compounds are used as food and food supplements. Moreover, it is clear that marine environment is an ideal source for exploring novel functional materials. In the search of novel biologically active metabolites, marine microorganism–derived functional ingredients have exhibited that they are unlike those found in terrestrial species. In order to develop advanced nutraceuticals with pinpointed health claims, these microbial metabolites seem to be good candidates. Applications of active metabolites of marine microbial origin in food and pharmaceutical products have just started the journey. The health claims of novel metabolites that have been mentioned in this chapter are acquired only through in vitro and in vivo studies. Thus, comprehensive studies on the mode of action, biological consequences, and possible side effects have to be conducted. Moreover, supplementation of diet with biologically active ingredients accompanies several complications. Therefore, legal controls have to be addressed when developing a novel ingredient such as a food supplement. Especially, microbial metabolites should come from strains that have been recognized as safe. Since marine microbiology is not a matured field, application of marine microbial products in food and pharmaceutical industries would be a challenge. Indeed, many novel techniques are applicable to reveal the biological and environmental role of these marine microbes and it would facilitate effective isolation and characterization of food-grade marine microbial species (Joint et al. 2010). Further, improved methodologies in fermentation technologies are necessary for the production of marine microbial metabolites in a cost-effective manner. Particularly, strains that have been identified from extreme environments will not be able to culture with conventional techniques. Recent advances of biotechnology in the field of microbiology and marine science would overcome these hurdles soon. We predict that the application of marine microorganism–derived novel metabolites in modern nutraceuticals is not so far.

References

Abe, K., Hattor, H., and Hiran, M. 2005. Accumulation and antioxidant activity of secondary carotenoids in the aerial microalga *Coelastrella striolata* var. *multistriata*. *Food Chemistry* 100:656–661.

Agyei, D. and Danquah, M.K. 2011. Industrial-scale manufacturing of pharmaceutical-grade bioactive peptides. *Biotechnology Advances* 29:272–277.

Ahmad, M.F., Ashraf, S.A., Ahmad, F.A., Ansari, J.A., and Siddiquee, M.R.A. 2011. Nutraceutical market and its regulation. *American Journal of Food Technology* 6:342–343.

Becker, E.W. 2007. Micro-algae as a source of protein. *Biotechnology Advance* 25:207–210.

FDA. 2004. FDA announces qualified health claims for omega-3 fatty acids (Press release). United States Food and Drug Administration. September 8, 2004. Retrieved 2006-07-10.

Gibson, G.R. 2007. Functional foods: Probiotics and prebiotics. *Culture* 28(2):1–7.

Guedes, A.C., Amaro, H.M., Barbosa, C.R., Pereira, R.D., and Malcata, F.X. 2011a. Fatty acid composition of several wild microalgae and cyanobacteria, with a focus on eicosapentaenoic, docosahexaenoic and α-linolenic acids for eventual dietary uses. *Food Research International* 44:2721–2729.

Guedes, A.C., Amaro, H.M., and Malcata, F.X. 2011b. Microalgae as sources of high added-value compounds—A brief review of recent work. *Biotechnology Progress* 27:597–613.

Gulati, O.P. and Ottaway, P.B. 2006. Legislation relating to nutraceuticals in the European Union with a particular focus on botanical-sourced products. *Toxicology* 221:75–87.

Gupta, A., Barrow, C.J., and Puri, M. 2012. Omega-3 biotechnology: Thraustochytrids as a novel source of omega-3 oils. *Biotechnology Advance*, doi:10.1016/j.biotechadv.2012.02.014.

Himaya, S.W.A., Ngo, D.H., Ryu, B., and Kim, S.K. 2012. An active peptide purified from gastrointestinal enzyme hydrolysate of Pacific cod skin gelatin attenuates angiotensin-1 converting enzyme (ACE) activity and cellular oxidative stress. *Food Chemistry* 132:1872–1882.

Imoff, J.F., Labes, A., and Wise, J. 2011. Bio-mining the microbial treasures of the ocean: New natural products. *Biotechnology Advance* 29:468–482.

Joint, I., Mühling, M., and Querellou, J. 2010. Culturing marine bacteria—An essential prerequisite for biodiscovery. *Microbial Biotechnology* 3:564–575.

Kathiresan, K. and Thiruneelakandan, G. 2008. Probiotics of lactic acid bacteria of marine origin. *Indian Journal of Biotechnology* 7:170–177.

Kim, S.K. and Kang, K.H. 2011. Medicinal effects of peptides from marine microalgae. *Advance Food and Nutritional Research* 64:313–323.

Laurienzo, P. 2010. Marine polysaccharides in pharmaceutical applications: An overview. *Marine Drugs* 8:2435–2465.

Logesh, A.R., Thillaimaharani, K.A., Sharmila, K., Kalaiselvam, M., and Raffi, S.M. 2012. Production of chitosan from endolichenic fungi isolated from mangrove environment and its antagonistic activity. *Asian Pacific Journal of Tropical Biomedicine* 140–143.

Mayfield, S.P., Manuell, A.L., Chen, S., Wu, J., Tran, M., Siefker, D., Muto, M., and Navarro, J.M. 2007. *Chlamydomonas reinhardtii* chloroplasts as protein factories. *Current Opinion in Biotechnology* 18:1–8.

Mortensen, A. 2006. Carotenoids and other pigments as natural colorants. *Pure and Applied Chemistry* 78:1477–1491.

Paulsen, S.B. 2002. Biologically active polysaccharides as possible lead compounds. *Phytochemistry Review* 1:379–387.

Plaza, M., Herrero, M., Cifuentes, A., and Ibanez, E. 2009. Innovative natural functional ingredients from microalgae. *Journal of Agricultural Food Chemistry* 57:7159–7170.

Poli, A., Anzelmo, G., and Nicolaus, B. 2010. Bacterial exopolysaccharides from extreme marine habitats: Production, characterization and biological activities. *Marine Drugs* 8:1779–1802.

Reddy, R.S., Swapna, PL.A., Ramesh, T., Singh, T.R., Vijayalaxmi, N., and Lavanya, R. 2011. Bacteria in oral health—Probiotics and prebiotics a review. *International Journal of Biological and Medical Research* 2:1226–1233.

Rodríguez, N.R., Beltrán, S., Jaime, I., Diego, S.M.D., Sanz, M.T., and Carballido, J.R. 2010. Production of omega-3 polyunsaturated fatty acid concentrates: A review. *Innovative Food Science and Emerging Technologies* 11:1–12.

Ryan, J., Farr, H., Visnovsky, S., Vyssotski, M., and Visnovsky, G. 2010. A rapid method for the isolation of eicosapentaenoic acid-producing marine bacteria. *Journal of Microbiological Methods* 82:49–53.

Ryu, B., Kang, K.H., Ngo, D.H., Qian, Z.J., and Kim, S.K. 2012. Statistical optimization of microalgae *Pavlova lutheri* cultivation conditions and its fermentation conditions by yeast, *Candida rugopelliculosa*. *Bioresource Technology* 107:307–313.

Sitanggang, A.B., Wu, H.S., and Wang, S.S. 2009. Determination of fungal glucosamine using HPLC with 1-napthyl isothiocyanate derivatization and microwave heating. *Biotechnology and Bioprocess Engineering* 14:819–827.

Soliev, A.B., Hosokawa, K., and Enomoto, K. 2011. Bioactive pigments from marine bacteria: Applications and physiological roles. *Evidence-Based Complementary and Alternative Medicine* 2011:1–17. doi:10.1155/2011/670349.

Ward, O.P. and Singh, A. 2005. Omega-3/6 fatty acids: Alternative sources of production. *Process Biochemistry* 40:3627–3652.

Waters, A.L., Hill, R.T., Place, A.R., and Hamann, M.T. 2010. The expanding role of marine microbes in pharmaceutical development. *Current Opinion in Biotechnology* 21:780–786.

Webster, N. and Hill, R. 2007. Vulnerability of marine microbes on the Great Barrier Reef to climate change. In *Climate Change on the Great Barrier Reef*. J.E. Johnso and P.A. Marshall, eds. pp. 96–120. Townsville, QLD, Australia: Great Barrier Reef Marine Park Authority and Australian Greenhouse Office.

Whitman, M. 2001. Understanding the perceived need for complementary and alternative nutraceuticals: Lifestyle issues. *Clinical Journal of Oncology Nursing* 5:190–194.

Potent Anticancer Actions of Omega-3 Polyunsaturated Fatty Acids of Marine Nutraceuticals

Kaipeng Jing and Kyu Lim

Contents

Abbreviations

ω3/ω6-PUFAs	omega-3/6 polyunsaturated fatty acids
AA	arachidonic acid
AIF	apoptosis-inducing factor
ALA	alpha-linolenic acid
ATGs	autophagy-related proteins
CDKs	cyclin-dependent kinases
C/EBPs	CCAAT/enhancer-binding proteins
CKIs	cyclin-dependent kinase inhibitors
COXs	cyclooxygenases
cPLA$_2$	cytosolic phospholipase A$_2$
CYPs	cytochrome P450 monooxygenases
cyt c	cytochrome c
DHA	docosahexaenoic acid
DNMTs	DNA methyltransferases
ECM	extracellular matrix
EPA	eicosapentaenoic acid
ER	estrogen receptor
HDACs	histone deacetylases
HIF	hypoxia-inducible factor
HPV	human papillomavirus
LA	linoleic acid
LOXs	lipoxygenases
LPS	lipopolysaccharide
LTs	leukotrienes
LXs	lipoxins
miRs	microRNAs
MMP	mitochondrial membrane potential
MMPs	matrix metalloproteinases
mTOR	mammalian target of rapamycin
NF-κB	nuclear factor κB
PGs	prostaglandins
PPARγ	peroxisome proliferator-activated receptor γ
ROS	reactive oxygen species
TIMPs	tissue inhibitor metalloproteinases
TNF-α	tumor necrosis factor-α
TXs	thromboxanes
VEGF	vascular endothelial growth factor

15.1 Introduction

Fish oil has not historically been a subject of study, and it is only because of the increasing recognition of its importance to human nutrition and health that it has become a topic of interest. The earliest studies on the health benefits of marine fish oil were conducted by Dyerberg and Bang who discovered that the deepwater fish, which contains omega-3 polyunsaturated fatty acids (ω3-PUFAs), was responsible for the low incidence of heart disease among Eskimo populations [1,2]. In the following 20 years after their finding, prospective epidemiological studies from the Netherlands, Chicago, and the Multiple Risk Factor Intervention Trial also confirmed the preventative effects of omega-3 fish oil on coronary heart disease [3–6]. These early observations led to increases in research examining the beneficial effects of ω3-PUFAs on numerous debilitating as well as common conditions, and up to now, it is almost impossible to find any human disorder where ω3-PUFAs have not been tested [7,8]. The findings from these studies lead health professionals to encourage the general population to consume more cold water fish, increasing omega-3 fish oil content in human diets [8–11].

Because of the minimal host cell toxicity and easy availability, cancer prevention and treatment using natural substances as dietary supplements have received tremendous interest in the past decades, and numerous food-derived natural products are currently under evaluation in oncology clinical trials [12,13]. Prompted by the observation that diet rich in fish oil has been associated with a lower incidence of common cancers in many human populations [14–17], dietary intake of ω3-PUFAs from fish oil for cancer prevention and treatment becomes the subject of intense study [18]. This is not totally surprising, considering that cancer is highly associated with dietary fat intake [19], and that 30%–40% of all cancers can be prevented by appropriate diets [20,21].

Increasing evidence from epidemiology, in vitro and animal studies suggests that ω3-PUFAs, particularly the long-chain ω3-PUFAs eicosapentaenoic acid (EPA) and docosahexaenoic acid (DHA), can suppress cancer development at different stages [14,22]. Although another shorter chain ω3-PUFA, alpha-linolenic acid (ALA) exists, and can be slowly desaturated and elongated to EPA and DHA in the human body, it is mainly found in green leafy vegetables and walnuts, which are also rich in linoleic acid (LA), the precursor of omega-6 polyunsaturated fatty acids (ω6-PUFAs) [23]. To deduce possible unique effects of ω3-PUFAs on cancer, the focus of this chapter is EPA and DHA, the highly unsaturated long-chain ω3-PUFAs from fish oil. Our objective is to provide a general background introduction of ω3-PUFAs including their chemistry and metabolism. In addition, the protective role of ω3-PUFAs against cancer development at certain stages and the potential mechanisms will be reviewed.

15.2 Nomenclature, metabolic conversion, and dietary sources of the ω3-PUFAS

15.2.1 Nomenclature and metabolic conversion

PUFAs are hydrocarbon chains with two or more double bonds situated along the length of the carbon chain, and depending on the location of the first double bond relative to the methyl end of the carbon chain, PUFAs can be classified as either ω3-PUFAs or ω6-PUFAs. ALA and LA, the precursors of ω3- and ω6-PUFAs, respectively, cannot be endogenously synthesized by mammals due to the lack of ω3- and ω6-desaturases required to place the double bond in position ω3 or ω6 in the PUFAs, and because they must be entirely obtained from the diet, ALA and LA are referred to as "essential fatty acids."

Both ALA and LA can be further metabolized to long-chain PUFAs, mainly in the liver, through a series of elongation and desaturation steps, which involve increases in chain length and degree of unsaturation [24]. ALA (18 carbon atoms and 3 double bonds; C18:3) can be metabolized to EPA (C20:5) and ultimately to DHA (C22:6); LA (C18:2) can be converted to arachidonic acid (AA; C20:4), the most important long-chain fatty acid of ω6 series [25]. However, competition for the same desaturases and elongases between ω3-PUFAs and ω6-PUFAs exists, with the enzymes having a greater affinity for the ω3-PUFAs (**Figure 15.1**). Indeed, increase in the dietary ω3-PUFAs intake reduces the rate of desaturation of LA and so, the production of AA, an inhibition which can be achieved by ALA, EPA, and DHA [26,27]. However, although the human body can produce EPA and DHA from ALA intake because of the existence of required enzymes, the very low conversion efficiency and wide variation in capacity for the conversion of ALA to EPA and DHA have been reported, even in healthy individuals [28–30]. Generally, women have a greater capacity for the conversion of ALA to DHA than men possibly due to a regulatory effect of estrogen, which may be important for demands of the fetus and neonate for DHA during pregnancy and lactation [31]. In addition, retroconversion of EPA from DHA occurs in mammals, but human beings struggle with this retroconversion as it is an inefficient process [32–34].

It is noteworthy, however, that the competition between the ω3- and ω6-PUFAs can also be extended to their other biological activities, such as incorporation into membrane phospholipids [35,36] and conversion into highly active

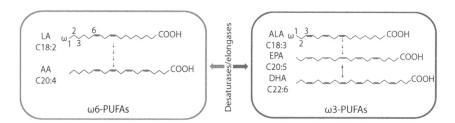

Figure 15.1 *Chemical structures of the major ω6- and ω3-PUFAs. The position of the first unsaturation counting from the methyl end of the fatty acid, the ω carbon, generated the name of the two classes. LA and ALA (short chain) are the precursors of ω6- and ω3-PUFAs, respectively, and they compete for the same desaturases as well as elongases in the synthesis of long-chain PUFAs. PUFAs, polyunsaturated fatty acids; ALA, alpha-linolenic acid; LA, linoleic acid; AA, arachidonic acid; EPA, eicosapentaenoic acid; DHA, docosahexaenoic acid.*

metabolites with them being substrates for cyclooxygenases (COXs) [37], lipoxygenases (LOXs) [38,39], and cytochrome P450 monooxygenases (CYPs) [40–42]. The enzymatic processing of PUFAs starts after their release from the membrane phospholipids by the action of cytosolic phospholipase A_2 (cPLA$_2$) following the binding of growth factors as well as hormones to membrane receptors and activation of signaling pathways [43–45], and generates a group of lipid mediators known as eicosanoids. COXs catalyze the conversion of AA into 2-series prostaglandins (PGs) and thromboxanes (TXs), and EPA into 3-series PGs and TXs; LOXs metabolize AA to form 4-series leukotrienes (LTs) and lipoxins (LXs), and EPA to form 5-series LTs; CYPs convert AA, EPA, and DHA into their corresponding hydroxy- and epoxy-fatty acids whose function remains unclear. DHA is not metabolized to a TX or to LT; however, it can competitively inhibit the eicosanoid biosynthesis from AA [46–48]. In addition to the aforementioned eicosanoids, another class of ω3-PUFAs-derived lipid mediators, resolvins, which are generated by the aspirin-acetylated COXs and LOXs, have recently been characterized and have attracted considerable interest because of their more powerful properties. These mediators include EPA-derived E-series resolvins, and DHA-derived D-series resolvins [49–51]. These lipid mediators involve a wide variety of physiological activities, such as fever and pain induction, platelet aggregation, smooth muscle contraction, and inflammation, which is a process tightly linked to cancer development [52,53]. It is widely believed that eicosanoids from ω6-PUFAs are pro-inflammatory, whereas those from ω3-PUFAs are weakly inflammatory or anti-inflammatory. However, this must be viewed as a generalization of their function, as AA-derived mediators, LXs, have also been reported to be anti-inflammatory [54–56].

Additionally, besides enzymatic oxidation, PUFAs can also be processed in a nonenzymatic pathway [41]. The nonenzymatic oxidation of PUFAs is triggered by the insult of reactive oxygen species (ROS), a heterogeneous group of highly reactive free radicals that oxidize targets in a biology system. Although ROS can be used as a second messenger in the intracellular transduction of some hormones and cytokines in cells, they intend to acquire electrons from neighboring molecules, such as DNA and proteins to reach their own stability, which often leads to the damage of corresponding molecules [57,58]. The presence of multiple double bonds (methylene group adjacent to a double bond) in PUFAs renders them extremely susceptible to ROS attack, and thus, the susceptibility of a PUFA is related to the number of double bonds that it contains, with one double bond increasing the vulnerability by a factor of 100 [59–61]. ROS can break any of the double bonds in PUFAs, and thus result in a bewildering array of products with either carbon structure preserved (isoprostanes, isofurans, and mono/dihydroxy fatty acids) or disintegrated (hydroxyhexenals and pentanes) [62,63]. Many of these resultant products are highly reactive, and are capable of propagating a chain reaction to generate additional ROS and/or interacting with other molecules [41,64–70] (**Figure 15.2**).

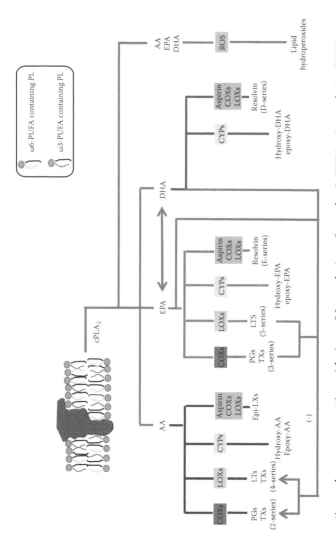

Figure 15.2 Enzymatic and non-enzymatic oxidation of long-chain ω6- and ω3-PUFAs. Long-chain PUFAs are incorporated into membrane phospholipids. Upon activation by various stimuli, cPLA₂ catalyzes the hydrolysis of cell membrane phospholipids and releases AA, EPA, and DHA. The released AA and EPA are converted by the same COXs and LOXs to their oxygenated metabolites. DHA is not metabolized to a TX or LT; but it inhibits the biosynthesis of lipid mediators from AA by competing/inhibiting COXs and LOXs. CYPs metabolize long-chain PUFAs into their corresponding hydroxyl and epoxy fatty acids. In the presence of aspirin, COXs and LOXs can convert long-chain PUFAs into their highly anti-inflammatory products. The released long-chain PUFAs can also be oxidized after the ROS attack in an enzyme-independent manner, generating a variety of lipid hydroperoxides. COXs, cyclooxygenases; LOXs, lipoxygenases; CYPs, cytochrome P450 monooxygenases; PGs, prostaglandins; TXs, thromboxanes; LTs, leukotrienes; LXs, lipoxins; ROS, reactive oxygen species; PL, phospholipids; cPLA₂, cytosolic phospholipase A₂.

15.2.2 Dietary sources

Dietary sources of ALA and LA are abundantly present in vegetable oils. However, the contents of ALA and LA in various vegetable oils differ widely. Corn, sunflower, cotton seed, safflower, soybean, and peanut oils are high in LA, while canola and linseed oils are rich sources of ALA [71]. The long-chain ω6-PUFA, AA, is present in small amounts in meat, poultry, and eggs and absent in vegetable oils [72], whereas long-chain ω3-PUFAs, EPA, and DHA are derived notably from marine fish and shellfish and to a lesser extent in egg yolk, meat, and some fruit [71,73]. The contents and ratios of EPA, and DHA in fish are dependent on the food that is available to the fish; in other words, the amount of phytoplankton and zooplankton from which ω3-PUFAs ultimately originate. Thus, it is not unexpected that fish oil from different sources contains a variable mixture of EPA and DHA, and that the contents of long-chain ω3-PUFAs in fish are influenced strongly by the species of fish, season of the year, and the geographical areas where the fish live in [74,75]. In general, fish such as mackerel, salmon, trout, and herring are higher in long-chain ω3-PUFAs than others [71,74].

15.3 Basic aspects of cancer development

Major advances have been made in understanding cancer development over the past century. It is now widely accepted that the development of cancer from normality to malignancy is a slow, complex multistep process, characterized by the acquisition of novel biological capabilities involving tumor cell aspects and modified microenvironment interactions [76–78]. This slow progression of cancer, which requires an average of 20–40 years in humans [12,79], accumulates genomic alterations that affect key molecular pathways in a stepwise fashion, and results in cell populations with unlimited proliferation potential based on evolutionary selection [80]. The acquired novel capabilities, also known as cancer hallmarks, constitute an organizing principle for rationalizing the complexities of cancer development, and regulate cancer progression through sustaining cell-proliferation signals, inhibiting growth suppressors and cell death, promoting angiogenesis and inflammation and activating metastasis [78,81,82]. In addition to these shared hallmarks of cancer, the development of each cancer also has its own unique set of risk factors, genetic and pathological characters. For instance, any form of the tobacco use is highly linked to lung cancer development [83]; persistent infection of high-risk human papillomavirus (HPV) is nearly responsible for all the premalignant and malignant epithelial lesions of the cervix [84]; the status of estrogen receptor (ER) plays a critical role in the progression and treatment of breast cancer [85].

In the following sections, we will review the role of ω3-PUFAs in genomic stability maintenance, preventing cancer cell-proliferation, inducing cancer cell death, inhibiting inflammation and angiogenesis, and reducing metastasis as well as the potential signaling pathways involved at each stage.

15.4 Anticancer actions of ω3-PUFAs during multistage of cancer progression

15.4.1 Role of ω3-PUFAs on genome instability: implications for epigenetic regulation

Genome instability refers to a set of events that result in an increased tendency of alterations within genome, and act as a major driving force for cancer development [86]. Studies from epidemiology and molecular biology have demonstrated that most, if not all, human cancers display some form of genomic instability, which can occur on the nucleotide level and/or chromosomal level [86,87]. Instability at the nucleotide level occurs due to faulty DNA repair pathways such as base excision repair and nucleotide excision repair and includes instability of repeat sequences caused by defects in the mismatch repair pathway; instability at the chromosomal level defines the existence of accelerated rate of chromosomal alteration that leads to gains or losses of whole chromosomes as well as inversions, deletions, duplications, and translocations of large chromosomal segments [87–89].

Because of the critical role played by genome instability in cancer development, it is rational to assume that maintaining genome stability is important for preventing the progression of cancer. Maintaining genome stability in the face of replication and recombination allows the proper passage of genomic information to subsequent generations, and requires a variety of different DNA-damage-response proteins involving DNA repair, cell cycle arrest, and epigenetic modifications of the genome [90–92]. As proteins involved in DNA repair and cell cycle arrest, such as p53 and p21, also regulate cancer cell proliferation and death, which will be reviewed in the subsequent sections, we here focus on how ω3-PUFAs maintain genome instability via regulating epigenetic modifications.

The term "epigenetics" refers to modifications in gene expression without altering the DNA sequence. Epigenetic regulation via DNA methylation, chromatin modulation, and microRNAs (miRs), occurring in response to the environment such as diet, is heritable but reversible; thus it offers additional explanations for regulatory mechanisms involved in temporal and spatial gene expression [93–97].

15.4.1.1 DNA methylation and ω3-PUFAs

One of the most well-studied mechanisms of epigenetic transcriptional regulation is that of DNA methylation, controlled by the DNA methyltransferases (DNMTs). DNMTs confer covalent methyl modifications to CpG islands in the promoter regions of DNA [96]. This mechanism is thought to act through two ways: either inhibiting the binding of the transcriptional machinery, or acting as donor binding sites for methyl-CpG-binding domain proteins that may recruit histone modifiers [92,96]. Generally, the modification of DNA by DNMTs induces transcriptional repression, while hypomethylation of gene promoter regions is associated with increases in transcriptional activity. It is now believed that the promoter CpG island hypermethylation of tumor-suppressor genes and global DNA hypomethylation are two main types of methylation changes in cancer progression [98]. One study investigating the hypermethylation of tumor-suppressor CCAAT/enhancer-binding proteins (C/EBPs) and ω3-PUFAs found that EPA protected against promoter methylation of C/EBP δ and potently induced cell cycle arrest in U937 leukemia cells [99]. More recently, Kulkarni et al. [100] found a positive correlation between DHA amount and placental global DNA methylation level in a rat model. The mechanism by which DHA promotes the global DNA methylation level seems highly associated with the effect of DHA on one-carbon metabolism [100,101].

15.4.1.2 Chromatin modulation and ω3-PUFAs

Eukaryotic chromosomes consist of DNA packaged in a highly ordered way around histone proteins forming a nucleosome. These nucleosomes are further packaged into dense chromatin fibers, which exist as either transcriptionally silent heterochromatin, or transcriptionally active euchromatin. Nucleosome repositioning is the mechanism by which the DNA is made available for transcription, and allows the binding of polymerases to promoter sequences within the DNA [102]. The nucleosomal subunits of chromatin contain histone tail domains, which are unstructured and are the target sites of posttranslational modifications such as acetylation, methylation, phosphorylation, and ubiquitination [103]. It is thought that these modifications affect either the charges on the histone proteins, thereby causing repulsion and opening up the chromatin structure, or by permitting binding by other DNA-associated proteins. In this way, the repositioning of heterochromatin into euchromatin and nucleosome–nucleosome associations are regulated [104]. The placement of these modifications is orchestrated by a variety of site-specific enzymes such as histone methyltransferases [104,105]. ω3-PUFAs also cause changes in histone proteins and hence chromatin modulation is implicated greatly in ω3-PUFAs-induced epigenetic modification. For example, it has been reported that both EPA and DHA decrease the activity of histone 3 lysine 27 trimethylation in breast tumor cells via downregulating enhancer of zeste homologue 2, a histone methyltransferase [106]. In addition to histone methyltransferases, histone deacetylases (HDACs) also play an important role in chromatin modulation. Acetylation of lysine residues in the N-terminal tail domain of histones results in an allosteric change in the nucleosomal conformation and an increased accessibility to transcription factors by DNA, whereas deacetylation of histones is associated with transcriptional silencing [105]. A decreased level of acetylated histones with increases in HDAC7 expression in DHA-pretreated T cells (pro-inflammatory immune cells) had been reported [107]. Nevertheless, Suphioglu et al. [108] showed that DHA increased both the gene and protein levels of histone 3 and 4, attenuating HDAC1, 2, and 3 expression levels (unpublished data) in a human neuronal-derived M17 cell line, and prevented zinc deficiency-mediated neuronal degeneration. The different effects of DHA on histones function and HDACs, on the one hand, could be explained by the different targets. On the other hand, it may be related to the different cellular contexts. Indeed, numerous studies suggest that the effect of ω3-PUFAs appears to be dependent on the cell background. For instance, ω3-PUFAs selectively killed lung, breast, and prostate cancer cells without influencing human normal cells [109]; it has also been shown that DHA is toxic to neuroblastoma cells, whereas it protects healthy nerve cells [7,110].

15.4.1.3 miRs and ω3-PUFAs

miRs are naturally occurring, small, noncoding RNA sequences, which have been implicated at all stages of cancer from initiation to tumor progression [111]. One of the earlier reports on ω3-PUFAs-mediated miR alteration was from the study of carcinogen-induced colon cancer in rat models. A global perturbation in miR expression patterns was first identified in carcinogen-induced colonic tumors. Further examination of the differences in carcinogen treatment versus control miR expression across diet groups revealed that five known tumor-suppressor miRs (let-7d, miR-15b, miR-107, miR-191, and miR-324-5p) were selectively upregulated by ω3-PUFAs exposure [112]. A second study in which the ω3-PUFA, DHA was employed in three glioblastoma cell lines showed that most altered miRs, such as miR-20b and miR-145, and their target coding genes in all their DHA-treated brain tumor cells were associated with apoptotic cell death [113]. In addition to their effects on cancer, ω3-PUFAs have also been found to affect an obesity-related phenotype via mediating miR-522 in human beings [114]. Although it is still unclear how ω3-PUFAs regulate such a variety of miRs and trigger different effects, these findings clearly indicate that miR-mediated epigenetic modification is also involved in the anticancer actions of ω3-PUFAs.

15.4.2 Role of ω3-PUFAs in cancer cell proliferation

Noncancerous cells carefully control their division cycle by regulating proliferation signals to ensure a homeostasis of cell number and thus maintain the normal tissue structure and function, and arguably, the most fundamental trait of cancer cells involves their ability to proliferate infinitely [78]. The mammalian cell cycle consists of four main stages: G1, S (DNA synthesis), G2, and M (mitosis), with G1 and G2 being referred to as "gap" phases between the events of S and M, respectively. In addition, there is a fifth phase, referred to as G0, representing a state of quiescence outside the cell cycle in which cells stop cycling after division. Progress through the cycle is mediated by two types of cell cycle control mechanisms: the coordinated interaction of cyclins with their respective cyclin-dependent kinases (CDKs) that relay a cell from one stage to the next at the appropriate time, and a set of checkpoints that monitor completion of critical events and delay progression to the next stage if necessary [115].

15.4.2.1 Cyclin–CDK complex-controlled cell cycle and ω3-PUFAs

The cell cycle controlled by the specific cyclin–CDK complexes involves a diverse family of proteins, termed cyclin-dependent kinase inhibitors (CKIs), which bind and inactivate cyclin–CDK complexes. The four major mammalian CKIs could be classified into two groups: p21, p27, and p57 are related Cip/Kip family CKIs with a preference for CDK2– and CDK4–cyclin complexes, whereas p15, p16, p18, and p19 are INK4 family CKIs specific for CDK4– and CDK6–cyclin complexes [116–118]. DHA has been shown to increase the expression level of p21 and arrest cell cycle in leukemic cells [119], breast [120], and colon cancer cells [121]. Meanwhile, one study performed by Mernitz et al. showed that high intake of ω3-PUFAs significantly reduced tumor prevalence in a rodent model of carcinogen-induced lung carcinogenesis, and that the inhibitory effect of ω3-PUFAs on tumor prevalence was associated with p21 upregulation [122]. In addition to p21, an increased level of p27 protein expression with cycle arrest has been reported in DHA-treated melanoma [123] and mouse mammary cancer cells [124]. In a colon cancer cell line, Narayanan et al. [125] found that besides p21 and p27, DHA treatment also increased the expression levels of other CKIs (p57 and p19).

15.4.2.2 Checkpoint-controlled cell cycle and ω3-PUFAs

The main checkpoints involved in the cell cycle control are G1/S and G2/M [126], and since they allow cells to respond to DNA damage, checkpoints are important quality control of the cell cycle, protecting cells against genomic instability and ensuring the fidelity of cell division [127,128]. At the G1/S transition, cyclin D plays a critical role. It binds to CDK4 or CDK6 and promotes progression by phosphorylating the Rb protein, thus releasing the transcription factor E2F to transcribe genes required for cell cycle progression and DNA synthesis [129,130]. ω3-PUFAs, particularly DHA, have been shown to suppress the expression of cyclin D in various cancer cells such as colon [121,131,132], breast [120,133], and prostate cancer cells [134]. ω3-PUFAs also have an anti-proliferation influence on cancer cells by mediating the G2/M transition. Jourdan et al. [135] reported that in MDA-MB-231 breast cancer cells, EPA and DHA both markedly increased the duration of the G2/M phase with decreased activity of cyclin B1–CDK1 complex, an essential regulator involved in the progression from G2 to M.

15.4.3 Role of ω3-PUFAs in cancer cell death

Millions of cells commit suicide each second within the human body as a result of normal development and loss associated with pathology to preserve tissue homeostasis as well as eliminate damaged cells, and thus, failure to do so can lead to various human diseases, including cancer [136,137]. Although the types of cell death used to be classified by the different morphological criteria without a clear reference to precise biochemical mechanisms [138], it is now generally accepted that based on the measurable biochemical features, cell death could be classified into apoptosis, autophagic cell death (cell death by autophagy), and necrosis, at least three types, with each type occasionally overlapping with another [139]. Notably, given that autophagy is often observed in apoptosis [140,141] as well as necrosis [142], and the complicated role (i.e., pro-survival or pro-death of cells) it plays, a more encompassing description "autophagy-associated cell death" is proposed with "autophagic cell death" being a form of programmed cell death in which autophagy per se serves as a cell death mechanism (cell death by autophagy) [139,143–146]. In addition, while increasing numbers of studies show that ω3-PUFAs induce cancer cell death, research on the relationship between ω3-PUFAs and necrosis in cancer cell death is rare. One study using L929 murine fibrosarcoma cells showed that DHA attenuated the tumor necrosis factor-induced necrosis [147], suggesting that necrosis may not be the favorite cell death induction pathway that ω3-PUFAs choose to kill cancer cells. The reason for this is not clear. One possible explanation is the nature of necrotic cell death and ω3-PUFAs, as necrosis is a programmed cell death [148], characterized morphologically by cell membrane rupture and an induction of inflammation around the dying cell attributable to the release of cellular contents and pro-inflammatory molecules [149,150], and ω3-PUFAs have well-documented anti-inflammatory effects [52,151,152].

15.4.3.1 Apoptosis and ω3-PUFAs

Apoptosis is a series of dramatic perturbations to the cellular architecture, leading to not only cell death, but also the elimination of apoptotic cell corpses by phagocytes without unwanted immune responses [153]. Owing to the compelling functional studies in the past 20 years, the concept that apoptosis could serve as a natural barrier to cancer progression has been established [78,154,155]. The apoptosis machinery consists of two distinct pathways (intrinsic and extrinsic) with the intrinsic pathway being the primary death program responsive to the signals of survival factors, cell stress, and injury [154]. The major control center of the intrinsic pathway is mitochondria, which contain the pro- and anti-apoptotic members of regulatory proteins. These two groups of regulative proteins work together and maintain the mitochondrial membrane potential (MMP) [78,154]. It is now clear that MMP is determined by the balance between the pro-apoptotic Bax/Bak proteins and their anti-apoptotic Bcl2/BclXL cousins, and that upon apoptotic stimulation, the MMP decreases, releasing several pro-apoptotic regulators, including cytochrome c (cyt c) [156], smac/diablo [157], endonuclease G [158], and apoptosis-inducing factor (AIF) [159], to the cytosol and/or nucleus [160]. The released cyt c subsequently activates a cascade of caspases (caspase 9-caspase 3), which act through their proteolytic activities to trigger a variety of cellular changes associated with apoptosis; smac/diablo inhibits several members of anti-apoptotic protein and thus further promotes apoptosis; endonuclease G and AIF relocate to the nucleus where they induce massive DNA fragmentation independently of caspases [139,160]. Unlike the intrinsic pathway, which is highly dependent on mitochondria, the extrinsic apoptosis pathway is mediated by various death receptors located on the cell membrane. The binding of corresponding ligands to the receptors activates caspase 8-caspase 3 cascade directly, thereby triggering the executioner phase of caspase-dependent apoptosis. Notably, in some types of cells such as hepatocytes and pancreatic β cells, caspase 8 can also lead to the generation of a mitochondrion-permeabilizing fragment, which further downregulates MMP and triggers the extrinsic apoptosis associated with the mitochondria-mediated intrinsic pathway [139,161].

Both the intrinsic and extrinsic apoptosis-mediated cell death has been reported in a bewildering range of cancer cells treated with ω3-PUFAs [162–166]. In a study using three different cholangiocarcinoma cell lines, we reported that ω3-PUFAs activated the caspase 9, triggering the release of cyt c from mitochondria to cytosol, and suggested that the mitochondria-mediated intrinsic apoptosis is involved in the death induction of cancer cells treated with ω3-PUFAs. Further studies showed that the apoptosis induced by ω3-PUFAs was accompanied with the inactivation of Wnt/β-catenin signal pathway [167,168]. The reduced level of Wnt/β-catenin signaling molecules and increased level of apoptotic cell death were also observed by us in both ω3-PUFAs-treated pancreatic cancer cells and pancreatic cancer cells implanted into a Fat-1 transgenic mouse model, which expresses ω3-desaturases and results in elevated levels of ω3-PUFAs endogenously [169–171]. In another study using HCT116 and SW480 colon cancer cells, Calviello et al. reported that DHA reduced the activity of Wnt/β-catenin pathway and induced caspase 3-dependent apoptosis in both two cell lines as well, and that the β-catenin expression inhibition induced by DHA was posttranslationally modified in a proteasome-dependent manner [172]. Although, until now, we are not aware of any published studies that have investigated how exactly the inactivation of Wnt/β-catenin signaling caused by ω3-PUFAs is associated with the mitochondria-mediated intrinsic apoptosis, there is evidence suggesting that β-catenin can translocate to mitochondria, where it binds to the anti-apoptotic protein Bcl2 and promotes cancer cell survival [173,174]. Therefore, it is possible that the intrinsic apoptosis induction and the decreased Wnt/β-catenin signaling activity in ω3-PUFAs-treated cancer cells might have resulted from the decreased binding ability of β-catenin to anti-apoptotic protein Bcl2 in mitochondria.

Mitochondria as the key players in intrinsic apoptosis are able to sense a variety of cellular and stress signals, such as p53 [175,176]. Tumor-suppressor p53 can directly translocate to mitochondria and induce a transcription-independent apoptosis via several different mechanisms [177–179], and p53 has also been reported to play an important role in the apoptosis of the ω3-PUFAs-treated cancer cells. For example, hepatoma cells expressing wild-type p53 have been shown to be more sensitive than those with dysfunctional p53 to EPA-induced apoptosis [180]; Δ12-prostaglandin J3, an EPA-derived metabolite selectively activates p53 and induces p53-dependent apoptosis in mice leukemia stem cells [181]. It should be noted, however, that the effect of ω3-PUFAs on apoptosis induction is not extremely dependent on p53 [182–184], and that more than one single mechanism is involved in the apoptosis induced by ω3-PUFAs [41,185,186].

In addition to the mitochondria-mediated intrinsic apoptosis, studies suggest that ω3-PUFAs can also kill cancer cells via the extrinsic apoptosis pathway. DHA has been recently reported to activate caspase 8 and induce apoptosis in MCF-7 breast cancer cells, and inhibition caspase 8 using specific inhibitor or knockdown caspase 8 could prevent apoptosis induced by DHA [187]. Giros et al. [188] studied the apoptotic mode in colorectal cancer cells treated with ω3-PUFAs, and observed that besides the release of smac/diablo, cyt c from mitochondria into cytosol, the activity of caspase 8 increased remarkably, indicating the activation of both intrinsic and extrinsic apoptosis pathways. They further showed that the extrinsic apoptosis induction, however, was not dependent on the activation of known death receptors, but a direct effect of ω3-PUFAs on the FLICE-like inhibitory protein, a crucial extrinsic apoptosis pathway negative regulator. It is interesting to note that in another study, EPA was found to trigger apoptosis in HepG2 hepatoma cells along with an increase in the expression of cell death receptor Fas on cell surface, and the EPA-induced apoptosis could be almost completely rescued by inactivating Fas [180].

15.4.3.2 Autophagy-associated cancer cell death and ω3-PUFAs

Autophagy is an evolutionarily conserved process characterized by the sequestration of bulk cytoplasm and organelles in double- or multi-membrane autophagic vesicles (termed autophagosomes) and their following degradation [189,190]. The formation of autophagosome starts at phagophore and requires the conjugation of a series of autophagy-related proteins (ATGs), which regulate the phagophore expansion and completion of the sequestering vesicles [191]. The mammalian target of rapamycin (mTOR), a negative regulator of autophagy located upstream of ATGs, plays an important role in autophagy induction by integrating different sensory inputs from a variety of pathways [192] and modifying ATGs [193,194]. Upon completion, the autophagosome fuses with lysosome, becoming autolysosome, and the sequestered components are degraded by lysosomal hydrolases and released into the cytosol by lysosomal efflux permeases [195]. The protein degradation and organelle turnover controlled by autophagy is essential for cell survival, and the disruption of this process can result in abnormal cell growth or cell death, leading to various diseases [196].

Owing to the studies in the past decades, our knowledge regarding autophagy and its connection to various diseases including cancer has been vastly expanded [189,197–199]. Now it seems clear that autophagy has paradoxical roles during cancer development. Nearly all eukaryotic cells undergo autophagy at a basal level under normal physiological conditions to maintain cellular homeostasis [200,201], and at the early stage of cancer development, autophagy acts as an anticancer process to maintain genomic stability by eliminating damaged organelles/protein aggregates that produce genotoxic stresses as well as accelerating the degradation of proteins required for cancer cell growth [202]. Yet, at the advanced stage of cancer progression, cancer cells, in particular those located in the center of tumor mass, use autophagy as a survival mechanism against low-oxygen and low-nutrient stressful conditions [202,203]. Likewise, the role of autophagy during cancer cell death can also be complicated. Autophagy can either promote cancer cell survival in response to anticancer agents, or induce cancer cell death, depending on the different cellular contexts and stimuli [139,198,204].

Unlike numerous studies on the apoptosis induction effect of ω3-PUFAs, our understanding of autophagy in the ω3-PUFAs-treated cancer cells remains insufficient. In a recent study conducted by us, it has been revealed that autophagy may be also involved in the effect of ω3-PUFAs-induced cancer cell death [205]. When treating SiHa cervical cancer cells with DHA, we found a remarkable accumulation of numerous lamellar structures with cytosolic autophagic vacuoles in these cells under transmission electron microscopy. Further study showed that the autophagic vacuolization induced by DHA was associated with the mTOR inactivation and ATG8 augmentation. Meanwhile, the effect of DHA on autophagy induction seems not dependent on cancer cell types, as we confirmed this induction using several other types of cancer cells. The DHA-induced autophagy activation was also accompanied by the intrinsic apoptotic cell death in SiHa cells, and when inhibiting autophagy, the effect of DHA on apoptosis significantly reduced. These results indicate that instead of triggering autophagic cell death, autophagy contributes to apoptotic cell death in DHA-treated SiHa cells, in that DHA treatment induces the autophagy-associated apoptotic cell death. It is worth mentioning that we also found that the DHA-induced mTOR inactivation and autophagy was mediated by p53 loss in cancer cells expressing wild-type p53. This observation is consistent with several studies on the novel role of p53 in autophagy regulation [206–208], and reveals a novel mechanism for the cancer cell death induction effect of ω3-PUFAs. Obviously, further investigation is required to unravel the mechanism by which ω3-PUFAs downregulate p53 in these wild-type p53 expressing cancer cells, as well as whether autophagy activation is a universal effect of ω3-PUFAs on the death of various types of cancer cells, such as cancer cells carrying mutant p53.

15.4.4 Role of ω3-PUFAs in tumor-associated inflammation

The functional relationship between cancer and inflammation has been long recognized. Now it is clear that cell proliferation alone does not cause cancer, and it is the environment rich in inflammatory cells and activated stroma where the sustained cell proliferation occurs, potentates, and promotes the cancer progression [53]. Tumor cells produce various cytokines and chemokines that attract diverse inflammatory cells, for example, neutrophils, macrophages, and lymphocytes, which are capable of generating an assorted array of bioactive molecules. Inflammation contributes to multiple hallmark capabilities by providing the tumor microenvironment with the bioactive molecules, such as growth factors that sustain proliferative signaling, survival factors that limit cell death, pro-angiogenic factors, extracellular matrix (ECM)-modifying enzymes that facilitate angiogenesis and metastasis [78,209].

The key mediators and regulators of inflammation are eicosanoids generated from 20 carbon PUFAs (AA and EPA), and because inflammatory cells typically contain a high proportion of AA and low proportions of EPA, AA is generally the major substrate for eicosanoid synthesis [151]. As ω3-PUFAs compete with AA for the enzymatic conversion (see Section 15.2.1), it is therefore believed that ω3-PUFAs exert an anti-inflammatory action by inhibiting the production of AA-derived eicosanoids. Indeed, studies have shown that the levels of inflammation can be attenuated by promoting the proportion of ω3-PUFAs in inflammatory cell phospholipids, and that supplementing the diet of laboratory rodents and humans with omega-3 fish oil results in decreased production of a wide range of AA-derived eicosanoids [210–215]. However, it should be pointed out again that the eicosanoids derived from AA and EPA do not always have different effects on inflammation [151]. For instance, among AA-derived eicosanoids, COXs-synthesized PGE2, which plays an important role in inflammation [216,217], has an equivalent inhibitory effect as EPA-derived PGE3 on tumor necrosis factor-α (TNF-α) production in human mononuclear cells stimulated with lipopolysaccharide (LPS) [218], and as mentioned previously, AA-derived LXs are anti-inflammatory instead of being pro-inflammatory, indicating that the AA-derived lipid mediators may also play an important anti-inflammatory role. Thus, in addition to the absolute amounts of ω6- and ω3-PUFAs intake, the concept of dietary ω6/ω3-PUFAs ratio is introduced. Generally, a physiologic ratio of ω6-PUFAs to ω3-PUFAs is about 1–4:1, and the lower ratio is associated with the decreased inflammatory burden, although the optimal ratio may vary with the disease under consideration [219,220]. The high ω6/ω3-PUFAs ratio (15–16.7:1), as found in the current Western diets, is thus believed to be responsible for the increasing incidence of inflammation-associated diseases, including cancer [221].

Importantly, in addition to altering the profile of lipid mediators, ω3-PUFAs also affect the gene expression of peptide mediators (i.e., cytokines) through modifying the activity of transcription factors, most likely nuclear factor κB (NF-κB) and peroxisome proliferator-activated receptor γ (PPARγ) [151,222].

Mullen et al. [223] reported that ω3-PUFAs reduced the expression of pro-inflammatory interleukins, IL-6 and IL-1β, two downstream genes of NF-κB, in LPS-activated macrophages with decreases in NF-κB activity. MAD-MB-231 breast cancer cells treated with ω3-PUFAs also showed downregulated expression levels of NF-κB and pro-inflammation molecules [224]. These observations together, therefore, indicate a direct effect of ω3-PUFAs on inflammatory gene expression via inhibiting transcription factor NF-κB. Unlike NF-κB whose activation promotes inflammation, the second transcription factor, PPARγ acts in an anti-inflammation manner [151]. PPARγ can be activated by ω3-PUFAs and their metabolites [225]. Using immortalized human proximal tubular cell line HK-2, Li et al. [226] have reported that the anti-inflammation effects of EPA and DHA are indeed dependent on the activation of PPARγ. As such, this raises the possibility that ω3-PUFAs enhance the activity of PPARγ and lead to anti-inflammation effects during cancer development.

15.4.5 Role of ω3-PUFAs in tumor angiogenesis

Mammalian cells are located within 100–200 μm of blood vessels because of their requirement for oxygen and nutrients. To grow beyond this size, multicellular organisms must recruit new blood vessels by angiogenesis, a process regulated by a balance between pro- and anti-angiogenic molecules [227]. Prior to neovascularization, human tumors can remain dormant indefinitely until the balance is altered and the tumor switches to an angiogenic phenotype [228]. This is a critical step and signaling by the major pro-angiogenic factor vascular endothelial growth factor (VEGF) often represents a crucial rate-limiting step during tumor progression [229]. ω3-PUFAs have been shown to attenuate the VEGF-induced proliferation of several types of endothelial cells in vitro [230–232]. In vivo studies also demonstrated that ω3-PUFAs intake suppressed pathological retinal angiogenesis in mice, an effect induced by the bioactive ω3-PUFAs-derived lipid mediators, resolvins, and that this protective effect was regulated partly by inhibiting TNF-α, which is an inflammatory cytokine found in a subset of microglia closely associated with retinal vessels [233]. Subsequent studies from the same group further revealed that the beneficial effect on the inflammatory and angiogenic activation was mediated by the ω3-PUFAs-induced PPARγ activation [234]. Cotreatment with an inhibitor of PPARγ significantly abrogated the anti-angiogenic effect of ω3-PUFAs, and reversed the ω3-PUFAs-induced reduction in a group of pro-angiogenic molecules. Although VEGF did not involve the PPARγ-mediated anti-angiogenic effect of ω3-PUFAs in this mouse model of oxygen-induced retinopathy [234], several studies showed that ω3-PUFAs inhibited the expression of VEGF in cancer cells and in tumors derived from cells transplanted in ω3-PUFAs-fed mice [235–237]. These studies suggested that the reduced VEGF expression might be associated with the negative regulating effect of ω3-PUFAs on COX-2 and hypoxia-inducible factor (HIF), two key upstream regulators of VEGF during angiogenesis [238,239]. Indeed, studies have shown that ω3-PUFAs suppressed these two key angiogenesis molecules in a variety of cellular contexts and animal models [240–243]. Based on the data currently available on tumor angiogenesis, it is not clear whether other mechanisms are involved in the inhibitory effects of ω3-PUFAs on VEGF, and their detected effects on tumor vasculature deserve to be future investigated. In addition, effects of ω3-PUFAs on the activity of other pro- and anti-angiogenic molecules, such as platelet-derived growth factor, restin, and platelet-derived endothelial cell growth factor, have not been studied and represent an opportunity for original research [244].

15.4.6 Role of ω3-PUFAs in cancer invasion and metastasis

Tumor invasion and metastasis are the critical steps in determining the aggressive phenotype of cancers and the major contributor to death in cancer patients [245]. The term invasion indicates penetration of cancer cells into neighboring territories and their occupation; the invasion further permits these cells to enter into the circulation from where they can reach distant organs and eventually form secondary tumors, called metastasis [246]. This spread of cancer cells is a complicated process involving cross-talk between cancer cells and host cells as well as the alterations in genes associated with metastasis [246,247]. Not all the cancer cells can metastasize, and noninvasive tumors are benign because they are simply cured by surgical removal [246]. The successful metastasis is dependent on both the intrinsic properties of cancer cells and the factors derived from the tumor microenvironment that provides blood vessels and an inflammatory milieu consisting of immune cells and their secretory products, and a scaffold in form of the ECM [248,249].

Several studies have shown signs of promises in using ω3-PUFAs as a possible means of treating cancer metastasis. Prostate cancer cells and brain-metastatic melanoma cells exposed to ω3-PUFAs showed decreased invasiveness and migration in vitro, an effect caused by inhibiting inflammation molecule COX-2 activation [250,251]. Altenburg et al. reported that ω3-PUFAs inhibited the migration of MDA-MB-231 breast cancer cells mediated by CXCR4, a chemokine receptor widely expressed in invasive cancer cells and highly associated with inflammation [53,252,253]. In addition, a study investigating the effects of PUFAs on liver metastasis in experimental ductal pancreatic adenocarcinoma demonstrated that omega-3 fish oil feeding decreased number of pancreatic tumors and incidence of liver metastasis with reduced concentration of pro-inflammatory LTs and PGs in metastatic tissues in hamsters [254]. Observations like these indicate that at least partially, the anti-metastatic action of ω3-PUFAs is due to their effect on regulating inflammation. Furthermore, ω3-PUFAs have also been demonstrated to inhibit metastasis by directly modifying the expression of metastatic molecules in cancer cells. For example, ω3-PUFAs upregulated E-cadherin expression and decreased the invasion in breast cancer cells [106]. The expression of E-cadherin is important in maintaining the integrity of intercellular adhesion, and its loss is associated with an increased tendency for tumor metastasis [246]. Another group of metastasis-associated molecules, matrix metalloproteinases (MMPs), and their corresponding tissue inhibitor metalloproteinases (TIMPs) have also been shown to be involved in the direct anti-metastatic effects of ω3-PUFAs [172,255–260]. MMPs are endoproteinases, and TIMPs inhibit MMPs by forming tight-binding, noncovalent associations with their active sites. As cancer cells must break existing cell–matrix and cell–cell attachments to move through the ECM, MMPs whose enzymatic activity is directed against components of the ECM, remove physical barriers to invasion through degradation of ECM macromolecules such as collagens, laminins, and proteoglycans [261,262]. On the whole, the outcomes of the published studies seem all in agreement and demonstrate the effects of ω3-PUFAs on reducing tumor invasion and metastasis. However, given the scarcity of these studies, further investigation is required to definitely ascertain the anti-metastasis action of ω3-PUFAs and extensive work in this field is clearly needed to unveil the precise underlying signaling mechanisms.

15.5 Conclusions

Since the essential dietary role of ω3-PUFAs was first discovered in 1930s [263], ω3-PUFAs have been intensively studied and the beneficial effects have been linked with a bewildering array of human disorders such as arthritis [219], depression [264], neurodegenerative diseases [265], malaria [266], heart disease [267], and cancer [14]. In this chapter, we chose the model of cancer development as a frame and reviewed the considerable body of evidence for the effects of omega-3 fish oil on cancer progression as well as the potential mechanisms underlying these effects (**Figure 15.3**). Nevertheless, as described in Section 15.3, the development of certain cancer also involves some unique risk factors. It is highly possible that ω3-PUFAs may also target these unique features to inhibit the progression of certain cancer. This seems rational, because there have been already some studies showing that these unique factors indeed could be regulated by ω3-PUFAs. For instance, EPA and DHA reduced the proliferation of cervical keratinocytes immortalized with HPV [268], which is a necessary cause of cervical cancer [269], and DHA has been reported to induce ER proteasomal degradation and inhibit MCF-7 breast cancer cell proliferation [133]. Although more studies are needed to verify these effects, these observations together with the effects of ω3-PUFAs on the multistage of cancer development indicate that a variety of mechanisms are involved in the anticancer actions of ω3-PUFAs.

One aspect of the ω3-PUFAs study that we have scarcely touched upon is the differing effects of EPA and DHA. Indeed, studies have shown that in addition to the difference in their structure and metabolism in the body, DHA and EPA also have different effects on cell membrane properties [270], serum lipid metabolism [271], and blood pressure regulation [272]. Although most studies that used both EPA and DHA showed, generally, that the effect of DHA on cancer cell proliferation and cell death induction is more stronger

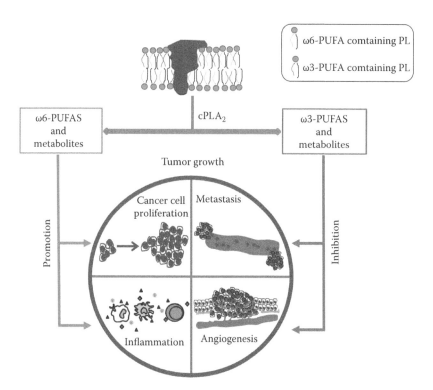

Figure 15.3 *Typical effects of ω3-PUFAs and their metabolites on cancer cell proliferation, tumor angiogenesis, inflammation, as well as metastasis. The cancer development is a complex process requiring a succession of hallmarks such as unstoppable proliferation, inflammation, new blood vessel formation, and spread of cancer cells to other sites of the body. ω3-PUFAs and their metabolites downregulate these acquired biological capabilities during the multistep development of tumors, and thus inhibit cancer progression. Of note, ω6-PUFAs and the metabolites derived from them, on the other hand, are generally believed to have contrasting effects on these hallmarks and promote the development of cancer. cPLA₂, cytosolic phospholipase A₂; PL, phospholipids.*

[163,169,187,273], EPA, but not DHA, has been shown to attenuate the loss of adipose tissue and preserve skeletal muscle in a murine model of cancer cachexia [274], indicating that these two PUFAs have different effects in the process of cancer development. Thus, it seems likely that the combination of EPA and DHA may have more potent effects on cancer treatment and prevention than either one alone.

Collectively, from the data discussed in this chapter, it is apparent that the anticancer effects of the omega-3 fish oil are likely mediated through influencing various pathways within tumors (**Figure 15.4**). Further investigation is required to study the molecular relationships between omega-3 fish oil and cancer progression, and how omega-3 fish oil interacts with endogenous systems as well as other exogenous factors. It is possible that continued studies on omega-3 fish oil with a high content of DHA and EPA will yield and

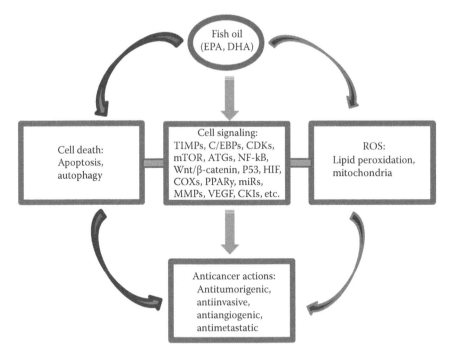

Figure 15.4 *Proposed anticancer action of omega-3 fish oil. Omega-3 fish oil possesses anticancer activities via their effects on a variety of biological pathways involved in gene expression, cell cycle and cell death regulation, angiogenesis, and metastasis. C/EBPs, CCAAT/enhancer-binding proteins; miRs, microRNAs; CDKs, cyclin-dependent kinases; CKIs, cyclin-dependent kinase inhibitors; mTOR, mammalian target of rapamycin; NF-κB, nuclear factor κB; PPARγ, peroxisome proliferator-activated receptor γ; VEGF, vascular endothelial growth factor; HIF, hypoxia-inducible factor; MMPs, matrix metalloproteinases; ATGs, autophagy-related proteins; TIMPs, tissue inhibitor metalloproteinases; COXs, cyclooxygenases. See text for further details.*

provide important implications in developing future strategies to use omega-3 fish oil in cancer prevention and therapy.

Acknowledgment

This work was supported by the National Research Foundation of Korea (NRF) grant funded by the Korean government (2012-0005767, 2012-0005456, and 2011-0003060).

References

1. Bang HO, Dyerberg J, Sinclair HM. The composition of the Eskimo food in north western Greenland. *Am J Clin Nutr* 1980, 33:2657–2661.
2. Bang HO, Dyerberg J, Nielsen AB. Plasma lipid and lipoprotein pattern in Greenlandic West-coast Eskimos. *Lancet* 1971, 1:1143–1145.

3. Dolecek TA, Granditis G. Dietary polyunsaturated fatty acids and mortality in the multiple risk factor intervention trial (MRFIT). *World Rev Nutr Diet* 1991, 66: 205–216.
4. Shekelle RB, Missel LV, Paul O, Shryock AM, Stamler J. Fish consumption and mortality from coronary heart disease. *N Engl J Med* 1985, 313:820–824.
5. Stone NJ. Fish consumption, fish oil, lipids, and coronary heart disease. *Circulation* 1996, 94:2337–2340.
6. Kromhout D, Bosschieter EB, de Lezenne Coulander C. The inverse relation between fish consumption and 20-year mortality from coronary heart disease. *N Engl J Med* 1985, 312:1205–1209.
7. Gleissman H, Johnsen JI, Kogner P. Omega-3 fatty acids in cancer, the protectors of good and the killers of evil? *Exp Cell Res* 2010, 316:1365–1373.
8. Riediger ND, Othman RA, Suh M, Moghadasian MH. A systemic review of the roles of n-3 fatty acids in health and disease. *J Am Diet Assoc* 2009, 109:668–679.
9. Gebauer SK, Psota TL, Harris WS, Kris-Etherton PM. n-3 fatty acid dietary recommendations and food sources to achieve essentiality and cardiovascular benefits. *Am J Clin Nutr* 2006, 83:1526S–1535S.
10. Abayasekara DR, Wathes DC. Effects of altering dietary fatty acid composition on prostaglandin synthesis and fertility. *Prostaglandins Leukot Essent Fatty Acids* 1999, 61:275–287.
11. Kolanowski W, Swiderski F, Lis E, Berger S. Enrichment of spreadable fats with polyunsaturated fatty acids omega-3 using fish oil. *Int J Food Sci Nutr* 2001, 52:469–476.
12. Kelloff GJ, Crowell JA, Steele VE, Lubet RA, Malone WA, Boone CW et al. Progress in cancer chemoprevention: Development of diet-derived chemopreventive agents. *J Nutr* 2000, 130:467S–471S.
13. Harvey AL. Natural products in drug discovery. *Drug Discov Today* 2008, 13:894–901.
14. Larsson SC, Kumlin M, Ingelman-Sundberg M, Wolk A. Dietary long-chain n-3 fatty acids for the prevention of cancer: A review of potential mechanisms. *Am J Clin Nutr* 2004, 79:935–945.
15. Caygill CP, Charlett A, Hill MJ. Fat, fish, fish oil and cancer. *Br J Cancer* 1996, 74:159–164.
16. Kaizer L, Boyd NF, Kriukov V, Tritchler D. Fish consumption and breast cancer risk: An ecological study. *Nutr Cancer* 1989, 12:61–68.
17. Tavani A, Pelucchi C, Parpinel M, Negri E, Franceschi S, Levi F, La Vecchia C. n-3 polyunsaturated fatty acid intake and cancer risk in Italy and Switzerland. *Int J Cancer* 2003, 105:113–116.
18. Siddiqui RA, Harvey KA, Xu Z, Bammerlin EM, Walker C, Altenburg JD. Docosahexaenoic acid: A natural powerful adjuvant that improves efficacy for anticancer treatment with no adverse effects. *Biofactors* 2011, 37:399–412.
19. Cohen LA. Lipids in cancer: An introduction. *Lipids* 1992, 27:791–792.
20. Glade MJ. Food, nutrition, and the prevention of cancer: A global perspective. American Institute for Cancer Research/World Cancer Research Fund, American Institute for Cancer Research, 1997. *Nutrition* 1999, 15:523–526.
21. Wiseman M. The second World Cancer Research Fund/American Institute for Cancer Research Expert Report. Food, nutrition, physical activity, and the prevention of cancer: A global perspective. *Proc Nutr Soc* 2008, 67:253–256.
22. Latham P, Lund EK, Johnson IT. Dietary n-3 PUFA increases the apoptotic response to 1,2-dimethylhydrazine, reduces mitosis and suppresses the induction of carcinogenesis in the rat colon. *Carcinogenesis* 1999, 20:645–650.
23. Berquin IM, Edwards IJ, Chen YQ. Multi-targeted therapy of cancer by omega-3 fatty acids. *Cancer Lett* 2008, 269:363–377.
24. de Gomez Dumm IN, Brenner RR. Oxidative desaturation of alpha-linoleic, linoleic, and stearic acids by human liver microsomes. *Lipids* 1975, 10:315–317.
25. Muskiet FA, Fokkema MR, Schaafsma A, Boersma ER, Crawford MA. Is docosahexaenoic acid (DHA) essential? Lessons from DHA status regulation, our ancient diet, epidemiology and randomized controlled trials. *J Nutr* 2004, 134:183–186.

26. Hagve TA, Christophersen BO. Effect of dietary fats on arachidonic acid and eicosap-entaenoic acid biosynthesis and conversion to C22 fatty acids in isolated rat liver cells. *Biochim Biophys Acta* 1984, 796:205–217.
27. Christiansen EN, Lund JS, Rortveit T, Rustan AC. Effect of dietary n-3 and n-6 fatty acids on fatty acid desaturation in rat liver. *Biochim Biophys Acta* 1991, 1082:57–62.
28. Stark AH, Crawford MA, Reifen R. Update on alpha-linolenic acid. *Nutr Rev* 2008, 66:326–332.
29. Kidd PM. Omega-3 DHA and EPA for cognition, behavior, and mood: Clinical findings and structural-functional synergies with cell membrane phospholipids. *Altern Med Rev* 2007, 12:207–227.
30. Brenna JT. Efficiency of conversion of alpha-linolenic acid to long chain n-3 fatty acids in man. *Curr Opin Clin Nutr Metab Care* 2002, 5:127–132.
31. Burdge GC, Wootton SA. Conversion of alpha-linolenic acid to eicosapentaenoic, docos-apentaenoic and docosahexaenoic acids in young women. *Br J Nutr* 2002, 88:411–420.
32. Gronn M, Christensen E, Hagve TA, Christophersen BO. Peroxisomal retroconversion of docosahexaenoic acid (22:6(n-3)) to eicosapentaenoic acid (20:5(n-3)) studied in isolated rat liver cells. *Biochim Biophys Acta* 1991, 1081:85–91.
33. Brossard N, Croset M, Pachiaudi C, Riou JP, Tayot JL, Lagarde M. Retroconversion and metabolism of [13C]22:6n-3 in humans and rats after intake of a single dose of [13C]22:6n-3-triacylglycerols. *Am J Clin Nutr* 1996, 64:577–586.
34. von Schacky C, Weber PC. Metabolism and effects on platelet function of the puri-fied eicosapentaenoic and docosahexaenoic acids in humans. *J Clin Invest* 1985, 76:2446–2450.
35. Fickova M, Hubert P, Cremel G, Leray C. Dietary (n-3) and (n-6) polyunsaturated fatty acids rapidly modify fatty acid composition and insulin effects in rat adipocytes. *J Nutr* 1998, 128:512–519.
36. Schley PD, Brindley DN, Field CJ. (n-3) PUFA alter raft lipid composition and decrease epidermal growth factor receptor levels in lipid rafts of human breast cancer cells. *J Nutr* 2007, 137:548–553.
37. Smith WL, Garavito RM, DeWitt DL. Prostaglandin endoperoxide H synthases (cyclo-oxygenases)-1 and -2. *J Biol Chem* 1996, 271:33157–33160.
38. Yamamoto S. Mammalian lipoxygenases: Molecular structures and functions. *Biochim Biophys Acta* 1992, 1128:117–131.
39. Radmark O, Samuelsson B. 5-Lipoxygenase: Mechanisms of regulation. *J Lipid Res* 2009, 50(Suppl):S40–S45.
40. Capdevila JH, Falck JR, Harris RC. Cytochrome P450 and arachidonic acid bioactiva-tion. Molecular and functional properties of the arachidonate monooxygenase. *J Lipid Res* 2000, 41:163–181.
41. Siddiqui RA, Harvey K, Stillwell W. Anticancer properties of oxidation products of docosahexaenoic acid. *Chem Phys Lipids* 2008, 153:47–56.
42. Rouzer CA, Marnett LJ. Endocannabinoid oxygenation by cyclooxygenases, lipoxygen-ases, and cytochromes P450: Cross-talk between the eicosanoid and endocannabinoid signaling pathways. *Chem Rev* 2011, 111:5899–5921.
43. Han C, Lim K, Xu L, Li G, Wu T. Regulation of Wnt/beta-catenin pathway by cPLA2al-pha and PPARdelta. *J Cell Biochem* 2008, 105:534–545.
44. Xu L, Han C, Lim K, Wu T. Activation of cytosolic phospholipase A2alpha through nitric oxide-induced S-nitrosylation. Involvement of inducible nitric-oxide synthase and cyclooxygenase-2. *J Biol Chem* 2008, 283:3077–3087.
45. Henderson RJ, Millar RM, Sargent JR. Effect of growth temperature on the positional distribution of eicosapentaenoic acid and trans hexadecenoic acid in the phospholip-ids of a Vibrio species of bacterium. *Lipids* 1995, 30:181–185.
46. Aveldano MI, Sprecher H. Synthesis of hydroxy fatty acids from 4, 7, 10, 13, 16, 19-[1–14C] docosahexaenoic acid by human platelets. *J Biol Chem* 1983, 258:9339–9343.
47. Rose DP, Connolly JM. Omega-3 fatty acids as cancer chemopreventive agents. *Pharmacol Ther* 1999, 83:217–244.

48. Corey EJ, Shih C, Cashman JR. Docosahexaenoic acid is a strong inhibitor of prostaglandin but not leukotriene biosynthesis. *Proc Natl Acad Sci USA* 1983, 80:3581–3584.

49. Arita M, Clish CB, Serhan CN. The contributions of aspirin and microbial oxygenase to the biosynthesis of anti-inflammatory resolvins: Novel oxygenase products from omega-3 polyunsaturated fatty acids. *Biochem Biophys Res Commun* 2005, 338:149–157.

50. Oh SF, Pillai PS, Recchiuti A, Yang R, Serhan CN. Pro-resolving actions and stereoselective biosynthesis of 18S E-series resolvins in human leukocytes and murine inflammation. *J Clin Invest* 2011, 121:569–581.

51. Serhan CN. Novel chemical mediators in the resolution of inflammation: Resolvins and protectins. *Anesthesiol Clin* 2006, 24:341–364.

52. Calder PC. n-3 polyunsaturated fatty acids, inflammation, and inflammatory diseases. *Am J Clin Nutr* 2006, 83:1505S–1519S.

53. Coussens LM, Werb Z. Inflammation and cancer. *Nature* 2002, 420:860–867.

54. Das UN. Essential fatty acids and their metabolites as modulators of stem cell biology with reference to inflammation, cancer, and metastasis. *Cancer Metastasis Rev* 2011, 30:311–324.

55. Serhan CN, Hamberg M, Samuelsson B. Lipoxins: Novel series of biologically active compounds formed from arachidonic acid in human leukocytes. *Proc Natl Acad Sci USA* 1984, 81:5335–5339.

56. Janakiram NB, Mohammed A, Rao CV. Role of lipoxins, resolvins, and other bioactive lipids in colon and pancreatic cancer. *Cancer Metastasis Rev* 2011, 30:507–523.

57. Blair AS, Hajduch E, Litherland GJ, Hundal HS. Regulation of glucose transport and glycogen synthesis in L6 muscle cells during oxidative stress. Evidence for cross-talk between the insulin and SAPK2/p38 mitogen-activated protein kinase signaling pathways. *J Biol Chem* 1999, 274:36293–36299.

58. Nose K. Role of reactive oxygen species in the regulation of physiological functions. *Biol Pharm Bull* 2000, 23:897–903.

59. Hart CM, Tolson JK, Block ER. Supplemental fatty acids alter lipid peroxidation and oxidant injury in endothelial cells. *Am J Physiol* 1991, 260:L481–L488.

60. Wagner BA, Buettner GR, Burns CP. Free radical-mediated lipid peroxidation in cells: Oxidizability is a function of cell lipid bis-allylic hydrogen content. *Biochemistry* 1994, 33:4449–4453.

61. Fukuzumi K. Relationship between lipoperoxides and diseases. *J Environ Pathol Toxicol Oncol* 1986, 6:25–56.

62. Niki E, Yoshida Y, Saito Y, Noguchi N. Lipid peroxidation: Mechanisms, inhibition, and biological effects. *Biochem Biophys Res Commun* 2005, 338:668–676.

63. Medina I, Satue-Gracia MT, German JB, Frankel EN. Comparison of natural polyphenol antioxidants from extra virgin olive oil with synthetic antioxidants in tuna lipids during thermal oxidation. *J Agric Food Chem* 1999, 47:4873–4879.

64. Gardner HW. Oxygen radical chemistry of polyunsaturated fatty acids. *Free Radic Biol Med* 1989, 7:65–86.

65. Pryor WA, Stanley JP, Blair E. Autoxidation of polyunsaturated fatty acids: II. A suggested mechanism for the formation of TBA-reactive materials from prostaglandin-like endoperoxides. *Lipids* 1976, 11:370–379.

66. Zhu X, Tang X, Anderson VE, Sayre LM. Mass spectrometric characterization of protein modification by the products of nonenzymatic oxidation of linoleic acid. *Chem Res Toxicol* 2009, 22:1386–1397.

67. Gao L, Yin H, Milne GL, Porter NA, Morrow JD. Formation of F-ring isoprostane-like compounds (F3-isoprostanes) in vivo from eicosapentaenoic acid. *J Biol Chem* 2006, 281:14092–14099.

68. Fam SS, Murphey LJ, Terry ES, Zackert WE, Chen Y, Gao L et al. Formation of highly reactive A-ring and J-ring isoprostane-like compounds (A4/J4-neuroprostanes) in vivo from docosahexaenoic acid. *J Biol Chem* 2002, 277:36076–36084.

69. Roberts LJ, II, Fessel JP, Davies SS. The biochemistry of the isoprostane, neuroprostane, and isofuran pathways of lipid peroxidation. *Brain Pathol* 2005, 15:143–148.

70. Yin H, Brooks JD, Gao L, Porter NA, Morrow JD. Identification of novel autoxidation products of the omega-3 fatty acid eicosapentaenoic acid in vitro and in vivo. *J Biol Chem* 2007, 282:29890–29901.
71. Kris-Etherton PM, Taylor DS, Yu-Poth S, Huth P, Moriarty K, Fishell V et al. Polyunsaturated fatty acids in the food chain in the United States. *Am J Clin Nutr* 2000, 71:179S–188S.
72. Trumbo P, Schlicker S, Yates AA, Poos M. Dietary reference intakes for energy, carbohydrate, fiber, fat, fatty acids, cholesterol, protein and amino acids. *J Am Diet Assoc* 2002, 102:1621–1630.
73. Simopoulos AP, Salem N, Jr. Egg yolk as a source of long-chain polyunsaturated fatty acids in infant feeding. *Am J Clin Nutr* 1992, 55:411–414.
74. Simopoulos AP. Omega-3 fatty acids in health and disease and in growth and development. *Am J Clin Nutr* 1991, 54:438–463.
75. Sargent JR. Fish oils and human diet. *Br J Nutr* 1997, 78(Suppl 1):S5–S13.
76. Barrett JC. Mechanisms of multistep carcinogenesis and carcinogen risk assessment. *Environ Health Perspect* 1993, 100:9–20.
77. Hanahan D, Weinberg RA. The hallmarks of cancer. *Cell* 2000, 100:57–70.
78. Hanahan D, Weinberg RA. Hallmarks of cancer: The next generation. *Cell* 2011, 144:646–674.
79. Kelloff GJ, Hawk ET, Crowell JA, Boone CW, Nayfield SG, Perloff M et al. Strategies for identification and clinical evaluation of promising chemopreventive agents. *Oncology (Williston Park)* 1996, 10:1471–1484; discussion 1484–1488.
80. Fisher JC. Multiple-mutation theory of carcinogenesis. *Nature* 1958, 181:651–652.
81. Yokota J. Tumor progression and metastasis. *Carcinogenesis* 2000, 21:497–503.
82. Merlo LM, Pepper JW, Reid BJ, Maley CC. Cancer as an evolutionary and ecological process. *Nat Rev Cancer* 2006, 6:924–935.
83. Steliga MA, Dresler CM. Epidemiology of lung cancer: Smoking, secondhand smoke, and genetics. *Surg Oncol Clin N Am* 2011, 20:605–618.
84. Castellsague X. Natural history and epidemiology of HPV infection and cervical cancer. *Gynecol Oncol* 2008, 110:S4–S7.
85. Sommer S, Fuqua SA. Estrogen receptor and breast cancer. *Semin Cancer Biol* 2001, 11:339–352.
86. Shen Z. Genomic instability and cancer: An introduction. *J Mol Cell Biol* 2011, 3:1–3.
87. Hoeijmakers JH. Genome maintenance mechanisms for preventing cancer. *Nature* 2001, 411:366–374.
88. Lengauer C, Kinzler KW, Vogelstein B. Genetic instabilities in human cancers. *Nature* 1998, 396:643–649.
89. Jefford CE, Irminger-Finger I. Mechanisms of chromosome instability in cancers. *Crit Rev Oncol Hematol* 2006, 59:1–14.
90. Esteller M. Cancer epigenomics: DNA methylomes and histone-modification maps. *Nat Rev Genet* 2007, 8:286–298.
91. Negrini S, Gorgoulis VG, Halazonetis TD. Genomic instability—An evolving hallmark of cancer. *Nat Rev Mol Cell Biol* 2010, 11:220–228.
92. Baylin SB, Herman JG. DNA hypermethylation in tumorigenesis: Epigenetics joins genetics. *Trends Genet* 2000, 16:168–174.
93. Djupedal I, Ekwall K. Epigenetics: Heterochromatin meets RNAi. *Cell Res* 2009, 19:282–295.
94. Gluckman PD, Hanson MA, Buklijas T, Low FM, Beedle AS. Epigenetic mechanisms that underpin metabolic and cardiovascular diseases. *Nat Rev Endocrinol* 2009, 5:401–408.
95. Egger G, Liang G, Aparicio A, Jones PA. Epigenetics in human disease and prospects for epigenetic therapy. *Nature* 2004, 429:457–463.
96. Jaenisch R, Bird A. Epigenetic regulation of gene expression: How the genome integrates intrinsic and environmental signals. *Nat Genet* 2003, 33(Suppl):245–254.
97. Lewin B. The mystique of epigenetics. *Cell* 1998, 93:301–303.

98. Hoffmann MJ, Schulz WA. Causes and consequences of DNA hypomethylation in human cancer. *Biochem Cell Biol* 2005, 83:296–321.
99. Ceccarelli V, Racanicchi S, Martelli MP, Nocentini G, Fettucciari K, Riccardi C et al. Eicosapentaenoic acid demethylates a single CpG that mediates expression of tumor suppressor CCAAT/enhancer-binding protein delta in U937 leukemia cells. *J Biol Chem* 2011, 286:27092–27102.
100. Kulkarni A, Dangat K, Kale A, Sable P, Chavan-Gautam P, Joshi S. Effects of altered maternal folic acid, vitamin B12 and docosahexaenoic acid on placental global DNA methylation patterns in Wistar rats. *PLoS ONE* 2011, 6:e17706.
101. Kale A, Naphade N, Sapkale S, Kamaraju M, Pillai A, Joshi S, Mahadik S. Reduced folic acid, vitamin B12 and docosahexaenoic acid and increased homocysteine and cortisol in never-medicated schizophrenia patients: Implications for altered one-carbon metabolism. *Psychiatry Res* 2010, 175:47–53.
102. Luger K, Mader AW, Richmond RK, Sargent DF, Richmond TJ. Crystal structure of the nucleosome core particle at 2.8 A resolution. *Nature* 1997, 389:251–260.
103. Marmorstein R. Protein modules that manipulate histone tails for chromatin regulation. *Nat Rev Mol Cell Biol* 2001, 2:422–432.
104. Zhang Y, Reinberg D. Transcription regulation by histone methylation: Interplay between different covalent modifications of the core histone tails. *Genes Dev* 2001, 15:2343–2360.
105. Wu J, Grunstein M. 25 years after the nucleosome model: Chromatin modifications. *Trends Biochem Sci* 2000, 25:619–623.
106. Dimri M, Bommi PV, Sahasrabuddhe AA, Khandekar JD, Dimri GP. Dietary omega-3 polyunsaturated fatty acids suppress expression of EZH2 in breast cancer cells. *Carcinogenesis* 2010, 31:489–495.
107. Yessoufou A, Ple A, Moutairou K, Hichami A, Khan NA. Docosahexaenoic acid reduces suppressive and migratory functions of CD4 + CD25 + regulatory T-cells. *J Lipid Res* 2009, 50:2377–2388.
108. Suphioglu C, Sadli N, Coonan D, Kumar L, De Mel D, Lesheim J et al. Zinc and DHA have opposing effects on the expression levels of histones H3 and H4 in human neuronal cells. *Br J Nutr* 2009, 103:344–351.
109. Begin ME, Ells G, Das UN, Horrobin DF. Differential killing of human carcinoma cells supplemented with n-3 and n-6 polyunsaturated fatty acids. *J Natl Cancer Inst* 1986, 77:1053–1062.
110. Gleissman H, Yang R, Martinod K, Lindskog M, Serhan CN, Johnsen JI, Kogner P. Docosahexaenoic acid metabolome in neural tumors: Identification of cytotoxic intermediates. *FASEB J* 2010, 24:906–915.
111. Calin GA, Croce CM. MicroRNA-cancer connection: The beginning of a new tale. *Cancer Res* 2006, 66:7390–7394.
112. Davidson LA, Wang N, Shah MS, Lupton JR, Ivanov I, Chapkin RS. n-3 Polyunsaturated fatty acids modulate carcinogen-directed non-coding microRNA signatures in rat colon. *Carcinogenesis* 2009, 30:2077–2084.
113. Farago N, Feher LZ, Kitajka K, Das UN, Puskas LG. MicroRNA profile of polyunsaturated fatty acid treated glioma cells reveal apoptosis-specific expression changes. *Lipids Health Dis* 2011, 10:173.
114. Richardson K, Louie-Gao Q, Arnett DK, Parnell LD, Lai CQ, Davalos A et al. The PLIN4 variant rs8887 modulates obesity related phenotypes in humans through creation of a novel miR-522 seed site. *PLoS ONE* 2011, 6:e17944.
115. Collins K, Jacks T, Pavletich NP. The cell cycle and cancer. *Proc Natl Acad Sci USA* 1997, 94:2776–2778.
116. Canepa ET, Scassa ME, Ceruti JM, Marazita MC, Carcagno AL, Sirkin PF, Ogara MF. INK4 proteins, a family of mammalian CDK inhibitors with novel biological functions. *IUBMB Life* 2007, 59:419–426.
117. Sherr CJ, Roberts JM. CDK inhibitors: Positive and negative regulators of G1-phase progression. *Genes Dev* 1999, 13:1501–1512.

118. Morgan DO. Principles of CDK regulation. *Nature* 1995, 374:131–134.
119. Siddiqui RA, Jenski LJ, Harvey KA, Wiesehan JD, Stillwell W, Zaloga GP. Cell-cycle arrest in Jurkat leukaemic cells: A possible role for docosahexaenoic acid. *Biochem J* 2003, 371:621–629.
120. Tsujita-Kyutoku M, Yuri T, Danbara N, Senzaki H, Kiyozuka Y, Uehara N et al. Conjugated docosahexaenoic acid suppresses KPL-1 human breast cancer cell growth in vitro and in vivo: Potential mechanisms of action. *Breast Cancer Res* 2004, 6:R291–R299.
121. Danbara N, Yuri T, Tsujita-Kyutoku M, Sato M, Senzaki H, Takada H et al. Conjugated docosahexaenoic acid is a potent inducer of cell cycle arrest and apoptosis and inhibits growth of colo 201 human colon cancer cells. *Nutr Cancer* 2004, 50:71–79.
122. Mernitz H, Lian F, Smith DE, Meydani SN, Wang XD. Fish oil supplementation inhibits NNK-induced lung carcinogenesis in the A/J mouse. *Nutr Cancer* 2009, 61: 663–669.
123. Albino AP, Juan G, Traganos F, Reinhart L, Connolly J, Rose DP, Darzynkiewicz Z. Cell cycle arrest and apoptosis of melanoma cells by docosahexaenoic acid: Association with decreased pRb phosphorylation. *Cancer Res* 2000, 60:4139–4145.
124. Khan NA, Nishimura K, Aires V, Yamashita T, Oaxaca-Castillo D, Kashiwagi K, Igarashi K. Docosahexaenoic acid inhibits cancer cell growth via p27Kip1, CDK2, ERK1/ERK2, and retinoblastoma phosphorylation. *J Lipid Res* 2006, 47:2306–2313.
125. Narayanan BA, Narayanan NK, Reddy BS. Docosahexaenoic acid regulated genes and transcription factors inducing apoptosis in human colon cancer cells. *Int J Oncol* 2001, 19:1255–1262.
126. Foijer F, te Riele H. Check, double check: The G2 barrier to cancer. *Cell Cycle* 2006, 5:831–836.
127. Bucher N, Britten CD. G2 checkpoint abrogation and checkpoint kinase-1 targeting in the treatment of cancer. *Br J Cancer* 2008, 98:523–528.
128. Callegari AJ, Kelly TJ. Shedding light on the DNA damage checkpoint. *Cell Cycle* 2007, 6:660–666.
129. Delston RB, Harbour JW. Rb at the interface between cell cycle and apoptotic decisions. *Curr Mol Med* 2006, 6:713–718.
130. Kato J, Matsushime H, Hiebert SW, Ewen ME, Sherr CJ. Direct binding of cyclin D to the retinoblastoma gene product (pRb) and pRb phosphorylation by the cyclin D-dependent kinase CDK4. *Genes Dev* 1993, 7:331–342.
131. Narayanan BA, Narayanan NK, Desai D, Pittman B, Reddy BS. Effects of a combination of docosahexaenoic acid and 1,4-phenylene bis(methylene) selenocyanate on cyclooxygenase 2, inducible nitric oxide synthase and beta-catenin pathways in colon cancer cells. *Carcinogenesis* 2004, 25:2443–2449.
132. Jakobsen CH, Storvold GL, Bremseth H, Follestad T, Sand K, Mack M et al. DHA induces ER stress and growth arrest in human colon cancer cells: Associations with cholesterol and calcium homeostasis. *J Lipid Res* 2008, 49:2089–2100.
133. Lu IF, Hasio AC, Hu MC, Yang FM, Su HM. Docosahexaenoic acid induces proteasome-dependent degradation of estrogen receptor alpha and inhibits the downstream signaling target in MCF-7 breast cancer cells. *J Nutr Biochem* 2010, 21:512–517.
134. Lu Y, Nie D, Witt WT, Chen Q, Shen M, Xie H et al. Expression of the fat-1 gene diminishes prostate cancer growth in vivo through enhancing apoptosis and inhibiting GSK-3 beta phosphorylation. *Mol Cancer Ther* 2008, 7:3203–3211.
135. Barascu A, Besson P, Le Floch O, Bougnoux P, Jourdan ML. CDK1-cyclin B1 mediates the inhibition of proliferation induced by omega-3 fatty acids in MDA-MB-231 breast cancer cells. *Int J Biochem Cell Biol* 2006, 38:196–208.
136. Golstein P, Kroemer G. A multiplicity of cell death pathways. Symposium on apoptotic and non-apoptotic cell death pathways. *EMBO Rep* 2007, 8:829–833.
137. Sun Y, Peng ZL. Programmed cell death and cancer. *Postgrad Med J* 2009, 85:134–140.
138. Kroemer G, Galluzzi L, Vandenabeele P, Abrams J, Alnemri ES, Baehrecke EH et al. Classification of cell death: Recommendations of the Nomenclature Committee on Cell Death 2009. *Cell Death Differ* 2009, 16:3–11.

139. Galluzzi L, Vitale I, Abrams JM, Alnemri ES, Baehrecke EH, Blagosklonny MV et al. Molecular definitions of cell death subroutines: Recommendations of the Nomenclature Committee on Cell Death 2012. *Cell Death Differ* 2012, 19:107–120.

140. Eisenberg-Lerner A, Bialik S, Simon HU, Kimchi A. Life and death partners: Apoptosis, autophagy and the cross-talk between them. *Cell Death Differ* 2009, 16:966–975.

141. Maiuri MC, Zalckvar E, Kimchi A, Kroemer G. Self-eating and self-killing: Crosstalk between autophagy and apoptosis. *Nat Rev Mol Cell Biol* 2007, 8:741–752.

142. Amaravadi RK, Thompson CB. The roles of therapy-induced autophagy and necrosis in cancer treatment. *Clin Cancer Res* 2007, 13:7271–7279.

143. Shen HM, Codogno P. Autophagic cell death: Loch Ness monster or endangered species? *Autophagy* 2011, 7:457–465.

144. Kroemer G, Levine B. Autophagic cell death: The story of a misnomer. *Nat Rev Mol Cell Biol* 2008, 9:1004–1010.

145. Levine B, Yuan J. Autophagy in cell death: An innocent convict? *J Clin Invest* 2005, 115:2679–2688.

146. Klionsky DJ, Abeliovich H, Agostinis P, Agrawal DK, Aliev G, Askew DS et al. Guidelines for the use and interpretation of assays for monitoring autophagy in higher eukaryotes. *Autophagy* 2008, 4:151–175.

147. Kishida E, Tajiri M, Masuzawa Y. Docosahexaenoic acid enrichment can reduce L929 cell necrosis induced by tumor necrosis factor. *Biochim Biophys Acta* 2006, 1761: 454–462.

148. Festjens N, Vanden Berghe T, Vandenabeele P. Necrosis, a well-orchestrated form of cell demise: Signalling cascades, important mediators and concomitant immune response. *Biochim Biophys Acta* 2006, 1757:1371–1387.

149. Zong WX, Thompson CB. Necrotic death as a cell fate. *Genes Dev* 2006, 20:1–15.

150. Edinger AL, Thompson CB. Death by design: Apoptosis, necrosis and autophagy. *Curr Opin Cell Biol* 2004, 16:663–669.

151. Calder PC. Polyunsaturated fatty acids and inflammatory processes: New twists in an old tale. *Biochimie* 2009, 91:791–795.

152. Mori TA, Beilin LJ. Omega-3 fatty acids and inflammation. *Curr Atheroscler Rep* 2004, 6:461–467.

153. Taylor RC, Cullen SP, Martin SJ. Apoptosis: Controlled demolition at the cellular level. *Nat Rev Mol Cell Biol* 2008, 9:231–241.

154. Lowe SW, Cepero E, Evan G. Intrinsic tumour suppression. *Nature* 2004, 432:307–315.

155. Adams JM, Cory S. The Bcl-2 apoptotic switch in cancer development and therapy. *Oncogene* 2007, 26:1324–1337.

156. Liu X, Kim CN, Yang J, Jemmerson R, Wang X. Induction of apoptotic program in cell-free extracts: Requirement for dATP and cytochrome c. *Cell* 1996, 86:147–157.

157. Verhagen AM, Ekert PG, Pakusch M, Silke J, Connolly LM, Reid GE et al. Identification of DIABLO, a mammalian protein that promotes apoptosis by binding to and antagonizing IAP proteins. *Cell* 2000, 102:43–53.

158. Li LY, Luo X, Wang X. Endonuclease G is an apoptotic DNase when released from mitochondria. *Nature* 2001, 412:95–99.

159. Susin SA, Lorenzo HK, Zamzami N, Marzo I, Snow BE, Brothers GM et al. Molecular characterization of mitochondrial apoptosis-inducing factor. *Nature* 1999, 397:441–446.

160. Wang X. The expanding role of mitochondria in apoptosis. *Genes Dev* 2001, 15:2922–2933.

161. Barnhart BC, Alappat EC, Peter ME. The CD95 type I/type II model. *Semin Immunol* 2003, 15:185–193.

162. Hu Y, Sun H, Owens RT, Gu Z, Wu J, Chen YQ et al. Syndecan-1-dependent suppression of PDK1/Akt/bad signaling by docosahexaenoic acid induces apoptosis in prostate cancer. *Neoplasia* 2010, 12:826–836.

163. Corsetto PA, Montorfano G, Zava S, Jovenitti IE, Cremona A, Berra B, Rizzo AM. Effects of n-3 PUFAs on breast cancer cells through their incorporation in plasma membrane. *Lipids Health Dis* 2011, 10:73.

164. Koumura T, Nakamura C, Nakagawa Y. Involvement of hydroperoxide in mitochondria in the induction of apoptosis by the eicosapentaenoic acid. *Free Radic Res* 2005, 39:225–235.
165. Chiu LC, Wong EY, Ooi VE. Docosahexaenoic acid from a cultured microalga inhibits cell growth and induces apoptosis by upregulating Bax/Bcl-2 ratio in human breast carcinoma MCF-7 cells. *Ann N Y Acad Sci* 2004, 1030:361–368.
166. Zand H, Rhimipour A, Bakhshayesh M, Shafiee M, Nour Mohammadi I, Salimi S. Involvement of PPAR-gamma and p53 in DHA-induced apoptosis in Reh cells. *Mol Cell Biochem* 2007, 304:71–77.
167. Lim K, Han C, Dai Y, Shen M, Wu T. Omega-3 polyunsaturated fatty acids inhibit hepatocellular carcinoma cell growth through blocking beta-catenin and cyclooxygenase-2. *Mol Cancer Ther* 2009, 8:3046–3055.
168. Lim K, Han C, Xu L, Isse K, Demetris AJ, Wu T. Cyclooxygenase-2-derived prostaglandin E2 activates beta-catenin in human cholangiocarcinoma cells: Evidence for inhibition of these signaling pathways by omega 3 polyunsaturated fatty acids. *Cancer Res* 2008, 68:553–560.
169. Song KS, Jing K, Kim JS, Yun EJ, Shin S, Seo KS et al. Omega-3-polyunsaturated fatty acids suppress pancreatic cancer cell growth in vitro and in vivo via downregulation of Wnt/beta-catenin signaling. *Pancreatology* 2011, 11:574–584.
170. Kang JX, Wang J, Wu L, Kang ZB. Transgenic mice: Fat-1 mice convert n-6 to n-3 fatty acids. *Nature* 2004, 427:504.
171. Kang JX. Fat-1 transgenic mice: A new model for omega-3 research. *Prostaglandins Leukot Essent Fatty Acids* 2007, 77:263–267.
172. Calviello G, Resci F, Serini S, Piccioni E, Toesca A, Boninsegna A et al. Docosahexaenoic acid induces proteasome-dependent degradation of beta-catenin, down-regulation of survivin and apoptosis in human colorectal cancer cells not expressing COX-2. *Carcinogenesis* 2007, 28:1202–1209.
173. Mezhybovska M, Yudina Y, Abhyankar A, Sjolander A. Beta-catenin is involved in alterations in mitochondrial activity in non-transformed intestinal epithelial and colon cancer cells. *Br J Cancer* 2009, 101:1596–1605.
174. Mezhybovska M, Wikstrom K, Ohd JF, Sjolander A. The inflammatory mediator leukotriene D4 induces beta-catenin signaling and its association with antiapoptotic Bcl-2 in intestinal epithelial cells. *J Biol Chem* 2006, 281:6776–6784.
175. Lee HC, Wei YH. Mitochondrial role in life and death of the cell. *J Biomed Sci* 2000, 7:2–15.
176. Shen Y, White E. p53-dependent apoptosis pathways. *Adv Cancer Res* 2001, 82:55–84.
177. Mihara M, Erster S, Zaika A, Petrenko O, Chittenden T, Pancoska P, Moll UM. p53 has a direct apoptogenic role at the mitochondria. *Mol Cell* 2003, 11:577–590.
178. Erster S, Mihara M, Kim RH, Petrenko O, Moll UM. In vivo mitochondrial p53 translocation triggers a rapid first wave of cell death in response to DNA damage that can precede p53 target gene activation. *Mol Cell Biol* 2004, 24:6728–6741.
179. Zhao Y, Chaiswing L, Velez JM, Batinic-Haberle I, Colburn NH, Oberley TD, St Clair DK. p53 translocation to mitochondria precedes its nuclear translocation and targets mitochondrial oxidative defense protein-manganese superoxide dismutase. *Cancer Res* 2005, 65:3745–3750.
180. Chi TY, Chen GG, Lai PB. Eicosapentaenoic acid induces Fas-mediated apoptosis through a p53-dependent pathway in hepatoma cells. *Cancer J* 2004, 10:190–200.
181. Hegde S, Kaushal N, Ravindra KC, Chiaro C, Hafer KT, Gandhi UH et al. Delta12-prostaglandin J3, an omega-3 fatty acid-derived metabolite, selectively ablates leukemia stem cells in mice. *Blood* 2011, 118:6909–6919.
182. Chiu LC, Wan JM. Induction of apoptosis in HL-60 cells by eicosapentaenoic acid (EPA) is associated with downregulation of bcl-2 expression. *Cancer Lett* 1999, 145:17–27.
183. Yonezawa Y, Hada T, Uryu K, Tsuzuki T, Nakagawa K, Miyazawa T et al. Mechanism of cell cycle arrest and apoptosis induction by conjugated eicosapentaenoic acid, which is a mammalian DNA polymerase and topoisomerase inhibitor. *Int J Oncol* 2007, 30: 1197–1204.

184. Kato T, Kolenic N, Pardini RS. Docosahexaenoic acid (DHA), a primary tumor suppressive omega-3 fatty acid, inhibits growth of colorectal cancer independent of p53 mutational status. *Nutr Cancer* 2007, 58:178–187.

185. Merendino N, Molinari R, Loppi B, Pessina G, D'Aquino M, Tomassi G, Velottia F. Induction of apoptosis in human pancreatic cancer cells by docosahexaenoic acid. *Ann N Y Acad Sci* 2003, 1010:361–364.

186. Hawkins RA, Sangster K, Arends MJ. Apoptotic death of pancreatic cancer cells induced by polyunsaturated fatty acids varies with double bond number and involves an oxidative mechanism. *J Pathol* 1998, 185:61–70.

187. Kang KS, Wang P, Yamabe N, Fukui M, Jay T, Zhu BT. Docosahexaenoic acid induces apoptosis in MCF-7 cells in vitro and in vivo via reactive oxygen species formation and caspase 8 activation. *PLoS ONE* 2010, 5:e10296.

188. Giros A, Grzybowski M, Sohn VR, Pons E, Fernandez-Morales J, Xicola RM et al. Regulation of colorectal cancer cell apoptosis by the n-3 polyunsaturated fatty acids docosahexaenoic and eicosapentaenoic. *Cancer Prev Res (Phila)* 2009, 2:732–742.

189. Levine B, Kroemer G. Autophagy in the pathogenesis of disease. *Cell* 2008, 132:27–42.

190. Ravikumar B, Futter M, Jahreiss L, Korolchuk VI, Lichtenberg M, Luo S et al. Mammalian macroautophagy at a glance. *J Cell Sci* 2009, 122:1707–1711.

191. Mizushima N, Yoshimori T, Ohsumi Y. The role of atg proteins in autophagosome formation. *Annu Rev Cell Dev Biol* 2011, 27:107–132.

192. He C, Klionsky DJ. Regulation mechanisms and signaling pathways of autophagy. *Annu Rev Genet* 2009, 43:67–93.

193. Ravikumar B, Vacher C, Berger Z, Davies JE, Luo S, Oroz LG et al. Inhibition of mTOR induces autophagy and reduces toxicity of polyglutamine expansions in fly and mouse models of Huntington disease. *Nat Genet* 2004, 36:585–595.

194. Jung CH, Jun CB, Ro SH, Kim YM, Otto NM, Cao J et al. ULK-Atg13-FIP200 complexes mediate mTOR signaling to the autophagy machinery. *Mol Biol Cell* 2009, 20:1992–2003.

195. Klionsky DJ, Cuervo AM, Seglen PO. Methods for monitoring autophagy from yeast to human. *Autophagy* 2007, 3:181–206.

196. Klionsky DJ, Emr SD. Autophagy as a regulated pathway of cellular degradation. *Science* 2000, 290:1717–1721.

197. Rosenfeldt MT, Ryan KM. The multiple roles of autophagy in cancer. *Carcinogenesis* 2011, 32:955–963.

198. Jing K, Lim K. Introduction: Why is autophagy important in human diseases? *Exp Mol Med* 2012, 44:69–72.

199. Hait WN, Jin S, Yang JM. A matter of life or death (or both): Understanding autophagy in cancer. *Clin Cancer Res* 2006, 12:1961–1965.

200. Hara T, Nakamura K, Matsui M, Yamamoto A, Nakahara Y, Suzuki-Migishima R et al. Suppression of basal autophagy in neural cells causes neurodegenerative disease in mice. *Nature* 2006, 441:885–889.

201. Ebato C, Uchida T, Arakawa M, Komatsu M, Ueno T, Komiya K et al. Autophagy is important in islet homeostasis and compensatory increase of beta cell mass in response to high-fat diet. *Cell Metab* 2008, 8:325–332.

202. Kondo Y, Kanzawa T, Sawaya R, Kondo S. The role of autophagy in cancer development and response to therapy. *Nat Rev Cancer* 2005, 5:726–734.

203. Cuervo AM. Autophagy: In sickness and in health. *Trends Cell Biol* 2004, 14:70–77.

204. Choi KS. Autophagy and cancer. *Exp Mol Med* 2012, 44(2):109–120.

205. Jing K, Song KS, Shin S, Kim N, Jeong S, Oh HR et al. Docosahexaenoic acid induces autophagy through p53/AMPK/mTOR signaling and promotes apoptosis in human cancer cells harboring wild-type p53. *Autophagy* 2011, 7:1348–1358.

206. Sui X, Jin L, Huang X, Geng S, He C, Hu X. p53 signaling and autophagy in cancer: A revolutionary strategy could be developed for cancer treatment. *Autophagy* 2011, 7:565–571.

207. Levine B, Abrams J. p53: The Janus of autophagy? *Nat Cell Biol* 2008, 10:637–639.

208. Tasdemir E, Maiuri MC, Galluzzi L, Vitale I, Djavaheri-Mergny M, D'Amelio M et al. Regulation of autophagy by cytoplasmic p53. *Nat Cell Biol* 2008, 10:676–687.
209. Grivennikov SI, Greten FR, Karin M. Immunity, inflammation, and cancer. *Cell* 2010, 140:883–899.
210. Calder PC. Long-chain fatty acids and inflammation. *Proc Nutr Soc* 2012:1–6.
211. Kelley DS, Taylor PC, Nelson GJ, Schmidt PC, Ferretti A, Erickson KL et al. Docosahexaenoic acid ingestion inhibits natural killer cell activity and production of inflammatory mediators in young healthy men. *Lipids* 1999, 34:317–324.
212. Trebble TM, Wootton SA, Miles EA, Mullee M, Arden NK, Ballinger AB et al. Prostaglandin E2 production and T cell function after fish-oil supplementation: Response to antioxidant cosupplementation. *Am J Clin Nutr* 2003, 78:376–382.
213. Gibney MJ, Hunter B. The effects of short- and long-term supplementation with fish oil on the incorporation of n-3 polyunsaturated fatty acids into cells of the immune system in healthy volunteers. *Eur J Clin Nutr* 1993, 47:255–259.
214. Endres S, Ghorbani R, Kelley VE, Georgilis K, Lonnemann G, van der Meer JW et al. The effect of dietary supplementation with n-3 polyunsaturated fatty acids on the synthesis of interleukin-1 and tumor necrosis factor by mononuclear cells. *N Engl J Med* 1989, 320:265–271.
215. Lee TH, Hoover RL, Williams JD, Sperling RI, Ravalese J, III, Spur BW et al. Effect of dietary enrichment with eicosapentaenoic and docosahexaenoic acids on in vitro neutrophil and monocyte leukotriene generation and neutrophil function. *N Engl J Med* 1985, 312:1217–1224.
216. Han C, Demetris AJ, Stolz DB, Xu L, Lim K, Wu T. Modulation of Stat3 activation by the cytosolic phospholipase A2alpha and cyclooxygenase-2-controlled prostaglandin E2 signaling pathway. *J Biol Chem* 2006, 281:24831–24846.
217. Xu L, Han C, Lim K, Wu T. Cross-talk between peroxisome proliferator-activated receptor delta and cytosolic phospholipase A(2)alpha/cyclooxygenase-2/prostaglandin E(2) signaling pathways in human hepatocellular carcinoma cells. *Cancer Res* 2006, 66:11859–11868.
218. Dooper MM, Wassink L, M'Rabet L, Graus YM. The modulatory effects of prostaglandin-E on cytokine production by human peripheral blood mononuclear cells are independent of the prostaglandin subtype. *Immunology* 2002, 107:152–159.
219. Simopoulos AP. Omega-3 fatty acids in inflammation and autoimmune diseases. *J Am Coll Nutr* 2002, 21:495–505.
220. Simopoulos AP. The importance of the omega-6/omega-3 fatty acid ratio in cardiovascular disease and other chronic diseases. *Exp Biol Med (Maywood)* 2008, 233:674–688.
221. Simopoulos AP. Evolutionary aspects of diet, the omega-6/omega-3 ratio and genetic variation: Nutritional implications for chronic diseases. *Biomed Pharmacother* 2006, 60:502–507.
222. Deckelbaum RJ, Worgall TS, Seo T. n-3 fatty acids and gene expression. *Am J Clin Nutr* 2006, 83:1520S–1525S.
223. Mullen A, Loscher CE, Roche HM. Anti-inflammatory effects of EPA and DHA are dependent upon time and dose-response elements associated with LPS stimulation in THP-1-derived macrophages. *J Nutr Biochem* 2010, 21:444–450.
224. Horia E, Watkins BA. Complementary actions of docosahexaenoic acid and genistein on COX-2, PGE2 and invasiveness in MDA-MB-231 breast cancer cells. *Carcinogenesis* 2007, 28:809–815.
225. Edwards IJ, O'Flaherty JT. Omega-3 fatty acids and PPARgamma in cancer. *PPAR Res* 2008, 2008:358052.
226. Li H, Ruan XZ, Powis SH, Fernando R, Mon WY, Wheeler DC et al. EPA and DHA reduce LPS-induced inflammation responses in HK-2 cells: Evidence for a PPAR-gamma-dependent mechanism. *Kidney Int* 2005, 67:867–874.
227. Carmeliet P, Jain RK. Angiogenesis in cancer and other diseases. *Nature* 2000, 407:249–257.

228. Ramanujan S, Koenig GC, Padera TP, Stoll BR, Jain RK. Local imbalance of proangiogenic and antiangiogenic factors: A potential mechanism of focal necrosis and dormancy in tumors. *Cancer Res* 2000, 60:1442–1448.
229. Tortora G, Melisi D, Ciardiello F. Angiogenesis: A target for cancer therapy. *Curr Pharm Des* 2004, 10:11–26.
230. Matesanz N, Park G, McAllister H, Leahey W, Devine A, McVeigh GE et al. Docosahexaenoic acid improves the nitroso-redox balance and reduces VEGF-mediated angiogenic signaling in microvascular endothelial cells. *Invest Ophthalmol Vis Sci* 2010, 51:6815–6825.
231. Yang SP, Morita I, Murota SI. Eicosapentaenoic acid attenuates vascular endothelial growth factor-induced proliferation via inhibiting Flk-1 receptor expression in bovine carotid artery endothelial cells. *J Cell Physiol* 1998, 176:342–349.
232. Tsuji M, Murota SI, Morita I. Docosapentaenoic acid (22:5, n-3) suppressed tube-forming activity in endothelial cells induced by vascular endothelial growth factor. *Prostaglandins Leukot Essent Fatty Acids* 2003, 68:337–342.
233. Connor KM, SanGiovanni JP, Lofqvist C, Aderman CM, Chen J, Higuchi A et al. Increased dietary intake of omega-3-polyunsaturated fatty acids reduces pathological retinal angiogenesis. *Nat Med* 2007, 13:868–873.
234. Stahl A, Sapieha P, Connor KM, Sangiovanni JP, Chen J, Aderman CM et al. Short communication: PPAR gamma mediates a direct antiangiogenic effect of omega 3-PUFAs in proliferative retinopathy. *Circ Res* 2010, 107:495–500.
235. Tevar R, Jho DH, Babcock T, Helton WS, Espat NJ. Omega-3 fatty acid supplementation reduces tumor growth and vascular endothelial growth factor expression in a model of progressive non-metastasizing malignancy. *JPEN J Parenter Enteral Nutr* 2002, 26:285–289.
236. Mukutmoni-Norris M, Hubbard NE, Erickson KL. Modulation of murine mammary tumor vasculature by dietary n-3 fatty acids in fish oil. *Cancer Lett* 2000, 150:101–109.
237. Calviello G, Di Nicuolo F, Gragnoli S, Piccioni E, Serini S, Maggiano N et al. n-3 PUFAs reduce VEGF expression in human colon cancer cells modulating the COX-2/PGE2 induced ERK-1 and -2 and HIF-1alpha induction pathway. *Carcinogenesis* 2004, 25:2303–2310.
238. Mizukami Y, Kohgo Y, Chung DC. Hypoxia inducible factor-1 independent pathways in tumor angiogenesis. *Clin Cancer Res* 2007, 13:5670–5674.
239. Huang SP, Wu MS, Shun CT, Wang HP, Hsieh CY, Kuo ML, Lin JT. Cyclooxygenase-2 increases hypoxia-inducible factor-1 and vascular endothelial growth factor to promote angiogenesis in gastric carcinoma. *J Biomed Sci* 2005, 12:229–241.
240. Hardman WE, Sun L, Short N, Cameron IL. Dietary omega-3 fatty acids and ionizing irradiation on human breast cancer xenograft growth and angiogenesis. *Cancer Cell Int* 2005, 5:12.
241. Tuller ER, Beavers CT, Lou JR, Ihnat MA, Benbrook DM, Ding WQ. Docosahexaenoic acid inhibits superoxide dismutase 1 gene transcription in human cancer cells: The involvement of peroxisome proliferator-activated receptor alpha and hypoxia-inducible factor-2alpha signaling. *Mol Pharmacol* 2009, 76:588–595.
242. Szymczak M, Murray M, Petrovic N. Modulation of angiogenesis by omega-3 polyunsaturated fatty acids is mediated by cyclooxygenases. *Blood* 2008, 111:3514–3521.
243. Bouwens M, van de Rest O, Dellschaft N, Bromhaar MG, de Groot LC, Geleijnse JM et al. Fish-oil supplementation induces antiinflammatory gene expression profiles in human blood mononuclear cells. *Am J Clin Nutr* 2009, 90:415–424.
244. Spencer L, Mann C, Metcalfe M, Webb M, Pollard C, Spencer D et al. The effect of omega-3 FAs on tumour angiogenesis and their therapeutic potential. *Eur J Cancer* 2009, 45:2077–2086.
245. Steeg PS. Metastasis suppressors alter the signal transduction of cancer cells. *Nat Rev Cancer* 2003, 3:55–63.
246. Mareel M, Leroy A. Clinical, cellular, and molecular aspects of cancer invasion. *Physiol Rev* 2003, 83:337–376.

247. Albini A, Mirisola V, Pfeffer U. Metastasis signatures: Genes regulating tumor-microenvironment interactions predict metastatic behavior. *Cancer Metastasis Rev* 2008, 27:75–83.
248. Mantovani A. Cancer: Inflaming metastasis. *Nature* 2009, 457:36–37.
249. Kim S, Takahashi H, Lin WW, Descargues P, Grivennikov S, Kim Y et al. Carcinoma-produced factors activate myeloid cells through TLR2 to stimulate metastasis. *Nature* 2009, 457:102–106.
250. Brown MD, Hart CA, Gazi E, Bagley S, Clarke NW. Promotion of prostatic metastatic migration towards human bone marrow stoma by omega 6 and its inhibition by omega 3 PUFAs. *Br J Cancer* 2006, 94:842–853.
251. Denkins Y, Kempf D, Ferniz M, Nileshwar S, Marchetti D. Role of omega-3 polyunsaturated fatty acids on cyclooxygenase-2 metabolism in brain-metastatic melanoma. *J Lipid Res* 2005, 46:1278–1284.
252. Altenburg JD, Siddiqui RA. Omega-3 polyunsaturated fatty acids down-modulate CXCR4 expression and function in MDA-MB-231 breast cancer cells. *Mol Cancer Res* 2009, 7:1013–1020.
253. Garber K. First results for agents targeting cancer-related inflammation. *J Natl Cancer Inst* 2009, 101:1110–1112.
254. Heukamp I, Kilian M, Gregor JI, Kiewert C, Schimke I, Kristiansen G et al. Impact of polyunsaturated fatty acids on hepato-pancreatic prostaglandin and leukotriene concentration in ductal pancreatic cancer—Is there a correlation to tumour growth and liver metastasis? *Prostaglandins Leukot Essent Fatty Acids* 2006, 74:223–233.
255. Shin SY, Kim YJ, Song KS, Kim NY, Jeong SY, Park JH et al. Mechanism of anti-invasive action of docosahexaenoic acid in SW480 human colon cancer cell. *J Life Sci* 2010, 20:561–571.
256. Yun E-J. Mechanisms of anticancer action of docosahexaenoic acid in breast cancer [doctor of medical science]. Daejeon, South Korea: Chungnam National University; 2008, p. 96.
257. Xia SH, Wang J, Kang JX. Decreased n-6/n-3 fatty acid ratio reduces the invasive potential of human lung cancer cells by downregulation of cell adhesion/invasion-related genes. *Carcinogenesis* 2005, 26:779–784.
258. Liu XH, Rose DP. Suppression of type IV collagenase in MDA-MB-435 human breast cancer cells by eicosapentaenoic acid in vitro and in vivo. *Cancer Lett* 1995, 92:21–26.
259. Wu MH, Tsai YT, Hua KT, Chang KC, Kuo ML, Lin MT. Eicosapentaenoic acid and docosahexaenoic acid inhibit macrophage-induced gastric cancer cell migration by attenuating the expression of matrix metalloproteinase 10. *J Nutr Biochem* 2012, 23:1434–1439.
260. McCabe AJ, Wallace JM, Gilmore WS, McGlynn H, Strain SJ. Docosahexaenoic acid reduces in vitro invasion of renal cell carcinoma by elevated levels of tissue inhibitor of metalloproteinase-1. *J Nutr Biochem* 2005, 16:17–22.
261. Kleiner DE, Stetler-Stevenson WG. Matrix metalloproteinases and metastasis. *Cancer Chemother Pharmacol* 1999, 43(Suppl):S42–S51.
262. Yoon SO, Park SJ, Yun CH, Chung AS. Roles of matrix metalloproteinases in tumor metastasis and angiogenesis. *J Biochem Mol Biol* 2003, 36:128–137.
263. Burr GO, Burr MM. Nutrition classics from *The Journal of Biological Chemistry* 82: 345–67, 1929. A new deficiency disease produced by the rigid exclusion of fat from the diet. *Nutr Rev* 1973, 31:248–249.
264. Liperoti R, Landi F, Fusco O, Bernabei R, Onder G. Omega-3 polyunsaturated fatty acids and depression: A review of the evidence. *Curr Pharm Des* 2009, 15:4165–4172.
265. Calon F, Cole G. Neuroprotective action of omega-3 polyunsaturated fatty acids against neurodegenerative diseases: Evidence from animal studies. *Prostaglandins Leukot Essent Fatty Acids* 2007, 77:287–293.
266. Fevang P, Saav H, Hostmark AT. Dietary fish oils and long-term malaria protection in mice. *Lipids* 1995, 30:437–441.
267. Mozaffarian D, Ascherio A, Hu FB, Stampfer MJ, Willett WC, Siscovick DS, Rimm EB. Interplay between different polyunsaturated fatty acids and risk of coronary heart disease in men. *Circulation* 2005, 111:157–164.

268. Chen D, Auborn K. Fish oil constituent docosahexaenoic acid selectively inhibits growth of human papillomavirus immortalized keratinocytes. *Carcinogenesis* 1999, 20:249–254.

269. Walboomers JM, Jacobs MV, Manos MM, Bosch FX, Kummer JA, Shah KV et al. Human papillomavirus is a necessary cause of invasive cervical cancer worldwide. *J Pathol* 1999, 189:12–19.

270. Gorjao R, Azevedo-Martins AK, Rodrigues HG, Abdulkader F, Arcisio-Miranda M, Procopio J, Curi R. Comparative effects of DHA and EPA on cell function. *Pharmacol Ther* 2009, 122:56–64.

271. Mori TA, Burke V, Puddey IB, Watts GF, O'Neal DN, Best JD, Beilin LJ. Purified eicosa-pentaenoic and docosahexaenoic acids have differential effects on serum lipids and lipoproteins, LDL particle size, glucose, and insulin in mildly hyperlipidemic men. *Am J Clin Nutr* 2000, 71:1085–1094.

272. Mori TA, Bao DQ, Burke V, Puddey IB, Beilin LJ. Docosahexaenoic acid but not eicosapentaenoic acid lowers ambulatory blood pressure and heart rate in humans. *Hypertension* 1999, 34:253–260.

273. Habermann N, Schon A, Lund EK, Glei M. Fish fatty acids alter markers of apoptosis in colorectal adenoma and adenocarcinoma cell lines but fish consumption has no impact on apoptosis-induction ex vivo. *Apoptosis* 2010, 15:621–630.

274. Tisdale MJ. Inhibition of lipolysis and muscle protein degradation by EPA in cancer cachexia. *Nutrition* 1996, 12:S31–S33.

16

Chitosan Application in Dentistry

Yoshihiko Hayashi, Kajiro Yanagiguchi, Zenya
Koyama, Takeshi Ikeda, and Shizuka Yamada

Contents

16.1 Introduction

Chitosan, an amino-polysaccharide obtained by alkaline deacetylation of chitin, is a natural and abundant polymer that is considered as an attractive candidate of biomaterial (Miyazaki et al., 1981; Madihally and Matthew, 1999). In dentistry, chitin–chitosan is applied as a dressing for oral mucous wound and a tampon following radical treatment of maxillary sinusitis. It is investigated as a root canal medicament (Ikeda et al., 2000) and an absorbing membrane for endodontic treatment and periodontal surgery, respectively. Furthermore, its application for oral care and hygiene has been proposed, in particular, to elderly people (Hayashi et al., 2007a). Bioactive natural organic materials originated from marine products now attract researchers' attention toward basic and clinical research for the development of alternative biomaterial, as amphixenosis such as BSE, avian, and swine influenzas occurs all over the world.

This chapter describes the characteristics and application of chitosan from marine sources in relation to especially dentistry (dento-oral fields). Among the resorbable polymers, chitosan has recently gained interest for its use in guided bone regeneration (GBR) membranes on account of its flexibility, biocompatibility, biodegradability, antibacterial properties, and low cost

(Kratz et al., 1997; Ueno et al., 1999; Teng et al., 2008). Furthermore, the scaffold innovation using chitosan complex is briefly reviewed in aspect of its importance as a potential source for GBR and tissue engineering (Hubbell, 1995).

16.2 Biological activities

The interesting biological properties of chitosan have led to various applications such as drug and gene delivery carriers, surgical thread, bone healing materials, and in particular wound dressing (Mi et al., 2001; Azad et al., 2004). In addition, chitosan exhibits myriad pharmacological actions, namely, hypocholesterolemic, hemostatic, antimicrobial (**Figure 16.1**), and wound healing properties (Ong et al., 2008). Therefore, it has been widely applied as a material for cell scaffolding, controlled release of pharmaceuticals, and wound dressing.

The term chitosan means a series of chitosan polymers with a different molecular weight (MW), viscosity, and degree of deacetylation (DD) (about 35%–100%). The MW and DD of chitosan influence the properties important for many applications, such as solubility of the product in dilute acids, viscosity of the obtained solutions, and their biological activity. The low MW and high DD chitosans are water soluble and can better incorporate the active molecule into the bacterial cell for the antimicrobial activity. The low MW and greater degree of quaternization have a positive influence on the antioxidant activity of chitosan. A low DD in the chitosan acetic acid physiological saline solution had a significant effect on the unusual aggregation and deformation

Streptococcus mutans
GS-5

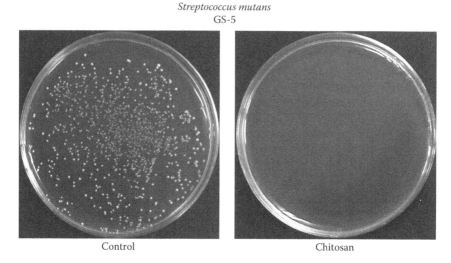

Control   Chitosan

Figure 16.1 *Antibacterial effect by 2% chitooligosaccharide application for typical cariogenic bacteria,* Streptococcus mutans. *Note almost complete inhibition of colony formation in chitosan group.*

of erythrocytes (hemostasis), compared with the effect of MW within a range between 10^5 and 10^6 (Yang et al., 2008). The higher DD of chitosan can accelerate wound healing in bone defect and biodegradation simultaneously progresses to solve the bone defect (Ikeda et al., 2002).

16.3 Special application for dental and oral fields

16.3.1 Soft tissue

Skin and oral mucosa dressing, and a tampon following radical therapy of maxillary sinusitis are popular to promote wound healing of soft tissue. Furthermore, endodontic application has been reported for pulp wound healing. Chitosan was used as a pulp capping medicament in rat pulp. However, the initial inflammatory reactions associated are severe and obvious in pulp tissue. This reaction is severer especially in large experimental animal. This type of tissue reaction needs to be overcome before it can be considered for clinical application (Yanagiguchi et al., 2001). This side effect by chitosan in soft tissue including gingiva could be also partially controlled by the pretreatment of tannic acid (Ishizaki et al., 2009).

Completely hydrolyzed chitin, D-glucosamine, which is almost completely deacetylated chitosan, promoted a good pulp wound healing with only initial slight inflammation after the application as a direct pulp capping agent (Matsunaga et al., 2006). Furthermore, it is thought that D-glucosamine has the effect of stabilization of cell membrane (Hayashi, 2010), besides the relief for osteoarthrosis per oral administration. The analgesic effect of D-glucosamine to peripheral nerve is investigated in relation to channels in membrane (personal communication).

Root canal treatment is a very popular dental treatment for conserving and maintaining the infected tooth structure and function that mainly originated from dental caries. Chitin–chitosan is also a useful medicament for the infected root canal treatment, especially in clinical cases where it is difficult to control inflammatory exudates and bleeding within root canal (**Figure 16.2**) (Ikeda et al., 2000; Yanagiguchi et al., 2002). These clinical studies revealed that symptoms and signs decreased in coincidence with the reduction of lysozyme and IL-1β concentrations. This means that inflammatory cells such as leukocyte and macrophage in the periapical regions would decrease in time by intracanal medication. The lysozyme and IL-1β concentrations could be used as a clinical marker for evaluating clinical conditions.

16.3.2 Hard tissue

GBR has been demonstrated to be a new method of filling the periodontal pocket or apical bone defect after apicotomy with neo-bone tissues by preventing the ingrowth of fibrous tissues. In order to promote effective bone

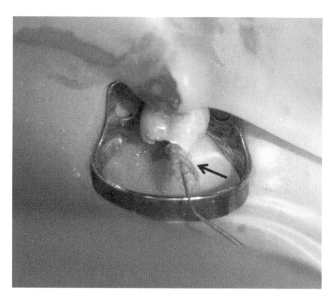

Figure 16.2 *Application of cotton-like chitin–chitosan (arrow) for root canal medicament (lower right first premolar, 71 year old male).*

regeneration using GBR procedure, the barrier membrane needs to have specific properties in terms of its bioactivity (osteoconductivity) and bioresorption, as well as space-maintaining ability, which is related to its mechanical stability during bone healing process (Ignatius et al., 2001). Bioabsorbable membranes, such as polylactic acid (Imbronito et al., 2002), polyglactin (Soncini et al., 2002), alginate (Ueyama et al., 2002; Wang et al., 2002; Hong et al., 2008), collagen (Ozmeric et al., 2000; Pek et al., 2008; Hayashi et al., 2012), have more interest than non-bioabsorbable membranes. Among the bioresorbable membranes, chitosan, a natural biodegradable cationic polymer with good biocompatibility, has also been widely used as GBR membrane (Mi et al., 2001; Kuo et al., 2003, 2006). However, the mechanical property is not satisfactory.

For the good mechanical property, chitosan–carboxymethyl cellulose complex membrane (Ryan and Sax, 1995; Bowler et al., 1999) and chitosan/poly-L-lysin composite membrane (Zheng et al., 2009) have been produced for the GBR application. These composite membranes properly functioned as mechanical stable barrier for GBR. However, as biodegradation of pure chitosan membrane generally takes a long time (over a few months), the secondary operation is needed for its removal from GBR site.

For regenerative medicine of hard tissue (Nakahara et al., 2004), it needs three interactive factors—responsive cells, supportive matrix (scaffold), and bioactive molecules—which correspond to seed, soil, and nutrient, respectively. The scaffold, which is used to create the three-dimensional organization needed for appropriate cell interactions (to serve as vehicles to deliver and retain the cells at a specific site), is a key element for regenerative medicine.

Pure chitosan scaffold lacks bioactivity to induce hard tissue formation, which limits its application in tissue engineering (Liao et al., 2010). For the improvement of biocompatibility and osteoconductivity, collagen/hydroxy-apatite (HA), and chitosan membrane (Teng et al., 2008), nanocomposite nanofiber of HA/chitosan (Zhang et al., 2008), chitosan/nanoHA composite membrane (Teng et al., 2009), membrane of hybrid chitosan–silica xerogel (Lee et al., 2009), chitosan–carboxymethyl cellulose polyelectrolyte com-plex membrane (Liuyun et al., 2009), and three-dimensional β-tricalcium phosphate chitosan scaffold (Liao et al., 2010; Tai et al., 2012) have been pro-posed and introduced as GBR membrane. The composite or hybrid scaffold showed higher proliferation rate of human periodontal ligament cells than the pure chitosan scaffold. Collagen and chitosan membranes, with their respective composites with different contents of HA, were successfully pre-pared by a newly developed dynamic filtration technique (Song et al., 2007; Teng et al., 2009). Collagen/HA composite membranes are known to have composition and structure similar to those of natural bone with excellent biocompatibility and osteoconductivity. The chitosan layer ensured high ten-sile strength and elastic modulus of the membrane, while the presence of the collagen/HA composite layers endowed it with good flexibility and bioactivity (Teng et al., 2008).

16.3.3 Oral care and hygiene

The studies pertaining to the chewing of gum in elderly populations have shown an improvement in oral health and acceptability within this age group (Markinen et al., 1996; Simons et al., 1997). Natural product such as mastic has been investigated to prevent plaque-related disease (Aksoy et al., 2006). Gum obtained from a native plant has been also reported to possess considerable in vitro antimicrobial activity in Iran (personal communication). Numerous antimicrobials and antibiotics, including chlorhexidine, spiramycin, and vancomycin, have been used against *Streptococcus mutans* to reduce plaque-mediated diseases, including dental caries (Cragg et al., 1997). However, in Japan since 1985, chlorhexidine has been prohibited for the use of mucous irrigation due to the occurrence of anaphylaxis.

The type of chitosan, MW, DD, and pH affect its bacteriostatic and bacteriocidal actions (Fujiwara et al., 2004). Recent studies have demonstrated that chewing the chitosan oligomer-containing gum (**Figure 16.3**) effectively inhibited the growth of the cariogenic (total bacteria, total *Streptococci*, mutant streptococci) (Hayashi et al., 2007b) and also periodontopathic bacteria (*Porphyromonas gingivalis*) in saliva (Hayashi, 2010). The chitosan oligomer-containing gum chewing has greater antibacterial effects and increases salivary secretion (Hayashi et al., 2007a). These findings strongly suggest that the application of such natural materials as chitosan is useful for both oral hygiene (especially in the outside activities) and the quality of life of elderly individuals through relatively high salivary secretion.

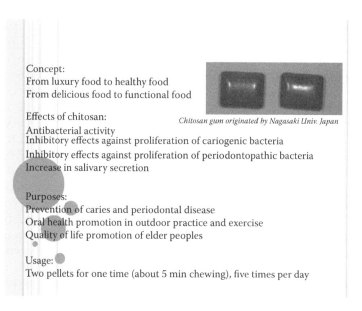

Concept:
From luxury food to healthy food
From delicious food to functional food

Effects of chitosan:
Chitosan gum originated by Nagasaki Univ. Japan
Antibacterial activity
Inhibitory effects against proliferation of cariogenic bacteria
Inhibitory effects against proliferation of periodontopathic bacteria
Increase in salivary secretion

Purposes:
Prevention of caries and periodontal disease
Oral health promotion in outdoor practice and exercise
Quality of life promotion of elder peoples

Usage:
Two pellets for one time (about 5 min chewing), five times per day

Figure 16.3 *Leaflet for chitosan-containing gum.*

16.4 Conclusions

Chitosan exhibits myriad pharmacological actions, namely, hypocholesterol-emic, hemostatic, antimicrobial, and wound healing properties. It has been widely applied as a material for cell scaffolding, controlled release of pharmaceuticals, and wound dressing.

Chitin–chitosan could be used as an effective medicament especially for periapical wounds, thanks to biodegradation and biocompatibility. The composite or hybrid scaffold showed higher proliferation rate of human periodontal ligament cells for bony wound healing than the pure chitosan scaffold. Furthermore, its application, such as chitosan-containing gum, is useful for both oral hygiene (especially in the outside activities) and the quality of life of elderly individuals.

References

Aksoy, A., Duran, N., and Koksal, F. (2006). In vitro and in vivo antimicrobial effects of mastic chewing gum against *Streptococcus mutans* and mutans streptococci. *Arch. Oral Biol.* 51, 476–481.

Azad, A. K., Sermsintham, N., Chandrkrachang, S., and Stevens, W. F. (2004). Chitosan membranes as a wound-healing dressing: Characterization and clinical application. *J. Biomed. Mater. Res.* 69B, 216–222.

Bowler, P. G., Jones, S. A., Davies, B. J., and Coyle, E. (1999). Infection control properties of some wound dressing. *J. Wound Care* 8, 499–502.

Cragg, G. M., Newman, D. J., and Snader, K. M. (1997). Natural products in drug discovery and development. *J. Nat. Prod.* 60, 52–60.

Fujiwara, M., Hayashi, Y., and Ohara, N. (2004). Inhibitory effect of water-soluble chitosan on growth of *Streptococcus mutans*. *Microbiologica* 27, 83–86.

Hayashi, Y. (2010). Applications of chitosan oligosaccharide and glucosamine in dentistry. In *Chitin, Chitosan, Oligosaccharides and Their Derivatives*, S.-K. Kim, Ed. CRC Press, Boca Raton, FL, pp. 447–460.

Hayashi, Y., Ohara, N., Ganno, T., Ishizaki, H., and Yanagiguchi, K., (2007a). Chitosan-containing gum chewing accelerates antibacterial effect with an increase in salivary secretion. *J. Dent.* 35, 871–874.

Hayashi, Y., Ohara, N., Ganno, T., Yanagiguchi, K., Ishizaki, H., Nakamura, T., and Sato, M. (2007b). Chewing chitosan-containing gum effectively inhibits the growth of cariogenic bacteria. *Arch. Oral Biol.* 52, 290–294.

Hayashi, Y., Yamada, S., Ikeda, T., and Yanagiguchi, K. (2012). Fish collagen and tissue repair. In *Marine Cosmeceuticals: Trends and Prospects*, S.-K. Kim, Ed. CRC Press, Boca Raton, FL, pp. 133–141.

Hong, H.-J., Jin, S.-E., Park, J.-S., Ahn, W. S., and Kim, C.-K. (2008). Accelerated wound healing by smad3 antisense oligonucleotides-impregnated chitosan/alginate polyelectrolyte-complex. *Biomaterials* 29, 4831–4837.

Hubbell, J. A. (1995). Biomaterials in tissue engineering. *Biotechnology (NY)* 13, 565–576.

Ignatius, A. A., Ohnmacht, M., Claes, L. E., Kreidler, J., and Palm, F. (2001). A composite polymer/tricalcium phosphate membrane for guided bone regeneration in maxillofacial surgery. *J. Biomed. Mater. Res. (Appl. Biomater.)* 58, 564–569.

Ikeda, T., Yanagiguchi, K., and Hayashi, Y. (2002). Application to dental medicine—In focus on dental caries and alveolar bone healing. *Bioindustry* 19, 22–30 (in Japanese).

Ikeda, T., Yanagiguchi, K., Viloria, I. L., and Hayashi, Y. (2000). Relationship between lysozyme activity and clinical symptoms following the application of chitin/chitosan in endodontic treatment. In *Chitosan per os: From Dietary Supplement to Drug Carrier*, R. A. A. Muzzarelli, Ed.. Atec Edizioni, Grottammare, Italy, pp. 275–292.

Imbronito, A. V., Todescan, J. H., Carvalho, C. V., and Arna-Chavez, V. E. (2002). Healing of alveolar bone in resorbable and non-resorbable membrane-protected defects. A histologic pilot study in dogs. *Biomaterials* 23, 4079–4086.

Ishizaki, H., Yamada, S., Yanagiguchi, K., Koyama, Z., Ikeda, T., and Hayashi, Y. (2009). Pre-treatment with tannic acid inhibits the intracellular IL-8 production by chitosan in a human oral epithelial cancer cell line. *Oral Med. Pathol.* 13, 135–141.

Kratz, G., Arnander, C., Swedenborg, J., Back, M., Falk, C., Gouda, I., and Larm, O. (1997). Heparin–chitosan complexes stimulate wound healing in human skin. *Scand. J. Plast. Reconstr. Surg. Hand. Surg.* 31, 119–123.

Kuo, S. M., Chang, S. J., Chen, T. W., and Kuan, T. C. (2006). Guided tissue regeneration for using a chitosan membrane: An experimental study in rats. *J. Biomed. Mater. Res.* A 76, 408–415.

Kuo, S. M., Chang, S. J., Lim, L.-C., and Chen, C. J. (2003). Evaluating chitosan/β-tricalcium phosphate/poly (methyl methacrylate) cement composites as bone-repairing materials. *J. Appl. Polym. Sci.* 89, 3897–3904.

Lee, E.-J., Shin, D.-S., Kim, H.-E., Kim, H.-W., Koh, Y.-H., and Jang, J.-H. (2009). Membrane of hybrid chitosan-silica xerogel for guided bone regeneration. *Biomaterials* 30, 743–750.

Liao, F., Chen, Y., Li, Z., Wang, Y., Shi, B., Gong, Z., and Cheng, X. (2010). A novel bioactive three-dimensional β-tricalcium phosphate/chitosan scaffold for periodontal engineering. *J. Mater. Sci.: Mater. Med.* 21, 489–496.

Liuyun, J., Yubao, L., and Chengdong, X. (2009). A novel composite membrane of chitosan-carboxymethyl cellulose polyelectrolyte complex membrane filled with nano-hydroxyapatite I. Preparation and properties. *J. Mater. Sci.: Mater. Med.* 20, 1645–1652.

Madihally, S. V. and Matthew, H. W. (1999). Porous chitosan scaffolds for tissue engineering. *Biomaterials* 20, 1133–1142.

Markinen, K. K., Pemberton, D., Makinen, P. L., Chen, C. Y., Cole, J., Hujoel, P., Lopatin, D., and Lambert, P. (1996). Poly-combinant saliva stimulants and oral health in veterans affairs patients—An exploratory study. *Spec. Care Dent.* 16, 104–115.

Matsunaga, T., Yanagiguchi, K., Yamada, S., Ohara, N., Ikeda, T., and Hayashi, Y. (2006). Chitosan monomer promotes tissue regeneration on dental pulp wounds. *J. Biomater. Res.* 76A, 711–720.

Mi, F. L., Shyu, S. S., Wu, Y. B., Lee, S. T., Shyong, J. Y., and Huang, R. N. (2001). Fabrication and characterization of sponge-like asymmetric chitosan membranes as a wound dressing. *Biomaterials* 22, 165–173.

Miyazaki, S., Ishii, K., and Nadai, T. (1981). The use of chitin and chitosan as drug carriers. *Chem. Pharm. Bull.* 29, 3067–3069.

Nakahara, T., Nakamura, T., Kobayashi, E., Kumemoto, K., Matsuno, T., Tabata, Y., Eto, K., and Shimizu, Y. (2004). In situ tissue engineering of periodontal tissues by seeding with periodontal ligament-derived cells. *Tissue Eng.* 10, 537–544.

Ong, S.-Y., Wu, J., Moochhala, S. M., Tan, M.-H., and Lu, L. (2008). Development of a chitosan-based wound dressing with improved hemostatic and antimicrobial properties. *Biomaterials* 29, 4323–4332.

Ozmeric, N., Bal, B., Oygur, T., and Balos, K. (2000). The effect of a collagen membrane in regenerative therapy of two-wall intrabony defects in dogs. *Periodontal Clin. Investig.* 22, 22–30.

Pek, Y. S., Gao, S., Mohamed Arshad, M. S., Leck, K.-J., and Ying, J. Y. (2008). Porous collagen-apatite nanocomposite foams as bone regeneration scaffolds. *Biomaterials* 29, 4300–4305.

Ryan, C. K. and Sax, H. C. (1995). Evaluation of a carboxymethyl cellulose sponge for prevention of postoperative adhesions. *Am. J. Surg.* 169, 154–159.

Simons, D., Kidd, E. A. M., Beighton, D., and Jones, B. (1997). The effect of chlorhexidine/xyliyol chewing gum on cariogenic salivary microflora: A clinical trial in elder patients. *Caries Res.* 31, 91–96.

Soncini, M., Rodriguez, R., Baena, R., Pietrabissa, R., Quaglini, V., Rizzo, S., and Zaffe, D. (2002). Experimental procedure for the evaluation of the mechanical properties of the bone surrounding dental implants. *Biomaterials* 23, 9–17.

Song, J.-H., Kim, H.-E., and Kim, H.-W. (2007). Collagen-apatite nanocomposite membranes for guided bone regeneration. *J. Biomed. Mater. Res.* 83B, 248–257.

Tai, H.-Y., Chou, S.-H., Cheng, L.-P., Yu, H.-T., and Don, T.-M. (2012). Asymmetric composite membranes from chitosan and tricalcium phosphate useful for guided bone regeneration. *J. Biomater. Sci., Polym. Ed.* 23, 1153–1170.

Teng, S.-H., Lee, E.-J., Wang, P., Shin, D.-S., and Kim, H.-E. (2008). Three-layered membranes of collagen/hydroxyapatite and chitosan for guided bone regeneration. *J. Biomed. Mater. Res.* 87B, 132–138.

Teng, S.-H., Lee, E.-J., Yoon, B.-H., Shin, D.-S., Kim, H.-E., and Oh, J.-S. (2009). Chitosan/nanohydroxyapatite composite membranes via dynamic filtration for guided bone regeneration. *J. Biomed. Mater. Res. A* 88, 569–580.

Ueno, H., Yamada, H., Tanaka, I., Kaba, N., Matsuura, M., Okumura, M., Kadosawa, T., and Fujinaga, T. (1999). Accelerating effects of chitosan for healing at early phase of experimental open wound in dogs. *Biomaterials* 20, 1407–1414.

Ueyama, Y., Ishikawa, K., Mano, T., Koyama, T., Nagatsuka, H, Suzuki, K., and Ryoke, K. (2002). Usefulness as guided bone regeneration membrane of the alginate membrane. *Biomaterials* 23, 2027–2033.

Wang, L., Khor, E., Wee, A., and Lim, L. Y. (2002). Chitosan-alginate PEC membrane as a wound dressing: Assessment of incisional wound healing. *J. Biomed. Mater. Res.* 63, 610–618.

Yanagiguchi, K., Ikeda, T., Takai, F., Ogawa, K., and Hayashi, Y. (2001). Wound healing following direct pulp capping with chitosan-ascorbic acid complex in rat incisors. In *Chitin and Chitosan*, T. Uragami, K. Kurita, and T. Fukamizo, Eds. Kodansha Scientific Ltd., Tokyo, Japan, pp. 240–243.

Yanagiguchi, K., Yamaguchi, K., Ohara, N., and Hayashi, Y. (2002). Relationship between clinical symptoms and concentration of inflammatory cytokine (IL-1β) in intracanal exudates during endodontic treatment. *Jpn. J. Conserv. Dent.* 45, 343–348 (in Japanese).

Yang, J., Tian, F., Wang, Z., Wang, Q., Zeng, Y.-J., and Chen, S.-Q. (2008). Effect of chitosan molecular weight and deacetylation degree on hemostasis. *J. Biomed. Mater. Res. B* 84, 131–137.

Zhang, Y., Venugopal, J. R., El-Turki, A., Ramakrishna, S., Su, B., and Lim, C. T. (2008). Electrospun biomimetic nanocomposite nanofibers of hydroxyapatite/chitosan for bone tissue engineering. *Biomaterials* 29, 4314–4322.

Zheng, Z., Wei, Y., Wang, G., Gong, Y., and Zhang, X. (2009). Surface characterization and cytocompatibility of three chitosan/polycation composite membranes for guided bone regeneration. *J. Biomater. Appl.* 24, 209–229.

Edible Marine Invertebrates
A Promising Source of Nutraceuticals

Se-Kwon Kim and S.W.A. Himaya

Contents

17.1 Introduction

The awareness and understanding of the health-promoting abilities of edible natural materials, beyond their nutritional benefits, paved the way to a highly growing market of functional foods and natural functional ingredients worldwide. The space for marine nutraceuticals in this market is continuously increasing with the understanding of unique features of marine natural ingredients with wide array of health benefits. Until recent times, the key players of marine nutraceutical market were fish oil, carotenoids, and chitosan products. However, intensive research on identification and characterization of value added ingredients from marine resources has revealed an enormous capacity of marine nutraceuticals. Identification of active ingredients from marine macro-algae, micro-algae, microbes, and invertebrates is widely researched. Among these groups the potential use of active ingredients, specifically functional polysaccharides and peptides from edible marine invertebrates, is discussed in this chapter.

17.1.1 Nutraceuticals

Even though the popularity for health-promoting foods both in the common society and in the research field has raised recently, the concept drove back to thousands of years, which was well explained by the Hippocrates 2500 years ago as "let food be thy medicine and medicine be thy food." Most of the Asian countries such as China, Japan, Korea, India, and Indonesia have used both terrestrial and marine resources as medicinal foods and traditional medicines. However, the awareness among the Western population has emerged much recently. Since the importance of medicinal foods has gained limelight in the field of food research, several different terminologies have appeared such as functional foods and nutraceuticals. The boundaries between functional foods and nutraceuticals are hard to identify except for their method of intake where functional foods are in a form of a food and nutraceuticals are in a form of a supplement such as capsules or pills (Espín et al. 2007).

The term nutraceutical is coined by Stephen De Felice, founder and chairman of the foundation for innovation in medicine, and he has defined it as "a nutraceutical is any substance that is a food or a part of a food and provides medical or health benefits, including the prevention and treatment of disease." As the name and definition suggests, nutraceuticals are unique metabolites with nutritional and pharmaceutical properties (Haller 2010). Most of the time continuous supplementation with nutraceuticals was found to be beneficial for health as they are able to prevent many adverse physiological conditions in human body such as cancer, heart diseases, inflammatory disorders, oxidative stress-related complications, neurodegeneration, and skin aging. The admiration toward nutraceuticals is increasing as they are safe materials to be consumed with significantly less or no side effects compared to synthetic drugs. And also nutraceuticals provide cost effectiveness over drugs as they are considerably cheaper materials (Bernal et al. 2011).

As most of the definitions suggest, the nutraceuticals should be of food origin. Marine resources have been used as food in most of the East Asian countries and they have identified the special healing power of these marine foods. Most of the marine foods have been used as traditional medicines in these countries. Among the marine resources, the medicinal value of marine invertebrates has gained more attention as most of the edible invertebrates have shown to possess functional ingredients that could be developed as nutraceuticals.

17.1.2 Edible marine invertebrates

Given the fact that ocean covers 70% of the earth surface, marine environment is far more diverse compared to the land. A large number of flora fauna has been identified in the marine environment and even larger numbers are to be discovered. Marine invertebrates are a distinct group of animals living in all the ecosystems of the marine environment where over 97% of all known species are under invertebrates. There are different phyla of marine invertebrates with specific characteristics acquired throughout the evolution for the survival. Some have shells or chitinous bodies, some have the ability to swim or move fast, and some are sessile and could produce toxic compounds as defense system (Faulkner 2000). Interestingly, it has been discovered that these defense metabolites or structures are responsible for the medicinal value of these invertebrates. This finding clarifies the specificity and effectiveness of marine invertebrate metabolites over the others, as they have been specially designed over years of evolution.

More than 100 species of marine invertebrates are utilized as foods, mainly in the Asia Pacific area for centuries. Some of them are even popular in the Western cuisines too. At present, marine invertebrates account for more than 40% of the seafood consumed worldwide (De Zoysa 2012). According to the taxonomical classification most of the major edible invertebrates belong to several prominent phyla, such as mollusca (oysters, abalones, clams, squids, cuttlefish, and octopus), arthropoda (crab, shrimp, and lobsters), Echinodermata (sea cucumbers). Moreover, some edible invertebrate groups such as tunicates are classified under the phylum chordata due to their embryonic characteristics. Most of these animals are used in Eastern medicine; present research has proved their medicinal values (Gopal et al. 2008). The medicinal properties of marine invertebrates are attributed to the presence of unique secondary metabolites, bioactive peptides, carbohydrates, and fatty acids. In the following sections of this chapter, efforts have been taken to discuss the nutraceutical potentials of these active metabolites from marine edible invertebrates.

17.2 Functional carbohydrates

17.2.1 Chitosan and chitosan oligosaccharides

The most studied functional polysaccharide of edible marine invertebrate origin is chitosan and chitosan derivatives. The exoskeleton of crustaceans and other arthropods are utilized as a source of chitin. Chitin is also abundant natural polysaccharide in terrestrial sources, microorganisms. However, mainly the shells of shrimp, crab, and lobsters are used in the commercial chitin production. As the crustacean shell is a processed by-product, it provides cheaper materials for chitin production. Furthermore, the utilization of the crustacean shell wall solves the problem of waste disposal in crustacean processing, as shell waste of crustaceans accounts for more than 45% by weight. Chitin is a long-chain polymer of N-acetylglucosamine with $\beta(1 \rightarrow 4)$ glycosidic bonds between each monomer (**Figure 17.1**). Over 60% of deacetylation of chitin with hot alkali produces chitosan, which is also a biopolymer composed of D-glucosamine monomers (**Figure 17.1**). The major limitation in chitin and chitosan usage is their poor solubility. Chitosan is more soluble in aqueous acid solutions such as formic acid and acetic acid compared to chitin. However, both fail to be soluble in water and therefore, the applications of chitins and chitosan are limited despite their promising functionality

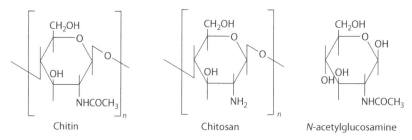

Figure 17.1 *Structures of Chitin, Chitosan and N-acetylglucosamine.*

in promoting health (Kim and Rajapakse 2005). To overcome this limitation, chemical (acid hydrolysis) (Il'ina and Varlamov 2004), physical (ultrasonic degradation, thermodynamic degradation) (Chen and Chen 2000), or enzymatic hydrolysis (Kuroiwa et al. 2002) of chitosan—in order to produce low molecular weight chitosan oligosaccharides (COS)—are used effectively. It has been reported that COS produced from chemical hydrolysis are not favorable as a bioactive material due to synthesis of other toxic compounds during the hydrolysis process (Kim and Rajapakse 2005). Therefore enzymatic hydrolysis for the production of bioactive COS is appreciated. COS have shown a wide array of biological activities. Evidently, COS is more potent and popular as a functional material due to its low molecular weight and water solubility. Recently COS derivatives have been synthesized to enhance and diversify the promising bioactivities of COS. These derivatives are prepared by the structural modification of COS by adding active chemical groups.

17.2.2 Biological activities of chitosan, COS, and COS derivatives

Both chitosan and its oligomers are used as nutraceuticals or active ingredients in functional food preparations (Xia et al. 2010). A large number of studies have been carried out in the investigation of biological activities of chitosan and COS. The popularity toward these functional carbohydrates is immense due to their abundance and low toxicity. They have shown promising results against intracellular oxidative stress, aging, photoaging, osteoarthritis, inflammation, asthma, allergy, diabetes, obesity, cancer, hypertension, and microbial infection. A list of biological activities of COS and COS derivatives is shown in **Table 17.1**. The biological activities shown by both COS and COS derivatives could be attributed to the basic nature with overall positive charge and the molecular structure having reactive hydroxyl and amino groups, respectively (Kim and Mendis 2006).

The mechanism of hypocholesterolemic and antiobesity activity of chitosan and COS is explained as they are able to scavenge fat and cholesterol in the digestive tract and remove it through excretion. The anticancer and the wound healing properties of chitosan and its derivatives are shown to stimulate

Table 17.1 Biological Activities of COS and COS Derivatives

COS/COS Derivative	Bioactivity	References
COS (<1 kDa)	Antioxidant	Mendis et al. (2007)
N-acetyl-COS	Antioxidant	Ngo et al. (2008a)
Gallate-COS	Antioxidant	Ngo et al. (2011a)
Aminoethyl-COS (1–3 kDa)	Antioxidant	Ngo et al. (2012a)
COS (<1 kDa)	Photoaging	Kim et al. (2012)
Carboxylated-COS	ACE-inhibitory	Huang et al. (2005)
Aminoethyl-COS	ACE-inhibitory	Ngo et al. (2008b)
COS (<5 kDa)	Prevention of negative mineral balance	Jung et al. (2006)
COS (3–5 kDa)	MMP-2 inhibitory activity	Kim and Kim (2006)
Carboxylated-COS	MMP-9 inhibitory activity	Rajapakse et al. (2006)
COS (3–5 kDa)	β-secratase inhibitory activity	Byun et al. (2005)
Dimethylaminoethyl-COS Diethylaminoethyl-COS	Acetylcholinesterase inhibitory activity	Yoon et al. (2009)
Aminoethyl-COS	Anti-inflammatory	Cho et al. (2010)
Aminoethyl-COS	Anti-neuroinflammatory	Ngo et al. (2012b)
COS (<1 kDa)	Anti-neuroinflammatory	Pangestuti et al. (2011)
Sulfated-COS	Antiarthritic	Ryu et al. (2012)
COS (1–3 kDa)	Antiallergy	Vo et al. (2012)
Dimethyl aminoethyl-COS Diethyl aminoethyl-COS	Antitumor	Karagozlu et al. (2012)
COS (3–5 kDa)	Antidiabetic	Karadeniz et al. (2010)
COS (3–5 kDa)	Anti-HIV	Artan et al. (2010)
COS (<10 kDa)	Antimicrobial	Jeon et al. (2001)

the immune system. The chitosan-mediated immune stimulation activates peritoneal macrophages and induces nonspecific host resistance followed by the removal of cancer cells (Kim and Mendis 2006; Suzuki et al. 1986). In the same manner the higher levels of macrophage production as a response to chitosan will release cytokines that are involved in the healing process (Okamoto et al. 2003). The degree of deacetylation and the molecular weight are key factors for the biological activities of COS. Low molecular weight chitosans (below 5 kDa) with higher degree of deacetylation (90%) are found to be potent radical scavengers and enzyme inhibitors (Je et al. 2004; Ngo et al. 2011b). This group of COS has effectively controlled the β-secretase enzyme and angiotensin-1 converting enzyme activity, which plays a major role in the progression of Alzheimer's disease and hypertension, respectively (Byun et al. 2005; Park et al. 2003).

All these studies have proved that chitosan and COS are potent regulators of adverse pathophysiological conditions of human body. Therefore, these molecules can be regarded as effective nutraceuticals to be used as supplements to regulate and prevent from wide array of diseases.

17.2.3 Glucosamine

N-acetylglucosamine (**Figure 17.1**) is the monomer unit of chitin and chitosan. The extensive hydrolysis of chitin will produce N-acetylglucosamine. However glucosamine and N-acetylglucosamine are readily synthesized in the human body from glucose. Higher glucosamine levels are found in bone tissues of humans. Therefore in osteoarthritis conditions, supplementation of glucosamine is used to relieve from arthritic symptoms (Anderson et al. 2005). Clinical studies have also proved this hypothesis (Block et al. 2010). Other than the pain-relieving ability of glucosamine, it was found that it has cartilage building and lubricating properties for the joints (Gorsline and Kaeding 2005). Other than its positive contribution to the joint pain relief, it has shown anti-inflammatory activity by suppressing inflammatory activation of synovial cells, endothelial cells, and intestinal epithelial cells (Nagaoka et al. 2011). Therefore, glucosamine as a chondroprotective and anti-inflammatory agent is a highly demanded nutraceutical in the market. Most of the commercially available glucosamine products—glucosamine hydrochloride, glucosamine sulfate, and N-acetylglucosamine—are derived from hydrolysis of chitin.

17.3 Bioactive peptides

At present there is an increasing interest in the utilization of food-derived biologically active peptides as nutraceuticals or nutritional supplements since they show potential benefits for the promotion of human health by reducing the risk of exposure to many diseases. Bioactive peptides exert physiological hormone-like effects on humans beyond their nutritional value (Erdmann et al. 2008). Generally the bioactive peptides remain latent within the parent protein molecule and are released by hydrolysis. The means of hydrolysis could be gastrointestinal digestion, enzymatic hydrolysis, fermentation (microbial hydrolysis) or physical methods of hydrolysis such as use of acids and sonication. By far enzymatic hydrolysis at optimal temperature and pH remains the most popular method to generate bioactive peptides. The enzymes used in the hydrolysis are either gastric enzymes (trypsin, chymotrypsin, pepsin), microbial enzymes (microbial proteases), or plant-based enzymes (papain) from marine sources. However, the fermentation method has also been used frequently to obtain bioactive peptides from marine invertebrates. The size of the peptide is of utmost importance for its activity. Most of the bioactive peptides are 2–50 amino acids long. Small-size peptides (less than 1 kDa) are considered to be more effective nutraceuticals as they are readily absorbed into the body. Therefore in the purification of bioactive peptides, membrane bioreactors equipped with ultrafiltration membranes are used to separate low molecular weight peptides. These bioreactors facilitate sequential hydrolysis of the protein sample with different enzymes at their optimal conditions. Most of the edible marine invertebrates are studied as potential sources of bioactive peptides and have resulted in promising peptides with a range of biological activities, mainly antioxidant, antihypertensive, and antimicrobial activities (**Table 17.2**).

Marine invertebrate-derived bioactive peptides have shown wide spectrum of biological activities such as antioxidant, antihypertensive, and antimicrobial activities. The protein content and the amino acid compositions are key factors to be considered when selecting a source for the purification of bioactive peptides as the activity is correlated with the structure and composition of the peptide. The amino acid composition of the sequence plays a critical role in its bioactivity. The size, hydrophobicity, charges, and microelement binding properties of the peptide are some of the key factors for the biological activities of the peptides.

Oxidative stress is a common risk factor in a number of chronic diseases, such as arthritis, diabetes, neurodegenerative diseases, and cardiovascular complications (Bernardini et al. 2011). Reactive oxygen species (ROS) generated during oxidative stress conditions cause adverse effect to cells by oxidizing major cellular components such as membrane lipids and proteins. Therefore, the production of ROS should be tightly regulated. Effective antioxidant peptides have been isolated from marine invertebrates with ROS scavenging properties. The structural features of the peptide are directly proportional to their activity. The antioxidant activity of the peptide was found to be more potent when its molecular weight is between 500 and 1500 Da compared to peptides above 1500 Da and below 500 Da (Li et al. 2008). Hydrophobicity is an important feature of antioxidant enzyme in accessing hydrophobic targets such as the cell membrane (Hsu 2010). The anti-lipid-peroxidation activity of the peptide is higher when the hydrophobicity increases as it can solubilize into lipid for more effective radical scavenging activities. Furthermore, positioning of leucine at the N terminal of the sequence is thought to be important for its antioxidant activity (Ranathunga et al. 2006). Cardiovascular diseases are responsible for about 30% of deaths worldwide according to World Health Organization (WHO) (2010) statistics. Increase in blood pressure (hypertension) is the mos t common risk factor of heart diseases. The regulation of blood pressure is associated with the renin-angiotensin system, where angiotensin-I converting enzyme (ACE) ultimately leads to hypertension (Hong et al. 2008). Therefore, inhibition of ACE has become the main target in the treatment of hypertension. Most of the identified marine invertebrate peptides have shown ACE inhibitory activity. These peptides or their hydrolysates are potential candidates of antihypertensive nutraceuticals. The ACE inhibitory peptides have also shown special structural features. The binding of inhibitory peptide to the ACE is strongly supported by the presence of hydrophobic amino acids at the carboxyl terminal and branched-chain aliphatic amino acids at the amino terminal (Himaya et al. 2012). Furthermore it

Table 17.2 Active Peptides Isolated from Edible Marine Invertebrates

Source	Method of Hydrolysis	Peptide Sequence	Activity	References
Oyster, *Pinctada fucata martencii*	Denazyme AP	Leu-Phe	ACE inhibitory	Matsumoto et al. (1994)
Oyster, *Crassostrea talienwhanensis* Crosse	Pepsin	Val-Val-Tyr-Pro-Trp-Tyr-Glu-Arg-Phe	ACE inhibitory	Wang et al. (2008)
Oyster, *Crassostrea gigas*	Fermentation	Val-Lys-Lys	ACE inhibitory	Je and Kim (2005)
Clam, *Corbicula fluminea*	Protamex	Val-Lys-Lys	ACE inhibitory	Tsai et al. (2006)
		Val-Lys-Pro		
Blue mussel, *Mytilus edulis*	Fermentation	Glu-Val-Met-Ala-Gly-Asn-Leu-Tyr-Pro-Gly	ACE inhibitory	Je et al. (2005)
Antarctic krill	Pepsin and trypsin	Lys-Leu-Lys-Phe-Val	ACE inhibitory	Kawamura et al. (1992)
Shrimp, *Acetes chinensis*	Protease from Bacillus spp. SM98011	Phe-Cys-Val-Leu-Pro	ACE inhibitory	He et al. (2006)
		Ile-Phe-Val-Pro-Ala-Phe		
		Lys-Pro-Pro-Gln-Try-Val		
		Tyr-Leu-Leu-Phe		
		Ala-Phe-Leu		
Shrimp, *Pandalopsis dispar*	Alcalase	Tyr, Phe, Leu, Ile, Val, Lys	ACE inhibitory	Cheung and Li-Chan (2010)
Sea bream	Alkaline protease	Val-Ile-Tyr	ACE inhibitory	Fahmi et al. (2004)
		Val-Tyr		
Sea cucumber, *Acaudina molpadioidea*	Bromelain and Alcalase	Met-Glu-Gly-Ala-Gln-Glu-Ala-Gln-Gly-Asp	ACE inhibitory	Zhao et al. (2009)
Oyster, *C. gigas*	In vitro gastrointestinal digestion with pepsin, trypsin and α-chymotrypsin	Leu-Lys-Gln-Glu-Leu-Glu-Asp-Leu-Leu-Glu-Lys-Gln-Glu	Antioxidant	Qian et al. (2008)
Blue mussel, *M. edulis*	Fermentation	Phe-Gly-His-Pro-Tyr	Antioxidant	Jung et al. (2005)
		His-Phe-Gly-Asp-Pro-Phe-His		Rajapakse et al. (2005)

Source	Method	Peptide sequence	Activity	Reference
Sea mussel, *Mytilus coruscus*	In vitro gastrointestinal digestion with pepsin, trypsin, and α-chymotrypsin	Pro-Ala-Val-Cys-Val-Pro-Leu-Val-Gly-Asp-Glu-Gln-Ala-Val-	Antioxidant	Jung et al. (2007)
Rotifer, *Brachionus rotundiformis*	Pepsin	Leu-Leu-Gly-Pro-Gly-Leu-Thr-AsnHis-Ala	Antioxidant	Byun et al. (2009)
Rotifer, *Brachionus calyciflorus*	Neutrase	Asp-Leu-Gly-Leu-Gly-Leu-Pro-GlyAla-His Gly-His-Asp-Gly-Tyr-Glu-Pro-LeuSer-Ser	Antioxidant	Lee et al. (2010)
Jumbo squid, *Dosidicus gigas*	Trypsin	Phe-Asp-Ser-Gly-Pro-Ala-Gly-Val-Leu Asn-Gly-Pro-Leu-Gln-Ala-Gly-Gln-Pro-Gly-Glu-Arg	Antioxidant	Mendis et al. (2005)
Blue mussel, *M. edulis*	Physical extraction methods	Met-Glu-Ala-Pro	Anti-coagulant	Jung and Kim (2009)
Crab, *Panaeus vannemi*	Physical extraction methods	Tyr-Arg-Gly-Gly-Tyr-Thr-Gly-Pro-Ile-Pro-Arg-Pro-Pro-Ile-Gly-Arg-Pro-Pro-Leu-Arg-Leu-Val-Val-Cys-Ala-Cys-Tyr-Arg-Leu-Ser-Val-Ser-Asp-Ala-Arg-Asn-Cys-Ile-Lys-Phe-Gly-Ser-Cys-Cys-His-Leu-Val-Lys	Antimicrobial	Hancock et al. (2006)
Horseshoe crab, *Tachyplesin tridentatus*	Physical extraction methods	Lys-Trp-Cys-Phe-Arg-Val-Cys-Tyr-Arg-Gly-Ile-Cys-Tyr-Arg-Arg-Cys-Arg	Antimicrobial	Hancock et al. (2006)
Horseshoe crab, *Limulus polyphemus*	Physical extraction methods	Arg-Arg-Trp-Cys-Phe-Arg-Val-Cys-Tyr-Arg-Gly-Phe-Cys-Tyr-Arg-Lys-Cys-Arg	Antimicrobial	Hancock et al. (2006)
Horseshoe crab, *T. tridentatus*	Physical extraction methods	Asn-Phe-Lys-Ile-Pro-Ala-Ile-Tyr-Ile-Gly-Ala-Thr-Val-Gly-Pro-Ser-Val-Trp-Ala-Tyr-Leu-Val-Ala-Leu-Val-Gly-Ala-Ala	Antimicrobial	Hancock et al. (2006)
Blue mussel, *M. edulis*	Physical extraction methods	Gly-Ala-Ser-Arg-Cys-Ala-Lys-Ala-Lys-Ala-Gly-Arg-Arg-Cys-Lys-Gly-Trp-Ala-Ser-Ala-Ser-Phe-Arg-Gly-Arg-Cys-Tyr-Lys-Cys-Phe-Arg-Cys	Antimicrobial	Hancock et al. (2006)
Bay mussel, *Mytilus galloprovincialis*	Physical extraction methods	Gly-Phe-Gly-Cys-Pro-Asn-Asn-Tyr-Gln-Cys-His-Arg-His-Cys-Lys-Ser-Ile-Pro-Gly-Arg-Cys-Gly-Gly-Thr-Cys-Gly-Gly-Cys-His-Arg-Leu-Arg-Cys-Thr-Cys-Tyr-Arg-Cys-Gly	Antimicrobial	Hancock et al. (2006)

has been suggested that proline, lysine, or arginine are the preferred amino acids at the C-terminal for increased ACE inhibitory potency of the peptide.

As the number of pathogenic bacteria and infections related to them are increasing every year, effective natural methods of microbial attack are needed in day to day life. In this regard, nutraceutical supplementation with edible antimicrobial peptides would be a promising approach. It has been found that marine invertebrates are a rich source of antimicrobial peptides and has a higher potential to be used as effective antibiotics. Around 40 different antimicrobial peptides have been isolated from numerous marine invertebrates such as crustaceans, tunics, mollusks, and echinoderms (Sperstad et al. 2011). The key structural features of antimicrobial peptides are the charge attached to it. Cationic peptides isolated from marine invertebrates are effective antimicrobial agents. These cationic amphiphilic peptides are a component in the innate immune system of these invertebrates (Sperstad et al. 2011). Furthermore, these invertebrate cationic peptides have the ability to detoxify bacterial endotoxins such as lipopolysaccharides (LPS). These antimicrobial peptides comprise 12–50 amino acids and bear net positive charge of +2 to +9 (Hancock et al. 2006). The cationic nature is gained due to the absence or lack of acidic amino acids such as glutamate or aspartate and the higher number of cationic amino acids such as arginine, lysine, or histidine. The antimicrobial peptides affect microbes by interacting with the microbial membrane. The peptide folds the microbial membrane into three-dimensional structures and forms ion-permeable channels, which leads to increased permeability of the membrane (Hancock and Rozek 2002).

Potential use of bioactive peptides from marine invertebrates as nutraceuticals is practiced in pacific Asian countries as a part of their traditional medicine. Furthermore, their traditional foods also include fermented marine invertebrates such as oysters and crabs, which are a rich source of active peptides. Therefore larger scale applications of these peptides would be an effective method in combating disease conditions such as oxidative stress, hypertension, and microbial attack.

17.4 Concluding remarks

Collectively this chapter briefly points out the potential use of edible marine invertebrate-derived functional carbohydrates and peptides as nutraceuticals. Highly active chitosan and COS are prepared from crustacean processing waste. Therefore development of COS as nutraceuticals will solve the problem of waste treatment also. At present several companies, such as Weifang sea source biological products and Golden-shell biochemical, have started producing chitosan and oligomers at a larger scale to be used in food industry and as a health food. However, despite the proven biological activities and unique sequences, marine invertebrate-derived peptides except antimicrobial peptides have not been used commercially. The basic reason for this would be the challenges in purification of the peptides. Therefore, commercially feasible methods of peptide purification should be researched for the optimal utilization of this valuable nutraceutical. As described throughout the chapter, edible marine invertebrates serve a wide array of safe and unique materials with health-enhancing properties, which could be effectively used as nutraceuticals with the proper combination of economical production methods.

References

Anderson, J.W., Nicolosi, R.J., and Borzelleca, J.F. 2005. Glucosamine effects in humans: A review of effects on glucose metabolism, side effects, safety considerations and efficacy. *Food and Chemical Toxicology* 43: 187–201.

Artan, M., Karadeniz, F., Karagozlu, M.Z., Kim, M.M., and Kim, S.K. 2010. Anti-HIV-1 activity of low molecular weight sulfated chitooligosaccharides. *Carbohydrate Research* 345: 656–662.

Bernal, J., Mendiola, J.A., Ibánez, E., and Cifuentes, A. 2011. Advanced analysis of nutraceuticals. *Journal of Pharmaceutical and Biomedical Analysis* 55: 758–774.

Bernardini, R.D., Harnedy, P., Bolton, D., Kerry, J., O'Neill, E., and Mullen, A.M. 2011. Antioxidant and antimicrobial peptidic hydrolysates from muscle protein sources and by-product. *Food Chemistry* 124: 1296–1307.

Block, J.A., Oegema, T.R., Sandy, J.D., and Plaas, A. 2010. The effects of oral glucosamine on joint health: Is a change in research approach needed?. *Osteoarthritis and Cartilage* 18: 5–11.

Byun, H.G., Kim, Y.T., Park, P.J., Lin, X., and Kim, S.K. 2005. Chitooligosaccharides as a novel β-secretase inhibitor. *Carbohydrate Polymers* 61: 198–202.

Byun, H.G., Lee, J.K., Park, H.G., Jeon, J.K., and Kim, S.K. 2009. Antioxidant peptides isolated from the marine rotifer, *Brachionus rotundiformis*. *Process Biochemistry* 44: 842–846.

Chen, R.H. and Chen, J.S. 2000. Changes of polydispersity and limiting molecular weight of ultrasound-treated chitosan. *Advances in Chitin Science* 4: 361–366.

Cho, Y.S., Lee, S.H., Kim, S.K., Ahn, C.B., and Je, J.Y. 2010. Aminoethyl-chitosan inhibits LPS-induced inflammatory mediators, iNOS and COX-2 expression in RAW264.7 mouse macrophages. *Process Biochemistry* 46: 465–470.

De Zoysa, M. 2012. Medicinal benefits of marine invertebrates: Sources for discovering natural drug candidates. In *Advances in Food and Nutrition Research*, ed. S.K. Kim, Vol. 65, pp. 153–169. Elsevier, New York.

Erdmann, K., Cheung, B.W.Y., and Schröder, H. 2008. The possible roles of food derived bioactive peptides in reducing the risk of cardiovascular disease. *Journal of Nutritional Biochemistry* 19: 643–654.

Espín, J.C., García-Conesa, M.T., and Tomás-Barberán, F.A. 2007. Nutraceuticals: Facts and fiction. *Phytochemistry* 68: 2986–3008.

Faulkner, D.J. 2000. Marine pharmacology. *Antonie van Leeuwenhoek* 77: 135–145.

Gopal, R., Vijayakumaran, M., Venkatesan, R., and Kathiroli, S. 2008. Marine organisms in Indian medicine and their future prospects. *Natural Product Radiance* 7: 139–145.

Gorsline, R.T. and Kaeding, C.C. 2005. The use of NSAIDs and nutritional supplements in athletes with osteoarthritis: Prevalence, benefits, and consequences. *Clinics in Sports Medicine* 24: 71–82.

Haller, C.A. 2010. Nutraceuticals: Has there been any progress? *Clinical Pharmacology and Therapeutics* 87: 137–141.

Hancock, R.E.W., Brown, K.L., and Mookherjee, N. 2006. Host defense peptides from invertebrates—Emerging antimicrobial strategies. *Immunobiology* 211: 315–322.

Hancock, R.E.W. and Rozek, A. 2002. Role of membranes in the activities of antimicrobial cationic peptides. *FEMS Microbiology Letters* 206: 143–149.

Himaya, S.W.A., Ngo, D.H., Ryu, B., and Kim, S.K. 2012. An active peptide purified from gastro-intestinal enzyme hydrolysate of Pacific cod skin gelatin attenuates angiotensin-1 converting enzyme (ACE) activity and cellular oxidative stress. *Food Chemistry* 132: 1872–1882.

Hong, F., Ming, L., Yi, S., Zhanxia, L., Yongquan, W., and Chi, L. 2008. The antihypertensive effect of peptides: A novel alternative to drugs? *Peptides* 29: 1062–1071.

Hsu, K.C. 2010. Purification of antioxidative peptides prepared from enzymatic hydrolysates of tuna dark muscle by-product. *Food Chemistry* 122: 42–48.

Huang, R., Mendis, E., and Kim, S.K. 2005. Improvement of ACE inhibitory activity of chitool-igosaccharides (COS) by carboxyl modification. *Bioorganic and Medicinal Chemistry* 13: 3649–3655.

Il'ina, A.V. and Varlamov, V.P. 2004. Hydrolysis of chitosan in lactic acid. *Applied Biochemistry and Microbiology* 40: 300–303.

Je, J.Y. and Kim, S.K. 2005. Water-soluble chitosan derivatives as a β-secretase inhibitor. *Bioorganic and Medicinal Chemistry Letters* 13: 6551–6555.

Je, J.Y., Park, J.Y., Jung, W.K., Park, P.J., and Kim, S.K. 2005. Isolation of angiotensin I converting enzyme (ACE) inhibitor from fermented oyster sauce, *Crassostrea gigas*. *Food Chemistry* 90: 809–814.

Je, J.Y., Park, P.J., and Kim, S.K. 2004. Radical scavenging activity of heterochitooligosaccharides. *European Food Research and Technology* 219: 60–65.

Jeon, Y.J., Park, P.J., and Kim, S.K. 2001. Antimicrobial effect of chitooligosaccharides produced by bioreactor. *Carbohydrate Polymers* 44: 71–76.

Jung, W.K. and Kim, S.K. 2009. Isolation and characterisation of an anticoagulant oligopeptide from blue mussel, *Mytilus edulis*. *Food Chemistry* 117: 687–692.

Jung, W.K., Moon, S.H., and Kim, S.K. 2006. Effect of chitooligosaccharides on calcium bioavailability and bone strength in ovariectomized rats. *Life Sciences* 78: 970–976.

Jung, W.K., Qian, Z.J., Lee, S.H., Choi, S.Y., Sung, N.J., Byun, H.G., and Kim, S.K. 2007. Free radical scavenging activity of a novel antioxidative peptide isolated from in vitro gastrointestinal digests of *Mytilus coruscus*. *Journal of Medicinal Foods* 10: 197–202.

Jung, W.K., Rajapakse, N., and Kim, S.K. 2005. Antioxidative activity of a low molecular weight peptide derived from the sauce of fermented blue mussel, *Mytilus edulis*. *European Food Research and Technology* 220: 535–539.

Karadeniz, F., Artan, M., Kong, C.S., and Kim, S.K. 2010. Chitooligosaccharides protect pancreatic β-cells from hydrogen peroxide-induced deterioration. *Carbohydrate Polymers* 82: 143–147.

Karagozlu, M.Z., Karadeniz, F., Kong, C.S., and Kim, S.K. 2012. Aminoethylated chitooligo-mers and their apoptotic activity on AGS human cancer cells. *Carbohydrate Polymers* 87: 1383–1389.

Kim, J.A., Ahn, B.N., Kong, C.S., and Kim, S.K. 2012. Chitooligomers inhibit UV-A-induced photoaging of skin by regulating TGF-β/Smad signaling cascade. *Carbohydrate Polymers* 88: 490–495.

Kim, M.M. and Kim, S.K. 2006. Chitooligosaccharides inhibit activation and expression of matrix metalloproteinase-2 in human dermal fibroblasts. *FEBS Letters* 580: 2661–2666.

Kim, S.K. and Mendis, E. 2006. Bioactive compounds from marine processing byproducts—A review. *Food Research International* 39: 383.

Kim, S.K. and Rajapakse, N. 2005. Enzymatic production and biological activities of chitosan oligosaccharides (COS): A review. *Carbohydrate Polymers* 62: 357–368.

Kuroiwa, T., Ichikawa, S., Sato, S., Hiruta, O., Sato, S., and Mukataka, S. 2002. Factors affecting the composition of oligosaccharides produced in chitosan hydrolysis using immobilized chitosanases. *Biotechnology Progress* 18: 969–974.

Lee, J.K., Yun, J.H., Jeon, J.K., Kim, S.K., and Byun, H.G. 2010. Effect of antioxidant peptide isolated from *Brachionus calyciflorus*. *Journal of the Korean Society for Applied Biological Chemistry* 53: 192–197.

Li, X.X., Han, L.J., and Chen, L.J. 2008. In vitro antioxidant activity of protein hydrolysates prepared from corn gluten meal. *Journal of Science Food and Agriculture* 88: 1660–1666.

Matsumoto, K., Ogikubo, A., Yoshino, T., Matsui, T., and Osajima, Y. 1994. Separation and purification of angiotensin-I converting enzyme inhibitory peptide in peptic hydrolyzate of oyster. *Journal of Japanese Society of Food Science and Technology* 41: 589–594.

Mendis, E., Kim, M.M., Rajapakse, N., and Kim, S.K. 2007. An in vitro cellular analysis of the radical scavenging efficacy of chitooligosaccharides. *Life Sciences* 80: 2118–2127.

Mendis, E., Rajapakse, N., Byun, H.G., and Kim, S.K. 2005. Investigation of jumbo squid (*Dosidicus gigas*) skin gelatin peptides for their in vitro antioxidant effects. *Life Sciences* 77: 2166–2178.

Nagaoka, I., Igarashi, M., Hua, J., Ju, Y., Yomogida, S., and Sakamoto, K. 2011. Recent aspects of the anti-inflammatory actions of glucosamine. *Carbohydrate Polymers* 84: 825–830.

Ngo, D.N., Kim, M.M., and Kim, S.K. 2008a. Chitin oligosaccharides inhibit oxidative stress in live cells. *Carbohydrate Polymers* 74: 228–234.

Ngo, D.N., Kim, M.M., and Kim, S.K. 2012a. Protective effects of aminoethyl-chitooligosaccharides against oxidative stress in mouse macrophage RAW 264.7 cells. *International Journal of Biological Macromolecules* 50: 624–631.

Ngo, D.H., Ngo, D.N., Vo, T.S., Ryu, B., Ta, Q.V., and Kim, S.K. 2012b. Protective effects of aminoethyl-chitooligosaccharides against oxidative stress and inflammation in murine microglial BV-2 cells. *Carbohydrate Polymers* 88: 743–747.

Ngo, D.N., Qian, Z.J., Je, J.Y., Kim, M.M., and Kim, S.K. 2008b. Aminoethyl chitooligosaccharides inhibit the activity of angiotensin converting enzyme. *Process Biochemistry* 43: 119–123.

Ngo, D.H., Qian, Z.J., Vo, T.S., Ryu, B. Ngo, D.N., and Kim, S.K. 2011a. Antioxidant activity of gallate-chitooligosaccharides in mouse macrophage RAW264.7 cells. *Carbohydrate Polymers* 84: 1282–1288.

Ngo, D.H., Wijesekara, I., Vo, T.H., Ta, Q.V., and Kim, S.K. 2011b. Marine food-derived functional ingredients as potential antioxidants in the food industry: An overview. *Food Research International* 44: 523–529.

Okamoto, Y., Inooue, A., Miyatake, K., Ogihara, K., Shigemasa, Y., and Minami, S. 2003. Effects of chitin/chitosan and their oligomers/monomers on migrations of macrophages. *Macromolecular Bioscience* 3: 587–590.

Pangestuti, R., Bak, S.S., and Kim, S.K. 2011. Attenuation of pro-inflammatory mediators in LPS-stimulated BV2 microglia by chitooligosaccharides via the MAPK signaling pathway. *International Journal of Biological Macromolecules* 49: 599–606.

Park, P.J., Je, J.Y., and Kim, S.K. 2003. Angiotensin I converting (ACE) inhibitory activity of hetero-chitooligosacchatides prepared from partially different deacetylated chitosans. *Journal of Agricultural and Food Chemistry* 51: 4930–4934.

Qian, Z.J., Jung, W.K., Byun, H.G., and Kim, S.K. 2008. Protective effect of an antioxidative peptide purified from gastrointestinal digests of oyster, *Crassostrea gigas* against free radical induced DNA damage. *Bioresource Technology* 99: 3365–3371.

Rajapakse, N., Kim, M.M., Mendis, E., Huang, R., and Kim, S.K. 2006. Carboxylatedchitooligosaccharides (CCOS) inhibit MMP-9 express ion in human fibro sarcoma cells via down-regulation of AP-1. *Biochimica et Biophysica Acta* 1760: 1780–1788.

Rajapakse, N., Mendis, E., Jung, W.K., Je, J.Y., and Kim, S.K. 2005. Purification of a radical scavenging peptide from fermented mussel sauce and its antioxidant properties. *Food Research International* 38: 175–182.

Ranathunga, S., Rajapakse, N., and Kim, S.K. 2006. Purification and characterization of antioxidative peptide derived from muscle of conger eel (*Conger myriaster*). *European Food Research and Technology* 222: 310–315.

Ryu, B., Himaya, S.W.A., Napitupulu, R.J., Eom, T.K., and Kim, S.K. 2012. Sulfated chitooligosaccharide II (SCOS II) suppress collagen degradation in TNF-induced chondrosarcoma cells via NF-κB pathway. *Carbohydrate Research* 350: 55–61.

Sperstad, S.V., Haug, T., Blencke, H.M., Styrvold, O.B., Li, C., and Stensvåg, K. 2011. Antimicrobial peptides from marine invertebrates: Challenges and perspectives in marine antimicrobial peptide discovery. *Biotechnology Advances* 29: 519–530.

Suzuki, K., Mikami, T., Okawa, Y., Tokoro, A., Suzuki, S., and Suzuki, M. 1986. Antitumor effect of hexa-N-aetylchitohexaose and chitohexaose. *Carbohydrate Research* 151: 403–408.

Tsai, J.S., Lin, T.C., Chen, J.L., and Pan, B.S. 2006. The inhibitory effects of freshwater clam (*Corbicula fluminea*, Muller) muscle protein hydrolysate on angiotensin I converting enzyme. *Process Biochemistry* 41: 2276–2281.

Vo, T.S., Kim, J.A., Ngo, D.H., Kong, C.S., and Kim, S.K. 2012. Protective effect of chitosan oligosaccharides against FcRI-mediated RBL-2H3 mast cell activation. *Process Biochemistry* 47: 327–330.

Wang, J., Hu, J., Cui, Z., Bai, X., and Du, Y. 2008. Purification and identification of a ACE inhibitory peptide from oyster proteins hydrolysate and the antihypertensive effect of hydrolysate in spontaneously hypertensive rats. *Food Chemistry* 111: 302–308.

Xia, W., Liu, P., Zhang, J., and Chen, J. 2010. Biological activities of chitosan and chitooligosaccharides. *Food Hydrocolloids* 25: 170–179.

Yoon, N.Y., Ngo, D.N., and Kim, S.K. 2009. Acetylcholinesterase inhibitory activity of novel chitooligosaccharide derivatives. *Carbohydrate Polymers* 78: 869–872.

18

Chitosan and Its Derivatives
Potential Use as Nutraceuticals

Jae-Young Je and Se-Kwon Kim

Contents

18.1 Introduction

Chitosan (poly-D-glucosamine) is a natural polymer derived from chitin, the second most abundant polysaccharide after cellulose. Chitosan can be obtained by either chemical deacetylation by treating with 40% sodium hydroxide for several hours or enzymatic deacetylation by N-deacetylase (EC3.5.1.41) (Tokuyasu et al. 1996), and it generally contains less than 50% of N-acetylglucosamine units. Chitosan possesses special properties for use in pharmaceutical, biomedical, food industry, health, and agriculture due to biocompatibility, biodegradability, and a nontoxic nature. However, the high molecular weight and poor solubility at neutral pH values of chitosan limit its industrial applications. Therefore, chemical and enzymatic modifications were employed to improve water solubility at a wide range of pH values and to enhance biological activities. Chitooligosaccharides (COSs) by enzymatic hydrolysis of chitosan and substitution of functional groups by chemical modification on three reactive groups, a free amino group, and both primary and secondary hydroxyl groups at C6 and C3 in chitosan were developed in order to overcome most of limitations (Vinšová and Vavřiková 2011). This chapter, therefore, reviews some bioactivities of chitosan and its derivatives for potential use as nutraceuticals.

18.2 Hepatoprotective effects

Oxidative stress is defined as a condition in which cellular antioxidant defenses are insufficient to keep the reactive oxygen species (ROS) levels below a toxic threshold. Under oxidative stress in the cells, overproduction of ROS such as hydroxyl, superoxide, and nitric oxide has occurred by the depletion of antioxidant biomolecules such as catalase (CAT), superoxide dismutase (SOD), glutathione peroxidase (GPx) as well as glutathione (GSH), antioxidant vitamins, and carotenoids, ultimately leading to cellular injury. Recently, oxidative stress is evidenced to be a major health problem causing liver disease by impaired hepatic alcohol dehydrogenase and ethanol-induced hepatotoxicity by ROS (Qian et al. 2011). Further, oxidative stress decreased intracellular antioxidant capacity of the liver cells including GSH and antioxidant enzymes such as SOD, CAT, and GPx.

COSs with different molecular weight (MW) were prepared as COS I (5–10 kDa), COS II (1–5 kDa), and COS III (below 1 kDa) from 90% deacetylated chitosan for evaluating hepatoprotective effect against *tert*-butylhydroperoxide (*t*-BHP)-induced damage in human normal Chang liver cells (Senevirathne et al. 2011). The organic hydroperoxide, *t*-BHP, is a useful model compound for the study of mechanisms of oxidative cell injury (Altman et al. 1994; Coleman et al. 1989). It is shown that treatment of 150 µM *t*-BHP decreased the cell viability to around 50% compared to nontreatment cells. However, Senevirathne et al. found that pretreatment of COSs significantly ($p < 0.05$) increased the cell viability. They also demonstrated that pretreatment of COSs reduced intracellular ROS generation, lipid peroxidation, and enhanced the level of GSH under *t*-BHP-induced oxidative stress in Chang liver cells. Furthermore, they found that COSs ameliorated antioxidant enzyme activities including CAT, SOD, and GPx than those of the cells treated only with *t*-BHP. However, the effects of MW of COSs were not of significant difference ($p > 0.05$). Recently, Senevirathne et al. (2012) developed gallic acid-*grafted*-chitosans (GA-*g*-chitosans) with different grafting ratio of gallic acid into chitosan in order to improve water solubility and biological activities (**Figure 18.1**). They also investigated hepatoprotective effects of GA-*g*-chitosans against *t*-BHP-induced hepatic damage in Chang liver cells. They founded that GA-*g*-chitosan with the highest content of gallic acid showed the highest hepatoprotective effect against *t*-BHP-induced hepatotoxicity in Chang liver cells by attenuating the oxidative stress by inhibiting intracellular ROS generation, lipid peroxidation, and increasing levels of antioxidane enzymes.

In vivo hepatoprotective effects of COSs and its derivatives against carbon tetrachloride-induced liver damage in mice were also investigated (Yan et al. 2006). Administration of CCl_4 (20 mg/kg body weight, i.p.) induced marked increase in serum aspartate transaminase (AST) and alanine transaminase (ALT) activities, primed liver lipid peroxidation, impaired total antioxidant capabilities, and induced genotoxicity. However, pretreatment of COSs, D-glucosamine, and *N*-acetyl-D-glucosamine (1.5 g/kg body weight, i.g.) markedly restored

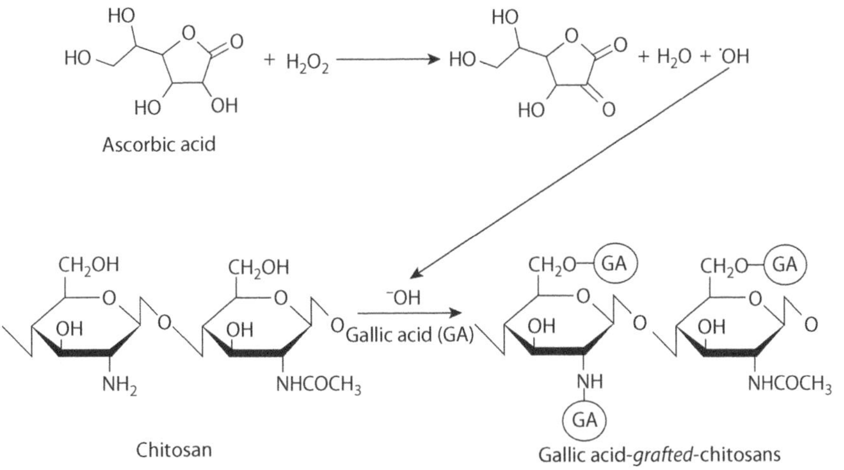

Figure 18.1 *Schematic diagram for the preparation of gallic acid-grafted-chitosans.*

antioxidant defense systems and serum AST and ALT activities. Hepatic malo-ndialdehyde formation caused by lipid peroxidation was also inhibited by pre-treatment of COSs, D-glucosamine, and N-acetyl-D-glucosamine. Additionally, pretreatment could also significantly decrease serum creatinine and uric acid levels and inhibit lipid peroxidation in kidney homogenate.

A number of studies have examined the hepatoprotective effects of chitosan against drugs-induced hepatotoxicity in rats. Rifampicin and isoniazid, anti-tubercular drugs, produce many metabolic and morphological aberrations in liver and induce hepatitis by a multiple-step mechanism. These drugs exhib-ited the cytotoxic action by increasing endogenous lipid peroxidation, and oxidative damage mediated by antitubercular drugs is generally the formation of the highly ROS, which act as stimulator of lipid peroxidation and source for destruction and damage to the cell membrane (Georgieva et al. 2004). Additionally, isoniazid and rifampicin-induced hepatotoxicity have resulted in alterations of various cellular antioxidant defense systems consisting of enzy-matic and nonenzymatic components such as reduced GSH (Tasduq et al. 2005). Santhosh et al. (2006, 2007) investigated effect of chitosan supplemen-tation on antitubercular drugs-induced hepatotoxicity in rats. Administration of isoniazid and rifampicin to rats was a significant elevation noticed in the levels of serum ALT, AST, lactate dehydrogenase (LDH), and alkaline phospha-tase (ALP); however, coadministration of chitosan (100 mg/kg body weight/day, MW 750,000 Da) significantly prevented the antitubercular drugs-induced elevation in the levels of serum diagnostic marker enzymes. It also exerted a significant antilipidemic effect against isoniazid- and rifampicin-induced hep-atitis by maintaining the levels of cholesterol, triglycerides, free fatty acids, and phospholipids in serum and liver. Isoniazid- and rifampicin-induced lipid peroxidation was also found to be prevented by the administration of chitosan.

18.3 Antioxidant activity

Antioxidants are compounds that significantly inhibit or delay oxidation of cellular oxidizable substrates when present at a low concentration compared to those of oxidizable substrates (Park et al. 2001; Vinšová and Vavřiková 2008). Antioxidants can help protect the body against ROS-mediated oxidative stress. It is well known that ROS is generated in normal metabolic processes; however, our body can detoxify ROS such as superoxide anion, hydroxyl radical, and hydrogen peroxide by using detoxifying antioxidant enzymes including SOD, CAT and GPx, as well as antioxidant molecules such as GSH, selenium vitamin C, and α-tocopherol (Wojcik et al. 2010). However, the generation of ROS and antioxidant balance in the human body can change with the progression of age and due to the other factors such as fatigue, sunlight, ultraviolet light, excessive caloric intake, and high fat diets (Samaranayaka and Li-Chan 2011). Excessive generated ROS can easily react with biomacromolecules such as DNA, membrane lipids, and proteins, resulting in the cell or tissue injury, and many diseases such as cancer, diabetes mellitus, AIDS, neurodegenerative, and inflammatory diseases are linked to ROS-mediated oxidative damage (Butterfield et al. 2002; Pryor 1982). Therefore, much attention had been directed toward the use of antioxidants to inhibit oxidative damage.

COSs were prepared from 89% deacetylated chitosan by enzymatic hydrolysis, and further fractionated into five MW ranges. Antioxidant effects of five kinds of COSs showed that COS with 1–3 kDa exhibited strong free radical scavenging activities (Park et al. 2003). Je et al. (2004) investigated free radical scavenging properties of hetero-COSs. They prepared nine kinds of hetero-COSs, and 90-MMWCOS (1–5 kDa, 90% deacetylated COS) showed the highest free radical scavenging activities than those of 75% and 50% deacetylated COSs and other MW COSs. Cellular antioxidant effects of COSs were also investigated in B16F1 melanoma cells, and COS (below 1 kDa) showed appreciable inhibition of intracellular ROS and ameliorated the intracellular GSH level (Mendis et al. 2007). A number of research studies have investigated the antioxidant abilities of chitosans from various sources. About 90%, 75%, and 50% deacetylated chitosan with average MW of 140–310 kDa were prepared from crab shell chitin, and free radical scavenging activities of chitosans were investigated using electron spin resonance (ESR) spectrometer (Park et al. 2004). All chitosans showed relatively high antioxidant activities, and 90% deacetylated chitosan showed the highest radical scavenging effects. Yen et al. (2008) also found that more free amino group in chitosan showed the highest antioxidant activity. These results indicate that free amino group at C-2 position and MW are major factors affecting antioxidant activities of chitosans and COSs.

Antioxidant activities of chitosan derivatives have also been developed. Je and Kim (2006) developed aminoderivatized chitosans with different degree of deacetylation by amino functional groups at a hydroxyl site in the chitosan

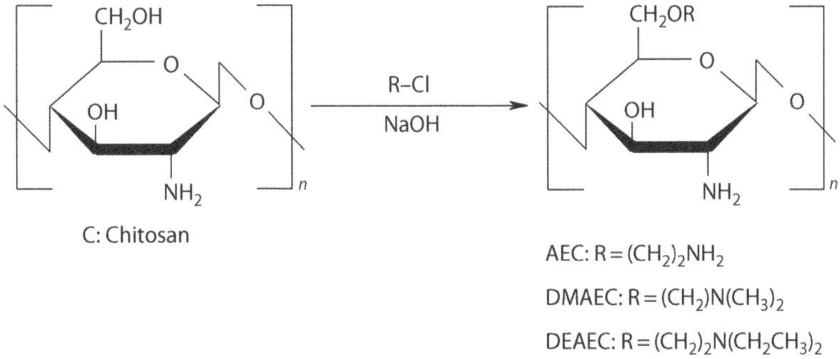

Figure 18.2 Schematic diagram for the preparation of aminoderivatized chitosans.

backbone (**Figure 18.2**). Antioxidant activities depend on the degree of deacetylation and the type of substituted group (aminoethyl, dimethylaminoethyl, and diethylaminoethyl group). Among them, aminoethyl-chitosan (AEC90) prepared from 90% deacetylated chitosan showed the highest hydroxyl, superoxide, and hydrogen peroxide scavenging activities. These results indicated that free amino group at C-2 position and introduction of amino group at C-6 position are major factors for ROS scavenging activity. In recent years, approaches for green and water-based chitosan conjugates were developed by grafting polyphenols onto chitosan. Pasanphan and Chirachanchai (2008) prepared chitosan gallate from a simple conjugating condition with gallic acid using carbodiimide. Chitosan gallate shows water-soluble property and exhibits significant antioxidant activity on free radicals. ESR studies revealed that chitosan gallate showed EC_{50} value of 137 µg/mL against DPPH radical, and also showed good thermal stability at 100°C for 1 h (Pasanphan et al. 2010). Aytekin et al. (2011) synthesized chitosan–caffeic acid conjugate to increase antioxidant activity compared to native chitosan by grafting with 1-ethyl-3-(3-dimethylaminopropyl) carbodiimide hydrochloride as covalent connector of amino group in native chitosan to carboxyl group in caffeic acid (**Figure 18.3**). The products were water-soluble in all pH and showed lower viscosity than native chitosan. The highest grafting ratio of caffeic acid was observed at 15%, and showed EC_{50} value of 64 µg/mL against DPPH radical.

A different method for the preparation of chitosan–polyphenols conjugate was also developed based on the use of an H_2O_2/ascorbic acid redox pair system. Crucio et al. (2009) performed covalent insertion of gallic acid and catechin on chitosan by a free radical grafting procedure. They investigated antioxidant capacity of chitosan–gallic acid and chitosan–catechin conjugates against DPPH, hydroxyl radical scavenging, and inhibition of lipid peroxidation. Chitosan–catechin conjugate enhanced 7.0-fold DPPH radical scavenging activity, 5.6-fold hydroxyl radical scavenging activity, and 4.3-fold lipid peroxidation inhibition activity compared to blank chitosan. Chitosan–gallic acid conjugate also augmented

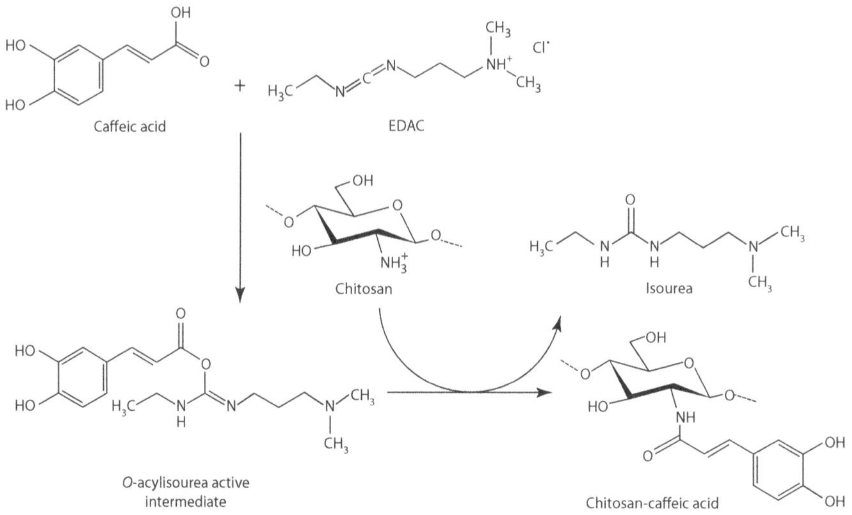

Figure 18.3 *Grafting reaction of caffeic acid with chitosan.*

DPPH, hydroxyl radical scavenging activity, and lipid peroxidation inhibition capacity compared to blank chitosan, but these activities were lower than those of chitosan–catechin conjugate. A possible mechanism of gallic acid and catechin insertion onto chitosan is related to the formation of macromolecular chitosan radicals by hydroxyl radicals that are generated by interactions between redox pair components (Cho et al. 2011a). Other research group also prepared GA-*g*-chitosans by the same method, but synthesized four kinds of GA-*g*-chitosans with different insertion ratios of gallic acid onto chitosan. All of GA-*g*-chitosans showed over 50% DPPH radical scavenging activity at the concentration of 50 μg/mL, and DPPH radical scavenging activities were enhanced with an increase of gallic acid content in chitosan. The best EC_{50} value of GA-*g*-chitosan against DPPH radical scavenging was 17.9 μg/mL. GA-*g*-chitosans effectively quenched hydrogen peroxide (93.15%) at the concentration of 50 μg/mL. Cellular antioxidant capacity in RAW264.7 macrophage cells also revealed that GA-*g*-chitosan significantly inhibited lipid peroxidation and generation of intracellular ROS under oxidative stress (Cho et al. 2011b). Further, GA-*g*-chitosan protected DNA damage induced by hydroxyl radical, and increased the protein expression levels of SOD-1 and glutathione reductase (**Figure 18.4**).

Enzymatic grafting of polyphenols to chitosan was achieved using phenol oxidase. Sousa et al. (2009) prepared enzymatically functionalized chitosan conjugates with flavonoids such as catechin, epigallocatechin gallate, epicatechin, epigallocatechin, quercetin, fisetin, rutin, hesperidin, and daidzein. With the grafting of flavonoids onto chitosan, the antioxidant properties were enhanced when compared to the unmodified chitosan; however, it was observed that these enhancements are dependent on the

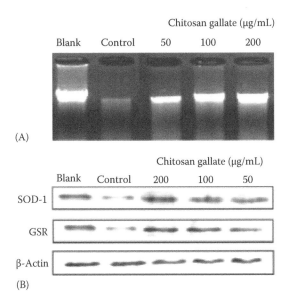

(A)

(B)

Figure 18.4 *(A) The protection effect of chitosan gallate on hydroxyl radical-induced DNA damage. (B) Effect of chitosan gallate on the levels of SOD-1 and glutathione reductase (GSR) under* H_2O_2*-mediated oxidative stress.*

antioxidant assay of the flavonoid used. Among them, chitosan–epigallo-catechin conjugate showed the highest enhancement, which is attributed to the 3′,4′,5′-trihydroxyl groups on B-ring, thus increasing the number of hydroxyl groups (Heim et al. 2002).

18.4 Enzyme inhibitors for Alzheimer's disease

Alzheimer's disease (AD) is a progressive neurodegenerative disease of the brain that is characterized by memory impairment, cognitive dysfunction, and personality changes (Bartolucci et al. 2001). Acetylcholine, neurotransmitter, plays an important role in the brain but deficiency of cholinergic neurotransmitters in the basal forebrains is predominantly involved in Alzheimer's disease; therefore, inhibiting acetylcholinesterase (AChE), which catalyzes the hydrolysis of acetylcholine to choline and acetate in both the peripheral nervous system and central nervous system, is a promising strategy to ameliorate symptoms of Alzheimer disease. AD is also caused by the progressive brain accumulation of β-amyloid (Aβ) peptides into fibrillar aggregates and insoluble plaques (Selkoe 2001). The Aβ peptides were generated by two proteolytic enzymes such as β-secretase and γ-secretase. The dominant role for the generation of Aβ peptides is β-secretase, and its activity is the rate-limiting step in Aβ peptides production in vivo (Vassar et al. 1999). Thus, the inhibition of β-secretase may reduce the production of Aβ peptides, thereby slowing or halting the progression of AD.

Byun et al. (2005) studied β-secretase inhibitory activities of COSs with different degree of deacetylation and MW prepared using an ultrafiltration membrane reactor system. Almost 90% deacetylated COSs showed more potent inhibitory activities than those of 75% and 50% deacetylated COSs. In addition, 90% deacetylated COS with 1–5 kDa exhibited the highest inhibitory activity compared to COS with 5–10 kDa and COS below 1 kDa. These results suggested that β-secretase inhibition was dependent on degree of deacetylation and MW of COSs. Lee et al. (2009) evaluated AChE inhibitory activity of COSs, and 90% deacetylated COSs showed more potent activity than low degree of deacetylated COSs, and 90% deacetylated COS with 1–5 kDa showed the highest inhibition activity. Cellular experiment revealed that 90% deacetylated COS with 1–5 kDa suppressed the level of AChE protein expression and AChE activity induced by $A\beta_{25-35}$ in PC12 cells. AChE inhibitory activities of GA-g-chitosans were also investigated (Cho et al. 2011a). Cho et al. found that GA-g-chitosan showed good AChE inhibitory activities in a dose-dependent manner, and IC_{50} value of GA-g-chitosan was 138.5 μg/mL. They also found that GA-g-chitosan acts as an AChE inhibitor by forming enzyme–substrate–inhibitor and enzyme–inhibitor complexes during the reaction to reduce the efficiency of catalysis. Cellular AChE inhibitory activities also revealed that GA-g-chitosan exhibited 35.3% AChE inhibitory activity in PC12 cells. These results further suggest that chitosan derivatives and COSs would be helpful for preventing neurodegenerative diseases.

18.5 Future prospects

Chitosan is natural mucopolysaccharides with unique biological properties including biocompatibility, biodegradability, and nontoxic nature. Chitosan and its derivatives have significant applications in the development of nutraceuticals and functional foods. However, their cytotoxicity and bioavailability is still unclear. Subacute toxicity and a safety evaluation of COSs suggested that COSs were generally deemed nontoxic. However, chitosan derivatives except COSs should be clarified of their cytotoxicity and bioavailability. Moreover, biological activities were dependent on their MW, degree of deacetylation, and substituted group onto chitosan; therefore, special types of chitosan derivatives displaying special bioactivities should be developed.

References

Altman, S.A., Zastawny, T.H., Randers, L., Lin, Z., Lumpkin, J.A., Remacle, J., Dizdaroglu, M., and Rao, G. 1994. *Tert*-butylhydroperoxide-mediated DNA base damage in cultured mammalian cells. *Mutat. Res.* 306:35–44.
Aytekin, A.O., Morimura, S., and Kida, K. 2011. Synthesis of chitosan-caffeic acid derivatives and evaluation of their antioxidant activities. *J. Biosci. Bioeng.* 111:212–216.
Bartolucci, C., Perola, E., Pilger, C., Fels, G., and Lambal, D. 2001. Three-dimensional structure of a complex of galanthamine (Nivalin®) with acetylcholinesterase from *Torpedo californica*: Implications for the design of new anti-Alzheimer drugs. *Proteins* 42:182–191.

Butterfield, D.A., Castenga, A., Pocernich, C.B., Drake, J., Scapagnini, G., and Calabrese, V. 2002. Nutritional approaches to combat oxidative stress in Alzheimer's diseases. *J. Nutr. Biochem.* 13:444–464.

Byun, H.G., Kim, Y.T., Park, P.J., Lin, X., and Kim, S.K. 2005. Chitooligosaccharides as a novel β-secretase inhibitor. *Carbohydr. Polym.* 61:198–202.

Cho, Y.S., Kim, S.K., Ahn, C.B., and Je, J.Y. 2011a. Preparation, characterization, and antioxidant properties of gallic acid-*grafted*-chitosans. *Carbohydr. Polym.* 83:1617–1622.

Cho, Y.S., Kim, S.K., Ahn, C.B., and Je, J.Y. 2011b. Inhibition of acetylcholinesterase by gallic acid-*grafted*-chitosans. *Carbohydr. Polym.* 84:690–693.

Cho, Y.S., Kim, S.K., and Je, J.Y. 2011c. Chitosan gallate as potential antioxidant biomaterial. *Bioorg. Med. Chem. Lett.* 21:3070–3073.

Coleman, J., Gilfor, D., and Farber, J.L. 1989. Dissociation of the accumulation of single-strand breaks in DNA from the killing of cultured hepatocytes by an oxidative stress. *Mol. Pharm.* 36:193–200.

Crucio, M., Puoci, F., Iemma, F., Parisi, O.I., Cirillo, G., Spizzirri, U.G., and Picci, N. 2009. Covalent insertion of antioxidant molecules on chitosan by a free radical grafting procedure. *J. Agric. Food Chem.* 57:5933–5938.

Georgieva, N., Gadjeva, V., and Tolekova, A. 2004. New isonicotinoylhydrazones with SSA protect against oxidative-hepatic injury of isoniazid. *Trakia J. Sci.* 2:37–43.

Heim, K.E., Tagliaferro, A.R., and Bobilya, D.J. 2002. Flavonoid antioxidants: Chemistry metabolism and structure–activity relationships. *J. Nutr. Biochem.* 13:572–584.

Je, J.Y. and Kim, S.K. 2006. Reactive oxygen species scavenging activity of aminoderivatized chitosan with different degree of deacetylation. *Bioorg. Med. Chem.* 14:5989–5994.

Je, J.Y., Park, P.J., and Kim, S.K. 2004. Free radical scavenging properties of hetero-chitooligosaccharides using an ESR spectroscopy. *Food Chem. Toxicol.* 42:381–387.

Lee, S.H., Park, J.S., Kim, S.K., Ahn, C.B., and Je, J.Y. 2009. Chitooligosaccharides suppress the level of protein expression and acetylcholinesterase activity induced by $A\beta_{25-35}$ in PC12 cells. *Bioorg. Med. Chem. Lett.* 19:860–862.

Mendis, E., Kim, M.M., Rajapakse, N., and Kim, S.K. 2007. An in vitro cellular analysis of the radical scavenging efficacy of chitooligosaccharides. *Life Sci.* 80:2118–2127.

Park, P.J., Je, J.Y., and Kim, S.K. 2003. Free radical scavenging activity of chitooligosaccharides by electron spin resonance spectrometry. *J. Agric. Food Chem.* 51:4624–4627.

Park, P.J., Je, J.Y., and Kim, S.K. 2004. Free radical scavenging activities of differently deacetylated chitosans using and ESR spectrometer. *Carbohydr. Polym.* 55:17–22.

Park, P.J., Jung, W.K., Nam, K.S., Shahidi, F., and Kim, S.K. 2001. Purification and characterization of antioxidative peptides from protein hydrolysate of lecithin-free egg yolk. *J. Am. Oil Chem. Soc.* 78:651–656.

Pasanphan, W., Buettner, G.R., and Chirachanchai, S. 2010. Chitosan gallate as a novel potential polysaccharide antioxidant: An EPR study. *Carbohydr. Res.* 345:132–140.

Pasanphan, W. and Chirachanchai, S. 2008. Conjugation of gallic acid onto chitosan: An approach for green and water-based antioxidant. *Carbohydr. Polym.* 72:169–177.

Pryor, W.A. 1982. Free radical biology: Xenobiotics, cancer, and aging. *Ann. N.Y. Acad. Sci.* 393:1–22.

Qian, Z.J., Zhang, C., Li, Y.X., Je, J.Y., Kim, S.K., and Jung, W.K. 2011. Protective effects of emodin and chrysophanol isolated from marine fungus *Aspergillus* sp. on ethanol-induced toxicity in HepG2/CYP2E1 cells. *Evid.-Based Complement. Alternat. Med.* 2011(452621):1–7.

Samaranayaka, A.G.P. and Li-Chan, E.C.Y. 2011. Food-derived peptidic antioxidants: A review of their production, assessment, and potential applications. *J. Funct. Food* 3:229–254.

Santhosh, S., Sini, T.K., Anandan, R., and Mathew, P.T. 2006. Effect of chitosan supplementation on antitubercular drugs-induced hepatotoxicity in rats. *Toxicology* 219:53–59.

Santhosh, S., Sini, T.K., Anandan, R., and Mathew, P.T. 2007. Hepatoprotective activity of chitosan against isoniazid and rifampicin-induced toxicity in experimental rats. *Eur. J. Pharmacol.* 572:69–73.

Selkoe, D.J. 2001. Alzheimer's disease: Genes, proteins, and therapy. *Physiol. Rev.* 81:741–766.

Senevirathne, M., Ahn, C.B., and Je, J.Y. 2011. Hepatoprotective effect of chitooligosaccharides against *tert*-butylhydroperoxide-induced damage in Chang liver cells. *Carbohydr. Polym.* 83:995–1000.

Senevirathne, M., Jeon, Y.J., Kim, Y.T., Park, P.J., Jung, W.K., Ahn, C.B., and Je, J.Y. 2012. Prevention of oxidative stress in Chang liver cells by gallic acid-*grafted*-chitosans. *Carbohydr. Polym.* 87:876–880.

Sousa, F., Guebitz, G.M., and Kokol, V. 2009. Antimicrobial and antioxidant properties of chitosan enzymatically functionalized with flavonoids. *Proc. Biochem.* 44:749–756.

Tasduq, S.A., Peerzada, K., Koul, S., Bhat, R., and Johri, R.K. 2005. Biochemical manifestations of anti-tuberculosis drugs induced hepatotoxicity and the effect of silymarin. *Hepatol. Res.* 31:132–135.

Tokuyasu, K., Ohnishi-Kameyama, M., and Hayashi, K. 1996. Purification and characterization of extracellular chitin deacetylase from *Colletotrichum lindemuthianum*. *Biosci. Biotechnol. Biochem.* 60:1598–1603.

Vassar, R., Bennett, B.D., Babu-Khan, S., Kahn, S., Mendiaz, E.A., and Denis, P. 1999. Beta-secretase cleavage of Alzheimer's amyloid precursor protein by the transmembrane aspartic protease BACE. *Science* 286:735–741.

Vinšová, J. and Vavříková, E. 2008. Recent advances in drugs and prodrugs design of chitosan. *Curr. Pharm. Des.* 14:1311–1326.

Vinšová, J. and Vavříková, E. 2011. Chitosan derivatives with antimicrobial, antitumor and antioxidant activities—A review. *Curr. Pharm. Des.* 17:3596–3607.

Wojcik, M., Burzynska-Pedziwiatr, I., and Wozniak, L.A. 2010. A review of natural and synthetic antioxidants important for health and longevity. *Curr. Med. Chem.* 17:3262–3288.

Yan, Y., Wanshun, L., Baoqin, H., Bing, L., and Chenwei, F. 2006. Protective effects of chitosan oligosaccharides and its derivatives against carbon tetrachloride-induced liver damage in mice. *Hepatol. Res.* 35:178–184.

Yen, M.T., Yang, J.H., and Mau, J.L. 2008. Antioxidant properties of chitosan from crab shells. *Carbohydr. Polym.* 74:840–844.

19

Marine Sulfated Polysaccharides with Unusual Anticoagulant Action through an Additional Unrelated-Natural Inhibitors Mechanism

Bianca F. Glauser, Paulo A.S. Mourão, and Vitor H. Pomin

Contents

19.1 Introduction to marine sulfated polysaccharides: chemistry and medical uses

The efforts in both structural and biofunctional studies of marine sulfated polysaccharides (MSPs) have been increasing significantly over the last two decades (Pomin 2009a; Pomin and Mourão 2008). This attention is a consequence of the large interest in novel sulfated polysaccharides (SPs) such as glycosaminoglycans (GAGs), sulfated fucans (SFs), and sulfated galactans (SGs) that may bring out structures as novel as therapeutic mechanisms for illness treatment, especially those involved in the cardiovascular deregulations. Hence, the sea already proved to be a new and rich environment for such intersectional research of glycobiology and pharmacology, thus offering with great perspectives potential molecular candidates concerning (1) differentiated structures and (2) new pharmacological effects. This chapter strikingly exemplifies these two topics at a single report.

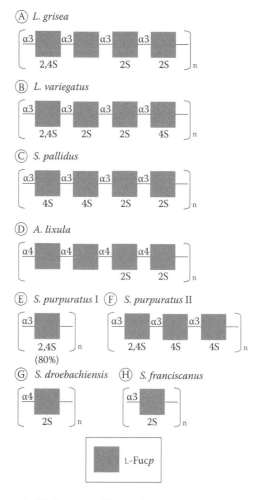

Bianca F. Glauser, Paulo A.S. Mourão, and Vitor H. Pomin

Glycosaminoglycans are of particular interest to researchers due to their extreme broad range of biological actions, especially those for clinical purposes exemplified by the therapeutic applications of heparin and chondroitin sulfates (CSs). The former is the most clinically exploited anticoagulant over the last 50 years (Fareed et al. 2000), and the latter is an effective supplement for cartilage regeneration and/or revigoration. Until the early 1990s, SFs, homopolymers composed of fucopyranosyl (Fuc$p$) units exclusively in α-L-form (Pomin 2009a; Pomin and Mourão 2008), and SGs, homopolymers of α-L- or α-D- and/or β-D-galactopyranosyl (Gal$p$) units (Pomin 2009b), were usually extracted and characterized from the cell walls of the three major groups of macroalgae. Phaeophyta (brown algae) expresses SFs while both Rhodophyta (red algae) and Chlorophyta (green algae) express SGs only. Historically, algal SFs and SGs are used mainly in clinical tests and large-scale extractions for industrial purposes. Actually, the greatest interest in the use and research of these molecules concerns their therapeutic effects (Pomin 2011). Recently, however, newer and interesting sources of these compounds have been found in the extracellular matrices of certain marine invertebrates (Mourão 2004, 2007; Mourão and Pereira 1999; Pomin 2009a,b; Pomin and Mourão 2008). In contrast with most algal SPs, the invertebrate polymers exhibit highly regular chemical structures (**Figures 19.1** and **19.2A** through **D**), which make the correlation of their biological functions to their respective structural features easier (Pomin 2009a; Pomin and Mourão 2008; Vilela-Silva et al. 2008). This is of tremendous benefit and contribution to the entire field of glycobiology.

Figure 19.1 *Chemical structures of repeating units of the SFs from the cell wall of the sea-cucumber (A) and from the egg jelly coat of sea urchins (B–H). These polysaccharides are composed of α-L-fucopyranosyl units (squares). The species-specific structures vary in sulfation patterns (exclusively 2- and/or 4-positions, in glycosidic linkages: α(1 → 3) (A–C, E, F, and H) and α(1 → 4) (D and G), and in number of residues of the repetitive units: tetrasaccharides (A–D), trisaccharides (F), and monosaccharides (E, G, and H), but they are all linear. The numbers and Greek letters over the black lines represent the glycosodic linkage type. The numbers before S represent the sulfation position. The structures are the following: (A) Ludwigothurea grisea [→3)-α- L-Fucp-2,4$\left(OSO_3^-\right)$-(1 → 3)-α- L-Fucp-(1 → 3)-α-L-Fucp-2$\left(OSO_3^-\right)$-(1 → 3)-α-L-Fucp-2$\left(OSO_3^-\right)$-(1 →]$_n$ (Mulloy et al. 1994); (B) Lytechinus variegatus [→3)-α-L-Fucp-2,4$\left(OSO_3^-\right)$-(1 → 3)-α-L-Fucp-2$\left(OSO_3^-\right)$-(1 → 3)-α-L-Fucp-2$\left(OSO_3^-\right)$-(1 → 3)-α-L-Fucp-4$\left(OSO_3^-\right)$-(1→]$_n$ (Mulloy et al. 1994); (C) Strongylocentrotus pallidus [→3)-α-L-Fucp-4$\left(OSO_3^-\right)$-(1 → 3)-α-L-Fucp-4$\left(OSO_3^-\right)$-(1 → 3)-α-L-Fucp-2$\left(OSO_3^-\right)$-(1 → 3)-α-L-Fucp-2$\left(OSO_3^-\right)$-(1→]$_n$ (Vilela-Silva et al. 2002); (D) Arbacia lixula [→4)-α-L-Fucp-2$\left(OSO_3^-\right)$-(1 → 4)-α-L-Fucp-2$\left(OSO_3^-\right)$-(1 → 4)-α-L-Fucp- (1 → 4)-α-L-Fucp-(1→]$_n$ (Alves et al. 1997); (E) Strongylocentrotus purpuratus-I ~80% [→3)-α-L-Fucp-2,4$\left(OSO_3^-\right)$-(1→]$_n$ and ~20% [→3)-α-L-Fucp-2$\left(OSO_3^-\right)$-(1→]$_n$ and (F) S. purpuratus-II [→3)-α-L-Fucp-2,4$\left(OSO_3^-\right)$-(1 → 3)-α-L-Fucp-4$\left(OSO_3^-\right)$- (1 → 3)-α-L-Fucp-4$\left(OSO_3^-\right)$-(1→]$_n$ (Alves et al. 1998); (F) Strongylocentrotus droebachiensis [→4)-α-L-Fucp-2$\left(OSO_3^-\right)$-(1→]$_n$ (Alves et al. 1997); and (G) Strongylocentrotus franciscanus [3)-α-L-Fucp-2$\left(OSO_3^-\right)$-(1→]$_n$ (Vilela-Silva et al. 1999).*

At one time it was relatively hard to accurately correlate the structure–activity relationship for the majority of SPs, especially those from mammalian sources. This advantage in use with marine sources is also observed for some marine invertebrate GAGs (**Figure 19.2F**). Unlike some marine GAGs, many mammalian GAGs exhibit a large variety of sulfation patterns (Gandhi and Mancera 2008) and consequentially demand much more effort to understand their biological activities, determine their specific structural features, and subsequently propose a reliable structure–function relationship.

The MSPs with well-defined chemical structures (SFs, SGs, and GAGs in **Figures 19.1** and **19.2**) can be extracted from echinoderms (Echinodermata) like sea cucumbers (Holothuroidea) (Mulloy et al. 1994), and sea urchins (Echinoidea) (Pomin and Mourão 2008; Vilela-Silva et al. 2008); or ascidians (Urochordata, Ascidiacea) (Mourão and Perlin 1987; Pavão 2002; Pavão et al. 1995, 1998; Santos et al. 1992; Vicente et al. 2001). The SFs (**Figure 19.1B** through H) and SGs (**Figure 19.2A** and **B**) are widely spread throughout the jelly coat that surrounds sea urchin eggs. Among all the MSPs with well-defined chemical structures described so far (**Figures 19.1** and **19.2**), there is

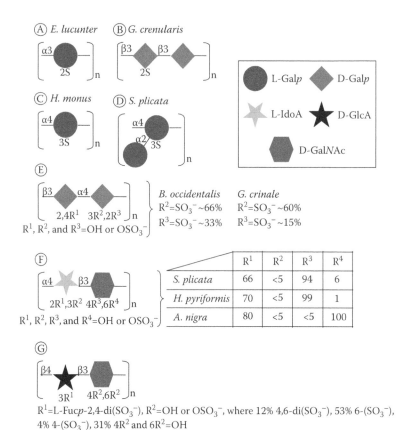

Bianca F. Glauser, Paulo A.S. Mourão, and Vitor H. Pomin

a description of a molecule from the body wall of the sea cucumber (**Figure 19.1A**) (Mulloy et al. 1994), a few examples of SGs isolated from the tunic of ascidians (**Figure 19.2C and D**) (Mourão and Perlin 1987; Pavão et al. 1998; Santos et al. 1992), and some reports from specific red algal cell walls (**Figure 19.2E**) (Fonseca et al. 2008; Pereira et al. 2005). All of the remaining structures (**Figures 19.1B** through **H** and **19.2A and B**) belong to the egg jelly coat of sea urchins (Alves et al. 1997, 1998; Mulloy et al. 1994; Vilela-Silva et al. 1999, 2002). The body of some species of sea cucumber and ascidians also contain peculiar GAGs that show distinct sulfation patterns or structures from mammalian GAGs (Mourão et al. 1996; Pavão et al. 1995, 1998; Vieira and Mourão 1988; Vieira et al. 1991) as exemplified with the tunicate dermatan sulfates (DSs) in **Figure 19.2F**, and the FucCS from the sea cucumber *L. grisea* in **Figure 19.2G**.

In this chapter, a differentiated anticoagulant action through an unrelated natural-inhibitors (serpin-independent) mechanism was found for two of these MSPs with well-defined oligomeric repetitive structures: the SG from the red alga *B. occidentalis* (**Figure 19.2E**), and the FucCS from the sea cucumber *L. grisea*. Hence, from now on, we will take a description of the methods of extraction, structural details, and anticoagulant properties of these two MSPs specifically. These two MSPs are great potential candidates in drug discovery and development in the clinical fields of hematology and cardiovascular system.

←──

Figure 19.2 Chemical structures of the repeating oligosaccharide units of SGs from the egg jelly coat of sea urchins (A and B), from the tunic of ascidians (C and D), from red algae (E); and the repeating GAG units of DSs from ascidians (F) and of FucCS from sea cucumber (G). These polysaccharides are composed of ʟ-galactopyranoses (circles), ᴅ-galactopyranoses (diamond), ʟ-iduronic acids (up-side-down star), ᴅ-glucuronic acids (star), and ᴅ-N-acetylgalactosamines (hexagons). The numbers and Greek letters over the black lines represent the glycosodic linkage type. The numbers before S and R represent the positions of sulfation or other radicals. The superscripted numbers after R represent the respective indication of radical. The structures are the following: (A) Echinometra lucunter [→3)-α-ʟ-Galp-2$\left(\text{OSO}_3^-\right)$-(1→]$_n$ (Alves et al. 1997); (B) Glyptocidaris crenularis [→3)-β-ᴅ-Galp-2$\left(\text{OSO}_3^-\right)$-(1 → 3)-β-ᴅ-Galp-1→]$_n$ (Castro et al. 2009); (C) Styela plicata {→4)-α-ʟ-Galp-2[→1)-α-ʟ-Galp-3$\left(\text{OSO}_3^-\right)$]-3$\left(\text{OSO}_3^-\right)$-(1→}$_n$ (Mourão and Perlin 1987); (D) Herdmania monus [→4)-α-ʟ-Galp-3$\left(\text{SO}_3^-\right)$-(1→]$_n$ (Santos et al. 1992); (E) both Botryocladia occidentalis and Gelidium crinali express [3-β-ᴅ-Galp-(1 → 4)-α-ʟ-Galp-1→]$_n$ with different sulfation contents (Fonseca et al. 2008; Pereira et al. 2005); (F) the DS from S. plicata, Halocynthia pyriformis, and Ascidian nigra are composed of [→4)-α-ʟ-IdoA-(1 → 3)-β-ᴅ-GalNAc-(1→]$_n$ with also different sulfation patterns (table in %) (Pavão et al. 1995, 1998); and (G) the FucCS from L. grisea composed of {→4)-β-ᴅ-GlcA-3[→1)-α-ʟ-Fucp-2,4-di$\left(\text{OSO}_3^-\right)$]-(1 → 3)-β-ᴅ-GalNAc-(1→]$_n$ (Fonseca et al. 2008; Mourão et al. 1996; Vieira and Mourão 1988; Vieira et al. 1991).

19.2 Fucosylated chondroitin sulfate from sea cucumber *L. grisea*

19.2.1 General structure and anticoagulant properties

FucCS is a MSP obtained from the body wall of the sea cucumber *L. grisea*. The backbone of this polysaccharide is made up of repeating disaccharide units of alternating 4-linked β-D-glucuronic acid (GlcA) units and 3-linked N-acetyl-β-D-galactosamine (GalNAc) (**Figure 19.2G**), the same structure of mammalian CS backbone. In the sea cucumber polysaccharide, however, the β-D-GlcA residues bear sulfated Fuc*p* branches bound at 3-position and the GalNAc units from the central core with also a complex sulfation pattern (**Figure 19.2G**). Approximately 12% are 4,6-di-sulfated, 53% are 6-mono-sulfated, 4% are 4-mono-sulfated, and 31% are non-sulfated GalNAcs (Fonseca et al. 2008; Mourão et al. 1996; Vieira and Mourão 1988; Vieira et al. 1991).

The physiological role of this SP in the sea cucumber is not completely understood but it is speculated that it plays a structural role in assembling the body wall of the invertebrate. FucCS is present on the connective tissue, and perhaps with its high negatively charged density and consequent capacity to retain water in the extracellular matrix, it might produce a "space" among muscle fibers, allowing the tissue to change its length rapidly and reversibly by more than 200% (Landeira-Fernandez et al. 2000).

L. grisea FucCS exhibits potential pharmacological actions in mammalian system such as anti-metastatic and anti-inflammatory activities (Borsig et al. 2007; Melo-Filho et al. 2010). However, this FucCS has been better explored as an anticoagulant and antithrombotic candidate (Fonseca and Mourão 2006). Unlike mammalian CSs, which do not show anticoagulant behavior, FucCS is an effective anticoagulant and antithrombotic molecule when administrated orally or by intravascular routes in rats (Fonseca and mourão 2006; Zancan and Mourão 2004). Because of its potency as an anticoagulant agent, FucCS could be an alternative drug candidate for the treatments of thrombosis, with many advantages over heparin, the most exploited anticoagulant GAG. Like heparin, FucCS has a serine-protease inhibitor (serpin)-dependent anticoagulant activity due to its capacity to increase thrombin (IIa)-inhibition more by heparin cofactor II (HCII) than by antithrombin (AT). However unlike heparin, FucCS also shows a serpin-independent anticoagulant activity by inhibiting tenase and prothrombinase procoagulant complexes (Glauser et al. 2008) as explained later. In this chapter we give some basis and details about these uncommon inhibiting mechanisms.

Anticoagulant and antithrombotic activities of FucCS are dependent on its molecular weight (MW) and can be attributable to some of its structural features. The native FucCS has an average MW of ~40 kDa and a very low polydispersity when compared with mammalian CSs (Mourão et al. 1996). Reduction on the MW diminishes its antithrombotic and anticoagulant efficacy (Zancan and Mourão 2004). Sulfated Fuc*p* units play a crucial role for

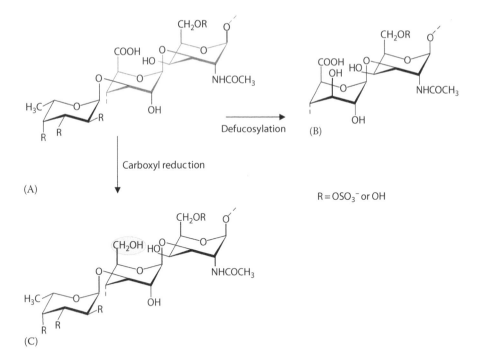

Figure 19.3 *Structure of the sea cucumber fucosylated chondroitin sulfate (FucCS) (A), its defucosylated (B), and carboxyl-reduced (cr-FucCS) (C) derivatives. The backbone of this polysaccharide is made up of repeating disaccharide units of alternating 4-linked β-D-glucuronic acid (GlcA) and 3-linked N-acetyl-β-D-galactosamine (GalNAc), the same structure as mammalian CS. But, in the sea cucumber polysaccharide, the β-D-glucuronic acid residues bear sulfated fucose branches at the three position. These branches are removed by mild acid hydrolysis while the hexuronic acid carboxyl groups can be reduced by 1-ethyl-3-(3-dimethylaminopropyl)carbodiimide/NaBH4 (Glauser et al. 2008; Mourão et al. 1996). Dashed line represents glycosidic bonds.*

the anticoagulant activity of FucCS. Removal of these branches by mild acid hydrolysis (**Figure 19.3B**), as well as desulfation reaction, reduces considerably its anticoagulant and antithrombotic activities to the same low levels of mammalian CS (Glauser et al. 2008; Mourão et al. 1996; Zancan and Mourão 2004). Moreover, sulfated Fuc*p* branches in FucCS molecules prevent its rapid digestions by enzymes, especially hyaluronidase that normally degrade GAGs in the gastrointestinal tract of vertebrates (Vieira and Mourão 1988; Vieira et al. 1991). Due to this fact FucCS was proved to be effective when administrated by oral routes, differently from heparin treatments (Fonseca and Mourão 2006).

Previous studies using MSPs from invertebrates have shown that the occurrence of 2,4-di-sulfated Fuc*p* units amplifies the effect of the AT-mediated anticoagulant activity of 3-linked α-L-fucans (**Figure 19.1A**, **B**, **E**, and **F**)

(Fonseca et al. 2009; Pereira et al. 2002a,b). Surprisingly, these 2,4-di-sulfated Fuc*p* units when occurring as branches did not favor interaction with AT in a similar way. Clearly, the anticoagulant effect of FucCS is mostly via potentiating HCII, as proved in the work of Fonseca et al. (2009).

The reduction of the carboxyl groups of the GlcA residues to glucose (**Figure 19.3C**, in gray ellipse) does not affect its anticoagulant activity (Glauser et al. 2008; Mourão et al. 1996; Zancan and Mourão 2004), meaning no such big influence of the negatively charged carboxyl groups for interaction. In contrast, carboxyl-reduced FucCS (cr-FucCS) are still able to retain the anticoagulant activity but with no traces of bleeding adverse effects as native FucCS and heparin have, indicating therefore an even more favorable antithrombotic action than the native unmodified molecule (Zancan and Mourão 2004).

cr-FucCS also has differences regarding its interaction with some plasma proteases. It possesses anticoagulant activity by potentiating inhibition of IIa through its natural inhibitor HCII while FucCS increases IIa inhibition by activating the HCII and, in a lesser extent, through AT as well (Glauser et al. 2008). Furthermore, in contrast with the native compound, cr-FucCS is proved to not activate factor XII from the contact system of coagulation (Fonseca and Mourão 2006). The importance of this activation has been highlighted by the observation that chemically oversulfated CS (OSCS), found as the main contaminant of pharmaceutical heparin preparations, activates factor XII and induces severe hypertension in patients associated with kallikrein release, when administered by intravenous injection (Blossom et al. 2008). Like OSCS, FucCS has also usually more sulfate groups than the mammalian CS. However, these two highly sulfated CSs differ significantly in their effects on coagulation and thrombosis (Fonseca et al. 2010).

These findings illustrate that the structural requirement for the interaction of these SPs with coagulation factors are very stereospecific and not only a mere consequence of sulfate content and increased charge density. Above all, while some nonspecific electrostatic interactions of SPs exist for binding to basic proteins, evidence continues to push the idea that many of the effects of SPs involve specific structural moieties (see Pomin 2009a). Taking the anticoagulant action of heparin as an example, approximately two-thirds of native heparin is unable to form a stable complex with AT. This is essentially due to the presence of specific binding sequence (discussed more in detail later) in only few specific regions of the heparin chain (Gray et al. 2008). Obviously, the identification of specific structural requirements in the invertebrate SP necessary for its interaction with coagulation cofactors is an essential step for more rational development of anticoagulant drugs (for more details, see Pomin 2009a, 2011).

19.2.2 Extraction methods

The body wall of sea cucumbers must be carefully separated from other tissues, immediately immersed in acetone and kept at low temperature for transportation or storage before handling for extraction. Acidic glycans can be extracted from the dry tissue by papain as previously described (Vieira and Mourão 1988; Vieira et al. 1991). About 20 mg of the crude acidic glycans from the body wall of *L. grisea* can be recovered and further dissolved in 1.5 mL of 0.3 M pyridinelacetate buffer (pH 6.0) for subjection into a gel-permeation chromatography on a Sepharose CL-4B column (115 × 1.5 cm), consequently eluted with the same buffer plus 4.0 M guanidine hydrochloride in 0.3 M pyridinelacetate buffer (pH 6.0). Columns can be successfully eluted at a flow rate of 6 mL h and aliquots of approximately 1.0 mL might be collected. The presence of sulfated glycans in each fraction can be easily detected by the DuBois reaction and/or by metachromatic property (Vieira and Mourão 1988; Vieira et al. 1991). Columns might be calibrated using blue dextran (2000 kDa) as a marker to V_0 and cresol red (~0.35 kDa) as a marker to V_t. After chromatography on Sepharose CL-4B column, the obtained crude polysaccharide must be further applied to a DEAE-cellulose column (3.5 × 2 cm) equilibrated with 0.1 M sodium acetate buffer (pH 6.0) and washed with 100 mL of the same buffer. The column is then developed by a linear gradient prepared by mixing 35 mL of 0.1 M sodium acetate buffer (pH 6.0), with 35 mL of 1.0 M NaCl and 35 mL of 2.0 M NaCl in the same buffer. The flow rate of the column could be set to 10 mL h, and fractions of 2.5 mL might be collected. The resultant fractions can be checked by the DuBois reaction, ultraviolet absorption, and conductivity. Fractions must be finally pooled, and dialyzed against distilled water, and lyophilized for future assays.

19.3 Sulfated galactan from *Botryocladia occidentalis*

19.3.1 General structure and anticoagulant properties

Sulfated galactan is an SP extracted from the cell wall of the red algae *B. occidentalis*. SG has a linear backbone made of alternating 3-linked β-D-Gal*p* and 4-linked α-D-Gal*p* residues (**Figure 19.2E**), showing a repeating pattern in disaccharide units like GAGs. The major structural variation in this polysaccharide is likely due to sulfation pattern. Two-thirds of its total α-units are 2,3-di-sulfated and one-third are 2-mono-sulfated (**Figure 19.2E**) (Farias et al. 2000).

This red algal SG also has a potent anticoagulant and antithrombotic activity when administered by intravascular route in rats. SG inhibits factor Xa and IIa by activating both serpins (AT and HCII). The occurrence of 2,3-di-sulfated α-Gal*p* units is a critical structural motif for promoting the interaction of the polysaccharide with the plasma proteases and serpins (Pereira et al. 2005). The 2,3-di-sulfated units have an amplifying effect on the anticoagulant activity, since other SGs that are either 2-*O*- or 3-*O*-sulfonated in α-Gal*p* residues (but not at both positions at same time) showed weak activity as compared to *B. occidentalis* SG. Again, the structural requirement for the interaction of these polysaccharides with coagulation cofactors and their target proteases seems to be stereospecific (Pereira et al. 2002). The antico-agulant activity of SG from *B. occidentalis* increased 15-fold when compared to a 2-*O*-sulfonated SG from the sea urchin *E. lucunter* (**Figure 19.2A**). In addition, highly sulfated dextrans (~3 sulfate esters/glucose unit) have lower anticoagulant potency when compared to the *B. occidentalis* SG (Farias et al. 2000; Pereira et al. 2005).

The native SG from *B. occidentalis* has an average MW of ~100 kDa (Farias et al. 2000). Low molecular weight (LMW) SG fractions can be prepared with less efficiency to induce factors IIa and Xa inactivation by serpins (AT e HCII) than the native compound. Using LMW SG fragments, it was proved that longer sugar chains than heparin are necessary for SG to achieve the same anticoagulant action of heparin (Melo et al. 2004). The *B. occidentalis* SG dif-fers from heparin in their way to interact with IIa and AT. The bulk structure of this SG allows it to interact more randomly with AT rather than a specific minor component-mediated like the pentasaccharide sequence in heparin (discussed later). In addition, *B. occidentalis* SG and heparin bind to different sites and/or induce distinct conformational activation of the serpin AT. This SG interacts predominantly with the heparin-binding exosite-II of IIa and catalyzes the formation of a covalent complex between AT and the protease, similar to heparin. Overall, this red algal SG has a higher affinity for throm-bin than for AT and probably the polysaccharide binds first to the protease through a high-affinity interaction (Melo et al. 2008).

This red algal SG has also a different serpin-independent activity by inhibiting tenase and prothrombinase complexes as seen for FucCS (Glauser et al. 2009; Melo et al. 2004). These mechanisms will be discussed in detail henceforth in this chapter, not only because of the novelty in the field but much more likely due to their very unusual way of preventing coagulation whose action was observed to be distinct for the case of heparin (Glauser et al. 2009).

19.3.2 Extraction methods

The marine red algae *B. occidentalis* have been collected at Pacheco beach, Caucaia, Ceará, Brazil, then separated from other species and sun-dried. The dried tissue (~5 g) was cut in small pieces, suspended in 250 mL of 0.1 M sodium acetate buffer (pH 6.0) containing 510 mg of papain (E. Merck, Darmstadt, Germany), 5 mM EDTA, and 5 mM cysteine, and incubated at 60°C for 24 h. The incubation mixture was then filtrated and the supernatant saved. The residue was washed with 140 mL of distilled water, filtered again, and the two supernatants were combined. SPs in solution were precipitated with 16 mL of 10% cetylpyridinium chloride solution. After standing at room temperature for 24 h, the mixture was centrifuged at $2 \times 10^3 g$ for 20 min at 5°C. The SPs in the pellet were washed with 610 mL of 0.05% cetylpyridinium chloride solu-tion, dissolved with 170 mL of a 2 M NaCl, ethanol (100/15, v/v) solution, and precipitated with 300 mL of absolute ethanol. After 24 h at 4°C, the precipitate was collected by centrifugation ($2 \times 10^3 g$ for 20 min at 5°C), washed twice with 300 mL of 80% ethanol, and once again with the same volume of absolute ethanol. The final precipitate was dried at 60°C overnight and 200 mg (dry weight) of crude polysaccharide was obtained after these procedures. The crude polysaccharide (10 mg) was applied to a Mono Q column FPLC2 (HR 5/5) (Amersham Pharmacia Biotech), equilibrated with 20 mM Tris-HCl buffer (pH 8.0). The column was developed by a linear gradient of 0–3.0 M NaCl in the same buffer. The flow rate of the column was 0.50 mL min and fractions of 0.5 mL were collected and assayed by metachromasia using 1,9-dimethylmethylene blue and by the phenol-H_2SO_4 reactions.

19.4 Most desirable biomedical use for MSP concerns interventions on blood coagulation

19.4.1 High incidence of cardiovascular diseases

Cardiovascular diseases (CVD) are the leading cause of death and illness throughout the world. It has been estimated that CVD are responsible for more than one million deaths annually in the United States (Lloyd-Jones et al. 2010; WHO 2004). After myocardial infarction and stroke, venous thromboembolism (VTE) is the third major cause of cardiovascular-associated death (Cushman 2007). VTE consists of deep vein thrombosis, which typically involves deep veins of the leg or pelvis, and its complication, pulmonary embolism (PE). VTE contributes to more than 250,000 hospital admissions/year. Thirty percent die within 30 days, one-fifth suffer from sudden death due to PE, and 30% developed episodic recurrence of VTE within 10 years in the United States (Lloyd-Jones et al. 2010; WHO 2004). It is estimated that 12% of the annual deaths occurring in the European Union are associated with VTE (Cohen et al. 2007). In the United States, PE causes about 300,000 deaths/year (Heit et al. 2005).

Arterial thromboembolism is also an important contributor to the growing burden of CVD (Lloyd-Jones et al. 2010). Atherosclerosis is a progressive disease characterized by the accumulation of lipids and fibrous elements in the large arteries that can grow sufficiently to block blood flow. The most important clinical complication is an acute occlusion due to the formation of a thrombus, resulting in myocardial infarction or stroke (Lusis 2000).

Anticoagulant drugs are highly effective for prevention of arterial thromboembolism and VTE (Gresele and Agnelli 2002). The conventional management of thrombotic and cardiovascular disorders is based on the use of heparin and its derivatives of low molecular weight (LMWH) (Hirsh et al. 2001). Heparin has been used in the clinic as an anticoagulant for more than 50 years (Fareed et al. 2000; Spyropoulos 2008). Anticoagulation treatment with heparin is necessary to prevent thrombus growth and reduce the risk of thromboembolism. Furthermore, any condition that predicts an increased risk of thrombosis requires the preventive use of anticoagulants.

Heparin is also required for extracorporeal circulation during cardiovascular surgeries and renal dialysis (Blossom et al. 2008).

Heparin is a mixture of heterogeneous SP chains varying from ~3 to ~30 kDa (Hirsh et al. 2001). It consists of a disaccharide repeating unit composed of an uronic acid (iduronic, IdoA, or glucuronic acid, GlcA) and a glucosamine (GlcN), with the majority of disaccharides units containing IdoA and di-sulfated GlcA (**Figure 19.4A**) (Li et al. 2004). Heparin acts as an anticoagulant by activating the physiological inhibitors of blood coagulation proteases serpins: AT and HCII. Serpins inhibit blood coagulation proteases, including IIa, the final protease generated in the blood coagulation cascade, responsible for the

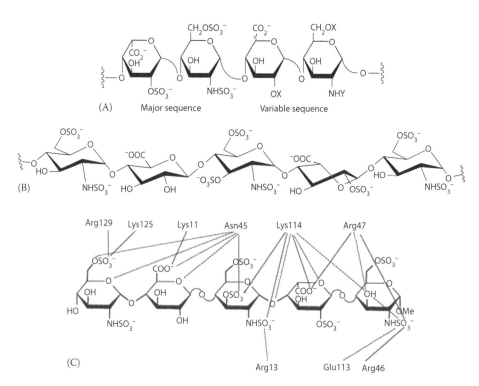

Figure 19.4 *Structure of heparin and the pentasaccharide motif with high affinity for AT. (A) Structure of heparin major and minor variable sequences (SO_3^- or SO_3^-, $COCH_3$, or H). (B) Heparin pentasaccharide antithrombin (AT) binding sequence. The sulfonate group at the C-3 position of the center glucosamine (GlcN) residue is critical for the interaction with AT. (From Kemp, M. and Linhardt, R.J., WIREs Nanomed. Nanobiotechnol., 2, 77, 2010.). (C) The AT residues (mostly cationic ones) that interact specifically with the heparin pentasaccharide are highlighted. The intermolecular interactions involved are also indicated (solid lines are ionic and dashed are hydrogen bonds). (From Carter, W.J. et al., J. Biol. Chem., 280, 2745, 2005.)*

cleavage of fibrinogen to form fibrin clot (Bourin and Lindahl 1993). In the presence of heparin, the rates of IIa-inhibition by AT is increased ~2000-fold (Ofosu et al. 1988). This interaction is specific via a pentasaccharide with high affinity for AT (**Figure 19.4B**), and the hydrogen bonds involved in such interaction of the heparin pentasaccharide with the positively charged aminoacid residues have been characterized (Carter et al. 2005).

A number of limitations associated with pharmacokinetic and biophysical properties of heparin have complicated its use. Heparin has a narrow therapeutic window and a highly variable dose–response relationship, necessitating frequent coagulation monitoring. The main side effects of this drug are bleeding, thrombocytopenia, and osteoporosis. Moreover, the heparin source is limited. It is obtained from pig intestine or bovine lung, and contamination of samples with pathogens is a serious concern in the extraction procedure (Mourao and Pereira 1999). The situation was further complicated recently due to contamination of heparin preparations with OSCS (Guerrini et al. 2008). As mentioned before, this contaminant is severe since it induces hypotension associated with kallikrein release when administered by intravenous injection (Kishimoto et al. 2008).

The incidence of CVD remains high (WHO estimates that by 2030 the number of annual deaths caused by CVD will rise from 17 to 23 million) and as a consequence, the use of heparin will increase strikingly. Hence, there is a current need for new anticoagulant drugs or alternative sources of heparin in anticoagulation. Fortunately, heparin is not the only SP capable of inhibiting blood coagulation proteases. Marine organisms are an abundant source of SPs with anticoagulant and antithrombotic activities as well (Blossom et al. 2008; Farias et al. 2000; Mourao and Pereira 1999; Pereira et al. 2002). In this chapter we presented in detail two of these MSPs (the *L. grisea* FucCS, **Figure 19.2G**, and the *B. occidentalis*, **Figure 19.2E**) with an unusual and promising serpin-independent anticoagulant action, which is not observed in the mechanisms of action of heparin.

19.4.2 Biochemical mechanisms of coagulation and thrombosis

Efficient functioning of the coagulation system is vital to human health. The maintenance of normal blood flow depends completely on the hemostatic system. Hemostasis must be tightly regulated and coordinated to prevent consequent diseases. Failure of its regulation and coordination leads to hemorrhagic disorders and thromboembolic events. The hemostasis system involves activation, regulation, and coordination of numerous proteases. Three catalytic complexes are involved in hemostasis and culminate in the generation of IIa, a key enzyme responsible for the cleavage of fibrinogen to ultimately form the fibrin clot. Each complex involves a vitamin K-dependent serine protease and a cofactor protein assembled on a phospholipid membrane surface provided by an activated or damaged cell, platelets, and endothelial cells (**Figure 19.5**; Mann et al. 1990).

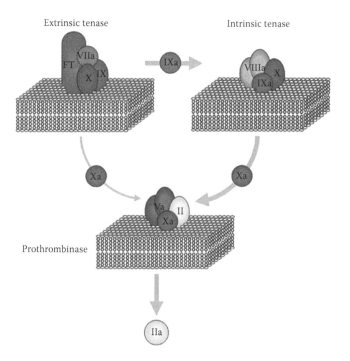

Figure 19.5 *Schematic representation of the vitamin K-dependent complexes of coagulation. Each serine protease is shown in association with the appropriate cofactor on the membrane surface.*

In normal physiologic conditions hemostasis is naturally initiated, controlled, and terminated. The response of the coagulation process is generally limited to the site of injury and is proportional to the extent of vascular damage. Coagulation response occurs somehow in a localized, amplified, and modulated manner. However, genetic or physiologic perturbations can lead to severe dysfunctions on one or more of these systems.

The physiologic response to vascular damage culminates in the rapid generation of precise and balanced amounts of thrombin at the site of injury. The blood coagulation cascade is triggered when subendothelial tissue factor (TF) is exposed to the blood flow as a consequence of a lesion of the vessel wall or activation of endothelium by chemicals, cytokines, or inflammatory and metastatic processes (Camera et al. 1999; Wilcox et al. 1989; Weiss et al. 1989). The contact of TF exposed with already circulating factor VIIa (Morrissey et al. 1993), in the presence of phospholipids and calcium ions, leads to the formation of extrinsic tenase complex, which can activate factor X, responsible for converting small amounts of prothrombin (II) to thrombin (IIa) (**Figures 19.5** and **19.6**) (Lawson and Mann 1991). Albeit rather inefficiently, this initial thrombin formation is necessary to accelerate coagulation process by activating platelets, factor V and factor VIII (Butenas et al. 1997;

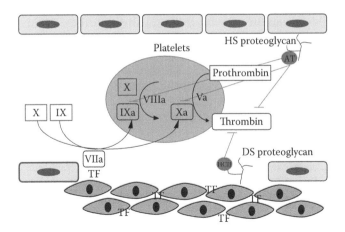

Figure 19.6 *Phospholipid-bound reactions that are involved in the activation of coagulation and the potentiation of antithrombin (AT) and heparin cofactor II (HCII) by glycosaminoglycans (GAGs) in vivo. Tissue factor (TF) is exposed on extravascular cells after vascular injury and binds circulating factor VIIa. The extrinsic tenase complex (FT-VIIa) accelerates factor X and factor IX activation. Factors IXa and Xa assemble with their protein cofactors (VIIIa and Va, respectively) on the surface of aggregated platelets to form the intrinsic tenase and prothrombinase complexes. This leads to local generation of large amounts of Xa and thrombin (IIa), followed by conversion of fibrinogen to fibrin. IIa is originated by a catalytic modification on prothrombin (II). AT inhibits coagulation factors IXa, Xa, and IIa when bound to heparan sulfate (HS) proteoglycans associated with vascular endothelial cells. After disruption of the endothelium, HCII is activated by dermatan sulfate (DS) proteoglycans in the vessel wall and inhibits IIa.*

Lawson et al. 1994). The extrinsic tenase complex can also activate factor IX. Factor IXa and cofactor VIIIa combine to form the intrinsic tenase complex assembled on a membrane surface, which activate large amount of factor X to supply the prothrombinase complex (**Figures 19.5** and **19.6**; Lawson and Mann 1991). The intrinsic tenase complex is the major activator of factor X. Fifty times more factor X is activated by intrinsic tenase complex than extrinsic tenase complex (Mann et al. 1992). The factor IXa-factor VIIIa complex is 10^5-fold more active than factor IXa alone as a factor X activator (Hockin et al. 2002). In the absence of factor VIIIa or factor IXa, the intrinsic tenase complex cannot be assembled; thus, no amplification of the factor Xa generation occurs. This is the principal defect observed in hemophilia A and B types (Cawthern et al. 1998). Factor Xa generated and cofactor Va assembled on a membrane surface to form the prothrombinase complex, which converts II to IIa (**Figures 19.5** and **19.6**). Prothrombinase is 300,000-fold more active than factor Xa alone in catalyzing II activation (Mann et al. 2003). Finally, IIa cleaves fibrinogen and factor XIII to form the insoluble cross-linked fibrin clot (Brummel et al. 1999).

Alternatively, coagulation may be initiated through the "intrinsic pathway" when factor XII is activated on a negatively charged surface by a process called contact activation (Gailani and Renné 2007; Renné and Gailani 2007). Activation of factor XII is followed sequentially by activation of factor XI and factor IX. The intrinsic and extrinsic pathways (factor VIIa-tissue factor) converge at the level of factor X activation (formation of Xa) to convert consequently fibrinogen to fibrin (Dahlbäck 2000; Davie and Kulman 2006; Furie and Furie 1992; Mann 1999). Typically, the intrinsic pathway is illustrated as a sequence of proteolytic reactions culminating in factor IXa (**Figures 19.6** and **19.7**). However, the hemorrhagic profiles of patients deficient in components of the intrinsic pathway suggest more complex interactions. The deficiency of factor IX or its cofactor (factor VIII) cause hemophilia B and hemophilia A, respectively, associated with severe hemorrhage into joints

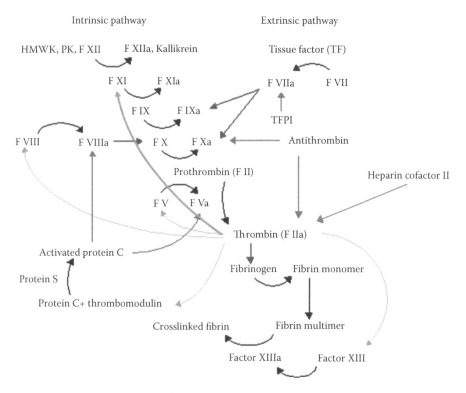

Figure 19.7 *The intrinsic and extrinsic coagulation pathway. The beginning of the coagulation cascade occurs after a vascular injury and exposure of tissue factor to the blood. This fact triggers the extrinsic pathway of coagulation (right side). The intrinsic pathway (left side) can be triggered in vitro when factor XII, prekallikrein, and high-molecular weight kininogen (HMWK) bind to kaolin, glass, or another artificial surface. Once bound, reciprocal activation of XII and prekallikrein occurs. Factor XIIa triggers clotting via the sequential activation of factors XI, IX, and X. The two pathways converge in the formation of thrombin (IIa).*

and muscles, and soft tissue bleeding that can be life threatening (Skinner 2011). In contrast, factor XI deficiency is associated with a milder disorder characterized by trauma or soft tissue-related hemorrhage (Salomon et al. 2006). In contrast, factor XII-deficient patients do not exhibit an abnormal bleeding tendency, even with surgery, despite having markedly prolonged clotting times on the activated partial thromboplastin time (aPPT) in assays used currently for clinical evaluation (Kaplan 1996; Lammle et al. 1991). These observations suggest that these proteases are not exclusively activated in a linear sequential way.

Although the relative contributions of the two mechanisms for factor IX activation are not clear, the activation by factor VIIa-TF is the more important, if not exclusively, mechanism in many instances. Factor XI may be activated by proteases other than factor XIIa, providing an explanation for the absence of a bleeding tendency in factor XII deficiency. Thrombin can convert factor XI to the active protease factor XIa (Gailani and Broze 1991; Naito and Fujikawa 1991) and it is postulated that the early thrombin generated in clot formation activates factor XI, creating a feedback loop that sustains coagulation (**Figure 19.7**) (von dem Borne et al. 1995). In animal models of hemostasis using deficient factor XII mouse, factor XII was not required for fibrin formation (Mackman 2004; Morrissey 2004; Rapaport and Rao 1995).

As deficiency of some coagulation factors causes hemorrhagic phenotypes, upregulated production of these proteases can also predict thrombotic diseases. Increased levels of factor VIII, IX, for example, are correlated with bigger incidence of VTE (Meijers et al. 2000; Vlieg et al. 2000). Moreover, there are some indications that the deficiency of factor VIII and IX provides some protection from arterial thrombosis events (Rosendaal et al. 1990; Triemstra et al. 1995). These findings are supported by the observation that venous thrombosis is rare in hemophiliacs. Because of the rarity of factor XI disorders and the lack of available data about factor XII deficiency, it is not clear whether these disorders contribute positively or negatively to thrombotic diseases.

Consequently, tight regulation of thrombin generation and/or thrombin activity is essential to prevent excessive thrombosis (Mann et al. 2003). Different anticoagulant principles are utilized such as proteolytic degradation of the enzyme cofactors factor Va and VIIIa and enzyme inhibition. The protein C anticoagulant system inhibits the procoagulant functions of factor VIIIa and Va, the cofactors in the tenase and prothrombinase complexes, respectively. Activated protein C (APC) cleaves a few peptide bonds in each of the phospholipid membrane-bound cofactors Va and VIIIa (**Figure 19.7**). The tissue factor pathway inhibitor (TFPI) regulates the initial steps of blood coagulation involving factor VIIa and TF (**Figure 19.7**). For further explanations see references (Broze 1995; Dahlbäck and Villoutreix 2005).

However, the most important regulators are the serine proteases inhibitors (serpins): antithrombin (AT) and heparin cofactor II (HCII) that inhibit procoagulant enzymes such as factor Xa and thrombin (IIa).

19.4.3 Classical mechanism of serpin-dependent inhibition

The principal procoagulant components of blood coagulation are serine proteases. Inasmuch, serpins are the predominant modulator of this system and control inappropriate, excessive, or mislocalized clotting in the blood flow. Two major serpins are involved in hemostasis: AT that inhibits several coagulation proteases, mostly important factor Xa and IIa; and HCII that inhibits exclusively IIa (**Figure 19.7**). Circulating AT and HCII are relatively inefficient inhibitors. GAGs such as heparin, heparan sulfate (HS), and DS have been found to significantly accelerate the interactions between serpins and coagulation proteases, providing maximal inhibitory activity (**Figure 19.6**; Bourin and Lindahl 1993).

AT is the most important serpin in hemostasis. Its importance is demonstrated by the high association of AT deficiency with venous thrombosis (Hirsh 1981), embryonic lethal phenotype in the mouse knockout model (Kojima 2002), and success of heparin therapy (Gray et al. 2008). The concentration of AT in plasma ($2-3\,\mu M$) greatly exceeds that of any of the target proteases generated during coagulation (Conard et al. 1983). AT is a single-chain, 58 kDa glycoprotein and inhibits several of the proteases involved in blood coagulation, including thrombin, factor Xa, IXa, XIa, and XIIa. Based on rates of inhibition, its primary targets are factor Xa and IIa (**Figure 19.7**) (Olson et al. 1993).

The physiological role of AT is to protect the circulation from liberated enzymes and to limit the coagulation process to sites of vascular injury. This is consistent with the observation that the free enzymes are readily inhibited by AT while clot bound IIa and factor Xa are protected from inactivation by AT (Rezaie 2001; Weitz and Buller 2002).

AT, as other serpins, inhibits serine proteases by forming a tight equimolar complex (**Figure 19.8**). The inhibition occurs through the cleavage of a peptide bond, present in an exposed loop of AT. The enzyme attacks the bond and remains attached to the inhibitor through a covalent bond. The cleavage causes a conformational change in AT, which irreversibly binds the protease to AT (Björk et al. 1992; Huntington et al. 2000; Kaslik et al. 1997; Olson et al. 1992; Ye et al. 2001). This interaction can be enhanced by heparin. The anticoagulant effect of commercially available heparin and its LMW derivatives is mediated predominantly through the activation of AT (Gray et al. 2008). Heparin can bind AT and provide the conformational change necessary in the inhibitor resulting in accelerated interaction with IIa or other proteases, and as consequence AT inhibition rate (Damus et al. 1973; Olson and Björk 1991). This binding occurs through a "heparin binding site" on AT that interacts with a unique pentasaccharide sequence found in one-third of the heparin chains (**Figure 19.4C**). The 3-O-sulfonated GlcN unit present in the middle of the pentasaccharide, is essential for this affinity with AT (Choay et al. 1983; Rosenberg et al. 1997; **Figure 19.4B**). Following complex formation with thrombin, AT loses its high affinity for heparin, which will be released and ready to activate another AT molecule (Olson and Shore 1986).

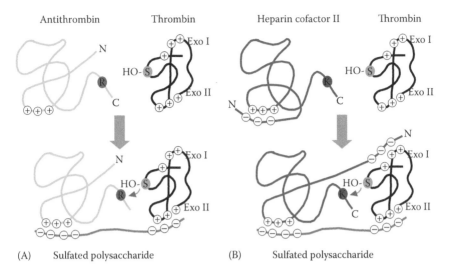

Figure 19.8 *Molecular schematic representation of (A) antithrombin and (B) heparin cofactor II activation by glycosaminoglycans. The thrombin (IIa) can be inhibited by serpins such as antithrombin (AT) or heparin cofactor II (HCII). In both cases, the glycosaminoglycans (GAGs) or other sulfated polysaccharides (SPs) bring together the serpins and the protease (IIa), mainly through electrostatic interactions of their opposite charges. In the thrombin, this charged cluster is the EXO II. Next, the hydroxyl groups of a serine (S) residue from thrombin will bind to the C-terminus of the serpins, actually to an arginine (R) residue of the AT, or to a lysine (K) residue in the case of the HCII. In the bind-states, a conformational change will occur in both serpins, although this change is more predominant and necessary at the HCII case. Note that the N-terminus of the HCII will interact also with the EXO I of IIa through also electrostatic contacts. With the examples described throughout the text, it is clear that the template mechanism between SPs, serpins (AT, HCII) and protease (IIa) has differential stabilities or formation kinetics directly related to the structural features of the SPs.*

The conformational change of the AT molecule is an important contributor to the rate enhancement, but a minimum chain length of ~18 saccharides is required to enhance the rates of thrombin inhibition (Shore et al. 1989). These findings are better explained by an additional mechanism, which predicted that heparin was acting as a surface or bridge of both proteins to the same heparin chain. So, heparin accelerates IIa inhibition rate through activation of AT by an allosteric modification of the structure of AT and by providing a template on which inhibitor and protease can interact (via exosite II on the protease) (**Figure 19.8**). The template mechanism accelerates the interaction of AT with IIa 1000-fold and with fXa 10,000-fold (Danielsson et al. 1986; Griffith 1982).

Although both factor Xa and IIa are capable of forming a ternary complex with AT and heparin, the inhibition of factor Xa by AT can be accelerated only by the pentasaccharide sequence (**Figure 19.4B** and **C**). The pentasaccharide

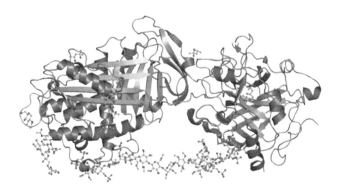

Figure 19.9 *Crystal structure of ternary Hep-AT-IIa complex reveals antithrom-botic template mechanism for heparin. (From Li, W. et al., Nat. Struct. Mol. Biol., 11(9), 857, 2004.) Left-hand side ribbon-model structure is AT, right-hand side is IIa, and ball-and-stick model structure represents the Heparin chain.*

alone accelerates the inhibition of factor Xa by 300-fold and of IIa by only twofold (Olson et al. 1992). Heparin enhancement of FXa inhibition is said to be through an "allosteric mechanism," whereas the enhancement of thrombin inhibition requires "template mechanism."

Recently, the crystal structure of heparin-catalyzed AT-templated within IIa has been shown. The bridging role of heparin is clear on the structure (**Figure 19.9**). The observation that the pentasaccharide sequence alone does not cata-lyze AT-mediated IIa-inhibition was confirmed. The pentasaccharide binding only facilitates the interaction of AT with IIa by increasing the flexibility of exposed loop.

Heparin is not the real physiological activator of AT. Presumably, heparin-like molecules, such as HS anchored to the surface of endothelial cells by its pro-teoglycan core, are responsible for the activation of AT and contribute to the non-thrombogenic properties of blood vessels in vivo (**Figure 19.6**) (Damus et al. 1973). The pentasaccharide sequence (**Figure 19.4B**) is also present on HS chains (Weitz 2003), although at much lower concentration than found in heparin chains. Nowadays, there is an intense debate whether HS is actually antithrombotic or not, since generated mice deficient in 3-O-sulfotransferase-1 failed to show accelerated thrombosis in injured arteries compared with wild-type mice (HajMohammadi et al. 2003; Weitz 2003).

HCII that occurs in plasma at μM concentration is a single-chain 66 kDa glycoprotein (Tollefsen et al. 1983) that also inactivates thrombin by for-mation of a stable complex (**Figure 19.8**). HCII has more restricted pro-tease specificity than AT. HCII does not react with factor Xa and other coagulation proteases (**Figure 19.7**; Tollefsen et al. 1982). In the absence of a GAG, the inhibition of IIa is very slow. In this case, DS is the more important activator of HCII. Heparin also stimulates HCII, but does not require the specific pentasaccharide structure shown in **Figure 19.4B** for stimulation by heparin (Hurst et al. 1983; Maimone and Tollefsen 1988). However, the affinity of HCII for DS is higher than for heparin and a 10-fold higher concentration of heparin is required to accelerate IIa inhibition by HCII (Tollefsen et al. 1982). Its relatively low affinity for heparin explains why the antithrombotic effect of therapeutic heparin is mediated mainly through activation of AT.

Like heparan sulfate, DS is a component of proteoglycans on the cell surface and in the extracellular matrix. DS is a repeating polymer of D-GlcA/L-IdoA and GalNAc units (Conrad 1989). O-sulfonation at IdoA residues in the C2 position and at GalNAc residues in the C4 and C6 positions occurs to a variable extent, yielding heterogeneous structures within the polymer. The high-affinity binding site for HCII in DS consists of three repeated 2-sulfated IdoA and 4-sulfated GalNAc disaccharide subunits. The 4-O-sulfonation of GalN is essential for activity with HCII (Halldórsdóttir et al. 2006; Pavão et al. 1995). The high-affinity hexasaccharide increases the rate of inhibition of thrombin by HCII about 50 times (Maimone and Tollefsen 1988), although DS chains containing up to 14 monosaccharide units are required for maximal stimulation (Tollefsen et al. 1986).

The stimulatory effect of heparin and DS appears to be allosteric and uses conformational activation of the serpin. The activation of HCII depends on the presence of an acidic polypeptide domain near the N-terminus of HCII (Van Deerlin and Tollefsen 1991). Heparin and DS displaces the N-terminal acidic domain to interact with exosite I of thrombin (Liaw et al. 1999; Myles et al. 1998; Ragg et al. 1990). Thrombin binds to the acidic domain of HCII and facilitates inhibition by bringing the active site of thrombin into approximation with the reactive site of HCII (**Figure 19.7**).

The function of HCII in vivo remains obscure. Individuals with inherited partial deficiency of HCII (~50% of normal) have been reported in association with histories of thrombotic disease (Tollefsen 2002). The AT deficiency is not compensated by HCII (Griffith et al. 1983). Some reports suggest that in vivo HCII assumes special relevance after vascular endothelial injury. DS is the predominant antithrombotic GAG in the vessel walls and is synthesized by cells from the subendothelial cells (**Figure 19.5**; Tovar et al. 2005).

HCII and AT could be activated by several MSPs. While HS and DS are physiological activators of AT and HCII, heparin and many different polyanions, including MSPs, are able to accelerate the serpin inhibition of some coagulation proteases (Glauser et al. 2008, 2009; Mourao and Pereira 1999; Sheehan and Walke 2006). SFs isolated from the brown algae as *Fucus vesiculosus* and *Ecklonia kurome* possess anticoagulant activity due to activation of HCII (Cumashi et al. 2007; Nishino et al. 1991). In contrast, the anticoagulant effect of the SF from *Ascophylum nodosum* is mediated mainly via AT (Mauray et al. 1995).

19.5 Complementary serpin-independent anticoagulant mechanism

Besides the common serpin-dependent anticoagulant activity described earlier for the sea cucumber FucCS and the red algal SG, these glycans possess a serpin-independent complementary mechanism that contributes to their anti-coagulant action. Both MSPs are able to inhibit the procoagulant tenase and prothrombinase complexes. Initially, its anticoagulant action was attributed only by its capacity of potentiate factor Xa and IIa inhibition by AT and HCII. Currently, FucCS and SG are known to inhibit the generation of factor Xa and IIa by these complexes. Factor Xa is activated mainly by the intrinsic tenase complex and IIa is converted from II by the prothrombinase complex (**Figure 19.5**). FucCS and SG were shown to inhibit the activation of these proteases by the complexes (Glauser et al. 2008, 2009). These are critical steps for the amplification of coagulation process. Protease activation by these complexes can be inhibiting through direct inactivation of the enzyme and/or the cofactor or by impairing assembling of all components of the complex on the phospholipid surface.

The contribution of serpins to the anticoagulant activity of FucCS and SG was determined in a plasmatic system using aPTT test. Basically, aPTT tests are used to evaluate the anticoagulant effect of the SPs. In this assay, the anticoagulant effect of the MSPs was evaluated by the measure of the time necessary to plasma clotting. Normal plasma and depleted plasma (free of serpins) were used in these tests. FucCS and SG were potent on the prolong clotting time, even when the serpins were not present (**Table 19.1**). This result suggested an anticoagulant activity that persists, independent of the inhibitory action based on serpins activity. More importantly, addition of FucCS and SG to either normal or serpin-free plasmas increased the clotting time with similar potency, demonstrating that the serpin-independent anticoagulant effect predominates over the serpin-dependent action. The result was unexpected because heparin and other GAGs with anticoagulant and antithrombotic effect are usually known to catalyze inhibition of factor Xa and thrombin by AT and HCII (Gray et al. 2008; Tollefsen et al. 1983). According to the literature, heparin has no effect on AT-depleted plasma (Anderson et al. 2001; Nagase et al. 1995).

Table 19.1 Concentrations of GAGs Required for Doubling APTT in Normal, AT- and/or HCII-free Plasmas

GAG	Plasma (Values in μg/mL)			
	Normal	AT-Free	HCII-Free	AT + HCII-Free
Heparin	0.60	NA	0.43	NA
Native FucCS	2.80	2.08	2.60	3.04
Carboxyl-reduced FucCS	7.31	3.06	4.62	3.60
SG	1.06	0.99	1.05	1.28

NA, not achieved.

Further evidence was demonstrated by a factor Xa and IIa generation tests where the generation of these proteases was quantified. In these assays, normal or serpin-free plasma was incubated with the SPs, and then activated to induce the coagulation. The assay is characterized by an initial lag phase, followed by an explosive generation of factor Xa or IIa until a plateau is reached. When the SPs were added to plasma a delay in the generation of the proteases and a decrease in the maximum amounts formed was observed (**Figure 19.10A** through **D**). Again, when similar assays were performed with AT + HCII-free plasma, the effect of heparin was totally abolished. In contrast, FucCS and SG still retain their activities.

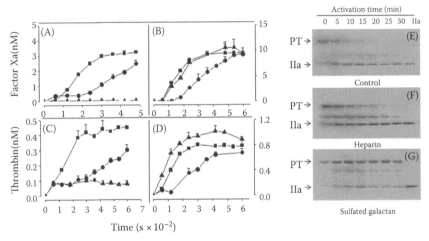

Figure 19.10 *Factor Xa (A and B) and thrombin (C and D) generation tests in normal (A and C) and serpin-free (B and D) plasmas. Defibrinated plasmas (50 μL) were incubated in the absence (■) or in the presence of 0.2 μg mL of heparin (▲) or sulfated galactan (SG) (●) with 50 μL of TS/PEG buffer and 10 μL of Cephalin reagent. After incubation for 2 min at room temperature, the protease generation reaction was started by addition of 100 μL 12.5 mM CaCl₂, and aliquots of 10 μL were removed each 15 s into microplate wells containing 40 μL TS/PEG buffer + 50 mM EDTA. The amounts of factor Xa or thrombin generated were determined using the chromogenic substrates S-2765 or S-2238, respectively. Substrate hydrolysis was detected using a Thermomax Microplate Reader. Reactions were recorded continuously at 405 nm for 15 min at 37°C. The panels show mean ± SD, n = 3. Effect of SPs on thrombin generation by the prothrombinase complex (E–G). The effect of SPs on prothrombinase complex was analyzed on 12% SDS-PAGE. The incubation mixtures contain: 1 nM plasma-derived factor Xa, 3 nM factor Va, 20 μM phospholipids and 10 μg mL sulfated polysaccharide in TS/PEG buffer containing 10 mM CaCl₂, final volume 500 μL. After incubation for 10 min at 37°C, the activation reaction was started by addition of 0.5 μM prothrombin. Aliquots from each reaction mixture were removed at the time point indicated in the panels and immediately quenched in SDS-PAGE loading buffer. The gel was stained with Coomassie blue and band intensities for prothrombin and thrombin were monitored by densitometric analysis. PT, prothrombin and IIa, thrombin. *p < 0.05 for vs. ●.*

Overall, these experiments clearly confirmed that FucCS and SG have an additional uncommon serpin-independent anticoagulant activity.

FucCS and SG interfere with the activation of factor X and II by the intrinsic tenase and prothrombinase complexes, respectively. In these assays, all the purified proteases of the complexes, calcium and phospholipids are used. Both glycans inhibit the generation of factor Xa and IIa by the complexes. The effect of SG on IIa generation was even demonstrated by monitoring II activation by SDS-PAGE assay (**Figure 19.10E** through **G**). As shown in **Figure 19.10E** through **G**, the conversion of II into IIa was almost complete after II incubation with the prothrombinase compounds. Heparin did not impair the conversion while SG abolishes the activation process (**Figure 19.10F** and **G**) (Glauser et al. 2008, 2009).

Additional experiments with FucCS demonstrated that the inactivation of the prothrombinase complex is not due to an interference with prothrombinase complex assembly on the phospholipid membrane. Probably, the invertebrate SP may make the assembly of factor Va on the complex difficult. It is possible that the predominant target of FucCS in the prothrombinase complex is related to the interaction of factor Va and Xa (**Figure 19.11**).

The main factor Xa residues responsible for its association with heparin were determined (Rezaie 2000). These same residues permit the binding of factor Xa with cofactor Va and/or the substrate in the prothrombinase complex (Rezaie 2001). These domains are conserved in vitamin K-dependent serine proteases, including factor IXa, and participate in the binding of factor IXa to cofactor VIIIa in the intrinsic tenase complex (Furie and Furie 1988; Mathur and Bajaj 1999; Stenflo 1991). Perhaps the site of action of FucCS and SG is the same in both systems. Recently, it was demonstrated that the SG binds to the heparin-binding exosite in factor Xa (Glauser et al. 2009). Possibly, FucCS and SG binds through this same region on factor IXa and factor Xa, since both proteases exhibit structural homology, thus hindering the binding of proteases to their cofactors in the formation of the complexes (**Figure 19.11**; Glauser et al. 2008). The carboxyl-reduced derivative of FucCS (cr-FucCS) (**Figure 19.3C**) reveals an effect on these systems very similar to the native compound. The potent effect of FucCS on inhibiting the tenase and prothrombinase complex is clearly associated with the presence of sulfated Fuc*p* branches and the pattern of sulfation of the molecule, since other GAGs such as DS and CS have no effect on these complexes (Barrow et al. 1994; Glauser et al. 2008; Nagase et al. 1996).

Similar results were found for two other SPs. One of these polysaccharides is pentosan polysulfate, obtained by chemical oversulfation of naturally occurring pentosan (Sie et al. 1986). The other is also obtained by an invertebrate and is denominated "depolymerized FucCS" (Nagase et al. 1995, 1996). Depolymerized FucCS also inhibits intrinsic tenase activity through interactions with the heparin-binding exosite present on factor IXa (Sheehan and Walke 2006). However, the absence of a clear description of its structure makes it difficult to trace a comparison with FucCS and SG. Pentosan polysulfate has

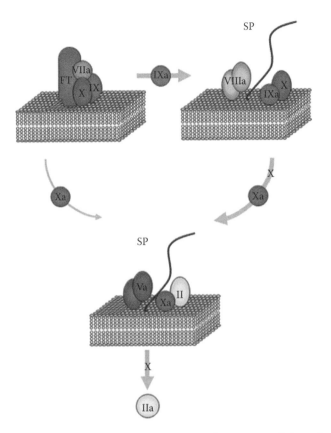

Figure 19.11 *Proposed mechanism of inhibition of factor Xa and thrombin activation by intrinsic tenase and prothrombinase complexes. Sulfated polysaccharides (SPs) may make the assembly of the cofactors, VIIIa and Va, on the complexes difficult. It is possible that the predominant target of SPs in procoagulant complexes are related to the interaction of the cofactors and the proteases, factor IXa and factor Xa.*

a critical limitation for its use as an antithrombotic drug due to a potent bleeding effect (Buchanan et al. 1986).

These findings demonstrate that FucCS and SG have an unusual serpin-independent anticoagulant activity in addition to their ability to potentiate the inhibitory action of AT and HCII. This effect is due to inhibition of factor Xa and IIa generation by the tenase and prothrombinase complexes, respectively.

19.6 Advantages in the medicinal use of FucCS and SG over other SPs

The amount of serine protease is amplified at each step in the coagulation system. Because of that, it is interesting that the selective inhibition of coagulation factors above IIa might be highly blocked as well. Furthermore, by not inhibiting all IIa activity directly, such anticoagulant agents might allow traces of IIa to escape neutralization, thereby facilitating hemostasis and leading to a favorable profile with respect to bleeding. Thus, there is a suggestion that the removal of all AT-dependent actions of heparin could further reduce the bleeding risk and increase the therapeutic action (Barrow et al. 1994). Given the critical role of tenase and prothrombinase enzyme complexes in hemostasis, it seems an antithrombotic drug candidate to act on these complexes. These observations led several authors to look for SPs, prepared from either new natural sources or by chemical modification of standard heparin, with serpin-independent anticoagulant action and, perhaps, with additional antithrombotic effects.

Heparin is not able to inhibit procoagulant complexes. The approach to obtain a derivative with predominant action on the intrinsic tenase complex, as FucCS and SG, involved periodate oxidation of LMWH, which reduces its affinity for AT followed by oversulfation of the saccharide chain. This chemically modified heparin became a potent inhibitor of the intrinsic tenase and prothrombinase complexes (Anderson et al. 2001). However, the serpin-independent anticoagulant effect of FucCS and SG does not require any additional chemical modification or laborious synthetic route. The structural requirement for its activity is clear, namely, sulfated Fuc$p$ branches linked to the polysaccharide core (Glauser et al. 2008).

Even, serpin-dependent anticoagulant drugs used nowadays are synthesized or chemically modified. For example, LMWH are also commonly used in the treatment of thromboembolic and CVDs. They are generally prepared by chemical or enzymatical depolymerization method of porcine unfractionated heparins (UFHs). In comparison to UFHs, LMWHs possess an antithrombotic property with fewer side effects (Glauser et al. 2011). Another strategy for the development of new antithrombotics was the synthetic heparin pentasaccharide (**Figure 19.4B**), called fondaparinux. This drug induces a conformational change in AT that increases the affinity of the serpin for factor Xa, potentiating the natural inhibitory effect of AT against factor Xa (Wienbergen and Zeymer 2007). Recently, a chemoenzymatically pentasaccharide was reconstructed in milligram quantities by following the heparin biosynthetic pathway (Xu et al. 2011). However, these methods are complex, expensive, increase the treatment costs, and are more time-consuming than compared to extraction-based procedures of natural sources.

In contrast with those observations, FucCS and SG occur at high concentrations in sea cucumber and red algal, respectively. Both organisms are abundant in nature. There is now more interest in therapeutics prepared from non-mammalian sources, which avoids the risk of contamination with pathogenic agents. The yield obtained from each sea cucumber specimen is ~7% per dry weight and about 35 mg of purified polysaccharide. FucCS and SG have clear advantages over heparin and other anticoagulant drugs currently used in the clinic of thrombosis. First of all, FucCS is absorbed after oral administration, and the peak in plasma is achieved about 2 h after oral administration in parallel with the antithrombotic effect (Fonseca and Mourão 2006). Its carboxyl-reduced derivative does not cause bleeding and retains the serpin-independent anticoagulant activity, indicating an even more favorable antithrombotic than the native polysaccharide (Zancan and Mourão 2004). It is noteworthy that the algal SG has no hemorrhagic effect even when tested at high doses (Fonseca et al. 2008).

These results concerning the serpin-independent anticoagulant activity of MSPs may help to design new drugs with specific actions on coagulation and thrombosis. For example, these new MSPs may be active in patients with inherited AT deficiency due to a mutation in the heparin-binding site (Patnaik and Moll 2008).

Obviously several points remain to be clarified before defining FucCS and SG as definitive antithrombotic drugs of practical use. Critical points are related to the bioavailability, half-life, and safety of these compounds. In addition, the magnitude of serpin-independent effect for the antithrombotic activity of these MSPs should be more carefully evaluated.

References

Alves, A. P., B. Mulloy, J. A. Diniz et al. 1997. Sulfated polysaccharides from the egg jelly layer are species-specific inducers of acrosomal reaction in sperms of sea urchins. *Journal of Biological Chemistry* 272: 6965–6971.

Alves, A. P., B. Mulloy, G. W. Moy et al. 1998. Females of the sea urchin *Strongylocentrotus purpuratus* differ in the structures of their egg jelly sulfated fucans. *Glycobiology* 8: 939–946.

Anderson, J. A. M., J. C. Fredenburgh, A. R. Stafford et al. 2001. Hypersulfated low molecular weight heparin with reduced affinity for antithrombin acts as an anticoagulant by inhibiting intrinsic tenase and prothrombinase. *Journal of Biological Chemistry* 276 (13): 9755–9761.

Barrow, R. T., E. T. Parker, S. Krishnaswamy, and P. Lollar. 1994. Inhibition by heparin of the human blood coagulation intrinsic pathway factor X activator. *Journal of Biological Chemistry* 269 (43): 26796–26800.

Björk, I., K. Ylinenjärvi, S. T. Olson, and P. E. Bock. 1992. Conversion of antithrombin from an inhibitor of thrombin to a substrate with reduced heparin affinity and enhanced conformational stability by binding of a tetradecapeptide corresponding to the P1 to P14 region of the putative reactive bond loop of the inhibitor. *Journal of Biological Chemistry* 267 (3): 1976–1982.

Blossom, D. B., A. J. Kallen, P. R. Patel et al. 2008. Outbreak of adverse reactions associated with contaminated heparin. *New England Journal of Medicine* 359 (25): 2674–2684.

von dem Borne, P., J. Meijers, and B. Bouma. 1995. Feedback activation of factor XI by thrombin in plasma results in additional formation of thrombin that protects fibrin clots from fibrinolysis. *Blood* 86 (8): 3035–3042.

Borsig, L., L. Wang, M. C. M. Cavalcante et al. 2007. Selectin blocking activity of a fucosylated chondroitin sulfate glycosaminoglycan from sea cucumber. *Journal of Biological Chemistry* 282 (20): 14984–14991.

Bourin, M. C. and U. Lindahl. 1993. Glycosaminoglycans and the regulation of blood coagulation. *Biochemical Journal* 289 (Pt 2): 313–330.

Broze, G. J. 1995. Tissue factor pathway inhibitor. *Thrombosis and Haemostasis* 74 (1): 90–93.

Brummel, K. E., S. Butenas, and K. G. Mann. 1999. An integrated study of fibrinogen during blood coagulation. *Journal of Biological Chemistry* 274 (32): 22862–22870.

Buchanan, M. R., F. A. Ofosu, F. Fernandez, and J. Van Ryn. 1986. Lack of relationship between enhanced bleeding induced by heparin and other sulfated polysaccharides and enhanced catalysis of thrombin inhibition. *Seminars in Thrombosis and Hemostasis* 12 (4): 324–327.

Butenas, S., C. van 't Veer, and K. G. Mann. 1997. Evaluation of the initiation phase of blood coagulation using ultrasensitive assays for serine proteases. *Journal of Biological Chemistry* 272 (34): 21527–21533.

Camera, M., P. L. A. Giesen, J. Fallon et al. 1999. Cooperation between VEGF and TNF-α is necessary for exposure of active tissue factor on the surface of human endothelial cells. *Arteriosclerosis, Thrombosis, and Vascular Biology* 19 (3): 531–537.

Carter W. J., E. Cama, and J. A. Huntington. 2005. Crystal structure of thrombin bound to heparin. *Journal of Biological Chemistry* 280: 2745–2749.

Castro, M. O., V. H. Pomin, L. L. Santos et al. 2009. A unique 2-sulfated {beta}-galactan from the egg jelly of the sea urchin *Glyptocidaris crenularis*: Conformation flexibility versus induction of the sperm acrosome reaction. *Journal of Biological Chemistry* 284: 18790–18800.

Cawthern, K. M., C. van 't Veer, J. B. Lock et al. 1998. Blood coagulation in hemophilia A and hemophilia C. *Blood* 91 (12): 4581–4592.

Choay, J., M. Petitou, J. C. Lormeau et al. 1983. Structure-activity relationship in heparin: A synthetic pentasaccharide with high affinity for antithrombin III and eliciting high anti-factor Xa activity. *Biochemical and Biophysical Research Communications* 116 (2): 492–499.

Cohen, A. T., G. Agnelli, F. A. Anderson et al. 2007. Venous thromboembolism (VTE) in Europe. The number of VTE events and associated morbidity and mortality. *Thrombosis and Haemostasis* 98 (4): 756–764.

Conrad, H. E. 1989. Structure of heparan sulfate and dermatan sulfate. *Annals of the New York Academy of Sciences* 556 (1): 18–28.

Conard, J., F. Brosstad, M. Lie Larsen, M. Samama, and U. Abildgaard. 1983. Molar antithrombin concentration in normal human plasma. *Pathophysiology of Haemostasis and Thrombosis* 13 (6): 363–368.

Cumashi, A., N. A. Ushakova, M. E. Preobrazhenskaya et al. 2007. A comparative study of the anti-inflammatory, anticoagulant, antiangiogenic, and antiadhesive activities of nine different fucoidans from brown seaweeds. *Glycobiology* 17 (5): 541–552.

Cushman, M. 2007. Epidemiology and risk factors for venous thrombosis. *Seminars in Hematology* 44 (2): 62–69.

Dahlbäck, B. 2000. Blood coagulation. *Lancet* 355 (9215): 1627–1632.

Dahlbäck, B. and B. O. Villoutreix. 2005. The anticoagulant protein C pathway. *FEBS Letters* 579 (15): 3310–3316.

Damus, P. S., M. Hicks, and R. D. Rosenberg. 1973. Anticoagulant action of heparin. *Nature* 246 (5432): 355–357.

Danielsson, A., E. Raub, U. Lindahl, and I. Björk. 1986. Role of ternary complexes, in which heparin binds both antithrombin and proteinase, in the acceleration of the reactions between antithrombin and thrombin or factor Xa. *Journal of Biological Chemistry* 261 (33): 15467–15473.

Davie, E. W. and J. D. Kulman. 2006. An overview of the structure and function of thrombin. *Seminars in Thrombosis and Hemostasis* 32 (Suppl 1): 3–15.

Farias, W. R. L., A. P. Valente, M. S. Pereira, and P. A. S. Mourão. 2000. Structure and anticoagulant activity of sulfated galactans. *Journal of Biological Chemistry* 275 (38): 29299–29307.

Fareed, J. W., D. Hoppensteadt, and R. L. Bick, 2000. An update on heparins at the beginning of the new millennium. *Seminars and Thrombosis Haemostasis* 26: 5–21.

Fonseca, R. J. C. and P. A. S. Mourão. 2006. Fucosylated chondroitin sulfate as a new oral antithrombotic agent. *Thrombosis and Haemostasis* 96: 822–829.

Fonseca, R. J. C., S. N. M. C. G. Oliveira, F. R. Melo et al. 2008. Slight differences in sulfation of algal galactans account for differences in their anticoagulant and venous antithrombotic activities. *Thrombosis and Haemostasis* 99: 539–545.

Fonseca, R. J. C., S. N. M. C. G. Oliveira, V. H. Pomin et al. 2010. Effects of oversulfated and fucosylated chondroitin sulfates on coagulation. Challenges for the study of anticoagulant polysaccharides. *Thrombosis and Haemostasis* 103: 994–1004.

Fonseca, R. J. C., G. R. C. Santos, and P. A. S. Mourão. 2009. Effects of polysaccharides enriched in 2,4-disulfated fucose units on coagulation, thrombosis and bleeding. Practical and conceptual implications. *Thrombosis and Haemostasis* 102: 829–836.

Furie, B. and B. C. Furie. 1988. The molecular basis of blood coagulation. *Cell* 53 (4): 505–518.

Furie, B. and B. C. Furie. 1992. Molecular and cellular biology of blood coagulation. *New England Journal of Medicine* 326 (12): 800–806.

Gandhi, N. S. and R. L. Mancera. 2008. The structure of glycosaminoglycans and their interactions with proteins. *Chemical Biology and Drug Design* 72: 455–482.

Gailani, D. and G. Broze. 1991. Factor XI activation in a revised model of blood coagulation. *Science* 253 (5022): 909–912.

Gailani, D. and T. Renné. 2007. The intrinsic pathway of coagulation: A target for treating thromboembolic disease? *Journal of Thrombosis and Haemostasis* 5 (6): 1106–1112.

Glauser, B. F., M. S. Pereira, R. Q. Monteiro et al. 2008. Serpin-independent anticoagulant activity of a fucosylated chondroitin sulfate. *Thrombosis and Haemostasis* 100: 420–428.

Glauser, B. F., R. M. Rezende, F. R. Melo et al. 2009. Anticoagulant activity of a sulfated galactan: Serpin-independent effect and specific interaction with factor Xa. *Thrombosis and Haemostasis* 102 (6): 1183–1193.

Glauser, B. F., B. C. Vairo, C. P. M. Oliveira et al. 2011. Generic versions of enoxaparin available for clinical use in Brazil are similar to the original drug. *Journal of Thrombosis and Haemostasis* 9 (7): 1419–1422.

Gray, E., B. Mulloy, and T. W. Barrowcliffe. 2008. Heparin and low-molecular-weight heparin. *Thrombosis and Haemostasis* 99 (5): 807–818.

Gresele, P. and G. Agnelli. 2002. Novel approaches to the treatment of thrombosis. *Trends in Pharmacological Sciences* 23 (1): 25–32.

Griffith, M. J. 1982. The heparin-enhanced antithrombin III/thrombin reaction is saturable with respect to both thrombin and antithrombin III. *Journal of Biological Chemistry* 257 (23): 13899–13902.

Griffith, M., T. Carraway, G. White, and F. Dombrose. 1983. Heparin cofactor activities in a family with hereditary antithrombin III deficiency: Evidence for a second heparin cofactor in human plasma. *Blood* 61 (1): 111–118.

Guerrini, M., D. Beccati, Z. Shriver et al. 2008. Oversulfated chondroitin sulfate is a contaminant in heparin associated with adverse clinical events. *Nature Biotechnology* 26 (6): 669–675.

HajMohammadi, S., K. Enjyoji, M. Princivalle et al. 2003. Normal levels of anticoagulant heparan sulfate are not essential for normal hemostasis. *Journal of Clinical Investigation* 111 (7): 989–999.

Halldórsdóttir, A. M., L. Zhang, and D. M. Tollefsen. 2006. N-Acetylgalactosamine 4,6-O-sulfate residues mediate binding and activation of heparin cofactor II by porcine mucosal dermatan sulfate. *Glycobiology* 16 (8): 693–701.

Heit, J., A. Cohen, F. Anderson, and V. T. E. I. A. G. on Behalf of the. 2005. Estimated annual number of incident and recurrent, non-fatal and fatal venous thromboembolism (VTE) events in the US. *Blood (ASH Annual Meeting Abstracts)* 106 (11): 910.

Hirsh, J. 1981. Blood tests for the diagnosis of venous and arterial thrombosis. *Blood* 57 (1): 1–8.

Hirsh, J., T. E. Warkentin, S. G. Shaughnessy et al. 2001. Heparin and low-molecular-weight heparin: Mechanisms of action, pharmacokinetics, dosing, monitoring, efficacy, and safety. *Chest* 119 (Suppl 1): 64S–94S.

Hockin, M. F., K. C. Jones, S. J. Everse, and K. G. Mann. 2002. A model for the stoichiometric regulation of blood coagulation. *Journal of Biological Chemistry* 277 (21): 18322–18333.

Huntington, J. A., R. J. Read, and R. W. Carrell. 2000. Structure of a serpin-protease complex shows inhibition by deformation. *Nature* 407 (6806): 923–926.

Hurst, R. E., M. C. Poon, and M. J. Griffith. 1983. Structure-activity relationships of heparin. Independence of heparin charge density and antithrombin-binding domains in thrombin inhibition by antithrombin and heparin cofactor II. *Journal of Clinical Investigation* 72 (3): 1042–1045.

Kaplan, A. P. 1996. *Intrinsic coagulation, thrombosis, and bleeding. Blood* 87 (5): 2090.

Kaslik, G., J. Kardos, E. Szabó et al. 1997. Effects of serpin binding on the target proteinase: Global stabilization, localized increased structural flexibility, and conserved hydrogen bonding at the active site. *Biochemistry* 36 (18): 5455–5464.

Kemp, M. and R. J. Linhardt. 2010. Heparin-based nanoparticles. *WIREs Nanomedicine and Nanobiotechnology* 2: 77–87.

Kishimoto, T. K., K. Viswanathan, T. Ganguly et al. 2008. Contaminated heparin associated with adverse clinical events and activation of the contact system. *New England Journal of Medicine* 358 (23): 2457–2467.

Kojima, T. 2002. Targeted gene disruption of natural anticoagulant proteins in mice. *International Journal of Hematology* 76 (Suppl 2): 36–39.

Lammle, B., W. A. Wuillemin, I. Huber et al. 1991. Thromboembolism and bleeding tendency in congenital factor XII deficiency—A study on 74 subjects from 14 Swiss families. *Thrombosis and Haemostasis* 65 (2): 117–121.

Landeira-Fernandez, A. M., K. R. M. Aiello, R. S. Aquino et al. 2000. A sulfated polysaccharide from the sarcoplasmic reticulum of sea cucumber smooth muscle is an endogenous inhibitor of the Ca2+-ATPase. *Glycobiology* 10 (8): 773–779.

Lawson, J. H., M. Kalafatis, S. Stram, and K. G. Mann. 1994. A model for the tissue factor pathway to thrombin. I. An empirical study. *Journal of Biological Chemistry* 269 (37): 23357–23366.

Lawson, J. H. and K. G. Mann. 1991. Cooperative activation of human factor IX by the human extrinsic pathway of blood coagulation. *Journal of Biological Chemistry* 266 (17): 11317–11327.

Li, W., D. J. D. Johnson, C. T. Esmon, and J. A. Huntington. 2004. Structure of the antithrombin-thrombin-heparin ternary complex reveals the antithrombotic mechanism of heparin. *Nature Structural and Molecular Biology* 11 (9): 857–862.

Liaw, P. C. Y., R. C. Austin, J. C. Fredenburgh, A. R. Stafford, and J. I. Weitz. 1999. Comparison of heparin- and dermatan sulfate-mediated catalysis of thrombin inactivation by heparin cofactor II. *Journal of Biological Chemistry* 274 (39): 27597–27604.

Lloyd-Jones, D., R. J. Adams, T. M. Brown et al. 2010. Heart disease and stroke statistics-2010 update. *Circulation* 121: e-46–e-215.

Lusis, A. J. 2000. Atherosclerosis. *Nature* 407 (6801): 233–241.

Mackman, N. 2004. Role of tissue factor in hemostasis, thrombosis, and vascular development. *Arteriosclerosis, Thrombosis, and Vascular Biology* 24 (6): 1015–1022.

Maimone, M. M. and D. M. Tollefsen. 1988. Activation of heparin cofactor II by heparin oligosaccharides. *Biochemical and Biophysical Research Communications* 152 (3): 1056–1061.

Mann, K. G. 1999. Biochemistry and physiology of blood coagulation. *Thrombosis and Haemostasis* 82 (2): 165–174.

Mann, K. G., S. Butenas, and K. Brummel. 2003. The dynamics of thrombin formation. *Arteriosclerosis, Thrombosis, and Vascular Biology* 23 (1): 17–25.

Mann, K. G., S. Krishnaswamy, and J. H. Lawson. 1992. Surface-dependent hemostasis. *Seminars in Hematology* 29 (3): 213–226.

Mann, K., M. Nesheim, W. Church, P. Haley, and S. Krishnaswamy. 1990. Surface-dependent reactions of the vitamin K-dependent enzyme complexes. *Blood* 76 (1): 1–16.

Mathur, A. and S. P. Bajaj. 1999. Protease and EGF1 domains of factor IXa play distinct roles in binding to factor VIIIa. *Journal of Biological Chemistry* 274 (26): 18477–18486.

Mauray, S., C. Sternberg, J. Theveniaux et al. 1995. Venous antithrombotic and anticoagulant activities of a fucoidan fraction. *Thrombosis and Haemostasis* 74 (5): 1280–1285.

Meijers, J. C. M., W. L. H. Tekelenburg, B. N. Bouma, R. M. Bertina, and F. R. Rosendaal. 2000. High levels of coagulation factor XI as a risk factor for venous thrombosis. *New England Journal of Medicine* 342 (10): 696–701.

Melo, F. R., M. S. Pereira, D. Foguel, and P. A. S. Mourão. 2004. Antithrombin-mediated anticoagulant activity of sulfated polysaccharides. *Journal of Biological Chemistry* 279 (20): 20824–20835.

Melo, F. R., M. S. Pereira, R. Q. Monteiro, D. Foguel, and P. A. S. Mourão. 2008. Sulfated galactan is a catalyst of antithrombin-mediated inactivation of α-thrombin. *Biochimica et Biophysica Acta (BBA)—General Subjects* 1780 (9): 1047–1053.

Melo-Filho, N. M., C. L. Belmiro, R. G. Gonçalves et al. 2010. Fucosylated chondroitin sulfate attenuates renal fibrosis in animals submitted to unilateral ureteral obstruction: A P-selectin-mediated event? *American Journal of Physiology—Renal Physiology* 299 (6): 1299–12307.

Morrissey, J. H. 2004. Tissue factor: A key molecule in hemostatic and nonhemostatic systems. *International Journal of Hematology* 79 (2): 103–108.

Morrissey, J., B. Macik, P. Neuenschwander, and P. Comp. 1993. Quantitation of activated factor VII levels in plasma using a tissue factor mutant selectively deficient in promoting factor VII activation. *Blood* 81 (3): 734–744.

Mourão, P.A. 2004. Use of sulfated fucans as anticoagulant and antithrombotic agents: Future perspectives. *Current Pharmaceutical Design* 10: 967–981.

Mourão, P.A. 2007. A carbohydrate-based mechanism of species recognition in sea urchin fertilization. *Brazilian Journal of Medical and Biological Research* 40: 5–17.

Mourão, P.A. and A. S. Perlin. 1987. Structural features of sulfated glycans from the tunic of *Styela plicata* (Chordata-Tunicata). A unique occurrence of L-galactose in sulfated polysaccharides. *European Journal of Biochemistry*. 166: 431–436.

Mourao, P. A. and M. S. Pereira. 1999. Searching for alternatives to heparin: Sulfated fucans from marine invertebrates. *Trends in Cardiovascular Medicine* 9 (8): 225–232.

Mourão, P. A. S., M. S. Pereira, M. S. G. Pavão et al. 1996. Structure and anticoagulant activity of a fucosylated chondroitin sulfate from echinoderm. *Journal of Biological Chemistry* 271 (39): 23973–23984.

Mulloy, B., A. C. Ribeiro, A. P. Alves et al. 1994. Sulfated fucans from echinoderms have a regular tetrasaccharide repeating unit defined by specific patterns of sulfation at the 0–2 and 0–4 positions. *Journal of Biological Chemistry* 269: 22113–22123.

Myles, T., F. C. Church, H. C. Whinna, D. Monard, and S. R. Stone. 1998. Role of thrombin anion-binding exosite-I in the formation of thrombin-serpin complexes. *Journal of Biological Chemistry* 273 (47): 31203–31208.

Nagase, H., K. Enjyoji, K. Minamiguchi et al. 1995. Depolymerized holothurian glycosaminoglycan with novel anticoagulant actions: Antithrombin III- and heparin cofactor II-independent inhibition of factor X activation by factor IXa-factor VIIIa complex and heparin cofactor II-dependent inhibition of thrombin. *Blood* 85 (6): 1527–1534.

Nagase, H., K. Enjyoji, M. Shima et al. 1996. Effect of depolymerized holothurian glycosaminoglycan (DHG) on the activation of factor VIII and factor V by thrombin. *Journal of Biochemistry* 119 (1): 63–69.

Naito, K. and K. Fujikawa. 1991. Activation of human blood coagulation factor XI independent of factor XII. Factor XI is activated by thrombin and factor XIa in the presence of negatively charged surfaces. *Journal of Biological Chemistry* 266 (12): 7353–7358.

Nishino, T., Y. Aizu, and T. Nagumo. 1991. Antithrombin activity of a fucan sulfate from the brown seaweed *Ecklonia kurome*. *Thrombosis Research* 62 (6): 765–773.

Ofosu, F. A., L. M. Smith, N. Anvari, and M. A. Blajchman. 1988. An approach to assigning in vitro potency to unfractionated and low molecular weight heparins based on the inhibition of prothrombin activation and catalysis of thrombin inhibition. *Thrombosis and Haemostasis* 60 (2): 193–198.

Olson, S. T. and I. Björk. 1991. Predominant contribution of surface approximation to the mechanism of heparin acceleration of the antithrombin-thrombin reaction. Elucidation from salt concentration effects. *Journal of Biological Chemistry* 266 (10): 6353–6364.

Olson, S. T., I. Björk, R. Sheffer et al. 1992. Role of the antithrombin-binding pentasaccharide in heparin acceleration of antithrombin-proteinase reactions. Resolution of the antithrombin conformational change contribution to heparin rate enhancement. *Journal of Biological Chemistry* 267 (18): 12528–12538.

Olson, S. T., I. Bjork, and J. D. Shore. 1993. Kinetic characterization of heparin-catalyzed and uncatalyzed inhibition of blood coagulation proteinases by antithrombin. *Methods in Enzymology* 222: 525–559.

Olson, S. T. and J. D. Shore. 1986. Transient kinetics of heparin-catalyzed protease inactivation by antithrombin III. The reaction step limiting heparin turnover in thrombin neutralization. *Journal of Biological Chemistry* 261 (28): 13151–13159.

Patnaik, M. M. and S. Moll. 2008. Inherited antithrombin deficiency: A review. *Haemophilia* 14 (6): 1229–1239.

Pavão, M. S. 2002. Structure and anticoagulant properties of sulfated glycosaminoglycans from primitive Chordates. *Anais da Academia Brasileira Ciencias* 74: 105–112.

Pavão, M. S., K. R. Aiello, C. C. Werneck et al. 1998. Highly sulfated dermatan sulfates from Ascidians. Structure versus anticoagulant activity of these glycosaminoglycans. *Journal of Biological Chemistry* 273: 27848–27857.

Pavão, M. S., P.A. Mourão, B. Mulloy, and D. M. Tollefsen. 1995. A unique dermatan sulfate-like glycosaminoglycan from ascidian. Its structure and the effect of its unusual sulfation pattern on anticoagulant activity. *Journal of Biological Chemistry* 270: 31027–31036.

Pereira, M. G., N. M. B. Benevides, M. R. S. Melo et al. 2005. Structure and anticoagulant activity of a sulfated galactan from the red alga, Gelidium crinale. Is there a specific structural requirement for the anticoagulant action? *Carbohydrate Research* 340 (12): 2015–2023.

Pereira, M. S., F. R. Melo, and P. A. S. Mourão. 2002a. Is there a correlation between structure and anticoagulant action of sulfated galactans and sulfated fucans? *Glycobiology* 12 (10): 573–580.

Pereira, M. S., A.-C. E. S. Vilela-Silva, A.-P. Valente, and P. A. S. Mourão. 2002b. A 2-sulfated, 3-linked α-l-galactan is an anticoagulant polysaccharide. *Carbohydrate Research* 337 (21–23): 2231–2238.

Pomin, V. H. 2009a. Review: An overview about the structure-function relationship of marine sulfated homopolysaccharides with regular chemical structures. *Biopolymers* 91: 601–609.

Pomin, V. H. 2009b. Structural and functional insights into sulfated galactans: A systematic review. *Glycoconjugate Journal* 27: 1–12.

Pomin, V. H. 2011. Fucanome and galactanome: Marine glycomics contribution. *Journal of Glycobiology* 1: 101.

Pomin, V. H. and P. A. S. Mourão. 2008. Structure, biology, evolution, and medical importance of sulfated fucans and galactans. *Glycobiology* 18 (12): 1016–1027.

Ragg, H., T. Ulshöfer, and J. Gerewitz. 1990. On the activation of human leuserpin-2, a thrombin inhibitor, by glycosaminoglycans. *Journal of Biological Chemistry* 265 (9): 5211–5218.

Rapaport, S. I. and L. V. Rao. 1995. The tissue factor pathway: How it has become a "prima ballerina". *Thrombosis and Haemostasis* 74 (1): 7–17.

Renné, T. and D. Gailani. 2007. Role of Factor XII in hemostasis and thrombosis: Clinical implications. *Expert Review of Cardiovascular Therapy* 5 (4): 733–741.

Rezaie, A. R. 2000. Identification of basic residues in the heparin-binding exosite of factor Xa critical for heparin and factor Va binding. *Journal of Biological Chemistry* 275 (5): 3320–3327.

Rezaie, A. R. 2001. Prothrombin protects factor Xa in the prothrombinase complex from inhibition by the heparin-antithrombin complex. *Blood* 97 (8): 2308–2313.

Rosenberg, R. D., N. W. Shworak, J. Liu, J. J. Schwartz, and L. Zhang. 1997. Heparan sulfate proteoglycans of the cardiovascular system. Specific structures emerge but how is synthesis regulated? *Journal of Clinical Investigation* 99 (9): 2062–2070.

Rosendaal, F. R., E. Briét, J. Stibbe et al. 1990. Haemophilia protects against ischaemic heart disease: A study of risk factors. *British Journal of Haematology* 75 (4): 525–530.

Salomon, O., D. M. Steinberg, and U. Seligshon. 2006. Variable bleeding manifestations characterize different types of surgery in patients with severe factor XI deficiency enabling parsimonious use of replacement therapy. *Haemophilia* 12 (5): 490–493.

Santos, J. A., B. Mulloy, and P. A. Mourão. 1992. Structural diversity among sulfated alpha-L-galactans from ascidians (tunicates). Studies on the species *Ciona intestinalis* and *Herdmania monus. European Journal of Biochemistry* 204, 669–677.

Sheehan, J. P. and E. N. Walke. 2006. Depolymerized holothurian glycosaminoglycan and heparin inhibit the intrinsic tenase complex by a common antithrombin-independent mechanism. *Blood* 107 (10): 3876–3882.

Shore, J. D., S. T. Olson, P. A. Craig, J. Choay, and I. Björk. 1989. Kinetics of heparin actiona. *Annals of the New York Academy of Sciences* 556 (1): 75–80.

Sie, P., F. Ofosu, F. Fernandez et al. 1986. Respective role of antithrombin III and heparin cofactor II in the in vitro anticoagulant effect of heparin and of various sulphated polysaccharides. *British Journal of Haematology* 64 (4): 707–714.

Skinner, M. W. 2011. Haemophilia: Provision of factors and novel therapies: World federation of hemophilia goals and achievements. *British Journal of Haematology* 154 (6): 704–714.

Spyropoulos, A. C. 2008. Brave new world: The current and future use of novel anticoagulants. *Thrombosis Research* 123 (Suppl 1): S29–S35.

Stenflo, J. 1991. Structure-function relationships of epidermal growth factor modules in vitamin K-dependent clotting factors. *Blood* 78 (7): 1637–1651.

Tollefsen, D. M. 2002. Heparin cofactor II deficiency. *Archives of Pathology & Laboratory Medicine* 126 (11): 1394–1400.

Tollefsen, D. M., D. W. Majerus, and M. K. Blank. 1982. Heparin cofactor II. Purification and properties of a heparin-dependent inhibitor of thrombin in human plasma. *Journal of Biological Chemistry* 257 (5): 2162–2169.

Tollefsen, D. M., M. E. Peacock, and W. J. Monafo. 1986. Molecular size of dermatan sulfate oligosaccharides required to bind and activate heparin cofactor II. *Journal of Biological Chemistry* 261 (19): 8854–8858.

Tollefsen, D. M., C. A. Pestka, and W. J. Monafo. 1983. Activation of heparin cofactor II by dermatan sulfate. *Journal of Biological Chemistry* 258 (11): 6713–6716.

Tovar, A. M. F., D. A. de Mattos, M. P. Stelling et al. 2005. Dermatan sulfate is the predominant antithrombotic glycosaminoglycan in vessel walls: Implications for a possible physiological function of heparin cofactor II. *Biochimica et Biophysica Acta (BBA)—Molecular Basis of Disease* 1740 (1): 45–53.

Triemstra, M., F. R. Rosendaal, C. Smit, H. M. Van der Ploeg, and E. Briet. 1995. Mortality in patients with hemophilia: Changes in a Dutch population from 1986 to 1992 and 1973 to 1986. *Annals of Internal Medicine* 123 (11): 823–827.

Van Deerlin, V. M., and D. M. Tollefsen. 1991. The N-terminal acidic domain of heparin cofactor II mediates the inhibition of alpha-thrombin in the presence of glycosaminoglycans. *Journal of Biological Chemistry* 266 (30): 20223–20231.

Vicente, C. P., P. Zancan, L. L. Peixoto et al. 2001. Unbalanced effects of dermatan sulfates with different sulfation patterns on coagulation, thrombosis and bleeding. *Thrombosis and Haemostasis* 86: 1215–1220.

Vieira, R. P. and P. A. Mourão. 1988. Occurrence of a unique fucose-branched chondroitin sulfate in the body wall of a sea cucumber. *Journal of Biological Chemistry* 263 (34): 18176–18183.

Vieira, R. P., B. Mulloy, and P. A. Mourão. 1991. Structure of a fucose-branched chondroitin sulfate from sea cucumber. Evidence for the presence of 3-O-sulfo-beta-D-glucuronosyl residues. *Journal of Biological Chemistry* 266 (21): 13530–13536.

Vilela-Silva, A. C., A. P. Alves, A. P. Valente et al. 1999. Structure of the sulfated alpha-L-fucan from the egg jelly coat of the sea urchin *Strongylocentrotus franciscanus*: Patterns of preferential 2-O- and 4-O-sulfation determine sperm cell recognition. *Glycobiology* 9: 927–933.

Vilela-Silva, A. C., M. O. Castro, A. P. Valente et al. 2002. Sulfated fucans from the egg jellies of the closely related sea urchins *Strongylocentrotus droebachiensis* and *Strongylocentrotus pallidus* ensure species-specific fertilization. *Journal of Biological Chemistry* 277: 379–387.

Vilela-Silva, A. C., N. Hirohashi, and P. A. Mourão. 2008. The structure of sulfated polysaccharides ensures a carbohydrate-based mechanism for species recognition during sea urchin fertilization. *International Journal of Developmental Biology*. 52: 551–559.

Vlieg, A. v. H., I. K. van der Linden, R. M. Bertina, and F. R. Rosendaal. 2000. High levels of factor IX increase the risk of venous thrombosis. *Blood* 95 (12): 3678–3682.

Weiss, H., V. Turitto, H. Baumgartner, Y. Nemerson, and T. Hoffmann. 1989. Evidence for the presence of tissue factor activity on subendothelium. *Blood* 73 (4): 968–975.

Weitz, J. I. 2003. Heparan sulfate: Antithrombotic or not? *Journal of Clinical Investigation* 111 (7): 952–954.

Weitz, J. I. and H. R. Buller. 2002. Direct thrombin inhibitors in acute coronary syndromes. *Circulation* 105 (8): 1004–1011.

WHO. 2004. *The Global Burden of Disease: 2004 Update*. WHO, Geneva, Switzerland.

Wienbergen, H. and U. Zeymer. 2007. Management of acute coronary syndromes with fondaparinux. *Vascular Health and Risk Management* 3 (3): 321–329.

Wilcox, J. N., K. M. Smith, S. M. Schwartz, and D. Gordon. 1989. Localization of tissue factor in the normal vessel wall and in the atherosclerotic plaque. *Proceedings of the National Academy of Sciences* 86 (8): 2839–2843.

Xu, Y., S. Masuko, M. Takieddin et al. 2011. Chemoenzymatic synthesis of homogeneous ultralow molecular weight heparins. *Science* 334 (6055): 498–501.

Ye, S., A. L. Cech, R. Belmares et al. 2001. The structure of a Michaelis serpin-protease complex. *Nature Structural Biology* 8 (11): 979–983.

Zancan, P. and P. A. Mourão. 2004. Venous and arterial thrombosis in rat models: Dissociation of the antithrombotic effects of glycosaminoglycans. *Blood Coagulation & Fibrinolysis* 15 (1): 45–54.

Enzymatic Production of N-Acetyl-D-Glucosamine Using the Enzyme from the Liver of Squid

Masahiro Matsumiya

Contents

20.1 Introduction

Chitin is a polysaccharide formed from β-1,4 links of *N*-acetyl-ᴅ-glucosamine (GlcNAc). It is widely distributed in organisms, mainly in structural components such as arthropod exoskeletons, mollusk shells, and fungal cell walls. It is the second most abundant biomass next to cellulose and useful for an ingredient of many biomaterials [1,2]. Although chitin is an insoluble polymer, GlcNAc has water solubility and some physiological functions [3]. GlcNAc has been investigated intensively for the treatment of osteoarthritis [4–7] similar to ᴅ-glucosamine (GlcN) hydrochloride or sulfate, which have already been commercialized as food supplements [8–10], because GlcNAc and GlcN are components of proteoglycan in cartilage. GlcN salts have a bitter taste, but GlcNAc has a sweet taste, 55% compared to sucrose. For this reason, GlcNAc is a suitable functional material with sweetness for the food industry. Moreover, GlcNAc has recently been used in cosmetics for skin care [7], because it is a component of hyaluronic acid. GlcNAc is produced from GlcN by *N*-acetylation, but it is not approved as a natural material that can be used as a food additive in some countries. Some enzymatic methods of producing GlcNAc from chitin have been reported [11–17]. Most of those reports used commercial non-chitinase crude enzyme produced from microorganisms such as cellulase for hydrolysis of chitin, because commercial non-chitinase crude enzymes include both endo- and exo-type chitinolytic enzymes, chitinase (EC 3.2.1.14) and β-*N*-acetylhexosaminidase (EC 3.2.1.52) (NAHase), respectively. Moreover, non-chitinase crude enzyme is much cheaper than pure chitinolytic enzyme.

The current author previously reported the presence of endo- and exo-type chitinolytic enzymes in the liver of several species of squid [18]. This chapter therefore describes the purification and characterization of chitinase isozymes [19–21] and *N*-acetylhexosaminidase [22], and the production of GlcNAc [23] by crude enzyme prepared from the liver of Japanese common squid in order to effectively use the organs obtained during marine food processing.

20.2 Endo-type chitinolytic enzyme, chitinase: purification and characterization from the liver of Japanese common squid *Todarodes pacificus*

20.2.1 Method

20.2.1.1 Materials

Fresh *Todarodes pacificus* was purchased from Tokyo Wholesale Fish Market. Glycol chitin, N-acetylchitooligosaccharides ($GlcNAc_n$, $n = 1–6$), and p-nitrophenyl $GlcNAc_n$ (pNp-$GlcNAc_n$, $n = 1–4$) were purchased from Seikagaku Kogyo Co. Chitin (crab shell chitin) was obtained from Tokyo Kasei Co., Ltd. Colloidal chitin was prepared by the method of Ohtakara et al. [24]. All other chemicals were of high grade.

20.2.1.2 Measurement of chitinase activity

Chitinase activity was assayed according to the method described by Imoto and Yagishita [25] by measuring the amount of reducing sugar produced by an enzyme reaction with glycol chitin as substrate. One milliliter of 0.1 M sodium acetate buffer solution (pH 4.0), 0.5 mL of enzyme solution, and 1 mL of 0.05% (w/v) glycol chitin solution were incubated for 30 min at 37°C. After the enzyme reaction was completed, 2 mL of Schales' reagent (0.5 M sodium carbonate solution containing 0.05% potassium ferricyanide) was added to the solution to stop the reaction. The solution was then heated in a boiling water bath for 15 min. After cooling under running water, absorbency was measured at 420 nm. The value recorded was converted into an amount of GlcNAc by the standard curve prepared using GlcNAc. When colloidal chitin was used as the substrate, enzyme activity was assayed according to the method of Ohtakara [26] by measuring the amount of reducing sugar produced by an enzyme reaction. The reaction mixture, containing 2 mL of sodium acetate buffer (pH 4.0), 1 mL of enzyme solution, and 1.0 mL of 0.5% colloidal chitin, was incubated for 30 min at 37°C with shaking. The reaction was stopped by boiling for 3 min. After centrifugation, 1.5 mL of the supernatant fluid was mixed with 2 mL of Schales' reagent. Reducing sugar was measured in the same manner as for glycol chitin. The hydrolyzing activity of pNp-$GlcNAc_n$ was measured according to the method prescribed by Ohtakara [25] using pNp-$GlcNAc_n$ as substrate. Five-tenths of a milliliter of enzyme solution and 0.2 mL of 4 mM pNp-$GlcNAc_n$ ($n = 1–4$) were added to 0.5 mL of 0.2 M phosphate–0.1 M citrate buffer solution (pH 4.5). The mixture was then reacted for 10 min at 37°C. The reaction was stopped by adding 2 mL of 0.2 M sodium carbonate, and then absorbency was measured at 420 nm. The value recorded was converted into an amount of p-nitrophenol by the standard curve prepared using p-nitrophenol.

20.2.1.3 Determination of protein

Protein concentration was determined by the method of Bradford [27] with bovine serum albumin as the standard protein.

20.2.1.4 Gel electrophoresis

Sodium dodecyl sulfate–polyacrylamide gel electrophoresis (SDS–PAGE) was carried out in 12.5% polyacrylamide gel according to the manufacturer's instructions (Phast Gel, Amersham Biosciences). The proteins in the gels were stained with Coomassie brilliant blue R-250.

20.2.1.5 HPLC analysis

A 20 µL portion of the reaction mixture was injected into a Waters µ Bondasphere 5 µ NH$_2$ 100 Å column (3.9 × 150 mm) and eluted with 70% acetonitrile.

20.2.1.6 Amino acid sequence analysis

N-Terminal amino acid sequences of the chitinases were analyzed by a protein sequencer (PE Applied Biosystems 447/120A).

20.2.2 Purification and characterization of chitinase isozymes
20.2.2.1 Purification of chitinase isozymes

Two chitinase isozymes were purified from the Japanese common squid liver by ammonium sulfate fractionation and column chromatographies with Chitopearl Basic BL-03, CM-Toyopearl 650S, and Bio-Gel HTP [19,21].

20.2.2.2 Characterization of chitinase isozymes

As shown in **Table 20.1**, the purified enzymes were basic chitinases with molecular masses of 38 and 42 kDa. The N-terminal amino acid sequences of 38 and 42 kDa chitinases were YLLSXYFTNWSQYRPGAGKYFPQNI and EYRKVXYYTNWSQYREVPAKFFPEN, respectively. Although these were different from each other, they had homology with those of family 18 chitinase from higher organisms. Both enzymes showed two optimum pHs toward glycol chitin. On the other hand, both enzymes were most active at pH 3.0 for colloidal chitin. The optimum temperatures were 50°C and 60°C at 30 min incubation, respectively. Both enzymes were stable at acidic pH between pH 4.0 and 5.0 after incubation at 50°C for 10 min. K_m and k_{cat} of the 38 and 42 kDa chitinases toward a longer substrate, glycol chitin, were 0.071 mg/mL and 1.22/s, and 0.074 mg/mL and 0.196/s, respectively. This result means that the 38 kDa chitinase catalyzes 6.2 times faster than the 42 kDa chitinase in the reaction from

Table 20.1 Properties of Japanese Common Squid Chitinases

Properties	38 kDa Chitinase	42 kDa Chitinase
Isoelectric point	8.3	9.2
Optimum pH (glycol chitin)	1.5 and 8.5 (1.5 > 8.5)	3.5 and 7.0
Optimum pH (colloidal chitin)	3.0	3.0 and 9.0 (3.0 > 9.0)
Optimum temperature (°C)	50	60
Stable pH	4–6	4–5
K_m and k_{cat} (glycol chitin)	0.071 mg/mL, 1.22/s	0.074 mg/mL, 0.196/s

Table 20.2 Substrate Specificity of Japanese Common Squid Chitinases for Long Substrates

	Specific Activity (µmol/mg min)	
Substrate	38 kDa Chitinase	42 kDa Chitinase
Glycol chitin	2.01	0.347
Colloidal chitin	0.868	1.20
Powdered α-chitin[a]	0.734	0.403
Powdered β-chitin[a]	1.98	0.218
Chitosan (DA 95%)[b]	0.209	0.042

[a] Activity was measured by the same method of colloidal chitin.
[b] Activity was measured by the same method of glycol chitin.

Table 20.3 Substrate Specificity of Japanese Common Squid Chitinases for N-Acetylchitooligosaccharides

	Reaction	Relative Activity (%)	
Substrate	Pattern	38 kDa Chitinase	42 kDa Chitinase
GlcNAc$_2$ (II)	No reaction	ND	ND
GlcNAc$_3$ (III)	No reaction	ND	ND
GlcNAc$_4$ (IV)	IV → 2 II	100	16.2
GlcNAc$_5$ (V)	V → III + II	79.8	40.3
GlcNAc$_6$ (VI)	VI →	57.6	100
	→ IV + II	48.4	80.0
	→ 2 III	9.2	20.0

ND, not detected.

the enzyme–substrate complex to enzyme and product. As shown in **Table 20.2**, both chitinases hydrolyzed colloidal chitin, α-chitin, β-chitin, and chitosan (DA 95%), but not *Micrococcus lysodeikticus*. The cleavage pattern and reaction rate were investigated using GlcNAc$_n$ (n = 2–6) (**Table 20.3**). Both chitinases hydrolyzed GlcNAc$_n$ (n = 4–6). Release of GlcNAc was not observed.

Table 20.4 Substrate Specificity of Japanese Common Squid Chitinases for pNp-N-Acetylchitooligosaccharides

Substrate	Relative Activity (%)	
	38 kDa Chitinase	42 kDa Chitinase
pNp-GlcNAc (G-P)	ND	ND
pNp-GlcNAc$_2$ (G-G-P)	100	100
pNp-GlcNAc$_3$ (G-G-G-P)	79.1	64.3
pNp-GlcNAc$_4$ (G-G-G-G-P)	52.2	57.2

ND, not detected; G, GlcNAc; P, pNp.

The speed of the reaction was observed to be in the following order: GlcNAc$_4$ > GlcNAc$_5$ > GlcNAc$_6$ for the 38 kDa chitinase, and GlcNAc$_6$ > GlcNAc$_5$ > GlcNAc$_4$ for the 42 kDa chitinase. As shown in **Table 20.4**, both the chitinases released pNp from pNp-GlcNAc$_n$ (n = 2–4). Therefore, the cleavage site of both chitinases seems to be the second and third link from the terminus of GlcNAc$_n$.

20.3 Exo-type chitinolytic enzyme, NAHase: purification and characterization from the liver of Japanese common squid *Todarodes pacificus*

20.3.1 Method

The method was the same as that of Section 20.2.1 except for the following:

20.3.1.1 Measurement of NAHase activity

NAHase activity was measured by the same method of chitinase activity using pNp-GlcNAc as the substrate.

20.3.1.2 Measurement of anomer formation

The analysis was made according to the method prescribed by Koga et al. [28] using GlcNAc$_3$ as the substrate at pH 4.0. A 20 µL portion was analyzed by HPLC using a Tosoh TSK-Gel amide-80 column (0.46 ID × 25 cm).

20.3.2 Purification and characterization of NAHase
20.3.2.1 Purification of NAHase

NAHase was purified from the Japanese common squid liver by ammonium sulfate fractionation and column chromatographies with Butyl-Toyopearl 650S and Toyopearl HW-55S [22].

20.3.2.2 Molecular weight of NAHase

The molecular weight of the enzyme was estimated to be 125 kDa by gel filtration. The molecular weight of the enzyme was estimated to be 54 kDa by SDS–PAGE in the nonreducing condition and 33 kDa by SDS–PAGE in the reducing condition. These results mean that Japanese common squid β-*N*-acetylhexosaminidase is a homo-tetramer with monomeric molecular weight of 33 kDa. But, the monomer of 33 kDa was bound by S–S binding to make dimer, and the produced dimer was bound by non S–S binding to make tetramer.

20.3.2.3 Effects of pH and temperature on the activity and stability of NAHase

The enzyme activity was measured at 37°C after incubation for 10 min at various pH values. As shown in **Figure 20.1A**, the enzyme showed optimum pH at 4.0. The enzyme was preincubated at 37°C for 30 min at various pHs, then the remaining activity was measured. The enzyme was stable from pH 3.5 to 5.5. Regarding the temperature, as shown in **Figure 20.1B**, the enzyme activity was measured at various temperatures after incubation for 10 min at pH 4.5. The enzyme showed optimum temperature at 70°C. The enzyme was preincubated at pH 4.5 for 10 min at various temperatures, and then the remaining activity was measured. The enzyme was stable till 60°C and completely inactivated at 80°C after incubation for 10 min.

20.3.2.4 Cleavage patterns of p-nitrophenyl N-acetylchitooligosaccharides and tri-N-acetylchitotorioside

p-Nitrophenyl GlcNAc$_n$ and GlcNAc$_3$ were used to identify the cleavage pattern by Japanese common squid NAHase. As shown in **Figure 20.2**, a large

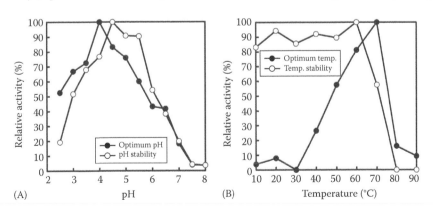

Figure 20.1 *Effects of pH and temperature on the activity and stability of Japanese common squid NAHase. (A) Effects of pH on the activity and stability of the enzyme. (B) Effects of temperature on the activity and stability of the enzyme.*

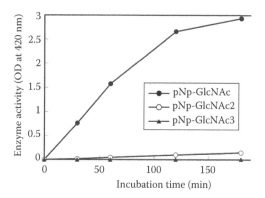

Figure 20.2 *p-Nitrophenyl GlcNAc$_n$ hydrolyzing activity of Japanese common squid NAHase.*

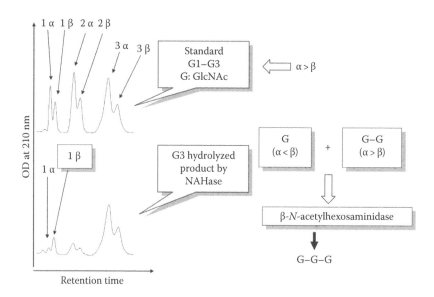

Figure 20.3 *HPLC analysis of the hydrolysis product of GlcNAc$_3$ by Japanese common squid NAHase.*

quantity of *p*-nitrophenol was released with time from pNp-GlcNAc, but little *p*-nitrophenol was observed in the reaction with pNp-GlcNAc$_2$. Furthermore, as shown in **Figure 20.3**, the enzyme produced GlcNAc with high β-anomer ratio compared to α-anomer from GlcNAc$_3$ by enzymatic hydrolysis. These results indicate that NAHase from the liver of Japanese common squid releases GlcNAc from the nonreducing end side.

20.4 Enzymatic production of *N*-acetyl-ᴅ-glucosamine

20.4.1 Method

20.4.1.1 Preparation of the crude chitinolytic enzyme from the liver

The liver was collected and ground uniformly in a mortar. Forty grams of ground liver were homogenized with 200 mL of 10 mM sodium acetate buffer (pH 6.0). The homogenate was centrifuged at 10,000 × g for 20 min. The supernatant was filtered through two layers of gauze to remove floating fat and used as the crude extract. Ammonium sulfate was added to the crude extract up to 65% saturation and left to stand for 1 day at 4°C. The precipitate was collected by centrifugation at 10,000 × g for 20 min, then was dissolved in a small volume of 10 mM sodium acetate buffer (pH 6.0) and dialyzed against the same buffer. The dialyzate was centrifuged at 10,000 × g for 20 min to remove insoluble material and the supernatant was used as the crude enzyme [18]. The chitinolytic activity in the crude enzyme was measured. The ratio of the activity of chitinase, 0.058 μmol/g min, and NAHase, 1.09 μmol/g min, in the crude enzyme was 1:19.

20.4.1.2 Hydrolysis of chitin by the crude enzyme

Almost 10 mL of 0.5% colloidal chitin suspension (50 mg of colloidal chitin), 20 mL of 0.1 M sodium acetate buffer (pH 4.0), and 10 mL of crude enzyme solution (corresponding to 2 g of liver) were incubated at 37°C with agitation. After the prescribed time, 3.5 mL of reaction mixture was removed, heated in a boiling water bath for 5 min, and filtered through filter paper. The amount of reducing sugar in the filtrate was measured by Schales' reagent described earlier, and converted into an amount of GlcNAc.

20.4.2 Enzymatic production of *N*-acetyl-ᴅ-glucosamine

Figure 20.4 shows the production of reducing sugar during the reaction. After 1 day of incubation, 22.3 mg of reducing sugar was produced by enzymatic reaction by crude enzyme. The amount of reducing sugar produced from days 1 to 5 increased slowly and reached 26.8 mg at day 5. To investigate the composition of the reaction product, HPLC analysis was done as described in Section 20.2.1. An HPLC chromatogram of the hydrolysate of colloidal chitin (50 mg) by chitinolytic enzyme from the liver of Japanese common squid at 37°C for 24 h is shown in **Figure 20.5**. The main product of the reducing sugar, with a retention time of 3.4 min, was GlcNAc. Colloidal chitin containing a small portion of glucosamine (GlcN) was produced by deacetylation of GlcNAc during the processes of chitin purification and colloidal chitin preparation. Other minor peaks may therefore be hetero oligomers, which contain GlcN. The 38 and 42 kDa chitinase isozymes from the liver of Japanese common squid cleaved $GlcNAc_n$ (n = 4–6) to $GlcNAc_2$ or

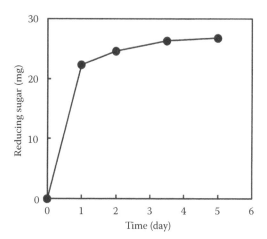

Figure 20.4 *Time course on the production of reducing sugar from colloidal chitin by chitinolytic enzyme from the liver of Japanese common squid.*

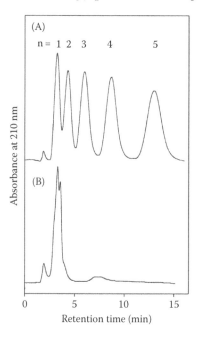

Figure 20.5 *HPLC chromatogram of the hydrolysate of colloidal chitin by chitinolytic enzyme from the liver of Japanese common squid. (A) Standard GlcNAc$_n$ (n = 1–5). (B) Hydrolysate of 50 mg colloidal chitin at 37°C for 24 h.*

GlcNAc$_3$. This result showed that dimeric or trimeric GlcNAc produced from colloidal chitin by hydrolysis of squid liver chitinase was in turn further hydrolyzed by NAHase to GlcNAc. The activity of NAHase was about 20 times higher than chitinase of the squid. This value was much higher than commercial bacterial enzymes like cellulases [15]. Though this ratio is very effective for the specific production of GlcNAc, the production amount is low. It is therefore concluded that squid liver can be a useful source of chitinolytic enzyme in combination with other chitinases for the enzymatic production of GlcNAc.

References

1. Muzzarelli, R. A. A. 2009. Chitins and chitosans for the repair of wounded skin, nerve, cartilage and bone. *Carbohydrate Polymers* 76: 167–182.
2. Jayakumar, R., Prabaharan, M., Sudheesh Kumar, P., Nair, S., and Tamura, H. 2011. Biomaterials based on chitin and chitosan in wound dressing applications. *Biotechnology Advances* 29: 322–337.
3. Chen, J. K., Shen, C. R., and Liu, C. L. 2010. *N*-Acetylglucosamine: Production and applications. *Marine Drugs* 8: 2493–2516.
4. Kikuchi, K. 2001. Characteristics and effects of natural *N*-acetylglucosamine "Marine Sweet." *Japan Fudo Saiensu* 40: 52–56 (in Japanese).
5. Matahira, K., and Yoshiharu, H. 1999. Characteristics and applications of natural *N*-acetylglucosamine as health food material. *New Food Industry* 41: 9–13.
6. Hatano, K., Miyakuni, Y., Hayashida, K., and Nakagawa, S. 2006. Effects and safety of soymilk beverage containing *N*-acetylglucosamine on osteoarthritis. *Jpn. Pharmacol. Ther.* 34: 149.
7. Matahira, Y. 2008. Physiological function and application of new material N-acetylglucosamine. *Nippon Suisan Gakkaishi* 74: 149–165 (in Japanese).
8. Kajimoto, O., Sakamoto, K., Takamori, Y., Kajitani, N., Imanishi, T., Matsuo, R., and Kajitani, Y. 1998. Therapeutic activity of oral glucosamine hydrochloride in osteoarthritis of the knee: A placebo-controlled, double-blind, cross-over study. *Nippon Rinsho Eiyo Gakkaishi* 20: 41–47.
9. Suguro, S., Minami, S., Kusuhara, S., Kumada, T., and Sakamoto, K. 2000. Effect of glucosamine hydrochloride on healing of cartilaginous injuries. *Food Style* 21: 67–73.
10. Nakamura, H. 2011. Application of glucosamine on human disease—Osteoarthritis. *Carbohydrate Polymers* 84: 835–839.
11. Sashiwa, H., Fujishima, S., Yamano, N., Kawasaki, N., Nakayama, A., Muraki, E., and Aiba, S. 2001. Production of *N*-acetyl-D-glucosamine from β-chitin by enzymatic hydrolysis. *Chemistry Letters* 30: 308–309.
12. Sashiwa, H., Fujishima, S., Yamano, N., Kawasaki, N., Nakayama, A., Muraki, E., Hiraga, K., Oda, K., and Aiba, S. 2002. Production of *N*-acetyl-D-glucosamine from α-chitin by crude enzymes from *Aeromonas hydrophila* H-2330. *Carbohydrate Research* 337: 761–763.
13. Pichyangkura, R., Kudan, S., Kuttiyawong, K., Sukwattanasinitt, M., and Aiba, S. 2002. Quantitative production of 2-acetamido-2-deoxy-D-glucose from crystalline chitin by bacterial chitinase. *Carbohydrate Research* 337: 557–559.
14. Sukwattanasinitt, M., Zhu, H., Sashiwa, H., and Aiba, S. 2002. Utilization of commercial non-chitinase enzymes from fungi for preparation of 2-acetamido-2-deoxy-D-glucose from β-chitin. *Carbohydrate Research* 337: 133–137.
15. Sashiwa, H., Fujishima, S., Yamano, N., Kawasaki, N., Nakayama, A., Muraki, E., Sukwattanasinitt, M., Pichyangkura, R., and Aiba, S. 2003. Enzymatic production of *N*-acetyl-D-glucosamine from chitin. Degradation study of *N*-acetylchitooligosaccharide and the effect of mixing of crude enzymes. *Carbohydrate Polymers* 51: 391–395.
16. Kuk, J. H., Jung, W. J., Hyun Jo, G., Ahn, J. S., Kim, K. Y., and Park, R. D. 2005. Selective preparation of *N*-acetyl-D-glucosamine and *N,N'*-diacetylchitobiose from chitin using a crude enzyme preparation from *Aeromonas* sp. *Biotechnology Letters* 27: 7–11
17. Kumar, S., Sharma, R., and Tewari, R. 2011. Production of *N*-acetylglucosamine using recombinant chitinolytic enzymes. *Indian Journal of Microbiology* 51: 319–325.

18. Matsumiya, M., Miyauchi, K., and Mochizuki, A. 1998. Distribution of chitinase and β-*N*-acetylhexosaminidase in the organs of a few squid and a cuttlefish. *Fisheries Science: FS* 64: 166–167.
19. Matsumiya, M. and Mochizuki, A. 1997. Purification and characterization of chitinase from the liver of Japanese common squid *Todarodes pacificus*. *Fisheries Science* 63: 409–413.
20. Matsumiya, M., Miyauchi, K., and Mochizuki, A. 2002. Characterization of 38 kDa and 42 kDa chitinase isozymes from the liver of Japanese common squid *Todarodes pacificus*. *Fisheries Science* 68: 603–609.
21. Matsumiya, M., Miyauchi, K., and Mochizuki, A. 2003. Purification and some properties of a chitinase isozyme from the liver of Japanese common squid *Todarodes pacificus*. *Fisheries Science* 69: 427–429.
22. Matsumiya, M., Suzuki, H., Tanaka, H., and Shigeo, M. 2007. Purification and characterization of β-*N*-acetylhexosaminidase from the liver of Japanese common squid *Todarodes pacificus*. In *Advances in Chitin Science*, Vol. IX. A. Domard, E. Guibal, and K. M. Vårum, Eds. European Chitin society, Montpellier, France, pp. 509–513.
23. Matsumiya, M. 2004. Enzymatic production of *N*-acetyl-ᴅ-glucosamine using crude enzyme from the liver of squids. *Food Science and Technology Research* 10: 296–299.
24. Ohtakara, A., Koga, D., and Hirano, S. 1991. Methods for measurement of chitinase activity. In *Experimental Manual of Chitin and Chitosan*, A. Ohtakara and M. Yabuki, Eds. Tokyo, Japan, Gihodoshuppan (in Japanese), pp. 116–120.
25. Imoto, T. and Yagishita, K. 1971. A simple activity measurement of lysozyme. *Agricultural and Biological Chemistry* 35: 1154–1156.
26. Ohtakara, A. 1988. Chitinase and β-*N*-acetylhexosaminidase from *Pycnoporus cinnabarinus*. In *Methods in Enzymology*. W. A. Wood and S. T. Kellogg, Eds. New York, Academic Press, Vol. 161, pp. 462–470.
27. Bradford, M. M. 1976. A rapid and sensitive method for the quantitation of microgram quantities of protein utilizing the principle of protein-dye binding. *Analytical Biochemistry* 72: 248–254.
28. Koga, D., Yoshioka, T., and Arakane, Y. 1998. HPLC analysis of anomeric formation and cleavage pattern by chitinolytic enzyme. *Bioscience, Biotechnology, and Biochemistry* 62: 1643–1646.

21

High-Density Chitin–Chitosan Production and Health Benefits

Siswa Setyahadi

Contents

21.1 Introduction

The potential use of chitin and chitosan is widely recognized, and many new applications have been developed. Currently, chitin, chitosan, and their derivatives are widely used in chemistry, medicine, pharmacy, cosmetics, food technology, water treatment, etc., but the use is restricted to specific applications because of the high price of technical-grade chitin and chitosan. A prerequisite for the greater use of chitin in industry is cheap manufacturing processes, or the development of profitable processes to recover chitin and byproducts such as protein pigments or carotenoproteins, if crustacean-processing waste products are used as starting material.

Marine byproducts rich in chitin and protein are renewable resources present in large amounts in many countries. Indonesia is one of the shrimp-producing and -exporting countries. According to the Stock Fishery Products, the total production of shrimp in Indonesia in August 2008 reached 290,000–470,000 tons.

If the prediction of shrimp production this year reached the target, the Indonesia will become the fourth highest producer of shrimp in the world after China, Thailand, and Vietnam, and it has the potential to develop the industry of extracting chitin from shrimp shell. Shrimp production in Indonesia rose from 290,000 MT year^{-1}, and nonedible wastes such as carapace and exoskeleton make up to 50%–60% of this volume. Due to their high availability and chemical composition (27% chitin, 40% protein, and 35% minerals), shrimp residues are used as a raw material for chitin–chitosan. Shrimp shell waste is partially used as animal feed but most of it is discarded, thereby causing serious ecological problems, and has also been used as an ingredient in culture media for extraction of pigments and proteins to formulate animal food, for reducing nematode populations in soils, to produce single-cell protein, and for enzyme immobilization.

Chitin, a linear copolymer of $\beta(1\rightarrow4)$-2-acetamido-2-deoxy-D-glucan, occurs widely in nature as a principal structural polymer in the integument of insects and crustaceans such as shrimp and crab shells and in the cell walls of many fungi. For chitin preparation, harsh chemical treatments are usually required to remove calcium carbonate and protein from raw chitinous material. This raw material has been most abundantly available in crab shells, shrimp shells, cuttle fish, and prawn shells. The raw material has been conventionally treated with a combination of HCl and NaOH (Hackman, 1954).

The traditional methods for commercial preparation of chitin from crustacean shell waste have remained essentially unchanged since first proposed (Axelsson, 1998) and generally involve mechanical grinding, demineralization with 1 N hydrochloric acid, and deproteinization with 3%–5% sodium hydroxide at 90°C–100°C. On a brief analysis of the main costs of process, transportation, and handling, chemical treatment and generation of potentially polluting wastes (which would have to be treated before being discarded) have been identified as the most important factor determining the high price for chitin and chitosan products. Such a situation is particularly important in areas such as food technology, in which an immense potential market will open once the FDA approves the use of chitin-based food additives. In this area, low-value widely accepted biopolymers such as cellulose, alginates, and starches dominate the market and will provide a tough competition for chitin–chitosan derivatives (George et al., 1999). Thus, it is evident that in order to increase chitin use and diminish emission of pollutants, a less expensive, environmentally friendly method for large-scale extraction of chitin is needed.

Recent investigations on the biodegradation of chitin in marine environments have demonstrated that over 90% of the annually produced chitin is degraded within 2 years. Different microorganism populations (with high densities and various hydrolytic potentials) take part in a sequential biodegradation, in which demineralization and deproteinization stages analogous to the industrial process play a decisive role. Additionally, cytological, biochemical, and structural studies on the moulting cycle of crustaceans (growth process) suggest that degradation of the old crustacean's cuticle in vivo is a highly efficient process.

After almost complete protein and chitin degradation, amino acids and amino sugars generated by hydrolysis of protein and chitin are quantitatively reabsorbed and reutilized in the synthesis of the materials for the new cuticle. Calcium carbonate is also solubilized, reabsorbed, and re-deposited in the new cuticle during the same cycle as a consequence of the action of the enzyme carbonic anhydrase and some phosphatides (Muzzarelli, 1990).

Thus, it is evident that the biological mechanisms for efficient demineralization, deproteinization, and hydrolysis of chitin are present in nature and, therefore, to integrate such mechanisms into the technology required for large-scale extraction of chitin can be highly rewarding. Based on these and some other evidences, the authors have proposed a whole continuous process shellfish waste utilization that considers microbial demineralization–deproteinization of shell waste with recovery of protein and pigment as byproducts, both with potential use as fish feed (Rigby, 1934).

21.2 Shrimp shell waste as a raw material chitin–chitosan

Indonesia is an archipelago surrounded by seas and oceans possessing a large potential supply of seafood products, including a variety of fish, shrimp, crab, and squid. Although seafood resources are relatively large, utilization of such resources has not been optimal. However, shrimp production is increasing due to the expansion of shrimp farms. Shrimp farms are typically low-lying impoundments along bays and tidal areas. Opportunities for shrimp products continue to grow with the increasing potential in export markets and the increase in the human population. Shrimp is becoming an alternative source of high-quality animal protein in Indonesia.

In 2005, shrimp-farming area was estimated to be 132,800 ha, with *Penaeus monodon* (*P. monodon*) accounting for 65% of total area and *Penaeus vannamei* (*P. vannamei*) accounting for 35%. The situation is expected to change in 2009. Total shrimp farming area will almost double. *P. monodon* shrimp farming will increase over 60%, but will only account for a little over half of total shrimp-farming area. *P. vannamei* shrimp-farming area will increase 160% and account for 47% of total farm area. Integrated shrimp development is on eastern and central Java, southern and northern Sumatera, and western Kalimantan. The larger shrimp farms tend to be located in the coastal regions of northern and southern Sumatera; the island of Java; western, southern, and eastern Kalimantan; southern, southern-eastern, and central Sulawesi; and western Nusa Tenggara (NTT).

Indonesia is one of the main shrimp-producing and -exporting countries. In 2008, 170,583 tons shrimps was exported and 90% of it was in the form of frozen headless-shelled ones. As a consequence, there has been a lot of shell waste from frozen shrimp industries. Shrimp shell waste consists of 45% of the whole shrimps (Dhewanto and Kresnowati, 2002); therefore in 2007, the amount of shell waste was about 100,188 tons, most of which was only used as

Table 21.1 Characteristics of the Shrimp Shell Waste from Indonesia

Shrimp Shell Waste Content	% w/w
Moisture	28.21
Ash	21.67
Protein	37.84
Mineral	36.03
Chitin	26.13

animal feed. Ash and chitin contents in shrimp shell waste from Indonesia were 21.67% to 26.13%, respectively. This content indicates that they were lower than the minced waste from coldwater areas (George et al., 1999). **Table 21.1** shows the characteristics of shrimp shell waste *P. monodon* obtained from the shrimp fishing industry in Lampung Indonesia (Setyahadi, 2007).

The shrimp shell waste content of *Crangon crangon* from the North Sea is different from *P. monodon* from the Indonesia Sea. The composition of *C. crangon* wastes as such and after the pretreatment for residual meat removal is shown in **Table 21.2**. For comparison of the proteolytic activity of enrichment

Table 21.2 Composition of Different Shrimp Shells

	P. monodon (Indonesia)	C. crangon (Büsum, North Sea)	
	Abdomen	Untreated[a]	Pre-treated[b]
TKN (mg g^{-1} shells[c])	73	76	47
Chitin-N (mg g^{-1} shells)	25	17	32
Chitin (mg g^{-1} shells)	363	240	459
Chitin (%)	36	24	46
Protein-N (mg g^{-1} shells)	48	61	16
Protein[d] (mg g^{-1} shells)	300	380	97
Protein (%)	30	38	10
Ca^{2+} (mg g^{-1} shells)	121	123	178
CaCO$_3$[e] (mg g^{-1} shells)	303	308	445
CaCO$_3$ (%)	30	31	44
Miscellaneous[f] (mg g^{-1} shells)	34	72	0

[a] Raw shrimp shells.
[b] Meat residues pressed off and shells then repeatedly washed.
[c] Dry weight.
[d] Protein-N × 6.25.
[e] Molecular weight of CaCO$_3$ is 100 g mol^{-1}.
[f] Lipids, pigments, etc. (not determined but calculated).

cultures, whole animals of *C. crangon* ("Siebkrabben"—small animals that could not be sold) were cooked and disrupted, and the aqueous extract used as a protein source (Xu et al., 2008).

21.3 Demineralization process

The most common type of fermentation of crustacean shell waste is carried out using lactic acid bacteria for demineralization. Lactic acid bacteria are a group of gram-positive, nonsporing, nonrespiring cocci or rods that produce lactic acid as the major endproduct during the fermentation of carbohydrates. Lactic acid bacteria can survive in low pH conditions and are resistant to inhibition by carbon dioxide. The most common production of organic acids were lactic acid and acetic acid.

Lactic acid fermentation combined with chemical treatments has been studied as an alternative method of chitin recovery that reduces the amount of alkali and acid. The removal of protein and calcium from shells is due to the enzymatic activity and mineral solubilization by organic acid produced by bacteria growth (Shirai et al., 1997).

Lactic acid fermentation combined with chemical treatment has been studied as an alternative method of chitin recovery that reduces the amount of alkali and acid needed. The removal of protein and calcium from shells is by enzymatic activity and mineral solubilization by organic acid produced in bacteria growth (Hsu and Wu, 2002; Luis et al., 2003; Shirai et al., 1997; Zakaria et al., 1998). Bacteria-producing lactic acid can be used for removing calcium from shrimp shell screened (Setyahadi, 2007) but the efficiency was low.

Legarreta et al. (1996) reported that adding *Lactobacillus* spp. into shrimp waste with carbohydrate addition results in production of lactic acid and various proteases. The study also showed that lactic acid caused low pH of the medium-suppressed spoilage bacteria and dissolved calcium carbonate presence in the shrimp shell creating precipitation of calcium lactate that was easily removed by washing. Another study (Fagbenro, 1996) stated that proteolytic enzyme produced by lactic acid bacteria hydrolyzed and solubilized the protein resulted in clean chitin and highly contained soluble peptide and amino acid liquor. The quantity of *Lactobacillus* inoculum, glucose, and type of acid added individually or in combination affected the efficiency of the lactic acid fermentation that included deproteination, demineralization, and spoilage of the shrimp waste (Rao et al., 2000).

The efficiency of fermentation using lactic acid bacteria depends on factors such as the quantity of inoculum, glucose, initial pH and pH during fermentation, the amount and type of acid used, and fermentation time. Shrimp shell was fermented with a fixed amount of inoculum (10% v/w)

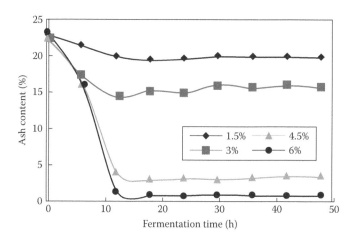

Figure 21.1 *Effect of glucose concentration (w/v) in demineralization process using* L. acidophilus *FNCC-116 with initial pH 7.0 at 37°C.*

and variable concentration of glucose (0%, 1.5%, 3.0%, 4.5%, and 6.0% w/v). The initial pH was adjusted to 7.0.

Rao et al., 2000 reported that chitin extraction process in a drum reactor using microbial *Lactobacillus plantarum* needed as much as 5% glucose, and Ling et al. (2006) for continuous production of lactic acid using *Lactobacillus rhamnosus* used glucose 4.0%–5.0%, and demineralization process using *Lactobacillus acidophilus* FNCC-116 needed 6.0% glucose to be optimal with pH 3.5 (**Figure 21.1**).

Figure 21.2 shows that microbial grown from the demineralization of shrimp shell processing containing 6% (w/v) glucose, exponential phase start from 0 to 12 h after cell culture into fermentation media, stationary phase at 16–30 h, and then phases of death from 35 to 48 h. In this phase, the supply of nutrients is necessary for growth and the addition of the cell, so that the viability and productivity remained high, in order to be maintained until the end of the process.

Glucose consumption increased for 24 h and glucose only for the remaining 0.3% of the initial conditions until the end of the process. Lactic acid production of *L. acidophilus* FNCC 116 increased and after 24 h of lactic acid production was relatively constant. Based on the analysis of patterns of microbial growth phase, glucose consumption, and lactic acid production, the demineralization process occurred between 16 and 32 h.

The presence of the lactic acid on mineral content in shrimp shell such as some previous research reported that mineral shrimp shell can react with lactic acid produced by bacteria, forming the mineral as byproducts lactate (Aye and Steven, 2004; Jung et al., 2005, 2007; Rao and Steven, 2006).

The lactic acid content during the process of demineralization is shown in **Figure 21.2**. The content of free lactic acid in fermentation media at the

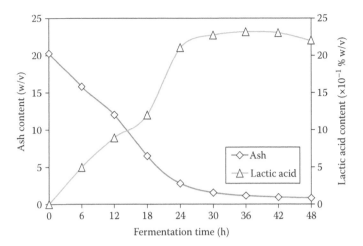

Figure 21.2 *Time course of lactic acid production and ash content of shrimp shell waste by using* L. acidophilus *FNCC-116 in demineralization process.*

beginning of the process was about 0.7%–1.5%, after 12 h of continuous process it rose to 1.6%–2.1% or increased by more than 50%, but for 24 h to the end of the process of decline in lactic acid content it was around 1.1%–1.4% or decreased by an average of nearly 33%. Ash content was 0.88% after 48 h.

21.4 Deproteination process

Deproteinization of crustacean shell wastes was reported using protease-producing microorganisms such as *Pseudomonas aeruginosa* K-187 (Wang and Chio, 1998), *Pseudomonas maltophilia* LC-102 (Shimahara et al., 1984), *Candida parapsilosis* (Chen et al., 2001), and *Bacillus subtilis* (Yang et al., 2000). Representatives of *Bacillus licheniformis* are major bacterial work-horses in the industry (Schallmey et al., 2004); their genome sequence was published already a couple of years ago (Rey et al., 2004; Veith et al., 2004), clarifying not only the close relationship to *B. subtilis* but also facilitating efficient genetic strain improvement. Indeed, a useful arsenal of genetic tools is available for improving the productivity of secreted enzymes and/or other desired features (Nahrstedt, 2005; Waldeck et al., 2007a,b). *B. licheniformis* DSM13 possesses an operon with two chitinase genes, *chiB* and *chiA*, which possibly hampers the recovery of high-molecular-mass chitin due to degradation. *B. licheniformis* strain F11 has a frameshift mutation in the chitinase A gene (*chiA*); it was efficiently used in shrimp shell protein hydrolysis, at the same time leaving the chitin largely unaffected (Waschkau et al., 2008). However, the other chitinase-encoding gene (*chiB*) localized immediately upstream of *chiA* as part of the operon may still have a negative impact on

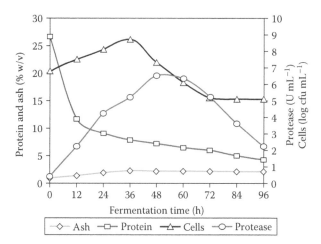

Figure 21.3 *Deproteination process of demineralized shrimp shells using* B. licheniformis *F11.1 at 55°C on 10L fermenter.*

chitin recovery. Thus, a clean *chiBA* operon deletion mutant of F11 was con-
structed (Hoffmann et al., 2010).

An optimum process of bio-demineralization has been established in laboratory scale. Bio-deproteinization using *B. licheniformis* F11 did not give a good
result. The isolate was replaced with genetically modified *B. licheniformis* F11
that designated as *B. licheniformis* F11.1 has given good results. Optimization
of bioconversion process in 10 L fermenter was done.

Deproteination process using *B. licheniformis* F11.1 was found at the most
after 24 h of fermentation at 55°C (1.3×10^8 cfu mL^{-1}) and protease activity was
4.23 U mL^{-1}; however, the maximum proteolytic enzyme production reached
after 48 h of fermentation (protease activity was 6.5 U mL^{-1}). During the deproteination process, the protein content of shrimp shells decreased from 27%
to 2.5%, whereas the ash content increased from 0.67% to 3% (**Figure 21.3**).

The deproteination stage plays a central role in the extraction of chitin for several reasons such as quality, and therefore the market price of chitin strongly
depends on the protein content of the final product that has to be below 5%.

21.5 Microbial process for chitin extraction

Commercial chitin and their derivatives originating from shrimp shells have
been produced in some countries such as the United States and Japan. The
traditional method of chitin production from crustacean shells involved hydrochloric acid treatment for removing calcium carbonate followed by alkali treatment for deproteination and bleaching stage with chemical reagent to get a
colorless product. Since such processes result in high pollution, it is necessary
to proceed with costly waste disposal treatment.

Research is ongoing in quest of alternative methods that give better quality products and environmentally friendly processes. Two types of bio process of chitin production have been suggested. The first process involved two stages, which were demineralization of crustacean shell with hydrochloric acid or acetic acid followed by deproteination step using microbial, plant, or animal protease (Bustos and Healy, 1994; Gagné and Simpson, 1993; Healy et al., 1994; Oh et al., 2000; Yang et al., 2000), and the other suggested process is demineralization and deproteination being done in one-step fermentation using lactic acid bacteria with or without pH control using organic acid such as citric acid, acetic acid, or lactic acid (Bautista et al., 2001; Rao et al., 2000; Shirai et al., 2001; Zakaria et al., 1998).

The experiment on process design of microbiological chitin extraction in laboratorium scale was respectively (Wahyuntari et al., 2011). The aim of this study was to see whether the process design scales up to 10 L. The process was started with demineralization of shrimp shells followed by deproteination. *L. acidophilus* FNCC 116 was used for demineralization and *B. licheniformis* F11.1, a proteolytic, chitinase-deficient bacterium, was used for the deproteination process (**Figure 21.4**).

Shrimp shells matrix is formed mainly of chitin and protein hardened by mineral salts, especially calcium carbonate (Beaney et al., 2005). *L. acidophilus* produced lactic acid by breaking down glucose-creating lactic acid, thereby lowering the pH of the fermentation broth and suppressing spoilage by microbial growth. The lactic acid reacted with calcium carbonate in the chitin fraction to form calcium lactate, which is soluble and could be removed by washing. This was followed by hydrolyzing protein in the chitin by fermentation of the proteolytic bacterium *B. licheniformis* F11.1. The proteolytic enzyme production increased along with the increase of cell concentration and reached the maximum activity after 48 h fermentation; however, the protein content was reduced drastically for the first 12 h of fermentation and decreased slowly for the rest of the fermentation time (Setyahadi et al., 2011). The removal of protein content in the shrimp shells in the second process

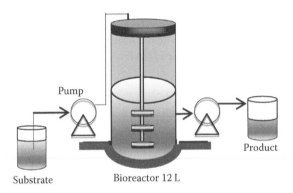

Figure 21.4 Schematic fermenter for chitin extraction process.

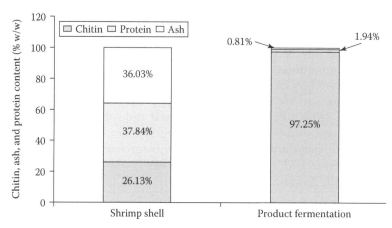

Figure 21.5 *Chitin extraction process by demineralization followed by deproteination.*

was higher since the calcium carbonate had been removed the proteolytic enzyme could contact more easily with the protein in the chitin fraction of the shells. The chitin extraction process by demineralization followed by deproteination was done for 96 h or 4 days. The protein content of the shrimp shells was reduced from 37.84% down to 1.94%. The ash content was reduced from 36.03% down to 0.81% (**Figure 21.5**).

The demineralization using *L. acidophilus* FNCC116 was done in a 10 L fermenter jar at 32°C with agitation 50 rpm for 48 h. The deproteination process was done in the same size of fermenter jar at 55°C, 250 rpm and 2.5 vvm aeration for 72 h using *B. licheniformis* F11.1 as a proteolytic bacterium. The results showed that the best condition for demineralization was 6.5% glucose. The chitin, ash, and protein contents of the chitin product were 95.4%, 3%, and 2.5%, respectively (Setyahadi, 2011).

For viscosity analysis of chitin products, samples were deacetylated under standard conditions with chemical treatment, as chitosan is easier to analyze. Microbiology combined with chemical process gives high-density chitosan. The viscosity values of biochitosan are higher than those that have been reported previously (Brugnerotto et al., 2001; Rinaudo et al., 1993; Yanagisawa et al., 2006), and also the analysis of chitosan produced from a different chitin source (*Gammarus lacustris*) led to lower values (Grigoryeva and Mezenova, 2007).

21.6 Application of high-density chitosan

21.6.1 Antimicrobial in dental field

Chitosan has antimicrobial activity against various bacteria and fungi (Liman et al., 2011). Antimicrobial chitosan against bacteria and fungi has the same mechanism, namely through the interaction between the positively charged group NH_3^+ on chitosan and the negative charge on cell membranes of microbes, resulting in electrostatic interactions that cause changes in permeability of the wall membrane and changes the balance of osmotic internally that can inhibit microbial growth, as it also causes the hydrolysis of peptidoglycan in the wall of microbes resulting in loss of intracellular electrolytes, proteins, nucleic acids, and glycated at mikroba (Tikhonov et al., 2006). Chitosan also has the ability to interact with compounds on the surface of bacterial cells and then absorbed to form a layer that can inhibit cell transport channels that resulted in cells deficient for substance developed resulting in cell death (Goy et al., 2009). Inhibitory power of chitosan against microbial leaching depends on the concentration of chitosan (Liu et al., 2006). In general, chitosan with a concentration of 0.1% can reduce the fungus growth (El Hadrami et al., 2010).

In the oral cavity, denture from nylon thermoplastic material can absorb specific molecules in saliva and form a thin organic layer called pellicle. Pellicle contains proteins that can bind to the adhesion of microbes so that the microbes can be attached to the denture surface and form plaques. Cleaning is done simply by flushing with water since plaque cannot be cleaned completely. Plaque cause denture stomatitis known as stomatitis is, one of them caused by *Candida albicans* (Abelson, 1981). *C. albicans* is fungus-shaped oval or round and is included in microgram-positive. The majority of *Candida* spp. have similar characteristic colonies such as white, shiny, and convex (Quinn et al., 2002). *C. albicans* is a normal microflora in the mouth (Arora et al., 2004). In a healthy mouth, the species is present in low concentrations so as not to cause disease, but when immunity decreases *C. albicans* can multiply quickly causing tissue damage (Jawetz, 1986).

In the denture using thermoplastic nylon material incubated in artificial saliva, *C. albicans* can be found attached to the surface. It is appropriate that the nature of thermoplastic nylon is hygroscopic, so that it can absorb specific molecules from the saliva and form a thin layer adhesion of *C. albicans* is able to bind (Billmeyer, 1984). High-density chitosan of 0.05% can inhibit the growth of *C. albicans*.

21.6.2 Fat absorption

Chitosan has many functions such as antitumor activity (Jeon and Kim, 2002; Qin et al., 2002; Suzuki et al., 1986; Tokoro et al., 1988), cholesterol-lowering effect (Gallaher et al., 2000; Ormrod et al., 1998), immuno-enhancing effect (Peluso et al., 1994), antidiabetic effect (Hayashi and Ito, 2002), wound-healing effect (Porporatto et al., 2003), antifungal activity and antimicrobial activity (Qin et al., 2006). Although numerous literatures are available on the aforementioned biological activities, the relationships of these activities with molecular weight and water solubility of chitosan deserve to be investigated. It can be easily hypothesized that the biological

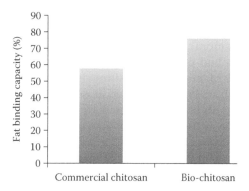

Figure 21.6 *Comparison between commercial chitosan and biochitosan on fat binding capacity.*

properties of chitosan may be closely related to the molecular weight and water solubility. As a preliminary study, in vivo absorption phenomena of different chitosan were investigated. Chitosan with high molecular weight was rarely absorbed in that it became gel and precipitated when the pH of the intestine was neutral or alkaline. The chitosan gel imbedded fat is intestinal so that fat could not be absorbed and greatly increases fecal fat excretion (Gades and Stern, 2003). High dose of chitosan with high-molecular weights can contribute to weight loss, but low dose of chitosan does not contribute to weight loss.

Chitosans with different physicochemical properties can be prepared under different reaction conditions. The degree of deacetylation and the molecular weight of a chitosan are two important characteristics that greatly affect its chemical and physiological properties. Our recent work studied the effects of the bio-chitosan (high-density chitosan) samples on their fat-binding, cholesterol-binding capacities in vitro. The results indicated that the fat-binding capacity of biochitosan (76.5%) is more effective than commercial chitosan (57.9%) producing chemical process (**Figure 21.6**).

21.7 Conclusions

Traditional methods for the commercial preparation of chitin from crustacean shell (exoskeleton) involve alternate hydrochloric acid and alkali treatment stages to remove calcium carbonate and proteins, respectively, followed by a bleaching stage with chemical reagents to obtain colorless product. Processing is expensive because wherever environmental controls are enforced, disposal cost must be added. The development of a process for cost-effective recovery of useful products would lead to a great commercial exploitation of chitin. For such a process to be successful, the cost of the reagents used in the process must be low.

The biotechnology process for extraction chitin from shrimp shell waste involves two-step process: demineralization and deproteination. Demineralization by using *L. acidophilus* FNCC 116 has a pH of 3 and can be used to remove calcium in shrimp shell waste. Next process is deproteination by using *B. licheniformis* F11.1 for removing protein content in shrimp shell. Deproteination is an important step in chitin extraction. The two-step chitin extraction processes are demineralization process using lactic acid bacterium followed by deproteination process using proteolytic-producing bacterium removal ash and protein. Chitin products were deacetylated under standard conditions with chemical treatment. Microbiology combined with chemical process gives high-density chitosan, called biochitosan. The possibility of gaining products of high-density chitosan offers new opportunities for special application products as they are not yet available in the market.

Chitosan has been noted for its application as a film-forming agent in cosmetics, a dye-binder for textiles, a strengthening additive in paper, and a hypolipidic material in diets. It has been used extensively as a biomaterial because of its immunostimulatory activities, anticoagulant properties, antibacterial and antifungal action, and for its action as a promoter of wound healing in the field of surgery. In the denture using thermoplastic nylon material incubated in artificial saliva, *C. albicans* can be found attached to the surface. It is appropriate that the nature of thermoplastic nylon is hygroscopic so that it can absorb specific molecules from the saliva and forms a thin layer adhesion of *C. albicans* that is able to bind. High-density biochitosan of 0.05% can inhibit the growth of *C. albicans*. The other application of biochitosan is cholesterol-binding capacities. As compared with chemical chitosan, biochitosan from shrimp shell waste is effective for fat-binding capacities.

Acknowledgments

This work was financially supported by an incentive research grant from the Ministry of Research and Technology (KNRT) from 2008 to 2011. Thanks to DAAD for granting the DAAD scholarship of the Special programme for Biosciences 2003, and also to Prof. Dr. Endang Sutriswati Rahayu, Faculty of Agricultural Technology, Gadjah Mada University, for giving *L. acidophilus* FNCC 116T. Thanks also to Prof. Bernward Bisping, Prof. Friedhelm Meinhardt, and Prof. Josef Winter for collaborating joint research between Germany and Indonesia in "Indonesia-Germany Biotechnology" Project.

References

Abelson, D.G. 1981. Denture plaque and denture cleansers. *J. Prosthet. Dent.*, 45: 376–379.
Arora, D.K., Khachatourians, G.G. 2004. *Applied Mycology and Biotechnology. Fungal Genomics*. Elsevier B.V., Amsterdam, the Netherlands, pp. 99–100.
Axelsson, L. 1998. Lactic acid bacteria: Classification and physiology. In *Lactic Acid Bacteria: Microbiology and Functional Aspects* (eds. Salminen, S. and von Wright, A.), Marcel Dekker Inc., New York.

Aye, K.N., Stevens, W.F. 2004. Improved chitin production by pretreatment of shrimp shells. *J. Chem. Technol. Biotechnol.* 79: 421–425.

Bautista, J., Jover, M., Gutierrez, J.F., Corpas, R., Cremades, O., Fontiveros, E., Iglesias, F., Vega, J. 2001. Preparation of crayfish chitin by in situ lactic acid production. *Process Biochem.* 37: 229–234.

Beaney, P., Lizardi-Mendoza, J., Healy, M. 2005. Comparison of chitins produced by chemical and bioprocessing methods. *J. Chem. Technol. Biotechnol.* 80: 145–150. doi: 10.1002/jctb.1164.

Billmeyer, F.W. 1984. *Textbook of Polymer Science*, John Wiley & Sons Inc., New York, pp. 407–411.

Brugnerotto, J., Desbrières, J., Heux, L., Mazeau, K., Rinaudo, M. 2001. Overview on structural characterization of chitosan molecules in relation with their behavior in solution. *Macromol. Symp.* 168: 1–20.

Bustos, R.O., Healy, M.G. 1994. Microbial/enzymatic deproteination of Prawn ShellWaste. In *Proceedings of the IChemE Research Event*, January 4–6, 1994, London, U.K., pp. 126–128.

Chen, H.C., Phang, K.A., Wu, S.D., Mau, W.J. 2001. Isolation of chitin from shrimp shells deproteinized by *Candida parapsilosis* CCRC20515. *Food Sci. Agric. Chem.* 3: 114–120.

Dhewanto, M., Kresnowati, M.T.A.P. 2002. Chitosan industry: An alternative for maritime industry in empowerment Indonesia. In *Proceedings of Indonesian Student Scientific Meeting*, October 4–6, 2002, Berlin, Germany, ISTEC-Europe, pp. 327–333.

El Hadrami, A., Adam, L.R., El Hadrami, I., Fouad, D. 2010. Chitosan in plant protection. *Mar. Drugs* 4: 968–987.

Fagbenro, O.A. 1996. Preparation, properties and preservation of lactic acid fermented shrimp head. *Food Res. Int.* 29: 595–599.

Gades, D.M., Stern, J.S. 2003. Chitosan supplementation and fecal fat excretion in men. *Obes. Res.* 11: 683–688.

Gagné, N., Simpson, B.K. 1993. Use of proteolytic enzyme to facilitate the recovery of chitin from shrimp waste. *Food Biotechnol.* 7(3): 253–263.

Gallaher, C.M., Munion, J., Hesslink, J.R., Wise, J., Gallaher, D.D. 2000. Cholesterol reduction by glucomannan and chitosan is mediated by changes in cholesterol absorption and bile acid and fat excretion in rats. *J. Nutr.* 130: 2753–2759.

George, M.H., Carole, L.R., Zainoha, Z. 1999. Fermentation of prawn waste by lactic acid bacteria. In *Proceedings of the Third International Conference of the European Chitin Society*, August 31–September 3, Potsdam, Germany, pp. 633–638.

Goy, R.C., Britto, D., Assis, O.B.G. 2009. A review of the antimicrobial activity of chitosan. *Polimeros: Ciencia e Tecnologia.* 3: 1–7.

Grigoryeva, E., Mezenova, O. 2007. Das Verfahren zur Chitosan-Gewinnung aus dem baltischen Krebs *Gammarus lacustris*. *Chem. Ing. Tech.* 79: 1189–1194.

Hayashi, K., Ito, M. 2002. Antidiabetic action of low molecular weight chitosan in genetically obese diabetic KK-Ay mice. *Biol. Pharm. Bull.* 25: 188–192.

Healy, M., Romo, R., Bustos, R. 1994. Bioconversion of marine crustacean shell waste. *Resour. Conserv. Recycl.* 11: 139–147.

Hoffmann, K., Daum, G., Köster, M., Kulicke, W.M., Meyer-Rammes, H., Bisping, B., Meinhardt, F. 2010. Genetic improvement of *Bacillus licheniformis* strains for efficient deproteinization of shrimp shells and production of high-molecular-mass chitin and chitosan. *Appl. Environ. Microbiol.* 76(24): 8211–8221.

Hsu, Y.L., Wu, W.T. 2002. A novel approach for scaling—Up a fermentation system. *Biochem. Eng. J.* 11: 123–130.

Jawetz, E.M. 1986. *Review of Medical Microbiology*, Lange Medical Publication, San Francisco, CA, pp. 143–148, 297–299.

Jeon, Y.J., Kim, S.K. 2002. Antitumor activity of chitosan oligosaccharides produced in ultrafiltration membrane reactor system. *J. Microbiol. Biotechnol.* 12: 503–507.

Jung, W.J., Jo, G.H., Kuk, J.H., Kim, Y.J., Oh, K.T., Park, R.D. 2007. Production of chitin from red crab shell waste by successive fermentation with *Lactobacillus paracasei* KCTC-3074 and *Serratia marcescens* FS-3. *J. Carbohydr. Polym.* 68(4): 746–750.

Jung, W.J., Kuk, J.H., Kim, K.Y., Park, R.D. 2005. Demineralization of red crab shell waste by lactic acid fermentation. *Appl. Microbiol. Biotechnol.* 67: 851–854.

Legarreta, G.I., Zakaria, Z., Hall, G.M. 1996. Lactic acid fermentation of prawn waste: Comparison of commercial and isolated starter culture. In *Advanced in Chitin Science* (eds. Domard, A., Jeuniaux, C., Mozzarelli, R.A.A., and Roberts, G.A.F.), Jacque s Andre, Lyon, France, Vol. 1, pp. 399–406.

Liman, Z., Selmi, S., Sadok, S., Abed, A. 2011. Extraction and characterization of chitin and chitosan from crustacean by products: Biological and physicochemical properties. *J. Biotechnol.* 4: 640–647.

Ling, L.S., Mohamad, R., Abdul Rahim, R., Wan, H.Y., Ariff, A.B. 2006. Improved production of live cells of *Lactobacillus rhamnosus* by continuous cultivation using glucose-yeast extract medium. *J. Microbiol.* 44(4): 439–446.

Liu, N., Chen, X., Park, H., Liu, C., Meng, X., Yu, L. 2006. Effect of MW and concentration of chitosan on antibacterial activity of *Escherichia coli*. *Carbohydr. Polym.* 64: 60–65.

Luis, S.J.T., Moldes, A.B., Alonso, J.L., Vazquez, M. 2003. Optimization of lactic acid production by *Lactobacillus delbrueckii* through response surface methodology. *J. Food Sci.* 68: 1454–1458.

Muzzarelli, R. 1990. Chitin and chitosan: Unique cationic polysaccharides. *Proceedings of the Symposium "Towards a Carbohydrate Based Chemistry,"* October 23–26, 1989, ECSC-EEC-EAEC, Amies, France, pp. 199–231.

Nahrstedt, H., Waldeck, J., Göne, M., Eichstädt, R., Feesche, J., Meinhardt, F. 2005. Strain development in *Bacillus licheniformis*: Construction of biologically contained mutants deficient in sporulation and DNA repair. *J. Biotechnol.* 119: 245–254.

Oh, Y.S., Shih, I.L., Tzeng, Y.M., Wang, S.L. 2000. Protease produced by K-187 and its application in the deproteinization of shrimp and crab shell waste. *Enzyme Microb. Technol.* 27(1–2): 3–10. doi:10.1016/S0141-0229(99) 00172-6.

Ormrod, D.J., Holmes, C.C., Miller, T.E. 1998. Dietary chitosan inhibits hypercholesterolaemia and atherogenesis in the apolipoprotein E-deficient mouse model of atherosclerosis. *Atherosclerosis* 138: 329–334.

Peluso, G., Petillo, O., Tanieri, M., Santin, M., Ambrosic, L., Calabro, D. 1994. Chitosan-mediated stimulation of macrophage function. *Biomaterials* 15: 1215–1220.

Porporatto, C., Bianco, I.D., Riera, C.M., Correa, S.G. 2003. Chitosan induces different l-arginine metabolic pathways in resting and inflammatory macrophages. *Biochem. Biophys. Res. Commun.* 304: 266–272.

Qin, C.Q., Du, Y.M., Xiao, L., Li, Z. 2002. Enzymic preparation of water-soluble chitosan and their antitumor activity. *Int. J. Biol. Macromol.* 31: 111–117.

Qin, C.Q., Li, H.R., Xiao, Q., Liu, Y., Zhu, J.C., Du, Y.M. 2006. Water-solubility of chitosan and its antimicrobial activity. *Carbohydr. Polym.* 63: 367–374.

Quinn, P.J., Markey, B.K., Donnnelly, W.J.C., Leonard, F.C. 2002. *Veterinary Microbiology and Microbial Disease*, Blackwell Science, Oxford, U.K., pp. 233–234.

Rao, M.S., Munoz, J., Stevens, W.F. 2000. Critical factors in chitin production by fermentation of shrimp biowaste. *Appl. Microbiol. Biotechnol.* 54(6): 808–813.

Rao, M.S., Stevens, W.F. 2006. Fermentation of shrimp biowaste under different salt concentration with amylolytic and non-amylolytic *Lactobacillus* strains for chitin production. *J. Food Technol. Biotechnol.* 44: 83–87.

Rey, M.W., Ramaiya, P., Nelson, B. A., Brody-Karpin, S. D., Zaretsky, E. J., Tang, M., Lopez de Leon, A. et al. 2004. Complete genome sequence of the industrial bacterium *Bacillus licheniformis* and comparisons with closely related *Bacillus* species. *Genome Biol.* 5: R77.

Rigby, G.W. 1934. Substantially undergraded deacetylated chitin and process for producing the same. Patent U.S. 2,040,879.

Rinaudo, M., Milas, M., Dung, P.L. 1993. Characterization of chitosan: Influence of ionic strength and degree of acetylation on chain expansion. *Int. J. Biol. Macromol.* 15: 281–285.

Schallmey, M., Singh, A., Ward, O.P. 2004. Developments in the use of *Bacillus* species for industrial production. *Can. J. Microbiol.* 50: 1–17.

Setyahadi, S. 2007. Screening of lactic acid bacteria for the purpose of chitin recovery processing. *Microbiol. Indonesia* 1(1): 48–50.

Setyahadi, S. 2011. Microbial processes for producing chitin from shrimp shell wastes. In *The 9th Asia-Pacific Chitin-Chitosan Symposium*, August 3–6, 2011, Nha Trang, Vietnam.

Setyahadi, S., Waltam, D.R, Hermansyah, H. 2011. Bioconversion of extraction chitin from *Penaeus vannamei* shell waste. In *Proceedings of the 12th International Conference on QiR (Quality in Research)*, July 4–7, 2011, Bali, Indonesia, ISSN 114-1284.

Shimahara, K., Yasuyuki, T., Kazuhiro, O., Kazunori, K., Osamu, O. 1984. Chemical composition and some properties of crustacean chitin prepared by use of proteolytic activity of *Pseudomonas maltophilia* LC102. In *Chitin, Chitosan and Related Enzymes* (ed. Zikakis, J.P.), Academic Press, Orlando, FL, pp. 239–255.

Shirai, K., Guerrero, I., Huerta, S., Saucedo, G., Rodriguez, G., Hall, G. 1997. Aspects in protein breakdown during the lactic acid fermentation. *Adv. Chitin Sci.* 2, 56–63.

Shirai, K., Guerrero, I., Huerta, S., Saucedo, G., Castillo, A., Gonzales, R.O., Hall, G.M. 2001. Effect of initial glucose concentration and inoculation level of lactic acid bacteria in shrimp waste ensilation. *Enzyme Microb Technol.* 28: 446–452.

Suzuki, K., Mikami, T., Okawa, Y., Tokoro, A., Suzuki, S., Suzuki, M. 1986. Antitumor effect of hexa-N-acetylchitohexaose and chitohexaose. *Carbohydr. Res.* 151: 403–408.

Tikhonov, V.E., Stepnova, E.A., Babak, V.G., Yamskov, I.A., Palma-Guerrero, J., Jansson, H., Lopez-Llorca, L.V. et al. 2006. Bacterial and antifungal activities of a low molecular weight chitosan and its N-2(3)-(Dodec-2-Enyl) succinoyl/-derivatives. *Carbohydr. Polym.* 64: 66–72.

Tokoro, A., Tatewaki, N., Suzuki, K., Mikami, T., Suzuki, S., Suzuki, M. 1988. Growth-inhibitory effect of hexa-N-acetylchitohexaose and chitohexaose against meth-A solid tumor. *Chem. Pharm. Bull.* 36: 784–790.

Veith, B., Herzberg, C., Steckel, S., Feesche, J., Maurer, K.H., Ehrenreich, P., Bäumer, S. et al. 2004. The complete genome sequence of *Bacillus licheniformis* DSM13, an organism with great industrial potential. *J. Mol. Microbiol. Biotechnol.* 7: 204–211.

Wahyuntari, B., Junianto, and Siswa, S. 2011. Process design of microbiological chitin extraction. *Microbiol. Indonesia* 5(1): 39–45.

Waldeck, J., Daum, G., Bisping, B., Meinhardt, F. 2006. Isolation and molecular characterization of chitinase-deficient *Bacillus licheniformis* strains capable of deproteinization of shrimp shell waste to obtain highly viscous chitin. *Appl. Environ. Microbiol.* 72: 7879–7885.

Waldeck, J., Meyer-Rammes, H., Nahrstedt, H., Eichstädt, R., Wieland, S., Meinhardt, F. 2007a. Targeted deletion of the *uvrBA* operon and biological containment in the industrially important *Bacillus licheniformis. Appl. Microbiol. Biotechnol.* 73: 1340–1347.

Waldeck, J., Meyer-Rammes, H., Wieland, S., Feesche, J., Maurer, K.H., Meinhardt, F. 2007b. Targeted deletion of genes encoding extracellular enzymes in *Bacillus licheniformis* and the impact on the secretion capability. *J. Biotechnol.* 130: 124–132.

Wang, S.L., Chio, S.H. 1998. Deproteinization of shrimp and crab shell with the protease of *Pseudomonas aeruginosa* K-187. *Enzyme Microb. Technol.* 22: 629–633.

Waschkau, B., Waldeck, J., Wieland, S., Eichstadt, R., Meinhardt, F. 2008. Generation of readily transformable *Bacillus licheniformis* mutants. *Appl. Microbiol. Biotechnol.* 78: 181–188.

Xu, Y., Gallert, C., Winter, J. 2008. Chitin purification from shrimp wastes by microbial deproteination and decalcification. *Appl. Microbiol. Biotechnol.* 79: 687–697.

Yanagisawa, M., Kato, Y., Yoshida, Y., Isogai, A. 2006. SEC-MALS study on aggregates of chitosan molecules in aqueous solvents: Influence of residual N-acetyl groups. *Carbohydr. Polym.* 66: 192–198.

Yang, J.-K., Shih, I.-L., Tzeng, Y.-M., Wang, S.-L. 2000. Production and purification of protease from a *Bacillus subtilis* that can deproteinize crustacean wastes. *Enzyme Microb. Technol.* 26: 406–413.

Zakaria, Z., Hall, G.M., Shama, G. 1998. Lactic acid fermentation scampi waste in a rotating horizontal bioreactor for chitin recovery. *Process Biochem.* 33: 1–6.

22

Antioxidant Effects of Marine Food-Derived Functional Ingredients

Se-Kwon Kim, Dai-Hung Ngo, and Thanh-Sang Vo

Contents

22.1 Introduction

Humans are impacted by many free radicals both from inside the body and surrounding environment, particularly reactive oxygen species (ROS) generated in living organisms during metabolism. It is produced in the forms of H_2O_2, superoxide radical, hydroxyl radical, and nitric oxide. In addition, oxidative stress may cause inadvertent enzyme activation and oxidative damage to cellular systems. Free radicals attack macromolecules such as DNA, proteins, and lipids, leading to many health disorders including hypertensive, cardiovascular, inflammatory, aging, diabetes mellitus, neurodegenerative and cancer diseases. Antioxidants may have a positive effect on human health as they can protect human body against deterioration by free radicals (Butterfield et al., 2002).

Oxidation in foods affects lipids, proteins, and carbohydrates. However, lipid oxidation is the main cause of deterioration of food quality, leading to rancidity and shortening of shelf-life. Oxidation of proteins in foods is influenced by lipid oxidation, where lipid oxidation products react with proteins causing their oxidation. Carbohydrates are also susceptible to oxidation, but they are less sensitive than lipids and proteins. To retard the oxidation and peroxidation processes in food and pharmaceutical industries, many synthetic commercial antioxidants such as butylated hydroxyanisole (BHA), butylated hydroxytoluene (BHT), tert-butylhydroquinone (TBHQ), and propyl gallate (PG) have been used. However, the use of these synthetic antioxidants must be under strict regulation due to potential health hazards (Blunt et al., 2006; Park et al., 2001). Therefore, the use of natural antioxidants available in food and other biological substances has attracted significant interest due to their presumed safety, nutritional, and therapeutic values.

Marine organisms are believed to be a potential source to provide not only novel biologically active substances for the development of pharmaceuticals but also essential compounds for human nutrition. Among them, marine algae represent one of the richest sources of natural antioxidants. It is believed that marine algae are protected against oxidative deterioration by certain antioxidant systems. In addition, marine food processing by-products including substandard muscles, viscera, skins, trimmings, and shellfish can be easily utilized to produce nutraceuticals and functional food ingredients with antioxidant activity. Most of the study on marine-derived antioxidants has focused on potency of crude extracts with effective antioxidant compounds remaining unisolated and unidentified. Only few researches have been carried out aiming at purification and characterization of antioxidants from marine sources (Shahidi, 2007). This chapter focuses on potential application of marine food-derived novel antioxidants such as bioactive peptides, chitooligosaccharide derivatives, sulfated polysaccharides, phlorotannins, and carotenoids in the food industry.

22.2 Development of antioxidant ingredients derived from marine food

There is a great potential in marine bioprocess industry to convert and utilize most of marine food products and marine food by-products as valuable functional ingredients. Apparently, there has been an increasing interest in utilization of marine products, and novel bioprocessing technologies are developing for isolation of some bioactive substances with antioxidative property from marine food products to be used as functional foods and nutraceuticals. Development of these functional ingredients involves certain bio-transformation processes through enzyme-mediated hydrolysis in batch reactors. Membrane bioreactor technology equipped with ultrafiltration membranes is recently emerging for the bioprocessing and development of functional ingredients and is considered as a potential method to utilize marine food products efficiently (Nagai and Suzuki, 2000).

This system has the main advantage that the molecular weight distribution of the desired functional ingredient can be controlled by the adoption of an appropriate ultrafiltration membrane (Jeon et al., 1999). Enzymatic hydrolysis of marine food products allows preparation of functional ingredients such as bioactive peptides and chitooligosaccharides. The physicochemical conditions of the reaction media, such as temperature and pH of the reactant solution, must then be adjusted in order to optimize the activity of the enzyme used. Proteolytic enzymes from microbes, plants, and animals can be used for the hydrolysis process of marine food products to develop bioactive peptides and chitooligosaccharide derivatives. Moreover, one of the most important factors in producing bioactive functional ingredients with desired functional properties is the molecular weight of the bioactive compound. Therefore, for the efficient recovery and to obtain bioactive functional ingredients with both a desired molecular size and functional property, a suitable method is the use of an ultrafiltration membrane system. In order to obtain functionally active peptides, it is a suitable method to use a three-enzyme system for sequential enzymatic digestion. Moreover, it is possible to obtain serial enzymatic digestions in a system using a multi-step recycling membrane reactor combined with an ultrafiltration membrane system to bioprocessing and development of marine food-derived bioactive peptides and chitooligosaccharide derivatives (Byun and Kim, 2001). This membrane bioreactor technology equipped with ultrafiltration membranes is recently emerging for the development of functional ingredients and is considered as a potential method to utilize marine food products as value-added nutraceuticals with beneficial health effects.

22.3 Antioxidants derived from marine food and their antioxidative property

22.3.1 Chitooligosaccharide derivatives

Chitosan (**Figure 22.1**) is a natural nontoxic biopolymer produced by the deacetylation of chitin, a major component of the shells of crustaceans such as shrimp, crab, and craw fish. Chitooligosaccharides (COS), partially hydrolyzed products of chitosan, are of great interest in pharmaceutical and medicinal applications due to their noncytotoxic and high-water-soluble properties.

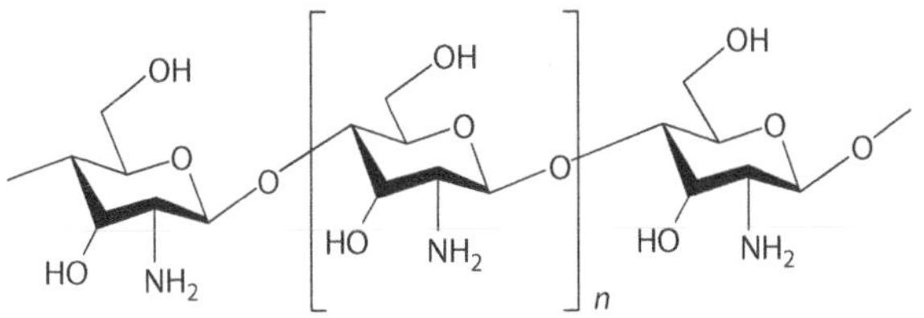

Figure 22.1 Structure of chitosan.

Recently, chitosan and COS have attracted considerable interest because of their numerous biological activities such as antioxidant, angiotensin-I converting enzyme inhibitory, antimicrobial, anticancer, antidiabetic, hypocholesterolemic, anti-Alzheimer's, anticoagulant, and adipogenesis inhibitory effects (Ngo et al., 2012; Park and Kim, 2010).

Various activities of COS are affected by degree of deacetylation (DD) and molecular weight (MW) or chain length. Chitin, an insoluble polymer in water, is the major limiting factor for its utilization in living systems. Therefore, it is important to produce soluble chitin or chitosan by several methods such as acidic and enzymatic hydrolysis. In recent years, antioxidant effects of chitin oligosaccharides (NA-COS) produced by acidic hydrolysis from crab shell chitin were evaluated in mouse macrophages cells (RAW 264.7). NA-COS can inhibit myeloperoxidase activity in human myeloid cells (HL-60) and decrease free-radical oxidation of DNA and membrane proteins. In addition, direct intracellular radical scavenging effect and intracellular glutathione level were significantly increased in the presence of NA-COS (Ngo et al., 2008, 2009).

Besides, COS (MW 1500, DD 90%) protect rat cortical neurons against copper-induced damage by attenuating intracellular ROS (Xu et al., 2010a). COS below 1 kDa decreased the formation of intracellular radical species in a murine melanoma cell line (B16F1), suggesting prevention of oxidative stress-related disease. The protective effect on DNA oxidation and intracellular glutathione (GSH) level was increased with the treatment of COS (Mendis et al., 2007). COS can effectively protect human umbilical vein endothelial cells (HUVEC, ECV304 cells) against H_2O_2-induced oxidative stress, which might be of importance in the treatment of cardiovascular diseases. COS exerted inhibitory effects on the formation of intracellular ROS and lipid peroxidation such as malondialdehyde, restoring activities of endogenous antioxidants including superoxide dismutase and GSH peroxidase, along with the capacity of increasing levels of nitric oxide (NO) and NO synthase (Liu et al., 2009). Moreover, COS of four different MW ranges (below 1, 1–3, 3–5, and 5–10 kDa) protect pancreatic β-cells from oxidative stress-induced cellular deterioration (Karadeniz et al., 2010). Xu et al. (2010b) investigated the protective effect of COS against H_2O_2-induced oxidative stress on human embryonic hepatocytes (L02 cells) and its scavenging activity against the DPPH radical in vitro, suggesting that COS might be useful in a clinical setting during the treatment of oxidative stress-related liver damage.

According to the electron spin resonance studies, 90% deacetylated medium MW hetero-COS have the highest free radical scavenging activity of all free radicals tested such as 1,1-diphenyl-2-picrylhydrazyl (DPPH), hydroxyl, superoxide and carbon centered radicals (Je et al., 2004). Inhibition of free radical mediated oxidation of cellular biomolecules such as lipids, proteins, and direct scavenging of ROS by carboxylated chitooligosaccharides (CCOS) has been reported (Rajapakse et al., 2007). On the other hand, low MW chitosans may inhibit neutrophil activation and oxidation of serum albumin commonly observed in patients undergoing hemodialysis, resulting in reduction of oxidative stress associated with uremia (Anraku et al., 2008). In addition, chitosan at an addition of 0.02% had antioxidant effects in lard and crude rapeseed oil but the activity was less than ascorbic acid. In the food industry, chitosan (edible chitosan, more than 83% DD) and COS have been used as dietary food additives and functional factors for their health beneficial effects as well as drug carriers (Xia et al., 2011). Therefore, the application of COS as antioxidants in the food industry is promising.

22.3.2 Bioactive peptides

Components of proteins in marine foods are containing sequences of bioactive peptides, which could exert a physiological effect in the body. Especially some of these bioactive peptides have been identified to possess nutraceutical potentials that are beneficial in human health promotion. Moreover, the possible roles of marine food-derived bioactive peptides in reducing the risk of diseases have been reported. Bioactive peptides usually contain from 3 to 40 amino acid residues, and their activities are based on amino acid composition and sequence. These short chains of amino acids are inactive within the sequence of the parent protein, but can be released during gastrointestinal digestion, food processing, or fermentation (Erdmann et al., 2008).

Marine bioactive peptides have been obtained widely by enzymatic hydrolysis of marine proteins such as algae, fish, mollusk, crustacean, and by-products including substandard muscles, viscera, skins, trimmings, and shellfish. Marine bioactive peptides based on their structural properties, amino acid composition, and sequences have been shown to display a wide range of biological functions including antioxidant, antihypertensive, antimicrobial, opioid agonistic, immunomodulatory, prebiotic, mineral binding, antithrombotic, and hypocholesterolemic effects (Betoret et al., 2011; Najafian and Babji, 2012). In fermented marine food sauces such as blue mussel sauce and oyster sauce, enzymatic hydrolysis has already been done by microorganisms, and bioactive peptides can be purified without further hydrolysis. In addition, marine processing by-products contain bioactive peptides with valuable functional properties. Recently, a number of studies have shown that peptides derived from various marine sources such as fish, blue mussel, conger eel, microalgae, and squid act as potential antioxidants (**Table 22.1**).

The antioxidant activity of bioactive peptides derived from marine has been determined by different in vitro methods, such as DPPH, carbon-centered, hydroxyl and superoxide anion radical scavenging activities which have been detected by ESR spectroscopy method as well as intracellular-free radical scavenging assays. The beneficial effects of antioxidant marine bioactive peptides are well known in scavenging ROS and free radicals or in preventing oxidative damage by interrupting the radical chain reaction of oxidation. In addition, peptides isolated from marine proteins have greater antioxidant properties than α-tocopherol in different oxidative systems. The peptide isolated from

Table 22.1 Antioxidant Peptides Derived from Marine Organisms

Source	Amino Acid Sequence	References
Tilapia	DPALATEPDPMPF	Ngo et al. (2010)
Pacific cod	TCSP, TGGGNV	Ngo et al. (2011)
Conger eel	LGLNGDDVN	Ranathunga et al. (2006)
Alga	VECYGPNRPQF	Sheih et al. (2009)
Squid	NADFGLNGLEGLA	Rajapakse et al. (2005a)
	NGLEGLK	
	FDSGPAGVL	Mendis et al. (2005)
	NGPLQAGQPGER	
Hoki	ESTVPERTHPACPDFN	Kim et al. (2007b)
Oyster	LKQELEDLLEKQE	Qian et al. (2008)
Blue mussel	HFGDPFH	Rajapakse et al. (2005b)
Tuna	VKAGFAWTANQQLS	Je et al. (2007)
Rotifer	LLGPGLTNHA	Byun et al. (2009)
	DLGLGLPGAH	
Prawn	IKK, FKK, and FIKK	Suetsuna (2000)
Yellowfin sole	RPDFDLEPPY	Jun et al. (2004)
Sardine	LQPGQGQQ	Suetsuna and Ukeda (1999)
Royal jelly	AL, FK, FR, IR, KF, KL, KY, RY, YD, YY, LDR, KNYP	Guo et al. (2009)
Alaska Pollack	GEOGPOGPOGPOGPOG	Kim et al. (2001)
	GPOGPOGPOGPOG	

the peptic hydrolysate of hoki (*Johnius balengeri*) frame protein has inhibited lipid peroxidation higher than that of α-tocopherol as positive control and efficiently quenched different sources of free radicals (Kim et al., 2007b). Moreover, the bioactive antioxidant peptide from oyster (*Crassostrea gigas*) exhibited higher protective activity against polyunsaturated fatty acid peroxidation than natural antioxidant, α-tocopherol (Qian et al., 2008). In addition, the bioactive peptide from jumbo squid inhibited lipid peroxidation in the linoleic acid model system and its activity was much higher than α-tocopherol and was close to highly active synthetic antioxidant, BHT. Antioxidant activities of bioactive peptides are mainly due to the presence of hydrophobic amino acids, some aromatic amino acids and histidine. Gelatin peptides are rich in hydrophobic amino acids, and the abundance of these amino acids favors a higher emulsifying ability. Hence, marine gelatin peptides possess higher antioxidant effects than peptides derived from other proteins because of the high percentage of glycine and proline. Therefore, antioxidant bioactive peptides derived from marine may have great potential for use as pharmaceuticals, nutraceuticals, and a substitute for synthetic antioxidants (Mendis et al., 2005). For example, Shahidi et al. (1995) clearly demonstrated that capelin fish protein hydrolysate when added to minced pork muscle at a level of 0.5%–3.0% reduced the formation of secondary oxidation products including thiobarbituric acid reactive substances (TBARS) in the product by 17.7%–60.4%. However, the bitter taste of protein hydrolysates prevents the use of bioactive peptides as food additives and the bioactivity may be reduced through molecular alteration during food processing or interaction with other food ingredients. As a treatment to this bitterness, Shahidi et al. (1995) treated fish protein hydrolysate with activated carbon, which removed bitter peptides. Also, the plastein reaction which can occur when a protein hydrolysate is incubated with a protease is able to debitter protein hydrolysates. Incorporation of a fish protein hydrolysate preparation made by autolysis of arrowtooth flounder protein into a coating of salmon fillets slowed down the lipid oxidation process. Furthermore, a brine solution containing salmon fish protein hydrolysate injected into smoked salmon fish fillets was shown to reduce lipid oxidation measured as TBARS during 6 weeks of cold storage (4°C) and 8 months of frozen storage (−18°C) (Samaranayaka and Li-Chan, 2011). The challenge for food technologists will be to develop functional foods and nutraceuticals without the undesired side effects of the added peptides.

22.3.3 Sulfated polysaccharides

Recently, various sulfated polysaccharides (SPs) isolated from marine algae have attracted much attention in nutraceutical/functional food, cosmetic/cosmeceutical and pharmaceutical applications. Marine algae are the most abundant source of nonanimal SPs in nature and their chemical structures (**Figure 22.2**) vary according to the species of algae. SPs are commonly found in three major groups of marine algae: red algae (Rhodophyceae), brown

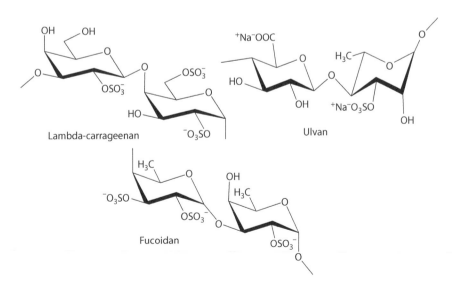

Figure 22.2 *Antioxidant sulfated polysaccharides from marine algae.*

algae (Phaeophyceae), and green algae (Chlorophyceae). The major SPs of red algae are galactans, commercially known as agar and carrageenan, and those of brown algae are fucans, including fucoidan, sargassan, ascophyllan, and glucuronoxylofucan. On the other hand, the major SPs of green algae are usually sulfated heteropolysaccharides that contain galactose, xylose, arabinose, mannose, glucuronic acid, or glucose. SPs include a complex group of macromolecules with a wide range of important biological activities such as antioxidant, anticoagulant, anticancer, antiviral, and anti-inflammation (Costa et al., 2010; Jiao et al., 2011).

In the last decade, SPs derived from seaweeds have appreciable antioxidant activity and can be used as potential antioxidants. Antioxidant activities of SPs have been determined by various methods such as ferric reducing antioxidant power (FRAP), lipid peroxide inhibition, DPPH, superoxide anion, hydroxyl, ABTS, and NO radicals scavenging assays (Wang et al., 2008). Xue et al. (1998) reported that several marine-derived SP have antioxidative activities in phosphatidylcholine-liposomal suspension and organic solvents. According to Kim et al. (2007a), the SPs of brown alga *Sargassum fulvellum* are more potent NO scavenger than commercial antioxidants such as BHA and α-tocopherol. Fucans from *Fucus vesiculosus* exhibited considerable FRAP (Ruperez et al., 2002) and superoxide radical scavenging property (Rocha de Souza et al., 2007). In addition, a positive correlation has been reported for sulfate content, superoxide and hydroxyl radicals scavenging assays in fucoidan fractions obtained from brown alga *Laminaria japonica* (Wang et al., 2009; Zhao et al., 2005). Furthermore, fucoidan has shown the highest antioxidant activity followed by alginate and laminaran from brown alga *Turbinaria conoides* according to FRAP and DPPH assays (Chattopadhyay et al., 2010). Besides, in vivo antioxidant activity of SP derived from marine red alga *Porphyra haitanensis* in aging mice has been reported (Zhang et al., 2003).

Antioxidant activity of SPs depends on their structural features such as sulfation level, distribution of sulfate groups along the polysaccharide backbone, molecular weight, sugar composition, and stereochemistry. Antioxidant properties of carrageenans (Rocha de Souza et al., 2007) and ulvans (Qi et al., 2005) also appeared related to sulfate content. In the latter study, low-molecular-weight and high-sulfate content derivatives of ulvans showed improved antioxidant activities (Sun et al., 2009). Interestingly, metal chelating, free radical and hydroxyl radical scavenging activities of fucan fractions appear to relate to their ratio of sulfate content/fucose (Wang et al., 2010). The rationale for this is low-molecular-weight SPs may incorporate into the cells more efficiently and donate proton effectively compared to high-molecular-weight SPs. Furthermore, SPs from marine algae are known to be important free radical scavengers and antioxidants for the prevention of oxidative damage, which is an important contributor in carcinogenesis. Collectively, these evidences suggest that among various naturally occurring substances, SPs prove to be one of the useful candidates in search for effective, nontoxic substances with potential antioxidant activity. Furthermore, seaweed processing by-products with bioactive SPs can be easily utilized for producing functional ingredients and could be used as a rich source of natural antioxidants with potential applications in the food industries and nutraceutical products.

22.3.4 Carotenoids

Carotenoids are linear polyenes that function as light energy harvesters and are a family of pigmented compounds that are synthesized by plants, algae, fungi, and microorganisms, but not animals. They are the most important pigments in nature that are responsible for various colors of different photosynthetic organisms. Carotenoids are thought to be responsible for the beneficial properties in preventing human diseases including cardiovascular diseases, cancer, and other chronic diseases (Agarwal and Rao, 2000). Carotenoid pigments have antioxidant properties by virtue of their highly unsaturated nature, which enable them to lend themselves to oxidation instead of other molecules. The antioxidant actions of carotenoids are based on their singlet oxygen quenching properties and their ability to trap free radicals, which mainly depends on the number of conjugated double bonds of the molecule and carotenoid end groups or the nature of substituents in carotenoids containing cyclic end groups. Fucoxanthin and astaxanthin (**Figure 22.3**) are known to be major ingredients of marine carotenoids and have also been recognized to possess excellent antioxidative potential (Miyashita and Hosokawa, 2008; Pangestuti and Kim, 2011).

In a recent study, fucoxanthin derived from the brown alga *Undaria pinnatifida* and fucoxanthinol which were prepared from fucoxanthin by hydrolysis with lipase are effective on scavenging of DPPH and 2,2′-Azinobis-3-ethylbenzo thizoline-6-sulphonate radicals (Sachindra et al., 2007). In addition, Nishida et al. (2007) demonstrated that carotenoids have stronger singlet

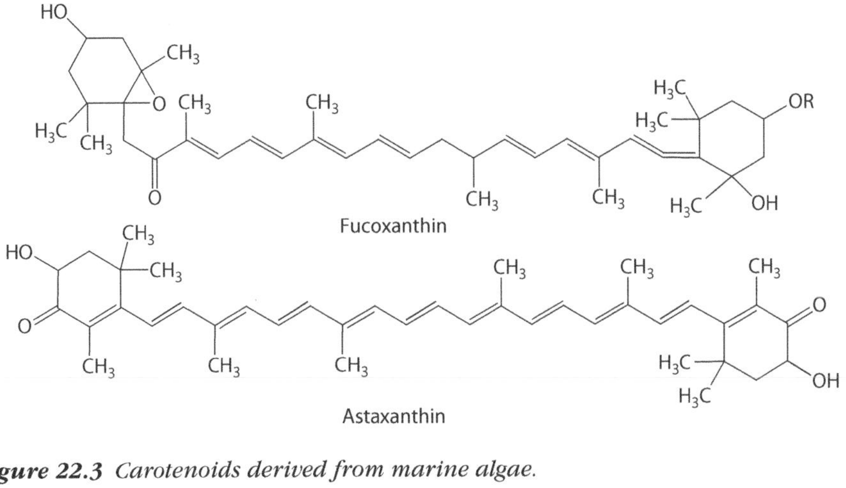

Figure 22.3 *Carotenoids derived from marine algae.*

oxygen quenching activities than α-tocopherol as well as α-lipoic acid and have reported that fucoxanthin from the brown algae *U. pinnatifida* and *L. japonica* was one of the active compounds. Moreover, the cytoprotective effect of fucoxanthin, from a brown alga *Sargassum siliquastrum*, against H_2O_2-induced cell damage has been reported by Heo et al. (2008). Fucoxanthin can effectively inhibit intracellular ROS formation, DNA damage, and apoptosis induced by H_2O_2. Noticeably, fucoxanthin also exhibited a strong enhance of cell viability against H_2O_2-induced oxidative damage. Furthermore, fucoxanthin has the potential to prevent UV-B-induced cell injury in human dermal fibroblasts (HDF cells) (Heo and Jeon, 2009).

Astaxanthin is effective as α-tocopherol in inhibiting free radical-initiated lipid peroxidation in rat liver microsomes (Palozza and Krinsky, 1992), and is 100 times higher than α-tocopherol in protecting rat mitochondria against Fe^{2+}-catalyzed lipid peroxidation in vivo and in vitro (Kurashige et al., 1990). Astaxanthin is also used not only as a source of pigment in the diet but also has potential clinical applications due to its higher and extremely versatile antioxidant activity than β-carotene and α-tocopherol (Miki, 1991). The higher antioxidant activity of astaxanthin than other carotenoids has been explained to be due to the presence of the hydroxyl and keto endings on each ionone ring in the structure of astaxanthin (Turujman et al., 1997). In addition, astaxanthin is effective against UVA-induced DNA alterations in human dermal fibroblasts, human melanocytes, and human intestinal cells (Lyons and O'Brien, 2002). Moreover, consumption of astaxanthin from marine animals leads to inhibit low-density lipoprotein oxidation (Iwamoto et al., 2000). These results suggest that marine-derived carotenoids such as fucoxanthin and astaxanthin are bioactive natural functional ingredients that may be important in human health as potential antioxidants.

22.3.5 Phlorotannins

Brown algae have been recognized as a rich source of phlorotannins, which are formed by the polymerization of phloroglucinol (1,3,5-tryhydroxybenzene) monomer units and biosynthesized through the acetate-malonate pathway. The phlorotannins are highly hydrophilic components with a wide range of molecular sizes ranging between 126 Da and 650 kDa. Marine brown algae accumulate a variety of phloroglucinol-based polyphenols, as phlorotannins of low, intermediate, and high MW containing both phenyl and phenoxy units could be used as functional ingredients in nutraceuticals with potential health effects. Based on the means of linkage, phlorotannins can be classified into four subclasses, including fuhalols and phlorethols (phlorotannins with an ether linkage), fucols (with a phenyl linkage), fucophloroethols (with an ether and phenyl linkage), and eckols (with a dibenzodioxin linkage). Among marine brown algae, *Ecklonia cava*, *Ecklonia stolonifera*, *Ecklonia kurome*, *Eisenia bicyclis*, *Ishige okamurae*, *Sargassum thunbergii*, *Hizikia fusiformis*, *U. pinnatifida* and *L. japonica* have been reported for phlorotannins with health beneficial biological activities. Notably, phlorotannins exhibited various bioactivities such as antioxidant, anticancer, antidiabetic, anti-HIV, matrix metalloproteinase enzyme inhibition, and antihypertensive activities (Vo and Kim, 2010; Wijesekara et al., 2010).

Many researchers have shown that phlorotannins derived from marine brown algae have strong antioxidant activities against free radical-mediated oxidation damage. The antioxidant activity can be the result of specific scavenging of radicals formed during peroxidation, scavenging of oxygen-containing compounds, or metal-chelating ability. The various phlorotannins can be observed that are able to overcome the sensitivity problem inherent in the detection of endogenous radicals in biological systems. According to the significant results of total antioxidant activity compared to tocopherol as positive control in the lineloic acid model system, the phlorotannins presented a greatly interesting potential against DPPH, hydroxyl, superoxide anion and peroxyl radicals in vitro, using ESR technique (Li et al., 2009). Phlorotannins derived from *H. fusiformis* are potential radical scavengers, thus a great source of natural antioxidative nutraceuticals (Siriwardhana et al., 2005). Furthermore, several phlorotannins purified from brown seaweeds such as *E. cava*, *E. kurome*, *E. bicyclis*, and *H. fusiformis* are responsible for potent antioxidant activities and show protective effects against hydrogen peroxide-induced cell damage. In addition, eckol, phlorofucofuroeckol A, dieckol, and 8,8'-bieckol have shown a potent inhibition of phospholipid peroxidation at $1\,\mu M$ in a liposome system and these phlorotannins have significant radical scavenging activities against free radicals effectively compared to ascorbic acid and α-tocopherol (Li et al., 2011). These findings suggest that phlorotannins, the natural antioxidant compounds found in edible brown algae, can protect food products against oxidative damage as well as preventing and treating free radical-related diseases.

22.4 Conclusions

Recent studies have provided evidence that marine-derived functional ingredients play a vital role in human health and nutrition. Moreover, the formation of cancer cells in human body can be directly induced by free radicals and natural anticancer drugs as chemopreventive agents have gained a positive popularity in treatment of cancer. Hence, radical scavenging compounds such as bioactive peptides, COS, SPs, phlorotannins, and carotenoids pigments including fucoxanthin and astaxanthin from marine foods and their by-products can be used indirectly as functional ingredients to reduce cancer formation in human body. Collectively, the wide range of biological activities associated with the antioxidative ingredients which were derived from marine food sources have the potential to expand its health beneficial value not only in the food industry but also in the pharmaceutical and cosmeceutical industries.

Acknowledgments

This study was supported by a grant from the Marine Bioprocess Research Center of the Marine Bio 21 Project funded by the Ministry of Land, Transport and Maritime, Republic of Korea.

References

Agarwal, S. and Rao, A. V. (2000). Carotenoids and chronic diseases. *Drug Metabolism and Drug Interactions*, *17*, 189–210.

Anraku, M., Kabashima, M., Namura, H., Maruyama, T., Otagiri, M., Gebicki, J., Furutani, N., and Tomida, H. (2008). Antioxidant protection of human serum albumin by chitosan. *International Journal of Biological Macromolecules*, *43*, 159–164.

Betoret, E., Betoret, N., Vidal, D., and Fito, P. (2011). Functional foods development: Trends and technologies. *Trends in Food Science and Technology*, *22*, 498–508.

Blunt, J. W., Copp, B. R., Munro, M. H. G., Northcote, P. T., and Prinsep, M. R. (2006). Marine natural products. *Natural Product Reports*, *23*, 26–78.

Butterfield, D. A., Castenga, A., Pocernich, C. B., Drake, J., Scapagnini, G., and Calabrese, V. (2002). Nutritional approaches to combat oxidative stress in Alzheimer's disease. *Journal of Nutritional Biochemistry*, *13*, 444–461.

Byun, H. G. and Kim, S. K. (2001). Purification and characterization of angiotensin I converting enzyme (ACE) inhibitory peptides from Alaska Pollack (*Theragra chalcogramma*) skin. *Process Biochemistry*, *36*, 1155–1162.

Byun, H. G., Lee, J. K., Park, H. G., Jeon, J. K., and Kim, S. K. (2009). Antioxidant peptides isolated from the marine rotifer, *Brachionus rotundiformis*. *Process Biochemistry*, *44*, 842–846.

Chattopadhyay, N., Ghosh, T., Sinha, S., Chattopadhyay, K., Karmakar, P., and Ray. B. (2010). Polysaccharides from *Turbinaria conoides*: Structural features and antioxidant capacity. *Food Chemistry*, *118*, 823–829.

Costa, L. S., Fidelis, G. P., Cordeiro, S. L., Oliveira, R. M., Sabry, D. A., Camara, R. B. G., Nobre, L. T. D. B. et al. (2010). Biological activities of sulfated polysaccharides from tropical seaweeds. *Biomedicine and Pharmacotherapy*, *64*, 21–28.

Erdmann, K., Cheung, B. W. Y., and Schroder, H. (2008). The possible roles of food-derived bioactive peptides in reducing the risk of cardiovascular disease. *Journal of Nutritional Biochemistry*, *19*, 643–654.

Guo, H., Kouzuma, Y., and Yonekura, M. (2009). Structures and properties of antioxidative peptides derived from royal jelly protein. *Food Chemistry*, *113*, 238–245.

Heo, S. J. and Jeon, Y. J. (2009). Protective effect of fucoxanthin isolated from *Sargassum siliquastrum* on UV-B induced cell damage. *Journal of Photochemistry and Photobiology B: Biology*, *95*, 101–107.

Heo, S. J., Ko, S. C., Kang, S. M., Kang, H. S., Kim, J. P., Kim, S. H., Lee, K. W., Cho, M. G., and Jeon, Y. J. (2008). Cytoprotective effect of fucoxanthin isolated from brown algae *Sargassum siliquastrum* against H_2O_2-induced cell damage. *European Food Research and Technology*, *228*, 145–151.

Iwamoto, T., Hosoda, K., Hirano, R., Kurata, H., Matsumoto, A., Miki, W., Kamiyama, M., Itakura, H., Yamamoto, S., and Kondo, K. (2000). Inhibition of low-density lipoprotein oxidation by astaxanthin. *Journal of Atherosclerosis and Thrombosis*, *7*, 216–222.

Je, J. Y., Park, P. J., and Kim, S. K. (2004). Free radical scavenging properties of heterochitooligosaccharides using an ESR spectroscopy. *Food and Chemical Toxicology*, *42*, 381–387.

Je, J. Y., Qian, Z. J., Byun, H. G., and Kim, S. K. (2007). Purification and characterization of an antioxidant peptide obtained from tuna backbone protein by enzymatic hydrolysis. *Process Biochemistry*, *42*, 840–846.

Jeon, Y. J., Byun, H. G., and Kim, S. K. (1999). Improvement of functional properties of cod frame protein hydrolysates using ultrafiltration membranes. *Process Biochemistry, 35,* 471–478.

Jiao, G., Yu, G., Zhang, J., and Ewart, H. S. (2011). Chemical structures and bioactivities of sulfated polysaccharides from marine algae. *Marine Drugs, 9,* 196–223.

Jun, S. Y., Park, P. J., Jung, W. K., and Kim, S. K. (2004). Purification and characterization of an antioxidative peptide from enzymatic hydrolysates of yellowfin sole (*Limanda aspera*) frame protein. *European Food Research and Technology, 219,* 20–26.

Karadeniz, F., Artan, M., Kong, C., and Kim, S. (2010). Chitooligosaccharides protect pancreatic β-cells from hydrogen peroxide-induced deterioration. *Carbohydrate Polymers, 82,* 143–147.

Kim, S. H., Choi, D. S., Athukorala, Y., Jeon, Y. J., Senevirathne, M., and Rha, C. K. (2007a). Antioxidant activity of sulfated polysaccharides isolated from *Sargassum fulvellum*. *Journal of Food Science and Nutrition, 12,* 65–73.

Kim, S. Y., Je, J. Y., and Kim, S. K. (2007b). Purification and characterization of antioxidant peptide from hoki (*Johnius belengerii*) frame protein by gastrointestinal digestion. *The Journal of Nutritional Biochemistry, 18,* 31–38.

Kim, S. K., Kim, Y. T., Byun, H. G., Nam, K. S., Joo, D. S., and Shahidi, F. (2001). Isolation and characterization of antioxidative peptides from gelatin hydrolysate of *Alaska pollack* skin. *Journal of Agricultural and Food Chemistry, 49,* 1984–1989.

Kurashige, M., Okimasu, E., Inoue, M., and Utsumi, K. (1990). Inhibition of oxidative injury of biological membranes by astaxanthin. *Physiological Chemistry and Physics and Medical NMR, 22,* 27–38.

Li, Y., Qian, Z. J., Ryu, B. M., Lee, S. H., Kim, M. M., and Kim, S. K. (2009). Chemical components and its antioxidant properties in vitro: An edible marine brown alga, *Ecklonia cava*. *Bioorganic and Medicinal Chemistry, 17,* 1963–1973.

Li, Y. X., Wijesekara, I., Li, Y., and Kim, S. K. (2011). Phlorotannins as bioactive agents from brown algae. *Process Biochemistry, 46,* 2219–2224.

Liu, H., Li, W., Xu, G., Li, X., Bai, X., Wei, P., Yu, C., and Du, Y. (2009). Chitosan oligosaccharides attenuate hydrogen peroxide-induced stress injury in human umbilical vein endothelial cells. *Pharmacological Research, 59,* 167–175.

Lyons, N. M. and O'Brien, N. M. (2002). Modulatory effects of an algal extract containing astaxanthin on UVA-irradiated cells in culture. *Journal of Dermatological Science, 30,* 73–84.

Mendis, E., Kim, M. M., Rajapakse, N., and Kim S. K. (2007). An in vitro cellular analysis of the radical scavenging efficacy of chitooligosaccharides. *Life Sciences, 80,* 2118–2127.

Mendis, E., Rajapakse, N., Byun, H. G., and Kim, S. K. (2005). Investigation of jumbo squid (*Dosidicus gigas*) skin gelatin peptides for their in vitro antioxidant effects. *Life Science, 77,* 2166–2178.

Miki, W. (1991). Biological functions and activities of animal carotenoids. *Pure and Applied Chemistry, 63,* 141–146.

Miyashita, K. and Hosokawa, M. (2008). Beneficial health effects of seaweed carotenoid, fucoxanthin. In C. Barrow and F. Shahidi (Eds.), *Marine Nutraceuticals and Functional Foods* (pp. 297–319). Boca Raton, FL: CRC Press.

Nagai, T. and Suzuki, N. (2000). Isolation of collagen from fish waste material—Skin, bone, and fins. *Food Chemistry, 68,* 277–281.

Najafian, L. and Babji, A. S. (2012). A review of fish-derived antioxidant and antimicrobial peptides: Their production, assessment, and applications. *Peptides, 33,* 178–185.

Ngo, D. N., Kim, M. M., and Kim, S. K. (2008). Chitin oligosaccharides inhibit oxidative stress in live cells. *Carbohydrate Polymers, 74,* 228–234.

Ngo, D. N., Lee, S. H., Kim, M. M., and Kim, S. K. (2009). Production of chitin oligosaccharides with different molecular weights and their antioxidant effect in RAW 264.7 cells. *Journal of Functional Foods, 1,* 188–198.

Ngo, D. H., Ngo, D. N., Vo, T. S., Ryu, B. M., Ta, Q. V., and Kim, S. K. (2012). Protective effects of aminoethyl-chitooligosaccharides against oxidative stress and inflammation in murine microglial BV-2 cells. *Carbohydrate Polymers*, *88*, 743–747.

Ngo, D. H., Qian, Z. J., Ryu, B. M., Park, J. W., and Kim, S. K. (2010). In vitro antioxidant activity of a peptide isolated from Nile tilapia (*Oreochromis niloticus*) scale gelatin in free radical-mediated oxidative systems. *Journal of Functional Food*, *2*, 107–117.

Ngo, D. H., Ryu, B. M., Vo, T. S., Himaya, S. W. A., Wijesekara, I., and Kim, S. K. (2011). Free radical scavenging and angiotensin-I converting enzyme inhibitory peptides from Pacific cod (*Gadus macrocephalus*) skin gelatin. *International Journal of Biological Macromolecules*, *49*, 1110–1116.

Nishida, Y., Yamashita, E., and Miki, W. (2007). Quenching activities of common hydrophilic and lipophilic antioxidants against singlet oxygen using chemiluminescence detection system. *Carotenoid Science*, *11*, 16–20.

Palozza, P. and Krinsky, N. (1992). Astaxanthin and canthaxanthin are potent antioxidant in a membrane model. *Archives of Biochemistry and Biophysics*, *297*, 291–295.

Pangestuti, R. and Kim, S. K. (2011). Biological activities and health benefit effects of natural pigments derived from marine algae. *Journal of Functional Foods*, *3*, 255–266.

Park, P. J., Jung, W. K., Nam, K. D., Shahidi, F., and Kim, S. K. (2001). Purification and characterization of antioxidative peptides from protein hydrolysate of lecithin-free egg yolk. *Journal of American Oil Chemists Society*, *78*, 651–656.

Park, B. K. and Kim, M. M. (2010). Applications of chitin and its derivatives in biological medicine. *International Journal of Molecular Sciences*, *11*, 5152–5164.

Qi, H., Zhang, Q., Zhao, T., Chen, R., Zhang, H., Niu, X., and Li, Z. (2005). Antioxidant activity of different sulfate content derivatives of polysaccharide extracted from *Ulva pertusa* (Chlorophyta) *in vitro*. *International Journal of Biological Macromolecules*, *37*, 195–199.

Qian, Z. J., Jung, W. K., Byun, H. G., and Kim, S. K. (2008). Protective effect of an antioxidative peptide purified from gastrointestinal digests of oyster, *Crassostrea gigas* against free radical induced DNA damage. *Bioresource Technology*, *99*, 3365–3371.

Rajapakse, N., Kim, M. M., Mendis, E., and Kim, S. K. (2007). Inhibition of free radical-mediated oxidation of cellular biomolecules by carboxylated chitooligosaccharides. *Bioorganic and Medicinal Chemistry*, *15*, 997–1003.

Rajapakse, N., Mendis, E., Byun, H. G., and Kim, S. K. (2005a). Purification and in vitro antioxidative effects of giant squid muscle peptides on free radical-mediated oxidative systems. *The Journal of Nutritional Biochemistry*, *16*, 562–569.

Rajapakse, N., Mendis, E., Jung, W. K., Je, J. Y., and Kim, S. K. (2005b). Purification of a radical scavenging peptide from fermented mussel sauce and its antioxidant properties. *Food Research International*, *38*, 175–182.

Ranathunga, S., Rajapakse, N., and Kim, S. K. (2006). Purification and characterization of antioxidative peptide derived from muscle of conger eel (*Conger myriaster*). *European Food Research and Technology*, *222*, 310–315.

Rocha de Souza, M. C., Marques, C. T., Dore, C. M. G., Ferreira da Silva, F. R., Rocha, H. A. O., and Leite, E. L. (2007). Antioxidant activities of sulphated polysaccharides from brown and red seaweeds. *Journal of Applied Phycology*, *19*, 153–160.

Ruperez, P., Ahrazem, O., and Leal, A. (2002). Potential antioxidant capacity of sulphated polysaccharides from the edible marine brown seaweed *Fucus vesiculosus*. *Journal of Agricultural and Food Chemistry*, *50*, 840–845.

Sachindra, N. M., Sato, E., Maeda, H., Hosokawa, M., Niwano, Y., Kohno, M., and Miyashita, K. (2007). Radical scavenging and singlet oxygen quenching activity of marine carotenoid fucoxanthin and its metabolites. *Journal of Agricultural and Food Chemistry*, *55*, 8516–8522.

Samaranayaka, A. G. P. and Li-Chan, E. C. Y. (2011). Food-derived peptidic antioxidants: A review of their production, assessment, and potential applications. *Journal of Functional Foods*, *3*, 229–254.

Shahidi, F. (2007). *Maximising the Value of Marine By-Products*. Boca Raton, FL: CRC Press.

Shahidi, F., Han, X. Q., and Synowiecki, J. (1995). Production and characteristics of protein hydrolysates from capelin (*Mallotus villosus*). *Food Chemistry*, 53, 285–293.

Sheih, I. C., Wu, T. K., and Fang, T. J. (2009). Antioxidant properties of a new antioxidative peptide from algae protein waste hydrolysate in different oxidation systems. *Bioresource Technology*, 100, 3419–3425.

Siriwardhana, N., Lee, K. W., and Jeon, Y. J. (2005). Radical scavenging potential of hydrophilic phlorotannins of *Hizikia fusiformis*. *Algae*, 20, 69–75.

Suetsuna, K. (2000). Antioxidant peptides from the protease digest of prawn (*Penaeus japonicus*) muscle. *Marine Biotechnology*, 2, 5–10.

Suetsuna, K. and Ukeda, H. (1999). Isolation of an octapeptide which possesses active oxygen scavenging activity from peptic digest of sardine muscle. *Nippon Suisan Gakkaishi*, 65, 1096–1099.

Sun, L., Wang, C., Shi, Q., and Ma, C. (2009). Preparation of different molecular weight polysaccharides from *Porphyridium cruentum* and their antioxidant activities. *International Journal of Biological Macromolecules*, 45, 42–47.

Turujman, S. A., Wamer, W. G., Wei, R. R., and Albert, R. H. (1997). Rapid liquid chromatographic method to distinguish wild salmon from aquacultured salmon fed synthetic astaxanthin. *Journal of AOAC International*, 80, 622–632.

Vo, T. S. and Kim, S. K. (2010). Potential anti-HIV agents from marine resources: An overview. *Marine Drugs*, 8, 2871–2892.

Wang, J., Liu, L., Zhang, Q., Zhang, Z., Qi, H., and Li, P. (2009). Synthesized oversulphated, acetylated and benzoylated derivatives of fucoidan extracted from *Laminaria japonica* and their potential antioxidant activity *in vitro*. *Food Chemistry*, 114, 1285–1290.

Wang, J., Zhang, Q., Zhang, Z., and Li, Z. (2008). Antioxidant activity of sulphated polysaccharide fractions extracted from *Laminaria japonica*. *International Journal of Biological Macromolecules*, 42, 127–132.

Wang, J., Zhang, Q., Zhang, Z., Song, H., and Li, P. (2010). Potential antioxidant and anticoagulant capacity of low molecular weight fucoidan fractions extracted from *Laminaria japonica*. *International Journal of Biological Macromolecules*, 46, 6–12.

Wijesekara, I., Yoon, N. Y., and Kim, S. K. (2010). Phlorotannins from *Ecklonia cava* (Phaeophyceae): Biological activities and potential health benefits. *Biofactors*, 36, 408–414.

Xia, W., Liu, P., Zhang, J., and Chen, J. (2011). Biological activities of chitosan and chitooligosaccharides. *Food Hydrocolloids*, 25, 170–179.

Xu, W., Huang, H., Lin, C., and Jiang, Z. (2010a). Chitooligosaccharides protect rat cortical neurons against copper induced damage by attenuating intracellular level of reactive oxygen species. *Bioorganic Medicinal Chemistry Letter*, 20, 3084–3088.

Xu, Q., Ma, P., Yu, W., Tan, C., Liu, H., Xiong, C., Qiao, Y., and Du, Y. (2010b). Chitooligosaccharides protect human embryonic hepatocytes against oxidative stress induced by hydrogen peroxide. *Marine Biotechnology*, 12, 292–298.

Xue, C., Yu, G., Hirata, T., Terao, J., and Lin, H. (1998). Antioxidative activities of several marine polysaccharides evaluated in a phosphatidylcholine-liposomal suspension and organic solvents. *Bioscience, Biotechnology and Biochemistry*, 62, 206–209.

Zhang, Q., Li, N., Zhou, G., Lu, X., Xu, Z., and Li, Z. (2003). In vivo antioxidant activity of polysaccharide fraction from *Porphyra haitanensis* (Rhodephyta) in aging mice. *Pharmacological Research*, 48, 151–155.

Zhao, X., Xue, C., Cai, Y., Wang, D., and Fang, Y. (2005). Study of antioxidant activities of fucoidan from *Laminaria japonica*. *High Technology Letters*, 11, 91–94.

23

Biological and Biomedical Applications of Marine Nutraceuticals

Janak K. Vidanarachchi, Maheshika S. Kurukulasuriya, and W.M.N.M. Wijesundara

Contents

23.1 Introduction

Lifestyle modifications including shifting to instant diets and lack of physical exercise have triggered the increase of health risks in many nations, especially in the developed world. In this context, the concept of "Nutraceutical," which is derived by combining the meaning of two words "nutrition" and "pharmaceutical," has a vital role in the improvement of human health. The nutraceuticals are substances that can be considered as a food or part of food that provides medical or human health benefits including prevention and treatment of diseases (Rajasekaran et al., 2008). The word "nutraceutical" is further elaborately described by the definition of Zeisel (1999), which considers nutraceuticals as a dietary supplement or a product in the form of a capsule, powder, softgel, or gelcap that delivers a concentrated form of a biologically active component of food in a non-food matrix to enhance the human health. Hence, the biologically active ingredients extracted from different food sources can be used as nutraceuticals in the prevention and treatment of human diseases. Since the natural food sources are recognized as safe for human consumption and the populations consuming a large proportion of diet with plant-based foods and seafood are known to have a lower incidence of diseases such as cardiovascular disease (CVD) and different types of cancers (Voutilainen et al., 2006), the active ingredients from the natural sources have wide acceptability in the nutraceutical industry.

Among the different natural nutraceuticals, applicability of marine-derived active compounds as nutraceuticals is a blooming aspect of the food industry. Marine environment is a biologically diverse source that covers approximately 71% of the earth surface (Baharum et al., 2010) with continuously varying extremely different environmental conditions. The heterogeneous groups of marine organisms have adapted to the environmental fluctuations through possessing structural compounds or by production of secondary metabolites which exhibit many biological properties. The important biological properties of marine-derived active compounds have the potential to be used for human health benefits. However, many of the marine-derived biologically active compounds and their important biological properties remain unexploited while the recent developments in marine research have increased the applicability of identified marine-derived compounds in the industrial scenario, including the nutraceutical industry. Some of the marine-derived compounds have been applied as nutraceuticals (Gulati, 2005; Dighe et al., 2009), while many of the isolated marine-derived compounds with important biological properties have the potential to be applied as nutraceuticals in the global food industry.

Therefore, this chapter was written with the objectives of reviewing the applications of marine-derived compounds as nutraceuticals with a special concern on the biological and biomedical applications of nutraceuticals derived from the marine origin.

23.2 Marine-derived compounds applied as biological and biomedical nutraceuticals

Marine environment consists of biologically diverse group of organisms, including marine macroalgae, microalgae, bacteria, cyanobacteria, fish species, and crustaceans. These marine organisms thrive in the changing, stressful marine environment either by having structural compounds within the organism or by producing secondary metabolites that have different biological properties such as antioxidant (Nomura et al., 1997; Yuan et al., 2009), antimicrobial (Kim et al., 2003; Hassan et al., 2004) or photo protection (Reimer et al., 2007) to overcome the surrounding stresses. Furthermore, due to anticancer (Jyonouchi et al., 2000; Jeon and Kim, 2002; Kong et al., 2009), anticoagulant (Carroll et al., 2004; Zhang et al., 2004), anti-inflammatory (Rajapakse et al., 2008), antihypertensive (Khan et al., 2003; Wijesinghe et al., 2011), and antilipogenic (Park et al., 2011) properties shown by many marine-derived compounds create many avenues for utilization of such compounds in various industrial applications. These biological properties have the potential to prevent or overcome the stressful conditions in humans also, which will have many health benefits in the prevention and treatment of diseases enabling the marine-derived compounds to be applied as nutraceuticals.

Despite the source of origin, the marine-derived bioactive substances that can be applied as nutraceuticals can be mainly categorized as polysaccharides, carotenoids, fatty acids, polyphenols, proteins or peptides, vitamins, and certain enzymes. The biological properties of these bioactive substances depend on the different functional groups present in the structure of these compounds (**Table 23.1**).

23.3 Biological and biomedical applications of marine nutraceuticals

The marine-derived polysaccharides, carotenoids, fatty acids, polyphenols, bioactive peptides, vitamins, and enzymes have different roles as nutraceuticals. The biological properties of these active compounds increase their applications as nutraceuticals, especially the biological and biomedical applications.

A majority of the marine natural products have been isolated from sponges, coelenterates (sea whips, sea fans, and soft corals), tunicates, *Opisthobranch molluscs* (nudibranchs, sea hares, etc.), echinoderms (starfish, sea cucumbers, etc.), and bryozoans (moss animals) and a wide variety of marine

Table 23.1 Marine-Derived Compounds Applied as Biological or Biomedical Nutraceuticals, Source of Origin, and the Functional Groups Present

Marine-Derived Compound	Type of Compound	Functional Groups	Source of Origin	References
Fucoidan	Sulfated polysaccharide	Sulfate group	*Undaria* sp.	Suetsuna and Nakano (2000)
			Focus vesiculosus	Patankar et al. (1993)
			L. japonica	Wang et al. (2008)
Chitin	Amino polysaccharide	Primary and secondary hydroxyl groups at C-3 and C-6 Amino group at C-2	Crustaceans	Vidanarachchi et al. (2010)
Chitooligosaccharides (COS)	Oligosaccharide	Primary and secondary hydroxyl groups at C-3 and C-6 Amino group at C-2	Crustaceans	Vidanarachchi et al. (2010)
Beta-carotene	Carotenoid	Double allenic bonds in the structure	*Dunaliella* sp.	Pisal and Lele (2005); Gomez et al. (2003)
Astaxanthin	Carotenoid	Hydroxyl and ketone groups at each β-ionine ring	*Haematococcus* sp.	Capelli and Cysewski (2006)
Fucoxanthin	Carotenoid	Double allenic bond and 5–6-monoepoxide	*Undaria* sp., *Laminaria* sp., *Sargassum* sp., *Hizika fusiformis*	Miyashita et al. (2011)
Docosahexanoic acid (DHA)	Omega-3 fatty acid		Marine fish	Rakshit et al. (2000)
Eicosapentaenoic acid (EPA)	Omega-3 fatty acid		Marine fish	Rakshit et al. (2000)
Phlorogucinol	Phenolic compounds	Phenolic hydroxyl groups	*E. cava*	Ahn et al. (2007); Shibata et al. (2008); Wijesinghe et al. (2011)
Eckol	Phenolic compounds	Phenolic hydroxyl groups	*E. cava*	Ahn et al. (2007); Shibata et al. (2008); Wijesinghe et al. (2011)
Dieckol	Phenolic compounds	Phenolic hydroxyl groups	*E. cava*	Ahn et al. (2007); Shibata et al. (2008); Wijesinghe et al. (2011)

Marine-Derived Compound	Type of Compound	Functional Groups	Source of Origin	References
Bioactive peptides	Peptides	Hydrophobic a.a. residue at N-terminus	Marine fish collagen	Byun and Kim (2001); Qian et al. (2007); Je et al. (2009); Zhu et al. (2010); Lee et al. (2010); Wijesekara et al. (2011)
		Positively charged a.a. at middle		
		Aromatic a.a. at C-terminus		
			Porphyra yezeoensis	Qu et al. (2010)
			U. pinnatifida	Suetsuna and Nakano (2000)
			Jelly fish — *Rhopilema esculentum*	Zhuang et al. (2012)
			Nemopilema nomurai	Kim et al. (2011)
			Marine shrimp — *Acetes chinensis*	Lun et al. (2006)

a.a., amino acid.

microorganisms. Marine natural nutraceutical products have yielded many bioactive compounds showing various pharmaceutical properties. Moreover, the marine-derived nutraceuticals such as polysaccharides, polyunsaturated fatty acids, polyphenols, and carotenoids have been identified as health beneficial compounds due to their characteristic biological properties including antioxidative, anti-inflammatory, anticancer, antiangiogenesis, antimicrobial, anticoagulant, prebiotic, and probiotic activities.

23.3.1 Antioxidant nutraceuticals

Antioxidants have become the subject of intensive investigation due to the ever-increasing demand by the food and pharmaceutical industries to develop natural, bioactive antiaging and anticarcinogenic compounds that produce measurable health benefits. Many of the synthetic antioxidants developed, however, produce side effects such as liver damage and carcinogenesis (Tezuka et al., 2001). Due to concerns regarding the toxicity of synthetic antioxidants, the search for alternatives from natural sources has received a great deal of interest. Recently, studies have focused on marine algae as a potential source of such activities, because they have been shown to contain many biologically active compounds with potential medicinal value, and their consumption has been associated with a reduced risk of several chronic diseases. Moreover, many bioactive molecules from marine algae produce antioxidant effects.

Marine-derived compounds such as β-carotene, astaxanthin, fucoxanthin, polyphenols, and fucoidans are antioxidants that are well known in industry. Many of these antioxidant compounds have the ability to prevent tissue damage in aging, atherosclerosis, and cancer through neutralizing reactive oxygen species (ROS). Therefore, there is significant attention on the use of marine-derived antioxidants in industry as nutraceuticals, which may have greater potential of protecting tissues from oxidative damages. Other than this, antioxidentability of certain marine antioxidants have the ability to relief inflammation (Guerin et al., 2003). According to Kochkina and Chirkov (2000), several mechanisms including hydroxyl radical scavenging, superoxide radical scavenging, erythrocyte hemolysis inhibiting, metal-chelating activities, and antilipid peroxidation have been studied to identify the antioxidant activity of marine origin compounds. Radical scavenging activity is one of the main modes of action of those marine-derived nutraceuticals. While superoxide radical is not so reactive and may not be able to cause any direct damage to cells, its reaction product hydrogen peroxide in the presence of trace metal ions is converted to more powerful hydroxyl radicals ($^{\cdot}$OH), which can oxidize most of the biomolecules. Thus, free radicals formed within the cells can induce multiple chemical changes in cellular organelles like membrane lipids, DNA, and proteins, which can eventually lead to cell death (Jing et al., 2011).

Recently, many studies have focused on marine algae as a potential source of antioxidant, because they have been shown to contain many biologically active compounds with potential medicinal value, and their consumption has been associated with a reduced risk of several chronic diseases.

Marine-derived bioactive peptides with antioxidant properties may have great potential for use as functional ingredients in functional foods and nutraceuticals instead of synthetic antioxidants. Shahidi et al. (1995) clearly demonstrated that Capelin fish protein hydrolysate when added to minced pork meat at a level of 0.5%–3.0% can reduce the formation of secondary oxidation products including thiobarbituric acid reactive substances in the product by 17.7%–60.4%.

Phlorotannins derived from marine brown algae could be used as nutraceutical with potential health effects (Wijesekara et al., 2010) including antioxidant function (Li et al., 2009b). Most of the phlorotannins purified from marine brown algae such as *Ecklonia cava* (Heo et al., 2005) have shown protective effects against hydrogen peroxide-induced cell damage (Kang et al., 2006) and act as free radical scavengers, reducing agents, and metal chelators, and thus effectively inhibit lipid oxidation (Ngo et al., 2011).

Exopolysaccharides (EPS) and intracellular polysaccharides (IPS) were isolated from broth and mycelia of *Armillaria mellea* by submerged culture results clearly demonstrating that EPS and IPS are highly effective in antioxidant property. Therefore, EPS and IPS from *A. mellea* by submerged culture can be developed as new antioxidants for potential applications in pharmaceutical and food industries (Lung and Chang, 2011).

Another marine-derived nutraceutical compound called chitooligosaccharide derivative also has been identified in pharmaceutical and medicinal applications such as antioxidant (Park et al., 2003b), antimicrobial (Kim et al., 2003; Hassan et al., 2004), anticancer (Jeon and Kim, 2002), antidiabetic (Liu et al., 2007), and anticoagulant (Park et al., 2004). Radical scavenging activity of COS is mainly responsible for the antioxidant ability. Rajapakse et al. (2007) reported that electron inhibition of free-radical-mediated oxidation of cellular biomolecules such as lipids, proteins, and direct scavenging of ROS by carboxylated chitooligosaccharides (CCOS). Moreover, Ngo et al. (2009) have reported the cellular antioxidant effects of chitin oligosaccharides produced from crab shell, which can inhibit myeloperoxidase activity and decrease free radical oxidation of DNA and membrane proteins.

It has been reported that some of the marine algae-derived sulfated polysaccharides (SPs) can be used as potential antioxidants (Kim et al., 2007; Wang et al., 2008). Furthermore, SPs from marine algae act as free radical scavengers and they prevent the oxidative damage, which is an important contributor in carcinogenesis. Structures of marine SPs (**Figure 23.1**) vary according

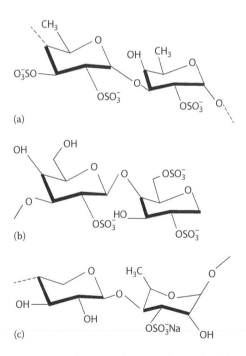

(a)

(b)

(c)

Figure 23.1 *Monemeric units of antioxidative polysaccharides from marine algae. (a) Fucoidan, (b) carrageenan, and (c) ulvan.*

to the species of algae such as fucoidan in brown algae (Phaeophyceae), carrageenan in red algae (Rhodophyceae), and ulvan in green algae (Chlorophyceae) (Costa et al., 2010). It has been found that those SPs are more effective than many commercial antioxidants such as butylated hydroxy anisole (BHA) and α-tocopherol (Ngo et al., 2011).

It is suggested that the antioxidant activity of the SP protects the red microalga *Porphyridium* sp. against ROS produced under high solar irradiation, possibly by scavenging the free radicals produced in the cell under stress conditions and transporting them from the cell to the medium (Cheng et al., 2005).

Carotenoids are also listed in marine-derived nutraceuticals, which have antioxidant properties. However, the modes of action, which is responsible for these antioxidant properties, are attributed by singlet oxygen quenching properties and their ability to trap free radicals (Britton, 1995; Stahl and Sies, 1996). Fucoxanthin (**Figure 23.2**) and astaxanthin (**Figure 23.3**) are known to be major marine carotenoids and have also been recognized to possess excellent antioxidative potential (Miyashita and Hosokawa, 2008).

Fucoxanthin derived from the brown alga *Undaria pinnatifida* and *Laminaria japonica* are effective on scavenging the free radicals (Sachindra et al., 2007), which leads to prevent oxidation of lipids. Furthermore, Heo et al. (2008) have reported the cytoprotective effect of fucoxanthin against H_2O_2-induced cell damage and can effectively inhibit intracellular ROS formation, DNA damage, and apoptosis induced by H_2O_2. Moreover, fucoxanthin has the potential to prevent UV-B-induced cell injury in human dermal fibroblasts (HDF cells) (Heo and Jeon, 2009).

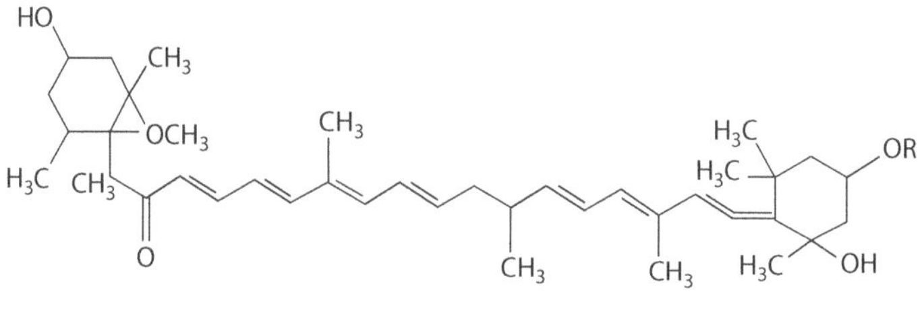

Figure 23.2 Molecular structure of marine-derived carotinoids, fucoxanthin.

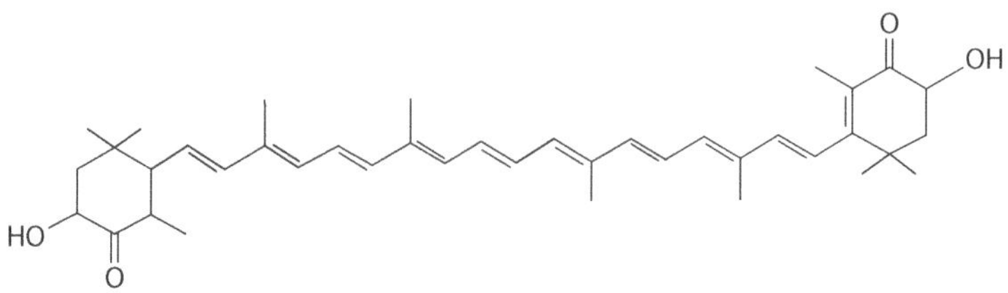

Figure 23.3 Molecular structure of marine-derived carotinoids, astaxanthin.

Many bioactive molecules from marine algae produce antioxidant effects with different modes of actions. For example, Phloroglucinol purified from *Conyza aegyptiaca* was shown to protect cells from oxidative damage by increasing cellular catalase activity and by regulating the extracellular signal-related kinase (ERK) signaling pathway (Kang et al., 2006). It was also shown to inhibit the production of nitrite, a known precursor of carcinogenic N-nitroso compounds (Choi et al., 1997; Park, 2005). Elsewhere, 5-hydroxymethyl-2-furfural isolated from marine red algae demonstrated antioxidant activities such as free radical scavenging and myeloperoxidase (MPO) inhibition, and increased the expression of superoxide dismutase (SOD) and glutathione (GSH) (Li et al., 2009a). Meanwhile, fucoidan and other polysaccharide-rich extracts have shown good superoxide anion scavenging ability (Qi et al., 2005; Xing et al., 2005).

23.3.2 Anti-inflammatory nutraceuticals

Inflammation is the first response of the immune system to infection or irritation and may be referred to as the innate cascade. However, some of the inflammatory reactions can affect to host own cells or tissues and chronic inflammation can lead to diseases such as arthritis, hepatitis, gastritis, periodontal disease, colitis, atherosclerosis, pneumonia, and neuro-inflammatory diseases (Pangestuti and Kim, 2004; Kim and Kim, 2010). Therefore, people are more attentive about natural anti-inflammatory substances especially on marine nutraceuticals, which gives fewer side effects. Potential anti-inflammatory nutraceuticals such as fucoidan, chitin, chitosan, COSs, and glucosamine play a significant role in the prevention of chronic inflammation either by inhibiting the production of nitric oxide (NO), prostaglandin E_2, inducible nitric oxide synthase (iNOS), cyclooxygenase-2 (COX-2), or proinflammatory cytokines (Park et al., 2003b; Rajapakse et al., 2008: Kim and Kim, 2010; Peerapornpisal et al., 2010).

According to Shin et al. (2011), algal extracts regulate inflammatory responses induced by oxidative stress. Moreover, they focused on toll-like receptor 4 (TLR4) signaling pathway, which is centrally involved in the inflammatory effects of lipopolysaccharide (LPS) that regulates the nuclear factor (NF)-κB and mitogen-activated protein kinase (MAPK) signaling pathways. Binding of LPS to TLR4 is important in the initiation and propagation of inflammatory responses (Hallenbeck, 2002; Janeway and Medzhitov, 2002). Extracted glycoprotein (PGP) from *Porphyra yezoensis* has antioxidant and anti-inflammatory effects, and assessed the potential contribution of TLR4 signaling to the anti-inflammatory mechanism of PGP. Similarly, Carlucci et al. (1997) also found that the antioxidant and anti-inflammatory effects of a glycoprotein isolated from the algae, *P. yezoensis*.

Ohgami et al. (2003) reported the effect of astaxanthin (**Figure 23.3**), a carotenoid found in crustacean cells, salmon, and sea stars on LPS-induced uveitis in rats both in vitro and in vivo. The mechanism of action determined for astaxanthin probably involved inhibition of NO, prostaglandin E_2, and TNF-α generation.

Anti-inflammatory effects of glucosamines in respect to prevention of osteoarthritis have been observed by Bruyere et al. (2004). Furthermore, Mayer et al. (2010) also have reported decreased symptoms of osteoarthritis from fucoidans found in the seaweed extract nutrient complex given orally, which will be more applicable as an anti-inflammatory nutraceutical. However, the omega-3 FAs such as EPA, which have been identified as a main active ingredient in the nutraceuticals, exert the anti-inflammatory action by inhibiting the metabolism of arachidonic acid by down regulation, thereby preventing the production of proinflammatory metabolites (Geering et al., 2006). Inflammatory bowel disease, lupus, and rheumatoid are treated with marine-derived n-3 fatty acids. Omega-3 and omega-6 are precursors for eicosanoids, a group of physiologically active molecules that have numerous biological effects.

Most anti-inflammatory action of those marine nutraceuticals are in the means of suppression of inflammatory cytokines and increasing epithelial cells internalization. However, there are few modes of actions that are specific to each substance. Furthermore, Lucas et al. (2003b) described that a detailed mechanistic study on the modulatory effect of bolinaquinone, a sesquiterpenoid isolated form a *Dysidea* sp. sponge, in several models of acute and chronic inflammation. Furthermore, marine-derived nutraceuticals with anti-inflammatory effects are listed in **Table 23.2**.

23.3.3 Anticoagulant activity and antithrombotic activities

Anticoagulant/antithrombotic, antiviral, immuno-inflammatory, antilipidemic, and antioxidant activities of SPs and their potential for therapeutic applications were found recently. For example, SPs synthesized by seaweeds including the galactans (e.g., agarans and carrageenans), ulvans, and fucans are recognized to possess a number of biological activities including anticoagulant, antiviral, and immuno-inflammatory activities that might find relevance in nutraceutical/functional food, cosmetic/cosmeceutical, and pharmaceutical applications.

Probably the most widely recognized and studied bioactivity of marine SPs is the heparin-like anticoagulant activity exhibited by fucoidans and other fucans of brown seaweeds. This was first reported for fucoidan isolated from *F. vesiculosus* by Springer and coworkers who found inhibition of fibrin clot formation and antithrombin activity (Bernardi and Springer, 1962).

Heparin is used extensively for the prevention of venous thrombosis and the treatment of other thromboembolic disorders due to its inhibition of thrombin and other enzymes in the coagulation system. To overcome the obvious potential side effects of bleeding, researchers have investigated means of reducing the anticoagulant activities of heparin while enhancing its antithrombotic

Table 23.2 Different Marine-Derived Anti-Inflammatory Compounds and Their Pharmacological Activities

Compound	Organisms	Chemistry	Pharmacological Activity	References
Astaxanthin	Salmon, sea stars	Tetraterpenee	Inhibition of endotoxin-induced uveitis in rats	Ohgami et al. (2003)
Bolinaquinone	Sponge, *Dysidea* sp.	Merosesquiterpenee	Inhibition of cytokine, iNOS and eicosanoids	Lucas et al. (2003b)
Cacospongionolide B	Sponge	Sesterterpenee	NO, PGE$_2$, and TNF-α inhibition in vitro and in vivo	Posadas et al. (2003a)
Clathriol B	Sponge	Sterole	Neutrophil superoxide inhibition	Keyzers et al. (2003)
Conicamin	Tunicate	Indole alkaloid	Histamine antagonist	Aiello et al. (2003)
Cycloamphilectene 2	Sponge	Diterpenee	NO inhibition	Lucas et al. (2003a)
Plakohypaphorine D	Sponge	Indole alkaloid	Histamine antagonist	Borrelli et al. (2004)
Pourewic acid A	Sponge	Diterpenese	Superoxide inhibition	Keyzers et al. (2004)
Cadlinolide C	Sponge	Diterpenee	Superoxide inhibition	Keyzers et al. (2004)
Petrocortyne A	Sponge	Polyacetylene	Macrophage inflammatory mediator inhibition	Hong et al. (2003)
Petrosaspongiolide M	Sponge	Sesterterpenee	NO, PGE$_2$, and TNF-α inhibition in vitro and in vivo	Posadas et al. (2003b)
Petrosaspongiolides M-R	Sponge	Sesterterpenese	Macrophage inflammatory mediator inhibition	Monti et al. (2004)
Pseudopterosin N	Sea whip	Diterpenee	Inhibition of mouse ear inflammation	Ata et al. (2003)
Pseudopterosin R	Sea whip	Diterpenee	Microglia thromboxane B$_2$ inhibition	Rodriguez et al. (2004)

activities including chemical modification and fractionation of native heparin to lower molecular forms (Barrow et al., 1994; Mourao and Pereira, 1999).

Carroll et al. (2004) found that three new peptides, dysinosins B, C, and D, isolated from the sponge *Lamellodysidea chlorea*, that inhibited the blood coagulation cascade serine proteases factor VIIa and thrombin. Furthermore, according to Zancan and Mourao (2004) and Mourao (2004), the two structural motifs of the dysinosins contributed to the binding of these compounds to factor VIIa and thrombin proteases.

A glycosaminoglycan called fucosylated chondroitin sulfate isolated from the Brazilian sea cucumber *Ludwigothurea grisea* showed antithrombotic activity. Furthermore, Melo et al. (2004) extended the pharmacology of anticoagulant sulfated galactans isolated from the red alga *Botryocladia occidentalis* and the sea urchin *Echinometra lucunter*.

Various alga contain significant amounts of anticoagulants. For example, carrageenans and agarans are derived from red algae. One of the best studied agarans is porphyran (Morrice et al., 1984) obtained from *Porphyra* species of red algae including *Porphyra capensis* and *Porphyra haitanensis* (Zhang et al., 2004). Furthermore, SPs from green algae such as Ulvan is the major water-soluble polysaccharide found in green seaweed of the order Ulvales (*Ulva* and *Enteromorpha* sp.) that has sulfate, rhamnose, xylose, iduronic, and glucuronic acids as main constituents (Cueto et al., 2001).

Pereira et al. (2002) reported a novel sulfated α-L-galactan isolated from the crude egg jelly of the sea urchinus *E. lucunter* and sulfated galactan had potential of anticoagulant activity as shown by the enhancement of thrombin or factor Xa inhibition by antithrombin.

23.3.4 Antimicrobial and anti-infection function of marine nutraceuticals

Marine-derived antimicrobial compounds play a significant role as nutraceuticals. Furthermore, the antimicrobial properties can improve the shelf-life of foods as well as prevent the effects of food-borne pathogens. The extracts from the marine algae, chitin derivatives, chitosan, COS, and certain bacteriocin have been identified as important antimicrobial compounds, which have the potential to prevent disease risks in humans (Vidanarachchi et al., 2010). Therefore, marine-derived antimicrobials have a high potential to be used as nutraceuticals in the food industry.

23.3.5 Antibacterial marine nutraceuticals

In view of the fact that resistance to current antibiotics remains a significant challenge for pathogenic bacterial infections, during past few years, scientists directed their studies to the search for novel antibacterial marine natural products that were identified as effective and less harmful to human body (Mayer and Lehmann, 2000; Mayer and Hamann, 2002, 2004, 2005).

Bugni et al. (2004) investigated a series of kalihinols, diterpenes isolated from the Philippine marine sponge *Acanthella cavernosa* as potential bacterial folate biosynthesis inhibitors.

A new dimeric bromopyrrole alkaloid, Nagelamide (**Figure 23.4**), was isolated from the Okinawan marine sponge *Agelas* sp. (Endo et al., 2004). Nagelamide G exhibited antibacterial activity against *Micrococcus luteus*, *Bacillus subtilis*,

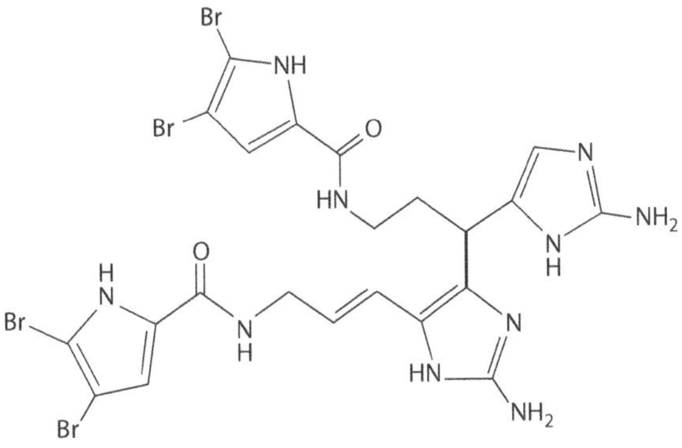

Figure 23.4 *Molecular structure of nagelamide.*

and *Escherichia coli*, but weakly inhibited protein phosphatase, thus suggesting that this enzyme may not be the main molecular target responsible for the antibacterial activity of this compound.

The different structures of marine-derived COS that contribute to the antibacterial activity have also been studied. COS-GTMAC was obtained, which exhibited 80% growth inhibition against *Streptococcus mutans*, whereas the COS showed 10% growth inhibition (Kim et al., 2003). Oligosaccharides can alter permeability characteristics of microbial cell membrane and further prevent the entry of materials or cause leakage of cell constituents that finally leads to bacterial death (Sudarshan et al., 1992). The exposure of *Aggregatibacter actinomycetemcomitans* to the COS resulted in the disruption of cell membranes and that could be considered for the treatment of periodontal diseases associated with this pathogen. Similar results were observed by Eaton et al. (2008), which have shown that COS had antimicrobial activity and were able to stimulate immune response. Lee et al. (2009) examined the antibacterial activity of water-soluble COS against *Vibrio vulnificus,* as well as their inhibitory effects against *V. vulnificus*-induced cell cytotoxicity, which proved the strong activity in intestinal epithelial cells in vitro.

Furthermore, several marine nutraceuticals were identified as antibacterial agents against certain bacteria (**Table 23.3**).

23.3.6 Antiviral marine nutraceuticals

The antiviral therapeutic efficacy of natural products can be certified than the synthetics, since they can easily convert their conventional use to therapeutic use against viral infection. Hence, their biological activity and safety for humans have undergone scientific evaluation worldwide and are recognized as having minimal side effects for the body. Currently there are nearly 40 antiviral substances that have been officially approved for clinical use. More than half of these are used for the treatment of human immunodeficiency virus (HIV) infection. Taking this situation into account, the

Table 23.3 Different Antimicrobial Marine Nutraceuticals and Their Pharmacological Activities

Compound/ Organism	Chemistry	Pharmacologic Activity	References
Spongosorites sp. alkaloids/ sponge	Alkaloid	*Streptococcs aureus* inhibition	Fennell et al. (2005)
Aurelin/jellyfish	Peptide	*E. Coli* inhibition	Fernandez-Busquets and Burger (2003)
Batzellaside A/sponge	Alkaloid	*S. epidermidis* inhibition	Frenz et al. (2004)
6-oxo-de-*O*-methyllasiodiplodi n/ fungus	Polyketide	*B. subtilis, S. aureus*, and *S. enteritidis* inhibition	Fujita et al. (2003)
Grammistins/fish	Peptide	*B. subtilis, S. aureus*, and *E. coli* inhibition	Gochfeld et al. (2003)
Marinomycins A–D/ bacterium	Polyketide	*S. aureus* inhibition	Goud et al. (2003)
Resistoflavin methyl ether/ bacteria	Polyketide	*B. subtilis* inhibition	Grant et al. (2004)
Streptomyces anthraquinones/bacterium	Polyketide	Methicillin-resistant *S. aureus* inhibition	Gustafson et al. (2004)
Xeniolide I/soft coral	Terpene	*E. coli* and *B. subtilis* inhibition	Hirono et al. (2003)

importance of developing new antiviral agents is clear. Therefore, scientists are mainly focusing on developing new compounds that can be used to treat viral diseases; hence, they are considered as difficult to control due to the nature of viruses. Therefore, it is necessary to develop new antiviral agents. Many of the natural marine derivates contain an abundance of biologically active substances with favorable pharmacological activities. Many marine algae, sponges, and other organisms contain antiviral compounds having potential antiviral properties.

Several investigations have described that the viral inhibitory effects of marine algal extracts from recent past. Algae contain large amounts of cell wall polysaccharides, peptides, fatty acids, and some phenolic compounds. SPs such as heparin, dextran sulfate, pentose polysulfate, and mannan sulfate cyclodextrans inhibit the replication of various enveloped viruses, including herpes simplex virus (HSV), human cytomegalo virus (HCMV), and HIV (Witvrouw et al., 1994).

Many investigators have reported the inhibitory effects of algal extracts and their constituents on the replication of HSVs. The HSVs are responsible for a broad range of human infectious diseases. Mazumder et al. (2002) found that the sulfated galactan derived from marine sources have antiviral activity against HSV-1 and -2 in bioassays, which is likely due to an inhibition of the initial virus attachment to the host cell.

Moreover, HSV infections were recognized as a risk factor for HIV infection (De Clercq, 2004). Spector et al. (1983) also found that the SP of the red microalgae *Porphyridium* sp. showed high order of antiviral activity against HSV-1 and -2 both in vitro and in vivo. According to Deig et al. (1974), aqueous extracts of *Farlowia mollis* and *Cryptosiphnia woodii* significantly interfered with HSV-1 reproduction. Ehresmann et al. (1977) studied the virus inhibitory activities of 28 extracts from marine algae and found that 10 species of Rhodophyta contained substances that significantly reduced HSV-1 infection. Carageenans type isolated from red seaweed *Gigartina skottsbergii* also potently and selectively inhibits HSV types 1 and 2 in plaque reduction assays in vitro (Carlucci et al., 1997).

Oligosaccharides are useful in preventing several phage infections. The mechanisms are described by Kochkina and Chirkov (2000) as inhibition of the bacteriophages' replication involved in structural change of phage particles or inhibition of the receptor-recognizing structures on them. Marine-derived COS are considered as an inactivated mucosal influenza vaccine by stimulating functional activity of macrophages and increasing generation of active oxygen species, which lead to viral destruction (Bacon et al., 2000). Another possible mechanism of antiviral activity of COS is explained in relation to the interactions between protein receptors on viral coat and blood leucocytes (Sosa et al., 1991). Furthermore, anionic derivatives of COS results to lower antiviral activity while increasing the degree of deacetylation improves antiviral activity.

AIDS remains one of the most dangerous viral diseases affecting a large number of people all over the world. The number of people suffering from AIDS is approximately 50 million and increases every day by 16,000. By early 2003, more than 150 highly active marine metabolites were found in the course of testing against HIV (Tzivelka et al., 2003; Gustafson et al., 2004). Edible algal polysaccharides, particularly fucoidans, carrageenans, and others, inhibit the penetration of HIV into human mononuclear cells. Some of them inhibit virus replication in very low concentrations and intensify the antiviral action of azidothymidine (Mayer et al., 2007).

Carrageenan is another substance derived by marine algae, which has been reported with anti-HIV activity. However, its strong anticoagulant activity is considered to be an adverse reaction according to Yuan and Song (2005). Moreover, Yamada et al. (2000) examined *O*-acylated carrageenan with different molecular weights and sulfate contents and the result showed that anti-HIV activity was increased by depolymerization and sulfation with low anticoagulant.

Great number of reports in literature shows that different pharmacological activities including antiviral activity of marine sponges (**Table 23.4**). One of the most important substances from sponges is the nucleoside ara-A isolated from the sponge *Cryptotethya crypta* that inhibits viral DNA synthesis (Bergmann and Feeney, 1951). Avarol inhibits HIV by blocking completely the synthesis of natural UAG suppressor glutamine transfer tRNA. Synthesis

Table 23.4 Antiviral Substances Derived from Marine Sponges

Substances	Class	Species	Action Spectrum	Reference
4-Methylaaptamine	Alkaloid	*Aaptos aaptos*	HSV-1	Souza et al. (2007)
Ara	Nucleoside	*C. crypta*	HSV-1, HSV-2, VZV	Bergmann and Feeney (1951)
Avarol	Sesquiterpene hydroquinone	*Dysidea avara*	HIV-1	Muller et al. (1987)
Dragmacidin	Alkaloid	*Halicortex* sp.	HSV-1	Cutignano et al. (2000)
Hamigeran		*Hamigera tarangaensis*	Herpes and polio viruses	Wellington et al. (2000)
Microspinosamide	Cyclic depsipeptide	*Sidonops microspinosa*	HIV-1	Rashid et al. (2001)
Mycalamide	Nucleosides	*Mycale* sp.	A59 coronavirus, HSV-1	Perry et al. (1990)
Papuamides	Cyclic depsipeptides	*Theonella* sp.	HIV-1	Ford et al. (1999)

of this tRNA is upregulated after viral infection, and it is important for the synthesis of a viral protease, which is necessary for viral proliferation (Muller and Schroder, 1991).

23.3.7 Antifungal marine nutraceuticals

Marine-derived COS could react with negatively charged groups of fungi and then exhibited the antifungal activity against both mold and yeast-like fungi. The inhibition mechanism of COS against fungi is similar to that against the bacteria. The reaction between COS and negatively charged groups on the cell surface directly interferes with the growth and normal physiological functions of fungi, suggesting that the antifungal activity is correlated with the charge distribution of COS (Hirano and Nagao, 1989). The inhibitory activities of isozyme and chitinase were lower than that of low molecular weight (LMW) chitosan and COS. This result strongly suggests that the depolymerized products of chitosan are effective for growth inhibition. Chitosan acts in the growth inhibition in several fungi (Kurosaki et al., 1986).

Yang et al. (2003) reported a new sterol sulfate isolated from a deep-water marine sponge of the family Astroscleridae, which exhibited antifungal activity against "supersensitive" *Saccharomyces cerevisiae*.

Nishimura et al. (2003) focused their research efforts in identifying inhibitors of the pathogenic fungus *Candida albicans*, geranylgeranyltransferase (GGTase), an enzyme that shares only 30% amino acid sequence homology with the human GGTase. Bioassay-guided fractionation resulted in the isolation of a novel alkaloid massadine from the marine sponge *Stylissa affmassa*, which inhibited fungal GGTase.

Similarly, another novel nutraceutical alkaloid, naamine, was reported from the Indonesian marine sponge *Leucetta chagosensis* that exhibited strong antifungal activity against the phytopathogenic fungus *Cladosporium herbarum* (Hassan et al., 2004). It remains to be determined if this compound will also be effective against fungi that infect mammalian hosts.

23.3.8 Anticancer/antitumor nutraceuticals

Tumor/cancer is one of the most dangerous threats against human health throughout the world. Although many drugs have been used in treating tumor, the side effect of those drugs exceeds the tolerance of many patients. Recently, many researchers have come to the realization that marine organisms hold immense potential as a source of novel molecules and new anticancer agents. Therefore, marine-derived nutraceuticals such as β-carotene, fucoidans, chitosan, chitosan oligosaccharides, astaxanthin, and phlorotanins are heavily used as anticancer agents.

Among all those agents, researchers pay more attention on marine oligosaccharides, which showed fewer side effects. These marine oligosaccharides could not only directly kill the tumor cells but also impair the tumor's nutritional support system by targeting the tumor blood vessel network, activating the immune system, etc. However, the mode of action of these marine-derived anticancer compounds varies and the cancer inhibitory pathways include activation of P53 anti-proliferative gene, induction of cell apoptosis through various pathways, inhibition of angiogenesis, and inhibition of matrix metalloproteinases (MMPs) (Vidanarachchi et al., 2011).

Over the last decade many cytotoxic compounds have been reported, some of which are now in clinical trials for anticancer, such as dehydrodidemnin B (**Figure 23.5**) and bryostatins (Rinehart et al., 1981).

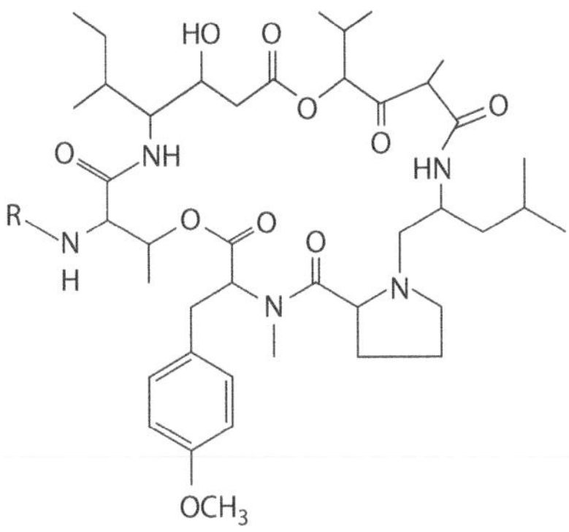

Figure 23.5 *Molecular structure of didemnin.*

Bryostatin-1 is a macrocyclic natural lactone isolated from the marine Bryozoan, *Bugula neritina*. It has shown both antitumor and immunomodulatory effects (Wender et al., 1999). It is a potent activator of the protein kinase C (PKC) family, lacking tumor-promoting activity and with antagonistic effects on tumor-promoting phorbol esters. Bryostatin-1 has cytotoxic activity against various leukemia and solid tumor cell lines in vitro (Hornung et al., 1992). It has also shown in vivo antitumor activity in various murine models, including leukemia, lymphoma, ovarian cancer, and melanoma.

The dolastatins are cytotoxic peptides, which can be cyclic or linear, derived from the sea hare, *Dolabella auricularia*, a mollusc from the Indian Ocean. It was shown to inhibit microtubule assembly, tubulin-dependent guanosine triphosphate (GTP) binding and inhibit vincristine and vinblastine binding to tubulin in vitro. Dolastatin 10 has activity against several human leukemia, lymphoma, and solid tumor cell lines in vitro. It has documented antitumor activity in various human solid tumor models, such as melanoma and ovarian carcinoma (Maki et al., 1996). Dehydrodidemnin B is a compound from Mediterranean tunicate. Surprisingly, it has shown to be six times more effective than didemnin B in animal tests (Jha and Zi-rong, 2004; Kijjoa and Sawangwong, 2004).

Parish et al. (1999) found that sulfated oligosaccharides could reduce tumor metastasis by inhibiting heparanase activity and/or by simultaneously inhibiting angiogenesis via blocking angiogenic growth factor action.

Mode of actions of many marine anticancer compounds is through directly killing tumor cells or inhibiting the growth of vessels. It is now well established that solid tumor growth is critically dependent on the growth of new vessels from pre-existing blood vessels surrounding the tumor, a process named angiogenesis (Douglas and Judah, 1996).

Usually the inhibition of cell apoptosis is a characteristic of cancers and the induction of apoptosis through nutraceuticals can play a major role in the destruction of cancer cells. For example, marine-derived fucoidans and phlorotanins enhance cell apoptosis through activating caspases which are enzymes involved in the apoptotic pathway (Kong et al., 2009; Kim et al., 2010) while κ-carageenan oligosaccharides and carotenoids stimulate the host immune system for the cancer cell apoptosis. Furthermore, some specific oligosaccharides on the tumor cell surface can interact with other marine-derived oligosaccharides, resulting in the death of the tumor cell. This antitumor activity has been proved by mild hydrochloric acid hydrolysis of carrageenan SPs from *Kappaphycus striatum* (Yuan and Song, 2005).

Beta-carotene is also responsible for the cancer cell apoptosis via expression of P53 antiproliferative gene in cancer cells and inhibition of angiogenesis, which prevents the blood supply to the developing cancer cells (Jawanda, 2009).

Moreover, the nutraceuticals that have the ability to inhibit MMPs are important for the prevention of tumor invasion to the surrounding tissues. Fucoidans, chitin, and chitin derivatives such as chitosan, COS, and glucosamines (Kim and Kim, 2006; Rajapakse et al., 2007) have been identified as important for the inhibition of MMPs, which degrades the extracellular matrix and allow the tissue cells to get invade. Therefore, the marine-derived compounds can play a greater role as antitumor nutraceuticals in the prevention and treatment of human cancers.

Similarly, the marine-derived biologically active compounds have important biological properties such as antioxidant, anti-inflammatory, anticoagulant, antithrombotic, and antimicrobial activities, which increase their application as nutraceuticals in the biomedical field. Similar to the biological applications, the application of marine nutraceuticals in biomedical field has attracted the attention of nutraceutical industry due to the high contribution of marine nutraceuticals in increase of lifespan of people and increase of quality of human life. The combined effects of several biological nutraceuticals such as antioxidant nutraceuticals, anti-inflammatory nutraceuticals, anticancer nutra-ceuticals, or anticoagulant nutraceuticals as well as the other important prop-erties of marine nutraceuticals such as antiobesity (Maeda et al., 2006; Kim et al., 2009b), antidiabetic (Hayashi and Ito, 2002; Lee et al., 2003; Liu et al., 2007), and hypocholesterolemic (Moon et al., 2007; Yao et al., 2008) effects can be used for the prevention and treatment of certain important disease conditions.

Marine nutraceuticals can play a vital role in photoprotection, neuroprotec-tion, and inflammatory conditions in the gut. Certain marine-derived nutra-ceutical compounds also promote the health of gut through probiotic or prebiotic nutraceuticals. Furthermore, marine bioactive compounds can be applied as nutraceuticals, which are important for the delivery of important drugs to the targeted sites without getting destroyed from the gastrointestinal enzymes. Moreover, nutraceuticals provide beneficial health effects for obe-sity, diabetes, and CVD, which have increased in the world during the recent past due to the lifestyle modifications of people.

23.3.9 Applications of marine nutraceuticals against obesity/antiobese marine nutraceuticals

Irrational dietary patterns and lack of physical exercise have triggered the condition of obesity around the world predisposing to many negative conse-quences such as increased risk of diabetes and CVD (Wallen and Hotamisligil, 2005; Dieterle et al., 2006). Obesity is a chronic metabolic disorder caused by an imbalance between energy intake and expenditure, which will ultimately result in accumulation of triglycerides in adipose tissue (Park et al., 2011).

Therefore, the inhibition of fat absorption while enhancing fecal excretion of triglycerides and cholesterol, suppression of adipogenesis and enhancement of lipolysis in adipose tissue may have a high contribution for the prevention of accumulation of triglycerides in the adipose tissue.

23.3.10 Marine-derived nutraceuticals for inhibition of dietary fat absorption

The prevention of fat absorption with the increase fecal fat excretion is the first step that can control obesity in humans. The marine-derived chitosans have an important function in the prevention of absorption of dietary lipids. According to the recent review of Prajapati (2009), chitosan inhibits fat absorption either by binding with fatty acids or bile acids in the human gastrointestinal tract. Chitosan has characteristic positively charged amino groups in their structure, which has the ability to bind with negatively charged fatty acids or bile acids. Since chitosan is indigestible by the gastrointestinal enzymes and nonabsorbable, fats bound to chitosan cannot be absorbed at the level of small intestine (Kaats et al., 2006). Furthermore, the solubilization of dietary fats is obstructed by the binding of chitosan with bile acids, which prevents the digestion and absorption of dietary fats (Kanauchi et al., 1995).

Although the mechanism is different from chitosan, certain other marine-derived compounds such as fucoidans and fucoxanthin also have an important role in the prevention of fat absorption. Increased fecal excretion of triglycerides and cholesterol due to the inhibitory effects on fat absorption has been observed from edible kelp, *Laminaria* sp., which contains alginate, fucoxanthin, and fucoidan as active substances (Miyashita et al., 2009). The observed low-fat absorbability from kelp may be mainly due to the biological properties of fucoxanthin found in *Laminaria* sp.

23.3.11 Marine-derived nutraceuticals for inhibition of adipogenesis

The absorbed fats are utilized for the body functions, whereas additional amounts are deposited in adipocytes of different body tissues, especially in white adipose tissues (WAT). Free fatty acids circulating in plasma are taken into the adipocytes and transported within the cells through binding to different proteins and undergo a cascade of reactions mediated by different enzymes such as acetyl CoA carboxylase to get converted into triglycerides, which will deposited finally in the adipocytes (Shi and Burn, 2004). This deposition of fat within body tissues has a high contribution for the development of obesity in humans. Thus the marine-derived nutraceuticals with the property of antiadipogenesis have a vital role in the prevention of human obesity.

The recent study of Kim et al. (2009b) has reported marked inhibition of adipogenesis from marine-derived fucoidans at a non-cytotoxic level of 1000 μg/mL mediating through inhibition of gene expression of fatty acid binding proteins (aP2), acetyl CoA carboxylase, and peroxisome proliferation-activated receptor γ (PPARγ). Since aP2, acetyl CoA carboxylase, and PPARγ are important for the differentiation of preadipocytes into adipocytes, fucoidans inhibit the final differentiation of adipocytes, thereby reducing the lipid accumulation in WAT.

Reduction of lipid accumulation in WAT through inhibition of PPARγ gene expression has been also reported from marine carotenoid; fucoxanthin. Fucoxanthin is converted into active form; fucoxanthinol in the small intestine and further converted to amaronciaxanthin A in the liver. Out of the two active forms of fucoxanthin, fucoxanthinol is mainly responsible for the inhibition of gene expression of PPARγ. Furthermore, fucoxanthinol inhibit glycerol-3-phosphate dehydrogenase activity, which is involved in the production of glycerol-3-phosphate within the 3T3-L1 cells in the adipogenesis pathway and obstruct the differentiation of adipocytes (Maeda et al., 2006). Similarly, Woo et al. (2009) also have observed inhibition of adipogenesis enzymes such as glycerol-3-phosphate dehydrogenase, malic enzyme, and fatty acid synthase from fucoxanthin favoring the application of fucoxanthin as antiadipogenesis agent.

Since the marine-derived fucoidans and fucoxanthin are able to inhibit the fatty acid transport proteins as well as the enzymes involved in adipogenesis, these marine materials can be applied as adipocyte differentiation inhibition nutraceuticals for the benefit of human health.

23.3.12 Role of marine-derived nutraceuticals in lipolysis and fat oxidation

Fats accumulated within the adipocytes are known to contribute for the metabolic thermogenesis through β-oxidation. Metabolic thermogenesis in brown adipose tissues (BAT) is mediated by uncoupling protein 1 (UCP-1) expression and dysfunction of this system contributes to the development of obesity (Miyashita et al., 2011). However, adult humans have low level of BAT whereas most of the fat is stored in WAT (Maeda et al., 2008). Therefore, the discovery of reduction of WAT weight through upregulation of expression of UCP-1 in WAT by marine-derived fucoxanthin is a positive factor in the development of nutraceuticals from the natural compounds (Maeda et al., 2005, 2007, 2008). Maeda et al. (2005) have observed a significant reduction of WAT weight from dietary *Undaria* lipids at a higher level such as 2%, whereas no significant reduction of WAT weight has been observed from 0.5% dietary *Undaria* lipids compared to the control group in an animal model. However, the same study has revealed significant reduction of WAT weight from incorporation of 0.4% marine fucoxantin in the diet reflecting the importance of fucoxanthin as an antiobesity agent. Similarly, Woo et al. (2009) and Jeon et al. (2010) have reported reduction of WAT weight from the doses of 0.2% and 0.05% fucoxanthin in the diet in a mice model. However, the inhibitory dose of fucoxanthin in UCP-1 mRNA expression in these two studies is controversial. Woo et al. (2009) have observed a higher induction of UCP-1 gene expression in WAT from a lower dose such as 0.05% dietary fucoxanthin whereas the study of Jeon et al. (2010) has not observed significant induction of UCP-1 mRNA expression from 0.05% fucoxanthin in the diet. According to Jeon et al. (2010), significant increase of UCP-1 mRNA expression has been reported from 0.2% dose of dietary fucoxanthin. Therefore, the determination of appropriated dose and the mode of action of fucoxanthin need further investigations prior to applying them as nutraceuticals. However, fucoxanthin has a greater potential for application as antiobese nutraceutical due to the presence of inhibitory effects on the visceral adipose tissue weight which reflects the prevention of accumulation of triglycerides within the adipocytes.

The accumulation of triglycerides within the adipocytes can be also prevented by enhancing the lipolysis process. Usually the relative rate of lipolysis is determined by the nutritional status and is regulated by endocrine factors such as catecholamines and insulin, which stimulates enzymes for lipolysis, especially the hormone-sensitive lipase (Shi and Burn, 2004). Marine-derived fucoidans enhance the lipolysis process in adiopocytes through increasing the protein expression of hormone-sensitive lipase increasing the applicability of fucoidans as an antiobese nutraceutical (Park et al., 2011).

Therefore, the marine-derived active compounds can play a vital role as antiobese nutraceuticals. The marine-derived chitosans have the nutraceutical applicability mainly due to the reduction of fat absorption while brown algal fucoidans and the carotenoid fucoxanthin have a higher potential to be applied as antiobese nutraceutical due to their involvement in inhibition of fat absorption as well as the inhibition of adipocyte differentiation and enhancement of fat oxidation or lipolysis in adipose tissues.

23.3.13 Applications of marine nutraceuticals as antidiabetic agents

Diabetes mellitus is a highly prevalent metabolic disorder that can be mainly classified as type I diabetes and type II diabetes, depending on the involvement of insulin in the development of disease. Type I or insulin-dependent diabetes occurs due to the insufficiency of insulin, which is usually characterized by destruction of insulin-producing β-cells in the pancreas, whereas ineffective utilization of insulin leads to the type II diabetes, which is also called noninsulin-dependent diabetes (Kong and Kim, 2010). The type II diabetes is further divided into lean type or obese type. However, currently obesity is the major risk factor for the development of noninsulin-dependent diabetes, especially in the people who are genetically predispose to inadequate β-cell compensation of insulin (Khan, 1998). In obesity, the excess accumulation of fat in WAT alters the endocrine function of WAT, deregulating the production of adipocytokines such as monocyte chemotactant protein (MCP-1), tumor necrosis factor α (TNF-α), interleukin-6 (IL-6), and adinopectin (Flier, 2004). Increase production of certain adipocytokines inhibit insulin-dependent glucose uptake leading to insulin resistance. Apart from the deregulation of production of adipocytokines, many inflammatory, neural, and cell-intrinsic pathways have been shown to be deregulated in obesity leading to insulin resistance (Qatanani and Lazar, 2007). Furthermore, fat accumulation in obesity results in the production of high amount of ROS, leading to systemic oxidative stress (Furukawa et al., 2004; Houstis et al., 2006) which can damage the insulin secretory β-cells of the pancreas aggravating the condition (Kim et al., 2009a). Since the regeneration of β-cells is known to be slow, the destruction of insulin secretory cells leads to insulin insufficiency, which ultimately manifest as diabetes. Therefore, the active substances that have the ability to prevent the destruction of pancreatic β-cells enhance the regeneration of damaged β-cells and enhance the efficient utilization of insulin having the potential to be applied as nutraceuticals in the prevention of diabetes (**Figure 23.6**).

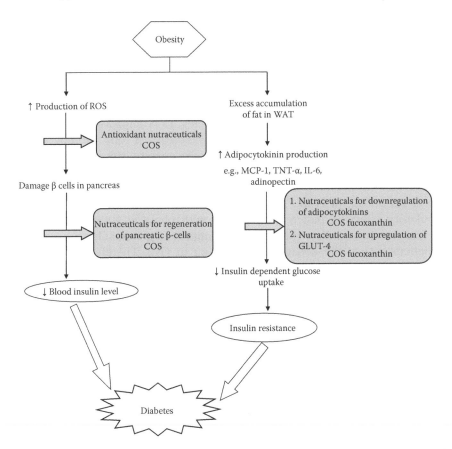

Figure 23.6 *Schematic diagram of development of diabetes and the nutraceuticals involved in the inhibition of different steps.*

Certain marine-derived polysaccharides have been identified as important compounds in the prevention of diabetes. Stimulatory effects on in vitro insulin secretion by rat RIN-SF cells have been observed from polysaccharides extracted from certain marine algae (Zhang et al., 2008). Furthermore, chitosan and COS have been identified as important antidiabetic compounds in many studies. Yao et al. (2008) have demonstrated the potential of high-molecular-weight chitosan in reducing hyperglycemia of diabetes rats, whereas plasma glucose level has not been reduced from the low-molecular-weight chitosan in the same rat study. In contrast, Hayashi and Ito (2002) have reported blood glucose level lowering effect from low-molecular-weight chitosan in a dose-dependent manner, which is favorable for the prevention and treatment of diabetes. The observed difference of the blood glucose lowering effect may have resulted from the differences in the chitosan structure and their mode of action in reducing hyperglycemia. The highly deacetylated COS with N-glucosamine units are known to have high biological properties including antioxidant activity (Je et al., 2004). The antioxidant property of COS can be applied for the minimization of the possible oxidative damage to the pancreatic β-cells from ROS (Ju et al., 2010). Yuan et al. (2009) have observed a dose-dependent pancreatic β-cell protective effect from COS at 10–500 mg/L in an in vitro study and significantly higher pancreatic β-cell protection effect from pre-treatment with COS at 500 mg/L in a mice model, reflecting the importance of antioxidant property of COS for antidiabetic effect.

Similarly, the regeneration of damaged pancreatic secretory cells is essential for the prevention of development of diabetes. COS improves the blood glucose metabolism through regeneration of pancreatic β-cells, thereby increasing the blood insulin needed for the glucose metabolism. The accelerated proliferation or neogenesis of pancreatic β-cells in streptozotocin-induced diabetic rats has been reported from the marine-derived COS (Liu et al., 2007; Kim et al., 2009a; Ju et al., 2010). Liu et al. (2007) have achieved maximum stimulatory effect on pancreatic β-cell growth phase at 100 mg/L concentration from in vitro studies while achieving improved general clinical symptoms in diabetic rats at the dosage of 500 mg/kg in vivo. Therefore, COS have a higher contribution as a nutraceutical in the prevention and treatment of diabetes through prevention of ROS-induced damage to the pancreatic β-cells and enhancing the regeneration of pancreatic secretory cells.

Even with normal pancreatic β-cell function, diabetes can occur due to insulin resistant, which is mediated by elevated levels of plasma adipocytokines. Kumar et al. (2009) have reported down-regulation of adipose tissue-specific TNF-α and IL-6 secretory molecules which belongs to the group of adipocytokines by treatment with COS. Furthermore, COS improves the insulin resistance directly by upregulating the expression of GLUT-4 (Ju et al., 2010), which is a major glucose transporter as well as an important regulatory factor in glucose metabolism (Shulman, 2000). Similar to COS, the marine carotenoid, fucoxanthin, is also known to have effects on down-regulation of TNF-α and IL-6. Furthermore, fucoxanthin has the ability to down-regulate

MCP-1 and plasminogen activator inhibitor, which are overexpressed in obese adipose tissue while upregulating GLUT-4 mRNA expression (Maeda et al., 2009). Therefore, fucoxanthin can be applied as antidiabetic nutraceutical due to its ability in improving the efficient utilization of insulin in glucose metabolism. Moreover, efficient utilization of glucose as well as prevention of destruction and enhancement of regeneration of pancreatic β-cells enables the application of COS in the nutraceutical industry.

23.3.14 Application of marine nutraceuticals in cardiovascular disease

Obesity as well as diabetes increases the risk of CVD, the consequences of which can be fatal in humans. Increased ROS produced in obesity contribute to the systemic oxidative stress, ultimately elevating the plasma lipid levels. Furthermore, ROS increases the insulin resistance leading to diabetes, which manifest as hyperglycemia (Wallen and Hotamisligil, 2005). Both hypercholesterolemia and hyperglycemia can damage the blood vessels, which are called microvascular disease, increasing the risk of CVD (Dieterle et al., 2006).

The term "cardiovascular disease" (CVD) is used for a group of disorders of the heart and blood vessels, which includes hypertension, coronary heart disease, heart failure, and peripheral vascular disease (Rajasekaran et al., 2008). The active compounds that can improve the different aspects of CVD have the potential to be applied as nutraceuticals for the improvement of human health (**Figure 23.7**). Thus the marine-derived compounds with the properties of reduction of progression of atherosclerosis, antithrombosis, antiarrythmic, antihypertensive, and hypocholesterolemic will have health benefits in the prevention and treatment of CVD.

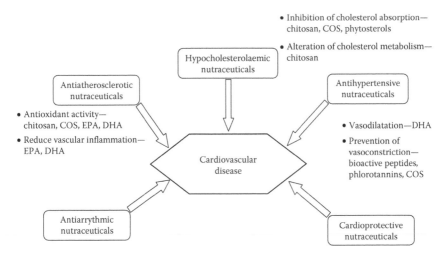

Figure 23.7 *Marine-derived nutraceuticals in the prevention of cardiovascular disease.*

23.3.15 Hypocholesterolemic marine-derived nutraceuticals

Hypercholesterolemia is known to have highly contributed to the development of CVD and thus the hypocholesterolemic nutraceuticals have received a greater attention during the recent past. The reduction of plasma cholesterol levels can be achieved either by inhibiting the cholesterol absorption and increasing the fecal excretion or by increasing the cholesterol metabolism in the body.

The dietary fibers are known to inhibit cholesterol absorption in the gastrointestinal tract of humans mainly by increasing the viscosity of digesta. The marine-derived polysaccharide, chitosan, has been identified as a compound having certain properties of dietary fiber. Though both dietary fiber and chitosan inhibit the intestinal absorption of cholesterol, the mechanism of action of chitosan is different from the other dietary fibers (Gallaher et al., 2002). There are several possible mechanisms of chitosan for the reduced fat absorption through formation of chitosan–cholesterol complexes in the digestive system (Kong and Kim, 2010). Chitosan dissolved in gastric HCl and entrapped the dietary lipid molecules. Chitosan–cholesterol complexes can be formed either by interaction between positively charged chitosan and negatively charged cholesterol molecules (Tai et al., 2000) or by hydrogen bonding between the OH$^-$ of cholesterol and amine groups of chitosan (Pavinatto et al., 2008). This chitosan lipid complex forms a gel in the small intestine making the chitosan–cholesterol complex unavailable for the intestinal absorption (Deuchi et al., 1994; Bokura and Kobayashi, 2003), thereby reducing the plasma cholesterol level. Since the presence of important functional groups such as hydroxyl and amine groups in the structure of chitosan has a greater contribution in the formation of chitosan–cholesterol complexes, the level of deacetylation as well as depolymerization of chitosan molecules is important for the hypocholesterolemic effects of chitosan.

Yao et al. (2008) have observed hypocholesterolemic effects in a rat model by high-molecular-weight chitosan, whereas low-molecular-weight chitosan had no effects on hypocholesterolemia. Similarly, an early study of Sugano et al. (1988) also reported poor hypocholesterolemic effects from COS, which have a low molecular weight. In contrast, high plasma cholesterol lowering ability has been observed from low-molecular-weight chitosan, especially from hydrophilic COS in many studies (Kim et al., 1998). This controversial results may have contributed from the difference of the level of deacetylation and depolymerization of the low-molecular-weight chitosan or COS used for the study. Thus the presence of beneficial functional groups in the structure of chitosan or COS enables them to be applied as hypocholesterolemic nutraceuticals due to the effects on lowering the dietary cholesterol absorption from the intestine.

Inhibition of cholesterol absorption has been also reported from certain marine algal extracts increasing the applicability of marine-derived compounds in the nutraceutical industry. Werman et al. (2003) have observed significant reduction of plasma cholesterol levels from the whole algal biomass and lipid extract of marine unicellular alga *Nanochloropsis* sp. in a rat model. Though the mode of action is uncertain, these authors have suggested that the observed hypocholesterolemic effect is exerted through the inhibition of intestinal cholesterol absorption by phytosterols present in alga. However, the high level of hypocholesterolemic effects that have been observed from whole algal biomass compared to the lipid extract is believed to be due to the insoluble fibers in algae that further reduce the cholesterol level by involving in cholesterol metabolism.

Though chitosan also can act as a dietary fiber, Yao et al. (2008) have reported minimum influence of chitosan on the bile acid. In contrast, increased fecal excretion of bile acids has been reported from feeding diet containing 5% chitosan into the hyperlipidemic rats (Xu et al., 2007). Murata et al. (2009) suggested that chitosan increases fecal excretion of bile acids through adsorbing either the primary bile acids such as taurocholate and glycocholate or secondary bile acid, taurodeoxycholate. The adsorption of bile acids from dietary chitosan prevents the reabsorption of bile acids leading to increased fecal excretion of cholesterol as bile acids, thereby reducing the plasma cholesterol level. Therefore, the marine-derived active compounds have the potential to be applied as nutraceuticals in reducing hypercholesterolemia through increasing the fecal excretion of cholesterol as well as by reducing the absorption of cholesterol in the human small intestine.

Similar to the reduction of cholesterol absorption, increased cholesterol metabolism in the liver has a greater role in hypocholesterolemia. The intestinally absorbed cholesterol circulates in the blood as a component of lipoproteins such as chylomicrones, VLDL, LDL, and HDL. Furthermore, cholesterol is biosynthesized from acetyl CoA after a cascade of reactions that occur within hepatocytes. Liver is also responsible for the conversion of cholesterol to bile acids and released into the duodenum with bile. However, primary bile acids as well as some of the secondary bile acids such as deoxycholic acid that have been produced in the lower small intestine and colon by the activity of bacteria are reabsorbed through the terminal ileum. The discovery of some key enzymes, receptors, and transporters in cholesterol biosynthesis and transfer has facilitated the development of drug for the regulation of cholesterol metabolism, which is favorable for the prevention of CVD (Menys and Durrington, 2006). Moreover, the recent discoveries in the biological properties of marine materials have found the marine-derived compounds that are able to inhibit the enzymes, receptors, and transporters in the cholesterol metabolism pathway. Therefore, these important biological properties of certain marine materials enable them to be used as hypocholesterolemic nutraceuticals in the prevention of CVD.

Chitosan, which has a wide range of biological properties, has been identified as an important marine-derived compound involved in the lowering of blood cholesterol level through hepatic cholesterol metabolism. The circulating LDL are subjected to LDL receptor mediated catabolism in the liver. Xu et al. (2007) have reported upregulation of LDL mRNA receptor expression in rat liver by 5% chitosan treatment, which increases the cholesterol catabolism, thereby decreasing the plasma cholesterol level. The upregulation of mRNA receptor expression of LDL receptor from chitosan has been confirmed by the recent study of Liu et al. (2011) who have observed significantly higher LDL mRNA receptor expression from molecular weight 10 kDa, 0.85° of deacetylated chitosan and O-carboxymethyl chitosan with a negative surface charge. In the same study, slight increase of LDL receptor mRNA expression has been observed from N-[(2-hydroxy-3-N,N-dimethylhexadecyl ammonium) propyl] chitosan chloride of 10 kDa with a positive surface charge. Furthermore, same authors have found increased mRNA expression of hepatic lipid metabolism enzymes such as lecithin cholesterol acyltransferase (LCAT), hepatic lipase and 3-hydroxy-3-methylglutaryl-CoA (HMG-CoA) reductase from 10 kDa chitosan and above two chitosan derivatives which increases the hepatic lipid metabolism.

Apart from the increased cholesterol catabolism and reduced synthesis of cholesterol, chitosan has the ability to enhance the activity of hepatic cholesterol 7α-hydroxylase, which is the rate-limiting enzyme in the conversion of cholesterol to bile acids. According to Moon et al. (2007), incorporation of 5% chitosan into the diet has significantly increased the activity of hepatic cholesterol 7α-hydroxylase, which can reduce the plasma cholesterol level. Therefore, chitosan as well as chitosan derivatives has the potential to alter the hepatic lipid metabolism, leading to hypocholesterolemia.

Since marine-derived compounds, especially the amino polysaccharide, chitosan, are able to reduce the intestinal cholesterol absorption, enhance hepatic cholesterol metabolism, and increase the fecal excretion of cholesterol as bile acids, chitosan and certain other marine algal extracts such as phycosterols have a greater potential to be applied as hypocholesterolemic nutraceuticals in the prevention and treatment of CVD.

23.3.16 Marine-derived nutraceuticals for prevention and development of atherosclerosis

Hypercholesterolemia can trigger the deposition of lipids, mainly cholesterol esters and cholesterol (Vijaimohan et al., 2006; Lamiaa, 2011), in the intimal layers of blood vessels leading to atherosclerosis lesions. Hence, the hypocholesterolemic nutraceuticals have a higher contribution for the prevention of development of atherosclerotic lesions. Similarly, inhibition of many other factors including ROS, endothelial adhesive property enhancing compounds, pro-inflammatory cytokines, and monocytes, which stimulates the accumulation of cholesterol, also have a special role in the prevention of atherosclerosis (**Figure 23.8**).

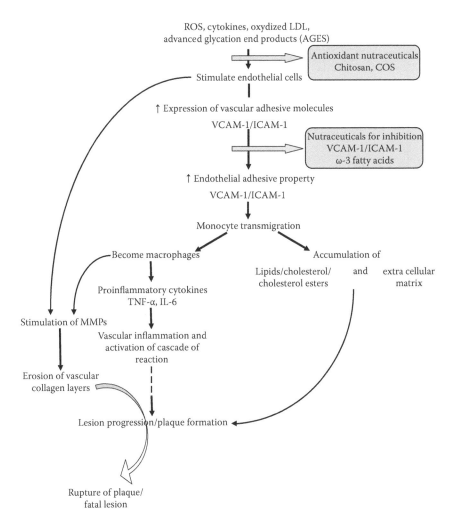

Figure 23.8 *Schematic diagram of the development of atherosclerosis and application of marine compounds as antiatherosclerotic nutraceuticals. (Adapted from Massaro, M. et al.,* Cardiovasc. Ther., *28, e13, 2010.)*

Since ROS have a special role in the initiation of cascade of atherosclerotic reactions, the antioxidant compounds, which have an ability to inhibit ROS, can prevent the development of atherosclerotic lesion at an early stage. Thus marine-derived compounds such as chitosan and omega-3 fatty acids, which have the ability to neutralize ROS, can be used as antiatherosclerotic nutraceuticals.

Reduction of the area of the aortic plaque has been reported from dietary chitosan (Ormrod et al., 1998) and marine omega-3 fatty acids (Erkkila et al., 2004; Wang et al., 2004a; Erkkila et al., 2006; Wan et al., 2010). Since the recent research of Lamiaa (2011) has observed a significant improvement of

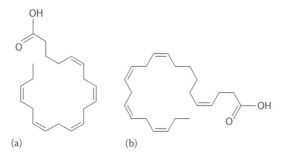

(a)   (b)

Figure 23.9 *Structure of (a) EPA and (b) DHA. (Adapted from Zulfakar, M.H. et al.,* Eur. J. Dermatol., *17(4), 284, 2007.)*

antioxidant enzymes such as GSH peroxidise by dietary chitosan, the observed antiatherogenic effects of dietary chitosan may have contributed from the antioxidant property of chitosan. Similarly, improvement of the antioxidant enzymes, especially SOD and catalase, has been observed from omega-3 fatty acids derived from the marine fish oils, reflecting the importance of omega-3 fatty acids in neutralizing ROS (Wang et al., 2004b).

Omega-3 fatty acids can be mainly found in marine fish oil and certain algal extracts (Rakshit et al., 2000; Fitton, 2003) in the form of eicosapentaenoic acid (EPA) and docosahexanoic acid (DHA) (**Figure 23.9**).

α-Linoleic acid (ALA), which is the precursor of EPA and DHA, is less effective in exerting anthero-protection (Degirolamo et al., 2010) due to the inefficient conversion of EPA and DHA in humans (Burdge and Calder, 2005). However, EPA and DHA can get incorporated into the cell membrane phospholipids in a dose-dependent manner (Robinson et al., 1993; Mebarek et al., 2009), especially in vascular endothelial cells, which are the target site for anti-atherosclerosis of these two compounds. Furthermore, marine-derived carotenoid, fucoxanthin, has the ability to enhance the DHA levels in body tissues, which will be favorable for antiatherosclerosis (Tsukui et al., 2009). The presence of EPA and DHA in the vascular endothelial cells has an association with the lower level of endothelial activation and vascular inflammation, which is beneficial for the prevention of atherosclerosis reaction cascade (He et al., 2009). The activated endothelial cells increase the expression of endothelial intercellular adhesion molecule-1 (ICAM-1) or vascular cell adhesion molecule-1 (VCAM-1), which will increase the endothelial adhesive property. Thus inhibition of these cell adhesive molecules is beneficial in the prevention of progression of atherosclerosis.

The inhibition of VCAM-1 has been reported from the dietary fish, which may have resulted from DHA and EPA present in the dietary fish (Erkkila et al., 2004). However, according to the same authors, there was no relationship between ICAM-1 and fish intake. In contrast, a recent study of Yamada et al. (2008) has observed a significant decrease in plasma concentrations of both ICAM-1 and VCAM-1 with 1.8 g of daily intake of highly purified EPA in both in vitro and in vivo studies. Similarly, He et al. (2009) also reported a significant negative association between ICAM-1 and fried fish consumption. Furthermore, the decreased tissue ration of n-6 or n-3 fatty acids has decreased the plasma concentration of ICAM-1 and monocyte chemotactic proteins (MCP-1) (Wan et al., 2010). Although different studies have reported contradictory results, the presence of inhibitory effects of cell adhesion molecules such as ICAM-1 and VCAM-1 from marine-derived omega-3 fatty acids allows them to be applied as nutraceuticals in the prevention of CVD.

Since the inhibition of ICAM-1 and VCAM-1 can alter the adhesive property of endothelium, the transmigration of monocytes and the inflammatory process will be prevented, obstructing the cascade of reaction of cell adhesive molecules, especially since DHA and EPA are known to have anti-inflammatory properties, and increasing their applicability as antiatherosclerosis nutraceuticals.

According to Ridker et al. (2000), C-reactive protein and IL-6 can be considered as two system inflammatory markers in the prediction of CVD. Omega-3 fatty acids, which are mainly derived from marine origin, have the property of decreasing the plasma concentrations of IL-6 and C-reactive proteins (He et al., 2009) as well as certain other inflammatory cytokines such as tumor necrosis factor α (TNF-α) (Wan et al., 2010). Furthermore, dietary long chain n-3 PUFA are also important for the inhibition of matrix metalloproteinase-3 (MMP-3) benefiting the prevention of rupture of aortic plaque, which can be fatal (He et al., 2009). Hence, omega-3 fatty acids have a greater potential to be applied as antiatherosclerotic nutraceutical due to the ability in the inhibition of cell adhesive molecules, inflammatory cytokines, and MMP-3s. Furthermore, neutralization of ROS by marine omega-3 fatty acids and chitosan enables these marine-derived compounds to be applied as antiatherosclerotic nutraceutical in the prevention of CVD.

23.3.17 Antihypertensive marine-derived nutraceuticals

Among the different cardiovascular risk factors, hypertension or high blood pressure is a major risk factor that accelerates the development of CVD. Thus the prevention of hypertension by vasodilatation or preventing vasoconstriction either by endothelial-dependent or independent pathway will be beneficial for the prevention of progression of CVD.

Antihypertensive effects via vasodilatation have been reported from marine green alga, *Cladophora patentiramea,* in an in vitro study. The pharmacological mechanism of action of this algal extract is suggested to be mediated through endothelium-dependent NO-cGMP pathway (Lim and Mok, 2010). Furthermore, the omega-3 fatty acids, especially DHA derived from marine fish oils, have exerted antihypertensive effects through vasodilatation (Khan et al., 2003). According to an early study of Bao et al. (1998), the antihypertensive effects of omega-3 fatty acids may depend on the vascular effects such as reduction of the reactivity of vascular smooth muscles. Therefore, certain marine-derived compounds can act as antihypertensive agents exerting their effects via an endothelium-dependent mechanism.

Apart from the endothelium-independent action, vasodilatory effects of DHA have been observed from endothelium-independent mechanism. Mori et al. (2000) have stated that vasodilator effects from DHA are predominantly by an endothelium-independent mechanism. This endothelium-independent vaso-dilator mechanism of DHA may have achieved from α_1 adrenergic receptor and angiotensin II receptor antagonists or prostaglandin synthesis inhibitors. Since angiotensin II can act as potent vasoconstrictor, the presence of angio-tensin II receptor antagonist activity in DHA increases the applicability of DHA as an antihypertensive nutraceutical compound. Furthermore, certain marine-derived compounds have the ability for the direct inhibition of angio-tensin II, increasing the applicability of marine-derived compounds in the nutraceutical industry.

The octapeptide, angiotensin II, is derived from the enzymatic conversion of angiotensin I, which is the inactive form through angiotensin-I convert-ing enzyme (ACE). Thus the angiotensin I converting enzyme has a greater contribution for the development of hypertension, especially with the activa-tion of rennin–angiotensin system due to the obesity-induced damage to the renal medulla (Zanella et al., 2001). Furthermore, the degradation of bradyki-nin, which is as a potent vasodilator by ACE, aggravates the development of hypertension in humans. Therefore, the active compounds with ACE inhibi-tory activity will have a greater potential to be applied as antihypertensives. Many active compounds including bioactive peptides, COS, and phlorotan-nins have been identified in diverse groups of marine organisms, which will be beneficial for the development of antihypertensive nutraceuticals and phar-maceuticals (**Table 23.5**) for the prevention of CVD.

Among the different antihypertensives, bioactive peptides derived from different marine sources have been studied extensively. Usually bioactive peptides are amino acid residues with 3–20 amino acids, and the amino acid composition and sequence in the peptide structure have contributed to their important bioactivities (Philanto, 2000). Therefore, certain bioac-tive peptides are found to be noncompetitive ACE inhibitors, which binds to the ACE site that is different from the substrate binding (Qian et al., 2007), whereas certain bioactive peptides are competitive ACE inhibitors that com-pete with the substrate for the active site of ACE (Lun et al., 2006; Kim et al., 2011). However, the recent study of Zhuang et al. (2012) has identified kidney as the target site for inhibitory activity of ACE by bioactive peptides. Although the structure–activity relationship of ACE inhibitory peptides has not yet been completely established, the presence of hydrophobic amino acid residues at N-terminus, positively charged amino acids at the middle, and aronatic amino acids at the C-terminus of peptides has been known to exert high ACE inhibition (Wijesekara et al., 2011). Moreover, the presence of

Janak K. Vidanarachchi, Maheshika S. Kurukulasuriya, and W.M.N.M. Wijesundara

Table 23.5 Marine-Derived Compounds with Antihypertensive Effects

Type of Marine Active Compound	Compound	Group of Organism	Organism	ID50	Reference
Oligosaccharide	COS — 50% deacetylated, 1000–5000 Da	Crustaceans	N/A	1.22 ± 0.13 mg/mL	Park et al. (2003b)
Oligosaccharide	COS — 90% deacetylated, 1000–5000 Da	Crustaceans	N/A	0.51 mg/mL	Park et al. (2008)
Oligosaccharide	Sulfated COS, 3–5 kDa	Crustaceans	N/A	0.25 mg/mL	Qian et al. (2010)
Oligosaccharide	Sulfated COS, 1–3 kDa	Crustaceans	N/A	0.325 mg/mL	Qian et al. (2010)
Oligosaccharide	Sulfated COS, 5–10 kDa	Crustaceans	N/A	0.775 mg/mL	Qian et al. (2010)
Oligosaccharide	Aminoethyl COS — 90% deacetylated, 800–3000 kDa	Crustaceans	N/A	0.8017 mg/mL	Ngo et al. (2008)
Peptide	Tensideal®	Marine fish	N/A	65 μg	Lahogue et al. (2009)
Phenolic compounds	Phlorotannins — Phlorogucinol, Triphlorethol-A, Eckol, Dieckol, Ecktolonol	Brown algae	E. cava	0.96 mg/mL	Wijesinghe et al. (2011)
Bioactive peptides	Thr-Phe-Pro-His-Gly-Pro (744 Da)	Marine fish	Syngnathus schegelli	0.62 mg/mL	Wijesekara et al. (2011)
Bioactive peptides	His-Trp-Thr-Gln-Arg (917 Da)	Marine fish	S. schegelli	1.44 mg/mL	Wijesekara et al. (2011)
Bioactive peptides	Gly, Pro, Glu, Ala, Asp (200–600 Da)	Jelly fish	Rhopilema esculentum	43 μg/mL	Zhuang et al. (2010)
Bioactive peptides	Gly-Pro-Leu (0.9 kDa)	Marine fish	Theragra chalcogramma	2.6 μM	Byun and Kim (2001)
Bioactive peptides	Gly-Pro-Met (0.9 kDa)	Marine fish	T. chalcogramma	17.13 μM	Byun and Kim (2001)
Bioactive peptides	Tyr-Ile (294.35 Da)	Jelly fish	Nemopilema nomurai	6.56 ± 1.12 μM (1.93 ± 0.33 μg/mL)	Kim et al. (2011)

(*continued*)

Type of Marine Active Compound	Compound	Group of Organism	Organism	ID50	Reference
Bioactive peptides	Ala-Ile-Tyr-Lys	Brown algae	*U. pinnatifida*	213 μM	Suetsuna and Nakano (2000)
Bioactive peptides	Tyr-Lys-Tyr-Tyr	Brown algae	*U. pinnatifida*	64.2 μM	Suetsuna and Nakano (2000)
Bioactive peptides	Lys-Phe-Tyr-Gly	Brown algae	*U. pinnatifida*	90.5 μM	Suetsuna and Nakano (2000)
Bioactive peptides	Tyr-Asn-Lys-Ley	Brown algae	*U. pinnatifida*	21 μM	Suetsuna and Nakano (2000)
Bioactive peptides	Gly-Asp-/ Leu-Gly-Lys-Thr-Thr-Thr-Val-Ser-Asn-Trp-Ser-Pro-Pro-Lys-Try-Lys-Asp-Thr-Pro (2482 Da)	Marine fish — tuna	N/A	11.28 μM	Lee et al. (2010)
Bioactive peptides	Trp-Pro-Glu-Ala-Ala-Glu-Leu-Met-Met-Glu-Val-Asp-Pro (1581 Da)	Marine fish — tuna	*Thunnus obesus*	21.6 μM	Qian et al. (2007)
Bioactive peptides	Met-Ile-Phe-Pro-Gly-Ala-Gly-Gly-Pro-Glu-Leu	Marine fish — yellowfin sole	*Limanda aspera*	28.7 μg/mL	Jung et al. (2006)
Bioactive peptides	Val-Glu-Cys-Tyr-Gly-Pro-Asn-Arg-Pro-Gln-Phe	Microalgae	*Chlorella vulgaris*	29.6 μM	Shei et al. (2009)
Bioactive peptides	Albumin, gliadin, glutelin	Red algae	*P. yezoensis*	1.6 g/L	Qu et al. (2010)

N/A, not available.

proline in the sequence of short peptides allows the peptide to be prevented from digestion by enzymes after ingestion and pass from the capillary into the circulation of blood. Since some peptides with potent ACE inhibitory activity found from in vitro get digested and inactivated after oral administration (Korhonen and Pihlanto, 2006), the development of gastrointestinal resistance due to the presence of proline in the peptide structure will be beneficial for the development of antihypertensive nutraceuticals from bioactive peptides.

Antihypertensive effects via ACE inhibitor activity of marine-derived bioactive peptides have been observed in many in vivo studies reflecting the ability of marine peptides to be resistant to the enzymatic digestion (Suetsuna and Nakano, 2000; Qian et al., 2007; Lee et al., 2010). However, according to Suetsana and Nakano (2000), the parent tetrapeptides administered orally get digested in the body to shorter fragments such as di- or tripeptides, which can easily get absorbed through the intestine. Moreover, these di- or tripeptides have more potent ACE inhibitor activity than the parent tetrapeptides. Therefore, the tetrapeptides extracted from the marine alga, *U. pinnatifida,* act as a prodrug that may possess health benefits by using them as nutraceuticals. Furthermore, the crude extracts of marine algae also contain fucoxanthin, phlorotannins, and phenolic compounds with ACE inhibitor activity, which increases the applicability of marine algae as antihypertensive nutraceuticals (Cha et al., 2006; Wijesinghe et al., 2011).

Apart from the algal-derived antihypertensives, ACE inhibitor activity has been observed from COS and COS derivatives extracted from the marine crustacean shell waste. ACE inhibitor activity of COS depends on the degree of deacetylation of the chitosan structure. An early study of Park et al. (2003a) has observed highest ACE inhibition from the chitosan with lowest degree of deacetylation that they have used for the study. In contrast, many studies have revealed increasing level of ACE inhibition with the increase of degree of deacetylation of chitosan in a dose-dependent manner (Huang et al., 2005; Ngo et al., 2008; Qian et al., 2010). Moreover, recent researchers have increased the ACE inhibitor activity of COS by synthesizing different COS derivatives by incorporating different functional groups into the structure of COS (**Table 23.5**). These COS derivatives compete with the substrate to the active site of ACE and inhibit ACE by binding to this active site. Similar to the binding activity of the synthetic antihypertensive drug, Catapril®, COS derivatives bind with the active site of enzyme via electrostatic interactions between negatively charged functional groups in the structure of COS derivatives such as sulfate groups or carboxyl groups with the positively charged sites of the enzyme (Huang et al., 2005; Qian et al., 2010), thereby COS derivatives inhibit the enzyme, ACE exerting their antihypertensive effects. The presence of antihypertensive properties in COS as well as in COS derivatives enables them to be applied as antihypertensive nutraceuticals in preventing CVD. Furthermore, the applicability of COS as antihypertensive nutraceutical increases by the presence of renin inhibitory activity of COS, which prevents the conversion of angiotensinogen to angiotensin I (Park et al., 2008). Hence, many marine-derived compounds such as COS, COS derivatives, bioactive peptides, phlorotannins, and omega-3 fatty acids are involved in acting as antihypertensive agents in preventing CVD by exerting different modes of actions. Irrespective of the mode of action, these marine materials have a greater role in the nutraceutical industry to be applied as antihypertensive nutraceuticals.

Similarly, the antiatherosclerotic, hypocholesterolemic, antiarrythmic, and cardioprotective marine-derived nutraceuticals will have a greater role in the prevention and treatment of human CVD.

23.4 Conclusion

Various compounds such as carotenoids, polysaccharides, fatty acids, polyphenols, proteins or peptides, vitamins, and certain enzymes derived from the diverse group of marine organisms have unique biological properties, which increases their applications in the industrial scenario including the nutraceutical industry. The characteristic biological properties of these marine materials include antioxidative, antimicrobial, anti-inflammatory, antiangiogenesis, anticoagulant, and antilipogenic ability. These special properties increase the applicability of marine-derived compounds as nutraceuticals in biological and biomedical fields, especially in the prevention and treatment of various important human disease conditions such as obesity, diabetes, various cancers, arthritis, and CVD. The recent discoveries of novel marine materials as well as novel biological properties in these marine materials create many paths for the development of novel nutraceuticals benefiting the human health. Moreover, these marine materials belong to the category of natural compounds and are considered as safe for human consumption; the active compounds derived from the marine sources have an expanding role in the future of the nutraceutical industry.

References

Ahn, G. N., Kim, K. N., Cha, S. H. et al. 2007. Antioxidant activities of phlorotannins purified from *Ecklonia cava* on free radical scavenging using ESR and H2O2-mediated DNA damage. *European Food Research and Technology* 226:71–79.

Aiello, R. J., Brees, D., and Francone, O. L. 2003. ABCA1-deficient mice: insights into the role of monocyte lipid efflux in HDL formation and inflammation. *Arteriosclerosis, Thrombosis, and Vascular Biology* 23:972–980.

Ata, A., Kerr, R. G., Moya, C. E. et al. 2003. Identification of anti-inflammatory diterpenes from the marine gorgonian *Pseudopterogorgia elisabethae. Tetrahedron* 59:4215–4222.

Bacon, A., Makin, J., Sizer, P. J. et al. 2000. Carbohydrate biopolymers enhance antibody responses to mucosally delivered vaccine antigens. *Infection and Immunity* 68(10):57–64.

Baharum, S. N., Beng, E. K., and Mokhtar, M. A. A. 2010. Marine microorganisms: Potential application and challenges. *Journal of Biological Sciences* 10:555–564.

Bao, D. Q., Mori, T. A., Burke, V. et al. 1998. Effects of dietary fish and weight reduction on ambulatory blood pressure in overweight hypertensives. *Hypertension* 32:710–717.

Barrow, R. T., Parker, E. T., Krishnaswamy, S. et al. 1994. Inhibition by heparin of the human blood coagulation intrinsic pathway factor X activator. *Journal of Biological Chemistry* 269:26796–26800.

Bergmann, W. and Feeney, R. J. 1951. Nucleosides of sponges: Discovery of the arabinose-based nucleotides/*Tethya crypta. Journal of Organic Chemistry* 16:981–987.

Bernardi, G. and Springer, G. F. 1962. Properties of highly purified fucan. *Journal of Biological Chemistry* 237:75–80.

Bokura, H. and Kobayashi, S. 2003. Chitosan decreases total cholesterol in women: A randomized double blind placebo-controlled trial. *European Journal of Clinical Nutrition* 57:721–725.

Borrelli, F., Campagnuolo, C., Capasso, R. et al. 2004. Iodinated indole alkaloids from *Plakortis simplex*—New plakohypaphorines and an evaluation of their antihistamine activity. *European Journal of Organic Chemistry* 1994(15):3227–3232.

Britton, G. 1995. Structure and properties of carotenoids in relation to function. *FASEB Journal* 9:1551–1558.

Bruyere, O., Pavelka, K., Rovati, L. C. et al. 2004. Glucosamine sulfate reduces osteoarthritis progression in ostmenopausal women with knee osteoarthritis: Evidence from two 3-year studies. *Menopause* 11:138–143.

Bugni, T. S., Singh, M. P., Chen, L. et al. 2004. Kalihinols from two *Acanthella cavernosa* sponges: Inhibitors of bacterial folate biosynthesis. *Tetrahedron* 60:6981–6988.

Burdge, G. C., and Calder, P. C. 2005. Conversion of alpha-linoleic acid to longer-chain poly-unsaturated fatty acids in human adults. *Reproduction, Nutrition, Development* 45(5): 581–597.

Byun, H. G. and Kim, S. K. 2001. Purification and characterization of angiotensin I convert-ing enzyme (ACE) inhibitory peptides from Alaska Pollack (*Theragra chalcogramma*) skin. *Process Biochemistry* 36:1155–1162.

Capelli, B. and Cysewski, G. (2006). *Natural Astaxanthin: King of the Carotenoids*. Cyanotech Corporation, San Fernando, CA.

Carlucci, M. J., Pujol, C. A., Ciancia, M. et al. 1997 Antiherpetic and anticoagulant properties of carrageenans from the red seaweed *Gigartina skottsbergii* and their cyclized deriva-tives: Correlation between structure and biological activity. *International Journal of Biological Macromolecules* 20:97–105.

Carroll, A. R., Buchanan, M. S., Edser, A. et al. 2004. Dysinosins B-D, inhibitors of factor VIIa and thrombin from the Australian sponge *Lamellodysidea chlorea*. *Journal of Natural Products* 67:1291–1294.

Cha, S. H., Lee, K. W., and Jeon, Y. J. 2006. Screening of extracts from red algae in Jeju for potentials marine angiotensin-I converting enzyme (ACE) inhibitory activity. *Algae* 21(3):343–348.

Cheng, S., Jian, W. W., Lei, F. et al. 2005. Characteristics of a polysaccharide from *Porphyra capensis* (Rhodophyta). *Carbohydrates Research* 340:2447–2450.

Choi, J. S., Lee, J. H., and Jung, J. H. 1997. The screening of nitrite scavenging effect of marine algae and active principle of *Ecklonia stolonifera*. *Journals of Korean Fish Society* 30:909–915.

Costa, L. S., Fidelis, G. P., Cordeiro, S. L. et al. 2010. Biological activities of sulfated polysac-charides from tropical seaweeds. *Biomedicine and Pharmacotherapy* 64:21–28.

Cueto, M., Jensen, P. R., Kauffman, C. et al. 2001. Pestalone a new antibiotic produced by a marine fungus in response to bacterial challenge. *Journal of Natural Products* 64:1444–1446.

Cutignano, A., Bifulco, G., Bruno, I. et al. 2000. Dactylolide, a new cytotoxic macrolide from the vanuatu sponge *Dactylospongia* sp. *Tetrahedron* 56:3743–3748.

De Clercq, E. 2004. Therapeutic potential of nucleoside/nucleotide analogues against poxvi-rus infections. *Journals of Clinical Virology* 30:73–89.

Degirolamo, C., Kelley, K. L., Wilson, M. D. et al. 2010. Dietary n-3 LCPUFA from fish oil but not α-linoleic acid-derived LCPUFA confers atheroprotection in mice. *Journal of Lipid Research* 51:1897–1905.

Deig, E. F., Ehresmann, D. W., Hatch, M. T. et al. 1974. Inhibition of herpesvirus replication by marine algae extracts. *Antimicrobial Agent Chemother* 61:525–528.

Deuchi, K., Kanauchi, O., Imasato, Y. et al. 1994. Decreasing effect of chitosan on the apparent fat digestibility by rats fed on a high-fat diet. *Bioscience, Biotechnology and Biochemistry* 58(9):1613–1616.

Dieterle, C., Bendel, M. D., Seissler, J. et al. 2006. Therapy of diabetes mellitus. Pancreas transplantation, islet transplantation, stem cell and gene therapy. *Internist (Berlin)* 47:489–496.

Dighe, N. S., Pattan, S. R., Musmade, D. S. et al. 2009. Recent synthesis of marine natu-ral products with antihypertensive activity: An overview. *Journal of Chemical and Pharmaceutical Research* 1(1):54–61.

Douglas, H. and Judah, F. 1996. Patterns and emerging mechanisms of the angiogenic switch during tumorigenesis. *Cell* 86:353–364.

Eaton, P., Fernandes, J. C., Pereira, E. et al. 2008. Atomic force microscopy study of the antibacterial effects of chitosans on *Escherichia coli* and *Staphylococcus aureus*. *Ultramicroscopy* 108(10):1128–1134.

Ehresmann, D. W., Deig, E. F., Hatch, M. T. et al. 1977. Antiviral substances from California marine algae. *Journal of Phycology* 13:37–40.

Endo, T., Tsuda, M., Okada, T. et al. 2004. New dimeric bromopyrrole alkaloids from marine sponge *Agelas* species. *Journal of Natural Products* 67:1262–1267.

Erkkila, A. T., Lichtenstein, A. H., Mozaffarian, D. et al. 2004. Fish intake is associated with a reduced progression of coronary artery atherosclerosis in postmenopausal women with coronary artery disease. *American Journal of Clinical Nutrition* 80:626–632.

Erkkila, A. T., Mattham, N. R., Herrington, D. M. et al. 2006. Reduced progression of coronary atherosclerosis in women with CAD. *Journal of Lipid Research* 47:2814–2819.

Fennell, B. J., Carolan, S., Pettit, G. R. et al. 2005. Effects of the antimitotic natural product dolastatin 10, and related peptides, on the human malarial parasite *Plasmodium falciparum*. *Journal of Antimicrobial Chemotherapy* 51:833–841.

Fernandez-Busquets, X. and Burger, M. M. 2003. Circular proteoglycans from sponges: First members of the spongican family. *Cellular and Molecular Life Sciences* 60:88–112.

Fitton, J. H. 2003. Brown marine algae: A survey of therapeutic potentials. *Alternative and Complementary Therapies* 9(1):39–41.

Flier, J. S. 2004. Obesity wars: Molecular progress confronts an expanding epidemic. *Cell* 116:337–350.

Ford, P. W., Gustafson, K. R., McKee, T. C. et al. 1999. Papuamides A-D, HIV-inhibitory and cytotoxic depsipeptides from the sponges *Theonella mirabilis* and *Theonella swinhoei* collected in Papua New Guinea. *Journal of the American Chemical Society* 121:5899–5909.

Frenz, J. L., Kohl, A. C., and Kerr, R. G. 2004. Marine natural products as therapeutic agents: Part 2. *Expert Opinion on Therapeutic Patents* 14:17–33.

Fujita, M., Nakao, Y., Matsunaga, S. et al. 2003. Callysponginol sulfate A, an MT1-MMP inhibitor isolated from the marine sponge *Callyspongia truncata*. *Journal of Natural Products* 66:569–571.

Furukawa, S., Fujita, T., Shimabukuro, M. et al. 2004. Increased oxidative stress in obesity and its impact on metabolic syndrome. *Journal of Clinical Investigation* 114:1752–1761.

Gallaher, D. D., Gallaher, C. M., Mahrt, G. J. et al. 2002. A glucomannan and chitosan fiber supplement decreases plasma cholesterol and increases cholesterol excretion in overweight normocholesterolemic humans. *Journal of the American College of Nutrition* 21(5):428–433.

Geering, R., Engelke, C., and Chandler, T. 2006. Polyunsaturated fatty acids—Are they alternative anti-inflammatories? *European Companion Animal Health*. http://www.touchbriefings.com/pdf/2398/Bionovate_tech.pdf (accessed on June 24, 2011).

Gochfeld, D. J., El Sayed, K. A., Yousaf, M. et al. 2003. Marine natural products as lead anti-HIV agents. *Mini-Reviews in Medicinal Chemistry* 3:401–424.

Gomez, P. I., Barriga, A., Cifuentes, A. S. et al. 2003. Effect of salinity on the quantity and quality of carotenoids accumulated by *Dunaliella salina* (strain CONC-007) and *Dunaliella bardawil* (strain ATCC 30861) Chlorophyta. *Biological Research* 36:185–192.

Goud, T. V., Srinivasulu, M., Reddy, V. L. et al. 2003. Two new bromotyrosine-derived metabolites from the sponge *Psammaplysilla purpurea*. *Chemical and Pharmaceutical Bulletin (Tokyo)* 51:990–993.

Grant, M. A., Morelli, X. J., and Rigby, A. C. 2004. Conotoxins and structural biology: A prospective paradigm for drug discovery. *Current Protein and Peptide Science* 5:235–248.

Guerin, M., Huntley, M. E., and Olaizola, M. 2003. Haematococcus astaxanthin: Applications for human health and nutrition. *Trends in Biotechnology* 21:210–216.

Gulati, O. P. 2005. The nutraceutical pycnogenol: Its role in cardiovascular health and blood glucose control. *Biomedical Reviews* 16:49–57.

Gustafson, K. R., Oku, N., and Milanowski, D. J. 2004. Antiviral marine natural products. *Current Protein and Peptide Science* 3:233–249.

Hallenbeck, J. M. 2002. The many faces of tumor necrosis factor in stroke. *Natural Medicine* 8:1363–1368.

Hassan, W., Edrada, R., Ebel, R. et al. 2004. New imidazole alkaloids from the Indonesian sponge *Leucetta chagosensis*. *Journal of Natural Products* 67:817–822.

Hayashi, K. and Ito, M. 2002. Antidiabetic action of low molecular weight chitosan in genetically obese diabetic KK-A^y mice. *Biological and Pharmaceutical Bulletin* 25(2):188–192.

He, K., Liu, K., Daviglus, M. L. et al. 2009. Associations of dietary long-chain n-3 polyunsaturated fatty acids and fish with biomarkers of inflammation and endothelial activation (from the multi-ethnic study of atherosclerosis [MESA]). *American Journal of Cardiology* 103(9):1238–1243.

Heo, S. J. and Jeon, Y. J. 2009. Protective effect of fucoxanthin isolated from *Sargassum siliquastrum* on UV-B induced cell damage. *Journal of Photochemistry and Photobiology B: Biology* 95:101–107.

Heo, S. J., Ko, S. C., Kang, S. M. et al. 2008. Cytoprotective effect of fucoxanthin isolated from brown algae *Sargassum siliquastrum* against H 2O2-induced cell damage. *European Food Research and Technology* 228:145–151.

Heo, S. J., Park, E. U., Lee, K. W. et al. 2005. Antioxidant activities of enzymatic extracts from brown seaweeds. *Bioresource Technology* 96:1613–1623.

Hirano, S. and Nagao, N. 1989. Effects of chitosan, pectic acid, lysozyme, and chitinase on the growth of several phytopathogens. *Agricultural and Biological Chemistry* 53(11):3065–3066.

Hirono, M., Ojika, M., Mimura, H. et al. 2003. Acylspermidine derivatives isolated from a soft coral, *Sinularia* sp, inhibit plant vacuolar H(+)-pyrophosphatase. *Journal of Biochemistry* 133:811–816.

Hong, S., Kim, S. H., Rhee, M. H. et al. 2003. In vitro anti-inflammatory and pro-aggregative effects of a lipid compound, petrocortyne A, from marine sponges. *Naunyn-Schmiedeberg's Arch* 368:448–456.

Hornung, R. L., Pearson, J. W., Beckwith, M. et al. 1992. Preclinical evaluation of bryostatin as an anticancer agent against several marine tumor cell lines: In vitro and in vivo activity. *Cancer Research* 52(1):101–107.

Houstis, N., Rosen, E. D., and Lander, E. S. 2006. Reactive oxygen species have a causal role in multiple forms of insulin resistance. *Nature* 440:944–948.

Huang, R., Mendis, E., and Kim, S. K. 2005. Improvement of ACE inhibitory activity of chitooligosaccharides (COS) by carboxyl modification. *Bioorganic and Medicinal Chemistry* 13:3649–3655.

Janeway, C. A. and Medzhitov, R. 2002. Innate immune recognition. *Annual Review of Immunology* 20:197–216.

Jawanda, M. K. 2009. Antitumour activity of antioxidant—Overview. *International Journal of Dental Clinics* 1:3–7.

Je, J. Y., Lee, K. H., Lee, M. H. et al. 2009. Antioxidant and antihypertensive protein hydrolysates produced from tuna liver by enzymatic hydrolysis. *Food Research International* 42:1266–1272.

Je, J. Y., Park, P. J., and Kim, S. K. 2004. Radical scavenging activity of hetero-chitooligosaccharides. *European Food Research and Technology* 219:60–65.

Jeon, Y. J. and Kim, S. K. 2002. Antitumor activity of chitosan oligosaccharides produced in ultrafiltrationmembrane reactor system. *Journal of Microbiology and Biotechnology* 12(3):503–507.

Jeon, S. M., Kim, H. J., Woo, M. N. et al. 2010. Fucoxanthin-rich seaweed extract suppresses body weight gain and improves lipid metabolism in high-fat-fed C57BL/6J mice. *Biotechnology Journal* 5:961–969.

Jha, R. K. and Zi-rong, X. 2004. Biomedical compounds from marine organisms. *Marine Drugs* 2:123–146.

Jing, J., Wang, L., Hao, W. et al. 2011. Bio-function summary of marine oligosaccharides. *International Journal of Biology* 3:74–86.

Ju, C., Yue, W., Yang, Z. et al. 2010. Antidiabetic effect and mechanism of chitooligosaccharides. *Biological and Pharmaceutical Bulletin* 33(9):1511–1516.

Jung, W. K., Mendis, E., Je, J. Y. et al. 2006. Angiotensin I-converting enzyme inhibitory peptide from yellowfin sole (*Limanda aspera*) frame protein and its antihypertensive effect in spontaneously hypertensive rats. *Food Chemistry* 94:26–32.

Jyonouchi, H., Sun, S., Iijima, K., and Gross, M. D. (2000). Antitumour activity of astaxanthin and its mode of action. *Nutrition and Cancer* 36, 59–65.

Kaats, G. R., Michalek, J. E., and Preuss, H. G. 2006. Evaluating efficacy of a chitosan product using a double-blinded, placebo-controlled protocol. *Journal of the American College of Nutrition* 25:389–394.

Kanauchi, O., Deuchi, K., Imasato, Y. et al. 1995. Mechanism for the inhibition of fat digestion by Chitosan and for the synergistic effect of ascorbate. *Bioscience Biotechnology and Biochemistry* 59(5):786–790.

Kang, K. A., Lee, K. H., Chae, S. et al. 2006. Cytoprotective effect of phloroglucinol on oxidative stress induced cell damage via catalase activation. *Journal of Cellular Biochemistry* 97:609–620.

Keyzers, R. A., Northcote, P. T., Berridge, M. V. et al. 2003. A new 14 beta marine sterol from the New Zealand sponge *Clathria lissosclera*. *Australian Journal of Chemistry* 56:279–282.

Keyzers, R. A., Northcote, P. T., and Zubkov, O. A. 2004. Novel anti-inflammatory spongian diterpenes from the New Zealand marine sponge *Chelonaplysilla violacea*. *European Journal of Organic Chemistry* 1994:419–425.

Khan, B. B. 1998. Type 2 diabetes: When insulin secretion fails to compensate for insulin resistance. *Cell* 92:593–596.

Khan, F., Elherik, K., Bolton-Smith, C. et al. 2003. The effect of dietary fatty acid supplementation on endothelial function and vascular tone of healthy subjects. *Cardiovascular Research* 59:955–962.

Kijjoa, A. and Sawangwong, P. 2004. Drugs and cosmetics from the sea. *Marine Drugs* 2:73–82.

Kim, J. N., Chang, I. Y., Kim, H. I. et al. 2009a. Long-term effects of chitosan oligosaccharide in streptozotocin-induced diabetic rats. *Islets* 1(2):111–116.

Kim, M. J., Chang, U. J., and Lee, J. S. 2009b. Inhibitory effects of fucoidan in 3T3-L1 adipocyte differentiation. *Marine Biotechnology* 11:557–562.

Kim, S. Y., Je, J. Y., and Kim, S. K. 2007. Purification and characterization of antioxidant peptide from hoki (*Johnius balengerii*) frame protein by gastrointestinal digestion. *Journal of Nutritional Biochemistry* 18:31–38.

Kim, S. J., Kang, S. Y., Park, S. L. et al. 1998. Effects of chitooligosaccharides on liver function in mouse. *Korean Journal of Food Science and Technology* 30(3):693–696.

Kim, M. M. and Kim, S. K. 2006. Chitooligosaccharides inhibit activation and expression of matrix mettaloproteinase-2 in human dermal fibroblast cells. *FEBS Letters* 580:2661–2666.

Kim, M. M. and Kim, S. K. 2010. Antiinflammatory activity of chitin chitosan and their derivatives. In *Chitin, Chitosan, Oligosaccharides and Their Derivatives*, S. K. Kim, Ed., pp. 215–221. Taylor & Francis Group, New York.

Kim, J. Y., Lee, J. K., Lee, T. S. et al. 2003. Synthesis of chitooligosaccharide derivative with quaternary ammonium group and its antimicrobial activity against *Streptococcus mutans*. *International Journal of Biological Macromolecules* 32(1–2):23–27.

Kim, Y. K., Lim, C. W., Yeun, S. et al. 2011. Dipeptide (Tyr-Ile) acting as an inhibitor of angiotensin-I-converting enzyme (ACE) from the hydrolysate of jelly fish *Nemopilema nomurai*. *Fisheries and Aquatic Sciences* 14(4):283–288.

Kim, E. J., Park, S. Y., Lee, J. Y. et al. 2010. Fucoidan present in brown algae induces apoptosis of human colon cancer cells. *BMC Gastroenterology* 10:96.

Kochkina, Z. M. and Chirkov, S. N. 2000. Effect of chitosan derivatives on the reproduction of coliphages T2 and T7. *Microbiology* 69(2):208–211.

Kong, C. S. and Kim, S. K. 2010. Antidiabetic activity and cholesterol-lowering effect of chitin, chitosan and their derivatives. In *Chitin, Chitosan, Oligosaccharides and Their Derivatives*, S. K. Kim, Ed., pp. 285–292. Taylor & Francis Group, New York.

Kong, C. S., Kim, J. A., Yoon, N. Y., and Kim, S. K. (2009). Induction of apoptosis by phloro-glucinol derivative from *Ecklonia cava* in MCF-7 human breast cancer cells. *Food and Chemical Toxicology* 47:1653–1658.

Korhonen, H. and Pihlanto, A. 2006. Bioactive peptides: Production and functionality. *International Dairy Journal* 16:945–960.

Kumar, S. G., Rahman, M. A., Lee, S. H. et al. 2009. Plasma proteome analysis for anti-obesity and anti-diabetic potentials of chitosan oligosaccharides in *ob/ob* mice. *Proteomics* 8:2149–2162.

Kurosaki, F., Tashiro, N., and Nishi, A. 1986. Induction of chitinase and phenylalanine ammo-nia-lyase in cultured carrot cells treated with fungal mycelial walls. *Plant and Cell Physiology* 27(8):1587–1591.

Lahogue, V., Vallee-Rahel, K., Kuart, M. et al. 2009. Isolation of angiotensin I converting enzyme inhibitory peptides from a fish by-products hydrolysate Tensideal®. *Peptides for Youth: The Proceedings of the 20th American Peptide Symposium*, S. Del Valle et al. Eds., p. 483. Springer, New York.

Lamiaa, A. A. B. 2011. Hypolipidemic and antiantherogenic effects of dietary chitosan and wheatbran in high fat-high cholesterol fed rats. *Australian Journal of Basic and Applied Sciences* 5(10):30–37.

Lee, B. C., Kim, M. S., Choi, S. H. et al. 2009. In vitro and in vivo antimicrobial activity of water-soluable chitosan oligosaccharides against Vibrio vulnificus. *International Journal of Molecular Medicine* 24:327–333.

Lee, H. W., Park, Y. S., Choi, J. W. et al. 2003. Antidiabetic effects of chitosan oligosaccharides in neonatal streptozotocin-induced noninsulin-dependant diabetes mellitus in rats. *Biological and Pharmaceutical Bulletin* 26(8):1100–1103.

Lee, S. H., Qian, Z. J., and Kim, S. K. 2010. A novel angiotensin I converting enzyme inhibi-tory peptide from tuna frame protein hydrolysate and its antihypertensive effect in spontaneously hypertensive rats. *Food Chemistry* 118:96–102.

Li, Q., Liu, H., Xin, X. et al. 2009a. Marine-derived oligosaccharide sulfate (JG3) suppresses heparanase-driven cell adhesion events in heparanase over-expressing CHO-K1 cells. *Acta Pharmacologica Sinica* 30:1033–1038.

Li, Y., Qian, Z. J., Ryu, B. M. et al. 2009b. Chemical components and its antioxidant proper-ties in vitro: An edible marine brown alga, *Ecklonia cava*. *Bioorganic and Medicinal Chemistry* 17:1963–1973.

Lim, Y. L. and Mok, S. L. 2010. In vitro vascular effects produced by crude aqueous extract of green marine algae, *Cladophora patentiramea* (Mont.) Kutzing, in aorta from normo-tensive rats. *Medical Principals and Practice* 19:260–268.

Liu, B., Liu, W. S., Han, B. Q. et al. 2007. Antidiabetic effects of chitooligosaccharides on pancreatic islet cells in streptozotocin-induced diabetic rats. *World Journal of Gastroenterology* 13(5):725–731.

Liu, X., Yang, F., Song, T. et al. 2011. Effects of chitosan, *O*-carboxymethyl chitosan and *N*-[(2-hydroxy-3-*N*,*N*-dimethylhexadecyl ammonium)propyl]chitosan chloride on lipid metabolism enzymes and low-density-lipoprotein receptor in a murine diet-induced obesity. *Carbohydrate Polymers* 85:334–340.

Lucas, R., Casapullo, A., Ciasullo, L. et al. 2003a. Cycloamphilectenes, a new type of potent marine diterpenes: Inhibition of nitric oxide production in murine macrophages. *Life Science* 72(22):2543–2552.

Lucas, R., Giannini, C., D'auria, M. V. et al. 2003b. Modulatory effect of bolinaquinone, a marine sesquiterpenoid, on acute and chronic inflammatory processes. *Journal of Pharmacology and Experimental Therapeutics* 304(3):1172–1180.

Lun, H. H., Lan, C. X., Yun, S. C. et al. 2006. Analysis o0f novel angiotensin-I-converting enzyme inhibitory peptides from protease-hydrolyzed marine shrimp *Acetes chinensis*. *Journal of Peptide Science* 12(11):726–733.

Lung, M. Y. and Chang, Y. C. 2011. In vitro antioxidant properties of polysaccharides from *Armillaria mellea* in batch fermentation. *African Journal of Biotechnology* 10(36):7048–7057.

Maeda, H., Hosokawa, M., Sashima, T. et al. 2005. Fucoxanthin from edible seaweed, *Undaria pinnatifida*, shows antiobesity effect through UCP1 expression in white adipose tissues. *Biochemical and Biophysical Research Communications* 332:392–397.

Maeda, H., Hosokawa, M., Sashima, T. et al. 2006. Fucoxanthin and its metabolite, fucoxanthinol, suppress adipocyte differentiation in 3T3-L1 cells. *International Journal of Molecular Medicine* 18:147–152.

Maeda, H., Hosokawa, M., Sashima, T. et al. 2007. Effect of medium-chain triglycerols on anti-obesity effect of fucoxanthin. *Journal of Oleo Science* 56(12):615–621.

Maeda, H., Hosokawa, M., Sashima, T. et al. 2009. Anti-obesity and anti-diabetic effects of fucoxanthin on diet-induced obesity conditions in a murine model. *Molecular Medicine Reports* 2:897–902.

Maeda, H., Tsukui, T., Sashima, T. et al. 2008. Seaweed carotenoid, fucoxanthin, as a multifunctional nutrient. *Asia Pacific Journal of Clinical Nutrition* 17(SI):196–199.

Maki, A., Mohammad, R., and Raza, S. 1996. Effect of dolastatin 10 on human non-Hodgkin lymphoma cell lines. *Anticancer Drugs* 7(3):344–350.

Massaro, M., Scoditti, E., Cariuccio, M. A. et al. 2010. Nutraceuticals and prevention of atherosclerosis: Focus on ω-3 polyunsaturated fatty acids and Mediterranean diet polyphenols. *Cardiovascular Therapeutics* 28:e13–e19.

Mayer, A. M. S. and Hamann, M. T. 2002. Compounds with antibacterial, anticoagulant, antifungal, anti-inflammatory, anthelmintic, anti-inflammatory, antiplatelet, antiprotozoal and antiviral activities; affecting the cardiovascular, endocrine, immune, and nervous systems; and other miscellaneous mechanisms of action. *Comparative Biochemistry and Physiology—Part C: Toxicology and Pharmacology* 132:315–339.

Mayer, A. M. S. and Hamann, M. T. 2004. Marine compounds with antibacterial, anticoagulant, antifungal, anti-inflammatory, antimalarial, antiplatelet, antituberculosis, and antiviral activities; affecting the cardiovascular, immune, and nervous systems and other miscellaneous mechanisms of action. *Marine Biotechnology (New York)* 6:37–52.

Mayer, A. M. S. and Hamann, M. T. 2005. Marine compounds with anthelmintic, antibacterial, anticoagulant, antidiabetic, antifungal, anti-inflammatory, antimalarial, antiplatelet, antiprotozoal, antituberculosis, and antiviral activities; affecting the cardiovascular, immune and nervous systems and other miscellaneous mechanisms of action. *Comparative Biochemistry and Physiology—Part C: Toxicology and Pharmacology* 140:265–286.

Mayer, A. M. S. and Lehmann, V. K. B. 2000. Marine compounds with antibacterial, anticoagulant, antifungal, anti-inflammatory, anthelmintic, antiplatelet, antiprotozoal, and antiviral activities; with actions on the cardiovascular, endocrine, immune, and nervous systems; and other miscellaneous mechanisms of action. *Pharmacologists* 42:62–69.

Mayer, A., Lidschreiber, M., Siebert, M. et al. 2010. Uniform transitions of the general RNA polymerase II transcription complex. *Nature Structural & Molecular Biology* 17(10):1272–1278.

Mayer, A. M., Rodriguez, A. D., Berlinck, R. G. et al. 2007. Marine compounds with anthelmintic, antibacterial, anticoagulant, antifungal, anti-inflammatory, antimalarial, antiplatelet, antiprotozoal, antituberculosis, and antiviral activities; affecting the Cardiovascular, immune and miscellaneous mechanisms of action. Comparative Biochemistry and Physiology—Part C: 145:553–581.

Mazumder, S., Ghosal, P. K., Pujol, C. A. et al. 2002. Isolation, chemical investigation and antiviral activity of polysaccharides from *Gracilaria corticata* (Gracilariaceae, Rhodophyta). *International Journal of Biological Macromolecules* 31(1–3):87–95.

Mebarek, S., Ermak, N., Benzaria, A. et al. 2009. Effects of increasing docosahexaenoic acid intake in human healthy volunteers on lymphocyte activation and monocyte apoptosis. *British Journal of Nutrition* 101(6):852–858.

Melo, F. R., Pereira, M. S., Foguel, D. et al. 2004. Antithrombin-mediated anticoagulant activity of sulfated polysaccharides: Different mechanisms for heparin and sulfated galactans. *Journal of Biological Chemistry* 279:20824–20835

Menys, V. C. and Durrington, P. N. 2006. Human cholesterol metabolism and therapeutic molecules. *Experimental Physiology* 93(1):27–42.

Miyashita, K. and Hosokawa, M. 2008. Beneficial health effects of seaweed carotenoid, fucoxanthin. In *Marine Nutraceuticals and Functional Foods*, C. Barrow and F. Shahidi Eds., pp. 297–319. Taylor & Francis Group, Boca Raton, FL.

Miyashita, M., Koyama, T., Kamitani, T. et al. 2009. Anti-obesity effect on rodents of the traditional Japanese food, Torokombu, shaved *Laminaria*. *Bioscience, Biotechnology and Biochemistry* 73(10):2326–2328.

Miyashita, K., Nishikawa, S., Beppu, F. et al. 2011. The allenic carotenoid fucoxanthin, a novel marine nutraceutical from brown seaweeds. *Journal of the Science of Food Agriculture* 91:1166–1174.

Monti, M. C., Casapullo, A., Riccio, R. et al. 2004. Further insights on the structural aspects of PLA(2) inhibition by gamma-hydroxybutenolide-containing natural products: A comparative study on petrosaspongiolides M-R. *Bioorganic & Medicinal Chemistry* 12:1467–1474.

Moon, M. S., Lee, M. S., Kim, C. T. et al. 2007. Dietary chitosan enhances hepatic CYP7A1 activity and reduces plasma and liver cholesterol concentrations in diet-induced hypercholesterolaemia in rats. *Nutrition Research and Practice* 1(3):175–179.

Mori, T. A., Watts, G. F., Burke, V. et al. 2000. Differential effects of eicosapentaenoic acid and docosahexaenoic acid on vascular reactivity of the forearm microcirculation in hyperlipidemic, overweight men. *Circulation* 102:1264–1269.

Morrice, L. M., McLean, M. W., Long, W. F. et al. 1984. Porphyran primary structure. *Life Sciences* 75(9):1063–1073.

Mourao, P. A. 2004. Use of sulfated fucans as anticoagulant and antithrombotic agents: Future perspectives. *Current Pharmaceutical Design* 10:967–981.

Mourao, P. A. S. and Pereira, M. S. 1999. Searching for alternatives to heparin: Sulfated fucans from marine invertebrates. *Trends Cardiovascular Medicine* 9:225–232.

Muller, W. E. G. and Schroder, H. C. 1991. Cell biological aspects of HIV-1 infection: Effects of the anti-HIV-1 agent avarol. *International Journal of Sports Medicine* 12:S43–S49.

Muller, W. E. G, Sobel, C., Diehl-Seifert, B. et al. 1987. Biochemical. *Pharmacology* 36:1489–1494.

Murata, Y., Kodama, Y., Harai, D. et al. 2009. Properties of an oral preparation containing a chitosan salt. *Molecules* 14:755–762.

Ngo, D. N., Lee, S. H., Kim, M. M. et al. 2009. Production of chitin oligosaccharides with different molecular weights and their antioxidant effect in RAW 264.7 cells. *Journal of Functional Foods* 1:188–198.

Ngo, D. N., Qian, Z. J., Je, J. Y. et al. 2008. Aminoethyl chitooligosaccharides inhibit the activity of angiotensin converting enzyme. *Process Biochemistry* 43:119–123.

Ngo, D. H., Wijesekara, I., Vo, T. S. et al. 2011. Marine food-derived functional ingredients as potential antioxidents in the food industry. *Food Research International* 44:523–529.

Nishimura, S., Matsunaga, S., Shibazaki, M. et al. 2003. Massadin, a novel geranylgeranyl-transferase type I inhibitor from the marine sponge *Stylissa* aff. massa. *Organic Letters* 5:2255–2257.

Nomura, T., Kikuchi, M., Kubodera, A. et al. 1997. Proton-donative antioxidant activity of fucoxanthin with 1,1-diphenyl-2-picrylhydrazyl (DPPH). *Biochemistry and Molecular Biology International* 42:361–370.

Ohgami, K., Shiratori, K., Kotake, S. et al. 2003. Effects of astaxanthin on lipopolysaccharide-induced inflammation in vitro and in vivo. *Investigative Ophthalmology & Visual Science* 44:2694–2701.

Ormrod, D. J., Holmes, C. C., and Miller, T. E. 1998. Dietary chitosan inhibits hypercholesterolaemia and atherogenesis in the apolipoprotein E-deficient mouse model of atherosclerosis. *Atherosclerosis* 138:329–334.

Pangestuti, R. and Kim, S. K. 2004. Neuroprotective effect of marine algae. *Marine Drugs* 9:803–818.

Parish, C. R., Freeman, C., Brown, K.J. et al. 1999. Identification of sulfated oligosaccharide-based inhibitors of tumor growth and metastasis using novel in vitro assays for angiogenesis and heparanase activity. *Cancer Research* 59(14):3433–3441.

Park, Y. B. 2005. Determination of nitrite-scavenging activity of seaweed. *Journal of the Korean Society of Food Science and Nutrition* 34:1293–1296.

Park, P. J., Ahn, C. B., Jeon, Y. J. et al. 2008. Renin inhibition activity by chitooligosaccharides. *Bioorganic and Medicinal Chemistry Letters* 18:2471–2474.

Park, P. J., Je, J. Y., and Kim, S. K. 2003a. Angiotensin I converting enzyme (ACE) inhibitory activity of heterochitooligosaccharides prepared from partially different deacetylated chitosans. *Journal of Agricultural and Food Chemistry* 51(17):4930–4934.

Park, P. J., Je, J. Y., and Kim, S. K. 2003b. Free radical scavenging activity of chitooligosaccharides by electron spin resonance spectrometry. *Journal of Agricultural and Food Chemistry* 51:4624–4627.

Park, M. K., Jung, U., and Roh, C. 2011. Fucoidan from marine brown algae inhibits lipid accumulation. *Marine Drugs* 9:1359–1367.

Park, P. J., Lee, H. K., and Kim, S. K. 2004. Preparation of hetero-chitooligosaccharides and their antimicrobial activity on Vibrio parahaemolyticus. *Journal of Microbiology and Biotechnology* 14(1):41–47.

Patankar, M. S., Oehninger, S., Barnett, T. et al. 1993. A revised structure for fucoidan may explain some of its biological activities. *Journal of Biological Chemistry* 268:21770–21776.

Pavinatto, F. J., Santos, D. S., and Olivera, O. N. 2008. Interaction between cholesterol and chitosan in Langmuir monolayers. *Polimeros: Ciencia e Tecnologia* 15:91–94.

Peerapornpisal, Y., Amornlerdpison, D., Jamjai, U. et al. 2010. Antioxidant and anti-inflammatory activities of brown marine alga, Padina minor Yamada. *Chiang Mai Journal of Science* 37:507–516.

Pereira, M. S., Vileila-Silva, A. C., Valente, A. P. et al. 2002. A 2-sulfated, 3-linked alpha-L-galactan is an anticoagulant polysaccharide. *Carbohydrate Research* 337:2231–2238.

Perry, N. B., Blunt, J. W., Munro, M. H. G. et al. 1990. Antiviral and antitumor agents from a New Zealand sponge, Mycale sp. 2. Structures and solution conformations of mycal-amides A and B. *Journal of Organic Chemistry* 55:223–227.

Philanto, L. 2000. Bioactive peptides derived from bovine whey proteins: Opioid and ace-inhibitory. *Trends in Food Science and Technology* 11:347–356.

Pisal, D. S. and Lele, S. S. 2005. Carotenoid production from microalgae, *Duneliella salina*. *Indian Journal of Biotechnology* 4:476–483.

Posadas, I., De Rosa, S., Terencio, M. C. et al. 2003a. Cacospongionolide B suppresses the expression of inflammatory enzymes and tumour necrosis factor-α by inhibiting nuclear factor-κB activation. *British Journal of Pharmacology* 138:1571–1579.

Posadas, I., Terencio, M. C., Randazzo, A. et al. 2003b. Inhibition of the NF-kappaB signaling pathway mediates the anti-inflammatory effects of petrosaspongiolide. *Medicinal Biochemistry & Pharmacology* 65:887–895.

Prajapati, B. G. 2009. Chitosan A marine medical polymer and its lipid lowering capacity. *Internet Journal of Health* 9(2):5. DOI: 10.5580/1045.

Qatanani, M. and Lazar, M. A. 2007. Mechanisms of obesity-associated insulin resistance: Many choices on the menu. *Genes and Development* 21:1443–1455.

Qi, H., Zhang, Q., Zhao, T. et al. 2005. Antioxidant activity of different sulfate content derivatives of polysaccharide extracted from *Ulva pertusa* (Chlorophyta) in vitro. *International Journal of Biological Macromolecules* 37:195–199.

Qian, Z. J., Eom, T. K., Ryu, B. et al. 2010. Angiotensin I-converting enzyme inhibitory activity of sulphated chitooligosaccgarides with different molecular weights. *Journal of Chitin and Chitosan* 15(2):75–79.

Qian, Z. J., Je, J. Y., and Kim, S. K. 2007. Antihypertensive effect of angiotensin I converting enzyme-inhibitory peptide from hydrolysates of Bigeye Tuna Dark Muscla, *Thunnus obesus*. *Journal of Agricultural and Food Chemistry* 55(21):8398–8403.

Qu, W., Ma, H., Pan, Z. et al. 2010. Preparation and antihypertensive activity of peptides from *Porphyra yezoensis*. *Food Chemistry* 123:14–20.

Rajapakse, N., Kim, M. M., Mendis, E. et al. 2007. Inhibition of free radicalmediated oxidation of cellular biomolecules by carboxylated chitooligosaccharides. *Bioorganic and Medicinal Chemistry* 15:997–1003.

Rajapakse, N., Kim, M. M., Mendis, E. et al. 2008. Inhibition of inducible nitric oxide synthase and cyclooxygenase-2 in lipopolysaccharide-stimulated RAW264.7 cells by carboxybutyrylated glucosamine takes place via down-regulation of mitogen-activated protein kinase-mediated nuclear factor-jB signalling. *Immunology* 123:348–357.

Rajasekaran, A., Sivagnanam, G., and Xavier, R. 2008. Nutraceuticals as therapeutic agents: A review. *Research Journal of Pharmacy and Technology* 1:328–340.

Rakshit, S. K., Vasuhi, R., and Kosugi, Y. 2000. Enrichment of polyunsaturated fatty acids from tuna oil using immobilized *Pseudomonas fluorescens* lipase. *Bioprocess and Biosystems Engineering* 23(3):251–255.

Rashid, M. A., Shigematsu, N., Gustafson, K. R. et al. 2001. Microspinosamide A, a new HIV-inhibitory depsipeptide from the marine sponge *Sidnopsis microspinosa*. *Journal of Natural Products* 64:117–121.

Reimer, U., Lamare, M. D., and Paeke, B. M. 2007. Temporal concentrations of sunscreen compounds (mycosporine-like amino acids) in phytoplankton and in the New Zealand krill, *Nyctiphanes australis* G.O. Sars. *Journal of Plankton Research* 29:1077–1086.

Ridker, P. M., Hennekens, C. H., Buring, J. E. et al. 2000. C-reactive protein and other markers of inflammation in the prediction of cardiovascular disease in women. *New England Journal of Medicine* 342:836–843.

Rinehart, K., Gloer, J. B., and Cook, J. C. 1981. Structures of the didemnins, antiviral and cytotoxic depsipeptides from a Caribbean tunicate. *Journal of the American Chemical Society* 103:1857–1859.

Robinson, D. R., Xu, L. L., Knoell, C. T. et al. 1993. Modification of spleen phospolipid fatty acid composition by dietary fish oil and by n-3 fatty acid ethyl esters. *Journal of Lipid Research* 34:1423–1434.

Rodriguez, I. I., Shi, Y. P., Garcia, O. J. et al. 2004. New pseudopterosin and seco-pseudopterosin diterpene glycosides from two Colombian isolates of *Pseudopterogorgia elisabethae* and their diverse biological activities. *Journal of Natural Products* 67:1672–1680.

Sachindra, N. M., Sato, E., Maeda, H. et al. 2007. Radical scavenging and singlet oxygen quenching activity of marine carotenoid fucoxanthin and its metabolites. *Journal of Agricultural and Food Chemistry* 55:8516–8522.

Shahidi, F., Han, X. Q., and Synowiecki, J. 1995. Production and characteristics of protein hydrolysates from capelin (*Mallotus villosus*). *Food Chemistry* 53:285–293.

Shei, I. C., Fang, T. J., and Wu, T. K. 2009. Isolation and characterization of a novel angiotensin I-converting enzyme (ACE) inhibitory peptide from the algae protein waste. *Food Chemistry* 115:279–284.

Shi, Y. and Burn, P. 2004. Lipid metabolic enzymes: Emerging drug targets for the treatment of obesity. *Nature Reviews Drug Discovery* 3:695–710.

Shibata, T., Ishimaru, K., Kawaguchi, S. et al. 2008. Antioxidant activities of phlorotannins isolated from Japanese Laminariaceae. *Journal of Applied Phycology* 20:705–711.

Shin, E. S., Hwang, H. J., Kim, I. H. et al. 2011. A glycoprotein from *Porphyra yezoensis* produces anti-inflammatory effects in liposaccharide-stimulated macrophages via the TLR4 signaling pathway. *International Journal of Molecular Medicine* 28:809–815.

Shulman, G. I. 2000. Cellular mechanisms of insulin resistance. *Journal of Clinical Investigation* 106(2):171–176.

Sosa, M. A. G., Fazely, F., Koch, J. A., Vercellotti, S. V. et al. 1991. *N*-Carboxymethylchitosan-*N*, *O*-sulfate as an anti-HIV-1 agent. *Biochemical and Biophysical Research Communications* 174(2):489–496.

Souza, T. M., Abrantes, J. L., Epifanio, R. et al. 2007. The alkaloid from 4 methylaaptamine isolated from the sponge *Aaptos aaptos* impairs herpes simplex type 1 penetration and immediate-early protein synthesis. *Planta Medica* 73(3):200–205.

Spector, I., Shochet, N. R., Kashman, Y. et al. 1983. *Latrunculins*: Novel marine toxins that disrupt microfilament organization in cultured cells. *Science* 219:493–495.

Stahl, W. and Sies, H. 1996. Lycopene: A biologically important carotenoid for humans Archives. *Journal of Biochemistry and Biophysics* 336:1–9.

Sudarshan, N. R., Hoover, D. G., and Knorr, D. 1992. Antibacterial action of chitosan. *Food Biotechnology* 6(3):257–272.

Suetsuna, K. and Nakano, T. 2000. Identification of an antihypertensive peptide from peptic digest of wakame (*Undaria pinnatifida*). *Journal of Nutritional Biochemistry* 11:450–454.

Sugano, M., Watanabe, S., Krishi, A. et al. 1988. Hypocholesterolemic action of chitosans with different viscosity in rats. *Lipids* 23:187–193.

Tai, T. S., Sheu, W. H. H., Lee, W. J. et al. 2000. Effect of chitosan on plasma lipoprotein concentrations in type-2 diabetic subjects with hypercholesterolemia. *Diabetes Care* 23:173–174.

Tezuka, Y., Irikawa, T., Kaneko, A. H. et al. 2001. Screening of Chinese herbal drug extracts for inhibitory activity on nitric oxide production and identification of an active compound of *Zanthoxylum bungeanum*. *Journal of Ethnopharmacology* 77:209–217.

Tsukui, T., Baba, N., Hosokawa, M. et al. 2009. Enhancement of hepatic docosahexaenoic acid and arachidonic acid contents in C57BL/6J mice by dietary fucoxanthin. *Fisheries Science* 75:261–263.

Tzivelka, L. A., Vagias, C., and Roussis, V. 2003. Natural products with Anti-HIV activity from marine organisms. *Current Topics in Medicinal Chemistry* 3:1512–1535.

Vidanarachchi, J. K., Kurukulasuriya, M. S., and Kim, S. K. 2010. Chitin, chitosan, and their oligosaccharides in food industry. In *Chitin, Chitosan, Oligosaccharides and Their Derivatives*, S. K. Kim, Ed., pp. 543–560. Taylor & Francis Group, New York.

Vidanarachchi, J. K., Kurukulasooriya, M. S., Samaraweera, A. M. et al. 2011. Applications of marine nutraceuticals in dairy products. In *Advances in Food and Nutrition Research*, Vol. 65, S. K. Kim, Ed., pp. 457–478. Elsevier Limited, Kidlington, Oxford, U.K.

Vijaimohan, K., Jainu, M., Sabitha, S. et al. 2006. Beneficial effect of alpha linoleic acid rich flaxseed oil on growth performance and hepatic cholesterol metabolism in high fat diet fed rats. *Life Science* 79:448–454.

Voutilainen, S., Nurmi, T., Mursu, J. et al. 2006. Carotenois and cardiovascular health. *American Journal of Clinical Nutrition* 83:1265–1271.

Wallen, K. E. and Hotamisligil, G. S. 2005. Inflammation, stress and diabetes. *Journal of Clinical Investigations* 115:1111–1119.

Wan, J. B., Huang, L. L., Rong, R. et al. 2010. Endogenously decreasing tissue n-6/n-3 fatty acid ratio reduces atherosclerosis lesions in *apolipoprotein-E* deficient mice by inhibiting systemic and vascular inflammation. *Atherosclerosis, Thrombosis, and Vascular Biology* 30:2487–2494.

Wang, H. H., Hung, T. M., Wei, J. et al. 2004a. Fish oil increases antioxidant enzyme activities in macrophages and reduces atherosclerosis lesions in apoE-knockout mice. *Cardiovascular Research* 61:169–176.

Wang, J., Jiang, X., Mou, H. et al. 2004b. Anti-oxidation of agar oligosaccharides produced by agarase from a marine bacterium. *Journal of Applied Phycology* 16(5):333–340.

Wang, J., Zhang, Q., Zhang, Z. et al. 2008. Antioxidant activity of sulphated polysaccharide fractions extracted from *Laminaria japonica*. *International Journal of Biological Macromolecules* 42:127–132.

Wellington, K. D., Cambie, R. C., Rutledge, P. S. et al. 2000. Chemistry of sponges. 19. Novel bioactive metabolites from *Hamigera tarangaensis*. *Journal of Natural Products* 63:79–85.

Wender, P. A., Hinkle, K. W., Koehler, M. F. et al. 1999. The rational design of potential chemotherapeutic agents: Synthesis of bryostatin analogues. *Medical Research Review* 19(5):388–407.

Werman, M. J., Sukenik, A., and Mokady, S. 2003. Effects of the marine unicellular alga *Nonnochloropsis* sp. to reduce the plasma and liver cholesterol levels in male rats fed on diets with cholesterol. *Bioscience Biotechnology and Biochemistry* 67(10):2266–2268.

Wijesekara, I., Qian, Z. J., Ryu, B. et al. 2011. Purification and identification of antihypertensive peptides from seaweed pipefish (*Syngnathus schlegeli*) muscle protein hydrolysate. *Food Research International* 44:703–707.

Wijesekara, I., Yoon, N. Y., and Kim, S. K. 2010. Phlorotannins from *Ecklonia cava* (Phaeophyceae): Biological activities and potential health benefits. *Biofactors* 36:408–414.

Wijesinghe, W. A. J. P., Ko, S. C., and Jeon, Y. J. 2011. Effect of phlorotannins isolated from *Ecklonia cava* on angiotensin I-converting enzyme (ACE) inhibitory activity. *Nutrition Research and Practice* 5(2):93–100.

Witvrouw, M., Desmyter, J., and De Clercq, E. 1994. Antiviral chemistry. *Chemotherapy* 5:345–359.

Woo, M. N., Jeon, S. M., Shin, Y. C. et al. 2009. Anti-obese property of fucoxanthin is partly mediated by altering lipid-regulating enzymes and uncoupling proteins of visceral adipose tissue in mice. *Molecular Nutrition & Food Research* 53:1603–1611.

Xing, R. G., Liu, S., Yu, H. H. et al. 2005. Preparation of high-molecular weight and high-sulfate content chitosans and their potential antioxidant activity in vitro. *Carbohydrate Polymers* 61:148–154.

Xu, G., Huang, X., Qiu, L. et al. 2007. Mechanism study of chitosan on lipid metabolism in hyperlipidemic rats. *Asia Pacific Journal of Clinical Nutrition* 16(Suppl 1):313–317.

Yamada, T., Ogamo, A., Saito, T. et al. 2000. Preparation of O-acylated low-molecular-weight carrageenans with potent anti-HIV activity and low anticoagulant effect. *Carbohydrate Polymers* 41(2):115–1120.

Yamada, H., Yoshida, M., Nakano, Y. et al. 2008. In vivo and in vitro inhibition of monocyte adhesion to endothelial cells and endothelial; adhesion molecules by Ecosapentaenoic acid. *Atherosclerosis, Thrombosis, and Vascular Biology* 28:2173–2179.

Yang, S. W., Chan, T. M., Pomponi, S. A. et al. 2003. Structure elucidation of a new antifungal sterol sulfate, Sch 575867, from a deep-water marine sponge (Family: Astroscleridae). *Journal of Antibiotics (Tokyo)* 56:186–189.

Yao, H. T., Huang, S. Y., and Chiang, M. T. 2008. A comparative study on hypoglycaemic and hypocholesterolemic effects of high and low molecular weight chitosan in streptozotocin-induced diabetic rats. *Food and Chemical Toxicology* 46:1525–1534.

Yuan, W. P., Liu, B., Liu, C. H. et al. 2009. Antioxidant activity of chito-oligosaccharides on pancreatic islet cells in streptozotocin-induced diabetic rats. *World Journal of Gastoenterology* 15(11):1339–1345.

Yuan, H. and Song, J. 2005. Preparation, structural characterization and in vitro antitumor activity of kappa-carrageenan oligosaccharide fraction from *Kappaphycus striatum*. *Journal of Applied Phycology* 17(1):7–13.

Zancan, P. and Mourao, P. A. 2004. Venous and arterial thrombosis in rat models: Dissociation of the antithrombotic effects of glycosaminoglycans. *Blood Coagulation & Fibrinolysis* 15:45–54.

Zanella, M. T., Kohlmann, O., and Rebeiro, A. B. 2001. Treatment of obesity hypertension and diabetes syndrome. *Hypertension* 38:705–708.

Zeisel, S. H. 1999. Regulation of "Nutraceuticals". *Science* 285(5435):1853–1855.

Zhang, D., Fugii, I., Lin, C. et al. 2008. The stimulatory activities of polysaccharide compounds derived from algae extracts on insulin secretion *in vitro*. *Biological and Pharmaceutical Bulletin* 31(5):921–924.

Zhang, Q., Li, N., Liu, X. et al. 2004. The structure of a sulfated galactan from *Porphyra haitanensis* and its in vivo antioxidant activity. *Carbohydrate Research* 339:105–111.

Zhu, C., Li, G. Z., Peng, H. B. et al. 2010. Therapeutic effect of marine collagen peptides on Chinese patients with type n diabetes mellitus and primary hypertension. *American Journal of the Medical Sciences* 340(5):360–366.

Zhuang, Y., Sun, L., and Li, B. 2010. Production of the angiotensin-I-converting enzyme (ACE)-inhibitory peptide from hydrolysates of Jellyfish (*Rhopilema esculentum*) collagen. *Food and Bioprocess Technology* 5(5):1622–1629. DOI: 10.1007/s11947-010-0439-9

Zhuang, Y., Sun, L., Zhang, Y. et al. 2012. Antihypertensive effect of long-term oral administration of jelly fish (*Rhopilema esculentum*) collagen peptides on renovascular hypertension. *Marine Drugs* 10:417–426.

Zulfakar, M. H., Edwards, M., and Heard, C. M. 2007. Is there a role for topically delivered ecosapenteinoid acid in the treatment of psoriasis. *European Journal of Dermatology* 17(4):284–291.

24

Fucoidans from Marine Brown Macroalgae
Isolation, Identification, and Potential Biological Activities

Yasantha Athukorala and Yvonne V. Yuan

Contents

24.1 Introduction

Marine macroalgae have long been known for their industrially important carbohydrates, including carrageenans (*Eucheuma* and *Kappaphycus*), agar (*Gelidium* and *Gracilaria*), alginates (*Laminaria, Macrocystis, Lessonia,* and *Ascophyllum*), and fucoidans (*Fucus* and *Undaria*), which are produced in large quantities around the world (Laurienzo, 2010). For example, alginates (linear polymers of 1,4-linked residues of β-D-mannuronic and α-L-guluronic acids) are the major carbohydrate produced from marine brown macroalgae (Kusaykin et al., 2008); approximately 32,000–39,000 MT of alginates are produced annually for food (e.g., pastry, canned meat, and ice cream), pharmaceutical (e.g., delivery of drugs and absorbents in dressings), and technical (e.g., dental impressions and wound dressings) applications mainly by Scotland, China, the United States, and Norway (Senanayake et al., 2011; Venugopal, 2011). Fucoidans, which are mainly composed of sulfated α-1,3 and 1,4-linked L-fucose (6-deoxy-L-galactose) polymers (**Figure 24.1A**), are bound to alginates and are thus primarily obtained as a cheap and economical byproduct from the process of alginate preparation (Kusaykin et al., 2008; Mabeau and Kloareg, 1987). As well, fucoidans frequently contain minor amounts of monosaccharides such as xylose, glucose, galactose, mannose, rhamnose, in addition to glucuronic acid, protein, phenolic compounds, phosphorus, and minerals including Na^+, Mg^{2+}, Ca^{2+}, K^+, $Fe^{2+/3+}$, and Al^{3+} (Shaklee et al., 2010). Compared to other marine carbohydrates, fucoidans are considered to be one of the most versatile, as they can be used in diverse applications in the food (e.g., nutritional beverages, soft drinks, functional drinks and hangover chasers, soups, yoghurts, and dietary supplements), cosmeceutic (e.g., anti-aging facial creams, moisturizers, soaps, and shampoos), and pharmacognosic

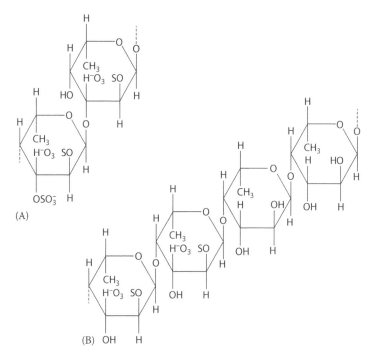

Figure 24.1 *Structures of fucoidan and sulfated fucan: (A)* Fucus vesiculosus, *disaccharide repeating unit [→4)-α-L-Fucp -(2,3-OSO₃⁻)-(1→3)-α-L-Fucp-(2OSO₃⁻)-(1→];* *(B)* Arbacia lixula, *homogeneous, unbranched tetrasaccharide sulfated fucan repeating unit. (From Berteau, O. and Mulloy, B.,* Glycobiology, *13, 29, 2003.)*

(e.g., anticancer, anticoagulant, and antithrombotic activities, stem cell mobilization; http://www.ebuenavista.co.kr/fucoidan) industries.

Fucoidans, which are also often referred to as fucoidin, fucan, fucan sulfates, fucosan, fucogalactan sulfates, xylofucoglycuronan (ascophyllans), and glycuronofucoglycan sulfates (sargassans), are only present in marine brown macroalgae (Phaeophyceae) and have not been identified in green (Chlorophyceae), red (Rhodophyceae), golden (Xanthophyceae), fresh water algae, or in terrestrial plants (Fitton et al., 2007; Kusaykin et al., 2008; Li et al., 2008). However, "sulfated fucans," which have similar structural and chemical features to fucoidans in marine brown macroalgae, have been reported in some marine invertebrates, that is, sea cucumber (*Ludwigothurea grisea*; Vilela-Silva et al., 1999) and sea urchin (*Arbacia lixula*; **Figure 24.1B**; Mulloy et al., 1994). While marine brown macroalgal fucoidans and sulfated fucans from sea invertebrates share some structural and functional similarities, fucoidans have been demonstrated to exhibit greater biological activities than sulfated fucans, despite the complex and heterogeneous nature of the former, compared to the latter. In the years after fucoidan was identified in 1913 (Berteau and Mulloy, 2003; Kusaykin et al., 2008; Kylin, 1913), research related to fucoidans has gained increasing interest. Since the beginning of the 1970s, there has been a dramatic increase in the amount of fucoidan-based research (Holtkamp et al., 2009). To date, over 800 individual research articles have been published focusing on the isolation, chemistry, and related biological activities of fucoidans: anticoagulant and antithrombotic activity (Irhimeh et al., 2009; Shanmugam and Mody, 2000; Wijesinghe et al., 2011); involvement in and control of cell-to-cell communication, adhesion and proliferation (Schumacher et al., 2011); antioxidant and free radical quenching properties (Kang et al., 2008; Thomes et al., 2010) as well as cosmeceutical effects, such as prevention of premature aging of the skin and amelioration of skin sensitivities (Mizutani et al., 2006).

In industry-linked fucoidan research, Takara Bio Inc. (Japan) identified three types of fucoidan from *Kjellmaniella crassifolia* (Nagakiri-kombu) with unique biological activities: U-fucoidan (containing approximately 20% glucuronic acid), F-fucoidan (composed mainly of sulfated fucose), and G-fucoidan (composed mainly of sulfated galactose; **Figure 24.2**; Mizutani et al., 2006). These co-workers developed an enzymatic digestion technique to produce fucoidans from marine brown macroalgae with tailored structures resulting in unique biological activities, such as anticancer and immunomodulating effects. Similarly, an Australian biotechnology company, Marinova, recently published a solvent-free, cold-water extraction technique for the purification of organically certified fucoidan from several marine brown macroalgae from Tasmania (*Undaria pinnatifida*; Australia), Nova Scotia (Canada), Tonga (South Pacific), and Patagonia (Chile; Fitton et al., 2007). One capsule of Maritech® 926 produced by Marinova contains 88.5 mg of *Undaria* fucoidan, equivalent to the amount of fucoidan typically consumed daily in the traditional Japanese diet (http://www.lef.org/Vitamins-Supplements/). In addition, several fucoidan-containing fruit juice products are also being marketed for nonspecific cardiovascular and anticancer health benefits, including Youth Juice® manufactured in Vancouver, British

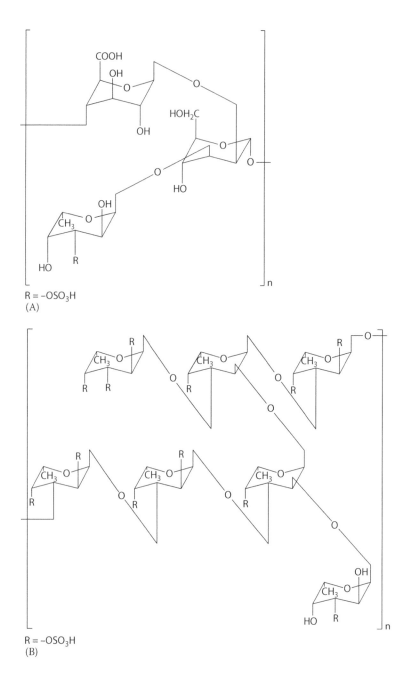

R = –OSO₃H

(A)

R = –OSO₃H

(B)

Figure 24.2 *Chemical structures for TaKaRa Kombu Fucoidan: (A) Kombu U-Fucoidan; (B) Kombu F-Fucoidan; (C) Kombu G-Fucoidan. (From Mizutani, S. et al., Fucoidan-containing cosmetics, U.S. Patent 0093566. http://www.google.co.il/patents/about/7678368_Fucoidan_containing_cosmetics.html?id=nanNAAAAEBAJ, 2006; Sakai, T. et al., Sulfated fucogalactan, U.S. Patent 6590097B1. http://www.google.com/patents/us 6590097, 2003).*

$R = -OSO_3H$
(C)

Figure 24.2 (continued) Chemical structures for TaKaRa Kombu Fucoidan: (C) Kombu G-Fucoidan. (From Mizutani, S. et al., Fucoidan-containing cosmetics, U.S. Patent 0093566. http://www.google.co.il/patents/about/7678368_Fucoidan_containing_cosmetics.html?id=nanNAAAAEBAJ, 2006; Sakai, T. et al., Sulfated fucogalactan, U.S. Patent 6590097B1. http://www.google.com/patents/us 6590097, 2003).

Columbia (Canada), containing a blend of berry juices and fucoidans from *Fucus gardneri* and *Cladosiphon* spp. (Ranatunga, 2007). As well, juice-based nutritional supplements, Limu Original®, Limu Lean®, and Limu Energy®, produced in Florida (Vitarich Laboratories, 2004) containing fucoidan extracts of Limu Moui (*Cladosiphon novae-caledoniae kylin*) are being marketed online. Historically, Limu Moui has been harvested and consumed for over 3000 years in the Kingdom of Tonga, whose people believe that this marine brown macroalga is responsible for their longevity, health, and vigor (Elkins, 2001). Thus, fucoidans and fucoidan-containing drinks and supplements as functional foods and nutraceuticals are being marketed increasingly for their potential health benefits around the globe. Academic and industry-based researchers in Asia and the West are investigating fucoidans for their potential biological activities and mechanisms of action. This chapter presents an overview of the rapidly expanding area of fucoidan research including (1) isolation and identification, (2) biological activities and mechanisms of action, (3) absorption and bioavailability, as well as the potential of fucoidans as pharmacognosial agents.

24.2 Occurrence and abundance of fucoidans in marine brown macroalgae

Fucoidans occur in almost all marine brown macroalgal species; in particular, the fucoidan constituents of the Fucales and Laminariales orders have been extensively studied, as well as others such as Ectocarpales, Chordariales,

Dictyosiphonales, Scytosiphonales, Desmarestiales, and Dictyotales (Black, 1954; Honya et al., 1999; Skriptsova et al., 2010; Usov et al., 2001). The fucoidan contents of marine brown macroalgae range from 0% to 24% (dry wt.; **Table 24.1**), with the highest levels recorded in extracts from *Pelvetia canaliculata* (18%–24%) and *Saundersella simplex* (20.4%; Nagumo and Nishino, 1996; Usov et al., 2001). The most highly consumed marine brown macroalgae, such as *Laminaria* and *Undaria*, contain approximately 4.76% and 12.75% (dry wt.) fucoidan, respectively (Honya et al., 1999; Skriptsova et al., 2010).

It is noteworthy that the fucoidan content and composition of individual marine brown macroalgal species have been reported to vary according to not only which part of the thallus was analyzed but also the developmental stage (juvenile or adult), as well as the season/time of year when the thalli were collected (**Table 24.1**). In the majority of studies, fucoidans were extracted from whole thalli; however, when an *Alaria fistulosa* thallus was divided into the blade, midrib, sporophylls, and rhizoids, not only did the fucoidan content vary greatly—0.7% in the blade, 0.6% in midrib, 7.8% in sporophylls, and 0.5% in the stipe—but also the monosaccharide composition varied—fucose (4.2%) and galactose (3%) contents of the sporophylls were greater than that of the midrib (1% and 0.5%), blade (2.1% and 1.1%) and stipe (1% and 0.4%), respectively (Usov et al., 2005). The sporophylls and tips of the marine brown macroalgal thalli have been reported to contain the greatest amounts of fucoidan. A fucoidan-based polysaccharide developed from the sprouts of *Undaria pinnatifida* has shown promising biological activities (Fujii et al., 2005; Skriptsova et al., 2010). Moreover, different biological activities have been reported for fucoidans extracted from the blade, midrib, stipe, sporophyll frill, sporophyll midrib, and whole thallus of *U. pinnatifida* (Khan et al., 2008). These findings show the importance of extracting fucoidans from different parts of the thallus. More recently, when fucoidan was detected using an enzyme-linked immunosorbent assay (ELISA) with a monoclonal antibody which would bind to purified fucoidan obtained from *Laminaria japonica* and *Kjellmaniella gyrate* (but not with laminarin, alginic acid, or purified polysaccharide from *U. pinnatifida*), fucoidan was observed to be localized to the outer cortical layer (approximately 0–150 µm from the surface of the thallus) of *L. japonica*, in a region characterized by smaller cells and mucilaginous secretions (Mizuno et al., 2009). Interestingly, while fucoidan appeared to be associated with cell wall structures, it could not be detected in the epidermal or inner cortical layers of the *L. japonica* thalli. Thus, it is likely that fucoidans are synthesized by marine brown macroalgal tissues prior to secretion into the outer cortical layer of tissues to form an extracellular matrix (Skriptsova et al., 2010).

Several studies have shown that tissue fucoidan content can vary with the depth of the harvested plant; greater fucoidan contents have been reported in marine brown macroalgae grown in shallower waters than those grown in deeper waters (Berteau and Mulloy, 2003; Evans, 1973). *P. canaliculata* (Channelled wrack) and *Fucus spiralis* (Spiral wrack), which both grow high up in the littoral zone and can thereby be exposed for prolonged periods

Table 24.1 Fucoidan Contents of Different Marine Brown Macroalgal Species

Brown Macroalgal Species	Order	Part of the Thallus	Fucoidan Content (% Dry Wt.)
Arthrothamnus bifidus P. et R.	Laminariales	Stipes	1.0
A. bifidus P. et R.	Laminariales	Blades	0.4
Alaria fistulosa P. et R.	Laminariales	Blades	1.1
A. fistulosa P. et R.	Laminariales	Midrib	0.5
Alaria marginata P. et R.	Laminariales	Midrib	0.6
A. marginata P. et R.	Laminariales	Sporophylls	6.7
A. marginata P. et R.	Laminariales	Stipes	0.6
A. marginata P. et R.	Laminariales	Blades	1.4
Agarum cribrosum Bory	Laminariales	Whole thallus	1.2
Chordaria gracilis Setch. et Gardn.	Chordariales	Whole thallus	9.0
Chordaria flagelliformis (Mull.) Ag.	Chordariales	Whole thallus	14.3
Desmarestia intermedia P. et R.	Desmarestiales	Whole thallus	0.4
Dictyosiphon foeniculaceus (Huds.) Grev.	Dictyosiphonales	Whole thallus	6.0
Fucus evanescens Ag.	Fucales	Whole thallus	7.7
Laminaria bongardiana P. et R.	Laminariales	Stipes	1.2
Laminaria dentigera Kjellm.	Laminariales	Blades	0.4
L dentigera Kjellm.	Laminariales	Stipes	0.7
Laminaria longipes Bory	Laminariales	Blades	2.4
L longipes Bory	Laminariales	Stipes	1.1
L. bongardiana P. et R.	Laminariales	Blades	1.0
Laminaria japonica	Laminariales	Blades	2–4.2
Pelvetia canaliculata	Fucales	—	18–24
Petalonia fascia (Mull.) Kuntze	Scytosiphonales	Whole thallus	0.6
Pilayella littoralis (L.) Kjellm	Ectocarpales	Whole thallus	1.2
Saundersella simplex (Saund.) Kylin	Chordariales	Whole thallus	20.4
Scytosiphon lomentaria (Lyngb.) J.Ag.	Scytosiphonales	Whole thallus	2.8
Thalassiophyllum clathrus (Gmel.) P. et R.	Laminariales	Whole thallus	1.2
Undaria pinnatifida	Laminariales	Sporophylls	3.2–16

Sources: Skriptsova, A.V. et al., *J. Appl. Phycol.,* 22, 79, 2010; Usov, A.I. et al., *Russ. J. Bioorg. Chem.,* 27, 395, 2001; Honya, M. et al., *Hydrobiology,* 398/399, 411, 1999; Black, W.A.P., *J.Sci. Food Agri.,* 5, 445, 1954.

during tidal fluctuations, exhibited large amounts of fucoidan (18%–24% dry wt.) while *Fucus serratus*, which grows lower down in the intertidal zone, contained much less fucoidan (ca. 13%; Evans, 1973). In earlier work, Black (1954) reported an inverse relationship between the depth of marine brown macroalgal growth and fucoidan content: thallus fucoidan contents decreased with increasing depths of immersion: Laminariaceae grown under submerged conditions exhibited a reduced amount of fucoidan (i.e., 2% in *Laminaria digitata*) compared to Fucaceae grown high up on the shore in the upper littoral zone (i.e., 20% in *P. canaliculata*). Tissue fucoidan levels and the monosaccharide composition of *U. pinnatifida* and *L. japonica* have also been reported to vary with the season or time of harvest; for example, fucoidan levels increased from 3.21% (dry wt.) in juvenile thalli collected in April to 16% in mature *U. pinnatifida* thalli harvested in July (Skriptsova et al., 2010). Moreover, from April to July, the proportion of fertile adult *U. pinnatifida* thalli increased from 0% to 95%. Honya et al. (1999) reported a similar increase in fucoidan levels of cultured *L. japonica* between April and September. Thus, the life cycle stages of the marine brown macroalgae have a significant effect on the accumulation of fucoidans, in that the greatest levels are found in reproductive plants; that is, fucoidan levels increased approximately fivefold in *U. pinnatifida* sporophylls (Skriptsova et al., 2010). Evidence indicating that macroalgal age or maturity influences fucoidan monosaccharide composition was obtained from studies demonstrating a greater amount of sulfated manno-galactofucan in juvenile *U. pinnatifida* in April (containing 19–28 mol% of mannose with approximately 20 mol% galactose), whereas sulfated galactofucan (containing >38 mol% galactose) predominated in plants harvested in June and July (Skriptsova et al., 2010); as well as greater molar ratios of galactose during April through June (0.24–0.37) compared to July and August (0.21–0.17) in the major fucan fractions of *L. japonica* (fucose molar ratio = 1.0; Honya et al., 1999). Taken together, these studies indicate that the yield and chemical composition of fucoidans vary between not only marine brown macroalgal species but also the different algal plant sections, age, maturity, and season of harvest of thalli.

24.2.1 Extraction and fractionation of fucoidans

Fucoidans are highly water soluble and fairly resistant to a broad range of temperatures and pH levels; therefore, the extraction of fucoidans is not difficult. Multiple extraction and fractionation techniques using a variety of solvents have been published for the separation and isolation of fucoidans from marine brown macroalgae: hot and cold water (Fitton et al., 2007; Li et al., 2006; Ozawa et al., 2006), diluted acids (e.g., HCl, H_2SO_4), aqueous $CaCl_2$ (Bilan et al., 2002, 2004), and aqueous ethanol (Mian and Percival, 1973; Ponce et al., 2003) have been used most often to extract crude fucoidans from marine brown macroalgae (**Table 24.2**). Enzymatic digestion (e.g., 4-α-D-glucosidase, carbohydrases, alginate lyase; Athukorala et al., 2006; Sakai et al., 2002), autoclaving (e.g., 120°C, 3 h; Wang et al., 2008), and ultrasonic

Table 24.2 Isolation Techniques and Structural Information of Fucoidans from Marine Brown Macroalgae and Sea Animals

Source of Fucoidan[a]	Main Steps of Fucoidan Isolation	Core Structural Information[b]
Chorda filum (A)	Defatted algal sample was extracted with 2% CaCl$_2$ (70°C, 20 min); applied onto DEAE-Sephadex® A-25, and fucoidans eluted step-wise with NaCl gradient (0.5, 1, 1.5, 2 M)	$[\rightarrow 3]$-L-α-Fuc$p$-$(1 \rightarrow 3)$-L-α-Fuc$p$-$(1 \rightarrow 3)$-L-α-Fuc$p$-$(1 \rightarrow 3)$-L-α-Fuc$p$-$(1 \rightarrow 3)$-L-α-Fuc$p$-$(1 \rightarrow]_n$ 2) ↑ L-α-Fuc$p$-(1 Mainly sulfated at C-4, and also sulfated and acetylated at C-2 positions. Composition of fucoidan fraction of *C. filum*: fucose: xylose: mannose: glucose: galactose (1:<0.01:<0.01:<0.01:0.02)
Cladosiphon okamuranus (A)	Fresh algae was extracted with 30% HCl (pH 3, 100°C, 15 min), and the supernatant treated with CaCl$_2$ (500 mg) and ethanol (1 L). Precipitate (200 mg) was dialyzed and applied to Econo-Pac® high Q cartridge and eluted 5 fold with CH$_3$COONa (50 mM), EDTA (10 mM) and NaCl (0.15 mM) containing buffer	$[\rightarrow 3]$-α-Fuc$p$-$(1 \rightarrow 3)$-L-α-Fuc$p$-$(4SO_3^-)$-$(1 \rightarrow 3)$-L-α-Fuc$p$-$(4SO_3^-)$-$(1 \rightarrow 3)$-L-α-Fuc$p$-$(1 \rightarrow]_n$2) ↑ GlcA Composition of fucoidan fraction of *C. okamuranus*: fucose: glucuronic acid: sulfate (6.1:1.0:2.9)
Ascophyllum nodosum (B)	Crude fucoidan was suspended in 1 N H$_2$SO$_4$ and hydrolyzed, 60°C, 90 min. Then, sample was ultra-filtered (5000 Da cut off), freeze-dried and eluted from a Sephacryl® S-300 HR column using 0.2 M NaCl	$[\rightarrow 3]$-L-α-Fuc$p$-$(2SO_3^-)$-$(1 \rightarrow 4)$-L-α-Fuc$p$-$(2,3SO_3^-)$-$(1 \rightarrow]_n$ 8–14 residue oligofucan of 3090 Da MW
Fucus evanescens (B)	Defatted algal sample extracted with aqueous CaCl$_2$, 85°C, followed by precipitation with Cetavlon. Crude fucoidan was fractionated with DEAE-Sephacel® using an aqueous NaCl gradient	$[\rightarrow 3]$-L-α-Fuc$p$-$(2SO_3^-)$-$(1 \rightarrow 4)$-L-α-Fuc$p$-$(2SO_3^-)$-$(1 \rightarrow]_n$ Composition of fucoidan fraction of *F. evanescens*: fucose: sulfate: acetate (1.0:1.23:0.36)

(continued)

Table 24.2 (continued) Isolation Techniques and Structural Information of Fucoidans from Marine Brown Macroalgae and Sea Animals

Source of Fucoidan[a]	Main Steps of Fucoidan Isolation	Core Structural Information[b]
Fucus serratus L (B)	Same as above	Fucoidan with a backbone of alternating [→3)-L-α-Fuc*p*-(1→4)-L-α-Fuc*p*-(1→]$_n$ with L-α-Fuc*p*-(1→4)-L-α-Fuc*p*-(1→3)-L-α-Fuc*p*-(1→]$_n$ Composition of fucoidan fraction of *F. serratus*: fucose: sulfate: acetate (1:1:0.1) Sulfated mainly at C-2 and sometimes at C-4
Ecklonia kurome (C)	Hot water extract of alga was precipitated with cetylpyridinium chloride and dissolved in 4 M NaCl. Then, extract, dissolved in water, was precipitated with 2% CaCl$_2$. Fucoidan was purified with a combination of both ion- and gel-filtration chromatography	[→3)-L-α-Fuc*p*-(1→3)-L-α-Fuc*p*-(1→]$_n$ Composition of fucoidan fraction of *E. kurome*: fucose, galactose, mannose, xylose, sulfate Sulfated at C-4 position
Hizikia fusiforme (D)	Hot water extract of alga was precipitated with ethanol and alginate removed by CaCl$_2$. Then, sample was fractionated with DEAE Sepharose® CL-6B and Sepharose® CL-6B columns to obtain fucoidan (Fraction 32)	Alternating units of →2)-α-D-Man (1→ and →4)-β-D-GlcA(1→ with a minor portion of →4)-β-D-Gal (1→]$_n$ Composition of fucoidan fraction of *H. fusiforme*: fucose, galactose, mannose, xylose, glucose, sulfate Molecular weight of fucoidan: 92.7 kDa

Kjellmaniella crassifolia (D)	Dry alga was extracted with 80% ethanol, 3 h; precipitate was filtered and washed. Residue was extracted with Na$_3$PO$_4$ (30 mM) buffer containing NaCl (100 mM), 95°C, 2 h. The sample was digested with 12,000 U of alginate-lyase and ultrafiltered	[→3)-L-α-Fuc*p*-(1→3)-L-α-Fuc*p*-(1→3)-L-α-Fuc*p*-(1→3)-L-α-Fuc*p*-(1→]$_n$ 2)
		
		        ↑
		   L-α-Fuc*p*-(1

| *Arbacia lixula* (sea urchin) | Crude polysaccharide was extracted from the egg jelly coat by papain digestion followed by cetylpyridinium chloride and ethanol precipitation. Then, sample was fractionated step-wise from DEAE-cellulose® and Sephacryl® S-400 columns with 50 mM sodium acetate, pH 5.0 and 3.0 M NaCl | $[\rightarrow 4]$-L-α-Fuc$p$-$(2\mathrm{SO}_3^-)$-$(1\rightarrow 4)$-L-α-Fuc$p$-$(2\mathrm{SO}_3^-)$-$(1\rightarrow 4)$-L-α-Fuc$p$-$(1\rightarrow 4)$-L-α-Fuc$p$-$(1\rightarrow]_n$

Composition of fucoidan fraction of *A. lixula*: fucose: galactose: sulfate $(1{:}{<}0.1{:}0.42)$ |
| *Ludwigothurea grisea* (sea cucumber) | Same as above | $[\rightarrow 3]$-L-α-Fuc$p$-$2{,}4$-(OSO_3^-)-$(1\rightarrow 3)$-L-α-Fuc$p$-$(1\rightarrow 3)$-L-α-Fuc$p$-2-(OSO_3^-)-$(1\rightarrow 3)$-L-α-Fuc$p$-2-(OSO_3^-)-$(1\rightarrow]_n$

Composition of the fucoidan fraction of *L. grisea*: fucose: galactose: sulfate/total sugars $(1{:}{<}0.01{:}1.10)$

Molecular weight of fucoidan: 30 kDa |

Sources: Chizhov, A.O. et al., *Carbohydr. Res.*, 320, 108, 1999; Nagaoka, M. et al., *Biotechnol. J.*, 3, 904, 2008; Chevolot, L. et al., *Carbohydr. Res.*, 330, 529, 2001; Chevolot, L. et al., *Carbohydr. Res.*, 319, 154, 1999; Bilan, M.I. et al., *Carbohydr. Res.*, 341, 238, 2006; Bilan, M.I. et al., *Carbohydr. Res.*, 337, 719, 2002; Nishino, T. et al., *Carbohydr. Res.*, 211, 77, 1991; Sakai, T. et al., *Mar. Biotechnol.*, 4, 399, 2002; Sakai, T. et al., *Mar. Biotechnol.*, 6, 335, 2004; Nagumo, T. and Nishino, T., Fucan sulfates and their anticoagulant activities, in *Polysaccharides in Medicinal Applications*, ed. S. Dumitriu, Marcel Dekker, New York, 1996; Li, B. et al., *Carbohydr. Res.*, 341, 1135, 2006; Alves, A.P. et al., *J. Biol. Chem.*, 272, 6965, 1997; Mulloy, B. et al., *J. Biol. Chem.*, 269, 22113, 1994; Vilela-Silva, A.C.E.S. et al., *Glycobiology*, 9, 927, 1999.

[a] (A) fucoidan with a backbone composed of C-3-linked residues of α-L-fucopyranose; (B) fucoidan with a backbone composed of alternating 3- and 4-linked α-L-fucopyranose residues; (C) sulfated galactofucose; (D) fucoidans of more complex composition.

[b] Fuc$p$, fucopyranose; GlcA, glucuronic acid; Man, mannose; Gal, galactose.

waves (Hagiwara, 2010) have also been used to extract fucoidans from marine brown macroalgae. For example, food grade enzymes such as AMG 300 L™ (an exo-1,4-α-D-Glucan glucohydrolase), which selectively digests 1,4 and 1,6-α bonds of plant cell wall polysaccharides, can convert water-insoluble polymers into water-soluble materials; thus, enzymatic digestion processes can release sulfated polysaccharides into the extraction media (Athukorala et al., 2006). Likewise, the application of ultrasonic waves (sonication) to *Nemacystus decipiens* (Itomozuku) in water for 4 h, pH 7, 20°C–60°C, extracts crude fucoidan for further purification (Hagiwara, 2010). This ultrasonic extraction protocol was reported to yield fucoidan with reduced amounts of alginic acid, arsenic, and heavy metals, as well as reduced seaweed smell and color compared to other hydrothermal extraction methods (Hagiwara, 2010). A clear benefit of both enzymatic and ultrasonic extractions of fucoidans is the absence of toxic chemicals, enabling direct use of products in the food or pharmaceutical industries.

Once fucoidan is extracted into the desired solvent, it can be precipitated with hexadecyl-(trimethyl)-ammonium bromide (Cetavlon, also known as Cetrimide; **Table 24.2**; Bilan et al., 2002), cetylpyridinium chloride (Nishino et al., 1991), or with ethanol (Athukorala et al., 2006). The crude fucoidan is subsequently washed or dialyzed, prior to centrifugation or lyophilization. Further purification of fucoidan requires a series of ion-exchange and gel-filtration steps. Since most crude fucoidan extracts contain a mixture of low- and high-molecular weight polysaccharides with varying degrees of sulfation, anion-exchange chromatography with a step-wise NaCl-containing buffer gradient elution has been widely used (**Table 24.2**; Athukorala et al., 2006; Chevolot et al., 1999). Different varieties of ion-exchange chromatography (DEAE-Sephadex®, DEAE Sephacel®, DEAE Sepharose®, Sephacryl® S-300 HR, DEAE cellulose®) and gel-filtration column materials (Sephadex® G50, Sepharose® CL-6B, Sepharose-4B®) have been utilized to fractionate and isolate biologically active fucoidans of interest from crude marine brown macroalgal polysaccharide extracts (**Table 24.2**; Athukorala et al., 2006; Béress et al., 1993; Bilan et al., 2002; Chevolot et al., 1999; Li et al., 2006; Nishino et al., 1994). However, fucoidans may not always bind to the anion-exchange material (i.e., QAE-Sephadex A25) at pH 8.0, despite the presence of strong negative charges present in the sulfated polysaccharides (Béress et al., 1993). This apparent electrostatically neutral behavior of the sulfated polysaccharide may be due to additional esterification of the $R-O-SO_2^-$ groups with the available hydroxyl groups of the polysaccharide; therefore, Béress et al. (1993) subsequently used a cation-exchange column (SP-Sephadex® C25) to separate the fucoidan fractions of *Fucus vesiculosus*. Following purification of fucoidan by anion-exchange chromatography, gel-filtration with Sephadex® G50 or Sephadex® G100 can be used to separate compounds based on their molecular size, which is particularly useful prior to structural analyses. The physical and chemical characteristics of fucoidans depend on the marine brown macroalgal species, extraction techniques, and sequence of extractions as well as the thallus structure analyzed; therefore, the chromatographic isolation of fucoidans will likely vary from species to species.

It is noteworthy that fucoidans isolated from the cell walls of different marine brown macroalgae (i.e., *P. canaliculata, F. vesiculosus, Sargassum muticum,* and *L. digitata*) by sequential fractionation varied considerably in their composition from initial mechanical disruption of thallus tissues (i.e., grinding and disruption in a French press) and solvent extraction to isolate cell walls and release free fucans, followed by Triton X-100 treatment, acid extraction (12 N H_2SO_4, 4°C, overnight), and finally alkali treatment (3% Na_2CO_3, 60°C, 8 h) with subsequent anion-exchange (i.e., DEAE Sepharose® A25, DEAE Sepharose® CL-6B) chromatography with linear NaCl gradients to release varied fucans (Mabeau and Kloareg, 1987; Mabeau et al., 1990). Each marine brown macroalgal species mentioned earlier and their subfractions had a unique but wide range of fucose-containing polysaccharides from highly sulfated homofucans to high-uronic acid, low fucose, and sulfate-containing polymers (Mabeau et al., 1990). Thus, the composition of fucoidan not only varies with the macroalgal species and extraction techniques but also the sequence of solvent extractions.

The removal of co-extractives such as phenolic compounds, smaller carbohydrates (i.e., mannitol, glucose) and other related polysaccharides (i.e., laminaran, alginic acid/alginates) and salts during the purification of fucoidans is quite difficult. Following ethanol extraction of marine brown macroalgal tissues to remove inorganic salts, mannitol, glucose, and myoinositol, Mian and Percival (1973) treated the macroalgal residue with 40% formaldehyde to polymerize contaminating phenolic constituents, followed by sequential extractions of the residue with reagents including aqueous 2% $CaCl_2$ to precipitate alginic acid/alginate. Of note, Usov et al. (2001, 2005) determined that the mannitol contents of the midribs of *A. fistulosa* and *Alaria marginata* thalli were 2–3 fold than elsewhere in the plant, indicative of the role in metabolite transport of this tissue. The presence of alginic acid/alginates in extracts in combination with fucoidans makes purification of the latter even more difficult due to the viscosity of the extracts; it has been observed that the release of fucoidans takes place upon the extraction of alginate; hence, fucoidan and alginate are believed to be closely associated in the intercellular spaces of marine brown macroalgae (Mabeau and Kloareg, 1987). Thus, the use of aqueous 2% $CaCl_2$ to precipitate alginate from crude fucoidan, followed by centrifugation, is commonplace in fucoidan isolation (**Table 24.2**; Chizhov et al., 1999; Ponce et al., 2003; Usov et al., 2005). Interestingly, the sporophylls of *A. fistulosa* contained greater amounts of fucoidan with less alginate than the other parts of the thallus (i.e., midrib and blade; Usov et al., 2005); thus, selective extraction of fucoidans from tissue subsections based on relative concentrations may be advantageous in streamlining the process.

24.2.2 Composition and structure of fucoidans

Polysaccharides with higher amounts of α-L-fucose with esterified sulfate groups are designated as fucoidans, as mentioned earlier. Moreover, fucoidans are noted to have irregular and branched structures, which are usually complex, heterogeneous, and species specific (**Table 24.2; Figures 24.1A and 24.2**). As discussed earlier, the composition of the minor compounds in fucoidans varies with the region of the thallus studied (Usov et al., 2001, 2005); in addition, geographical and climate conditions (temperature, sunlight intensity, day length, nutrient abundance, and tidal fluctuations), life cycle stage of growth or morphological changes, as well as time of the year, may have a significant influence on fucoidan composition (Black, 1954; Honya et al., 1999; Skriptsova et al., 2010; Usov et al., 2001, 2005). Therefore, it is quite difficult to isolate fucoidans of consistent chemical composition; for example, commercially available fucoidan, derived from *F. vesiculosus*, was reported to consist of 16 different polysaccharide fractions varying in charge density, molecular weight, composition and degree of sulfation, including typical fucoidans containing fucose and sulfate as well as minor amounts of low-sulfate heteropolysaccharide-like fucans containing neutral sugars other than fucose and uronic acids, suggesting that fucoidan was a mixture of structurally different polysaccharide components (Nishino et al., 1994). Moreover, the aforementioned subfractions of the commercial fucoidan exhibited different anticoagulant efficacies than that of the original sample, indicating their unique physical and chemical properties. These workers also identified a minor but novel polysaccharide among the subfractions of commercial fucoidan identified as a proteoglycan-like, amino sugar-containing fucan sulfate, composed of fucose, galactose, glucose, mannose, xylose, uronic acid, glucosamine, and sulfate (1:0.04:0.01:0.48:0.24:0.18:0.56:1.90; Nishino et al., 1994).

Different biosynthetic pathways of marine brown macroalgal species or families are likely responsible for the variations in fucoidan composition. In a recent review, Usov and Bilan (2009) assigned fucoidans to four main categories based on structural features: Group A, fucoidans with a backbone consisting of C-3-linked residues of α-L-fucopyranose (e.g., *Chorda filum, Cladosiphon okamuranus, Lessonia vadosa, Laminaria saccharina*, and *Chordaria flagelliformis*; **Table 24.2**); B, fucoidans with a backbone composed of alternating 1,3- and 1,4-linked α-L-fucopyranose residues (e.g., *Ascophyllum nodosum, Fucus evanescens, F. serratus, P. canaliculata*, and *Stoechospermum marginatum*); C, sulfated galactofucose (e.g., *Ecklonia kurome, A. fistulosa, U. pinnatifida, L. japonica, Laminaria cichorioides, Laminaria gurjanovae*, and *Sargassum patens*); and D, fucoidans of more complex composition (e.g., *Hizikia fusiforme, Padina gymnospora, Dictyota menstrualis, Spatoglossum schroederi*, and *Tagelus gibbus*). Despite the core structural data presented in **Table 24.2**, the detailed structures of fucoidans remain unclear mainly due to the random sulfation and acetylation of fucoidans (Li et al., 2008). Compared to the fucoidans of marine brown macroalgae, the sulfated fucans of sea animals have fairly consistent structures as linear polymers with repeated regular sequences (**Table 24.2**). For example, the sulfated fucan of the egg jelly coat of *A. lixula* (Black sea urchin) shares similar structural features with *Strongylocentrotus droebachiensis* (Green sea urchin) consisting of linear sulfated α-L-fucans with repeated tetrasaccharide units (Alves et al., 1997; Vilela-Silva et al., 2002). Thus, the sulfated fucans of sea animals are easier to extract, analyze, and identify than the fucoidans of marine brown macroalgae; hence, researchers have developed different extraction conditions to increase the extraction efficacy of distinct fucoidan forms.

24.2.3 Identification of fucoidan composition

According to the IUPAC standards of nomenclature and terminology, a sulfated polysaccharide containing 20%–60% L-fucose is considered to be a fucoidan. The L-fucose (6-deoxy-L-galactose) content of a sample can be determined by either colorimetry (e.g., after reaction with L-cysteine hydrochloride and concentrated H_2SO_4; Usov et al., 2001) or chromatography (e.g., CarboPac® PA1 columns with electrochemical detection (Athukorala et al., 2006), or gas chromatography (Mabeau and Kloareg, 1987; Usov and Bilan, 2009; Usov et al., 2005). It is essential to have purified fucoidan fractions to determine chemical composition accurately. In addition, the metachromatic staining activity of a sample can be used as an indicator of the separation of fucoidans; for example, a higher density of sulfation, or other anionic groups relatively close to each other, exhibits greater metachromatic activity with 1,9-dimethylmethylene blue (525 nm) compared to those with fewer sulfate groups (Athukorala et al., 2006; Baumann and Rys, 1999). Thus, metachromatic activity is indicative of the complex binding properties of fucoidans.

As appropriate, during the anion-exchange chromatography purification process, biologically active fucoidans may be identified using a suitable assay (e.g., activated partial thromboplastin time [APTT] assay for anticoagulant activity; Athukorala et al., 2006). Once the biologically active fucoidan fractions are collected, they can be pooled, dialyzed, and freeze-dried for re-chromatography with new column materials to obtain cleaner fucoidan isolates. Agarose gel electrophoresis of the sample followed by staining with toluidine blue is a good tool to determine fucoidan purity or the homogeneity. Once the desired purification level is achieved, gel filtration chromatography of the fucoidan, with known polysaccharide standards, can be used to determine fucoidan molecular weight. Alternatively, polyacrylamide gel electrophoresis of fucoidan, with known polysaccharide standards (i.e., dextran sulfates, pullulans, and chondroitin sulfates), has also been used to help determine fucoidan molecular weight (Athukorala et al., 2006; Holtkamp et al., 2009).

Fucoidans are often hydrolyzed with trifluoroacetic or sulfuric acid prior to analysis of the monosugar composition by HPLC, GC, or GC/MS (Athukorala et al., 2006; Mabeau and Kloareg, 1987; Merkle and Poppe, 1994; Mian and Percival, 1973; Skriptsova et al., 2010; Usov et al., 2005). HPLC analysis is rapid and can tolerate a wider range of concentrations; moreover, unlike GC, HPLC does not require derivatization of sample compounds prior to analyses. The structural analysis of fucoidans is normally followed by desulfation and methylation of the sample; methylation is used to analyze the bond linkages between the monosugar compounds, while desulfation combined with methylation and acetylation or carboxyreduction is used to determine the degree and relative position of sulfation of fucoidans (Jiao et al., 2011; Li et al., 2008). The presence of sulfate groups hinders the accurate structural analysis of fucoidans and cannot be determined by GC or GC/MS techniques. Thus, methylation analysis of both native and desulfated samples gives the clearest picture about the majority of the structural features of fucoidans; moreover, these data can be compared with 1D and 2D ^{1}H and ^{13}C NMR as well as IR spectra to confirm any assumptions made during the analyses. However, it can be difficult to analyze fucoidans with masses >20 kDa using NMR due to the complicated spectra with overlapped signals (Holtkamp et al., 2009); thus, higher molecular-weight fucoidans are usually hydrolyzed prior to NMR analysis, although this may not enable determination of the detailed structural features of the fucoidans under study.

24.2.4 Hydrolyzed or modified fucoidans

Due to the high molecular weight and structural complexity of the polymers, the structure of native fucoidans can be challenging to determine accurately; therefore, the majority of studies have been conducted on hydrolyzed or modified samples yielding lower molecular-weight fucoidans. Modification of fucoidan structures may be undertaken to produce "tailored" structures accompanied by specific biological activities such as anticoagulant or antiviral activities (Holtkamp et al., 2009). Thus, chemical (e.g., hydrolysis, desulfation, and deacetylation) and enzymatic techniques (e.g., fucoidanases) have been utilized to produce modified fucoidans (Mizutani et al., 2006); however, only the latter technique has shown promising results, as chemical hydrolysis is nonspecific in its reactions, which is an inherent disadvantage when elucidating structural composition. For example, hydrolysis of fucoidans with dilute acid at elevated temperatures may not only disrupt the sulfation pattern due to desulfation, as earlier, but also result in oligosaccharides or the release of monosaccharides, which may be undesirable in the isolation of biologically active fucoidan fractions (Holtkamp et al., 2009). Thus, enzymatic methods have gained momentum to hydrolyze fucoidans using fucoidan-degrading enzymes derived from bacteria (e.g., *Pseudoalteromonas atlantica*, *Pseudoalteromonas carrageenovora*, *Pseudoalteromonas citrea*, *Pseudoalteromonas issachenkonii*, *Pseudoalteromonas nigrifaciens*, and *Vibrio* sp. no. 5) and marine organisms (e.g., *Pecten maximus*, *Patinopecten yessoensis*, *Haliotis gigantea*, *Haliotis corrugata*, *Haliotis rufescens*, and *Littorina kurila*) to produce modified fucoidans (Berteau and Mulloy, 2003; Holtkamp et al., 2009; Kusaykin et al., 2008). These enzymes can be divided into two main groups: fucoidanases (also known as fucan sulfate hydrolases; EC 3.2.1.44; which cleave off oligosaccharides from the end or center of the polysaccharide chain generating fucoidans of lower molecular weights) and α-ʟ-fucosidases (EC 3.2.1.51; which cleave α-ʟ-fucosyl linkages at the non-reducing terminus of fucoidans) able to hydrolyze fucoidans in a reproducible and reliable manner, without damaging the original sulfation pattern with either an exo- or endo-mode of action, as mentioned earlier (Holtkamp et al., 2009).

Endo-fucoidanases isolated from bacteria, *Flavobacterium* sp. SA-0082 (FERM BP-5402) and *Alteromonas* sp. SN-1009 (FERM BP-5747), were utilized by Mizutani et al. (2006) to produce U-, G-, and F-fucoidans from *K. crassifolia* (Nagakiri-Kombu, Gagome Kombu; **Figure 24.2**) as mentioned earlier. The monosugar composition and the structures of the aforementioned three fucoidans vary as follows: U-fucoidan has greater amount (20%) of glucuronic acid; F-fucoidan is mainly composed of fucose and sulfate, as a homofucan; while G-fucoidan is a sulfated fucogalactan. These three types of fucoidan from *K. crassifolia* are commercially available (Takara-Bio Inc., Japan) with a focus on cosmeceutical bioactive properties including prevention of the effects of aging skin, skin sensitivities as well as hairloss (Mizutani et al., 2006). Chemically modified fucoidans including those which were oversulfated (by exposing dry fucoidan to SO_3-dimethylformamide [SO_3-DMF] or sulfur trioxide-trimethylamine-DMF; Cho et al., 2011; Koyanagi et al., 2003; Soeda et al., 1994; Wang et al., 2009), acetylated (by exposing dried fucoidan to acetic anhydride), or benzoylated (by exposing dried fucoidan to phthalic acid anhydride; Wang et al., 2009) have been evaluated for anticancer (Cho et al., 2011; Koyanagi et al., 2003; Soeda et al., 1994) and antioxidant (Wang et al., 2009) activities. If these modifications can be combined with enzyme hydrolytic methods, perhaps fucoidan derivatives with even greater efficacies may be possible.

24.3 Biological activities of fucoidans

Fucoidans have been reported to exhibit a range of biological activities both in vitro and in vivo (**Tables 24.3** and **24.4**); in particular, the anticoagulant activity of fucoidans has been well documented since the first report published in 1957 by Springer et al. (Athukorala et al., 2006; Church et al., 1989; Irhimeh et al., 2009; Jung et al., 2007; Nagumo and Nishino, 1996; Nishino et al., 1991; Shanmugam and Mody, 2000; Wijesinghe et al., 2011; Zvyagintseva et al., 2000). On the other hand, a fucoidan fraction isolated from *L. japonica* and *F. vesiculosus*, designated AV513, exhibited potent non-anticoagulant (i.e., hemostatic) activity in canine and murine models of hemophilia A (Prasad et al., 2008). Fucoidans have also been reported to exhibit a range of potential biological activities including anti-inflammatory (Angstwurm et al., 1995; Khan et al., 2008), anticarcinogenic (Boisson-Vidal et al., 2007; Park et al., 2011; Schumacher et al., 2011; Soeda et al., 1994), cardioprotective (Deux et al., 2002; Thomes et al., 2010), and antioxidant activities (Ajisaka et al.,

Table 24.3 Biological Activities of Fucoidans: In Vitro and Animal Model Studies

Biological Activity[a]	Source of Fucoidan	Structural Information[b]	Observed Effects
Cosmeceutic effect: Promotion of fibroblast-populated collagen gel contraction as an in vitro model of dermal tissue	Hibamata extract (from *Fucus vesiculosus*)	Crude polysaccharide extract containing fucoidan up to 30 kDa MW	Fucoidan extract promoted collagen gel volume contraction at concentrations as low as 0.0001%, associated with increased integrin α-2 and β-1 subunit expression on the surface of human fibroblasts (1×10^6 cells/mL)
Cardioprotective activity: in vivo model of myocardial infarction in male Wistar rats (140–160 g)	*C. okamuranus*	Fucoidan composed of 40% fucose, 3.7% uronic acid, 25% neutral monosaccharides and 18% sulfate; 380 kDa MW	Fucoidan given orally (150 mg/kg/day) for 7 days reduced the Isoproterenol induced (2 days; 150 mg/kg/day) myocardial infarction in rats by increasing the activities of enzymatic and non-enzymatic antioxidant molecules in cardiac tissue
Immunomodulatory, anti-gastric ulcer activity: in vivo assay with Wistar rats (130–160 g)	*F. vesiculosus*	Fucoidan (Sigma-Aldrich) composed predominantly of sulfated fucose and less than 10 EU/mg of endotoxin	Fucoidan pre-treatment (0.02 g/kg fucoidan, 14 days) reduced the Aspirin®-induced (400 mg/kg, on day 14) gastric mucosal ulcer formation; reduced levels of anti-inflammatory cytokines (e.g., interleukin 6 and 12), tumor necrosis factor alpha and collagen deposition were observed in rats pre-treated with fucoidan + Aspirin versus Aspirin alone
Inhibition of complement activation: Alternative pathway of complement (APC) activity; serum from healthy donors as source of complement; non-sensibilized rabbit RBCs used for reaction activation (RA) of APC	*F. evanescens*	Fucoidan 1: fucose, galactose, glucose, mannose, xylose, rhamnose, sulfate (150–500 kDa); Fucoidan 2: fucose, galactose, glucose, rhamnose, xylose, sulfate (150–200 kDa);	The highly sulfated α-L-fucan in fucoidan-2 of *L. cichorioides* exhibited the greatest APC I_{50} at 0.5–0.7 mg/mL. Complement-mediated RBC lysis was inhibited by 50% at 20 mg/L by the sulfated, heterogeneous fucoidan-1 of *L. japonica*. The other fucoidans exhibited APC I_{50} at 6–10 mg/mL
	L. japonica	Fucoidan 1: fucose, galactose, glucose, rhamnose, xylose, sulfate (22–39 kDa);	
	Laminaria cichorioides	Fucoidan 1: fucose, galactose, glucose, rhamnose, xylose (20–70 kDa); Fucoidan 2: fucose, sulfate (35 kDa)	

Table 24.3 (continued) Biological Activities of Fucoidans: In Vitro and Animal Model Studies

Biological Activity[a]	Source of Fucoidan	Structural Information[b]	Observed Effects
Inhibition of human sperm-zona pellucida tight-binding under hemizona assay (HZA) conditions in vitro	Fucoidan (Sigma-Aldrich, St. Louis, MO); F. vesiculosus	Six fractions of low MW fucoidan prepared by acid hydrolysis: 0.02 N HCl, 80°C, 1 h; column exclusion cutoff 1.8 kDa	At 1.0 mg/mL, all fucoidan fractions inhibited tight-binding of human sperm to human zona pellucida by 56%–94% under HZA conditions; fucoidan fractions had no effect on sperm motility. Inhibition of tight-binding may be mediated by a competitive receptor-ligand mechanism
Antibacterial activity: in vitro agar plate diffusion and minimal inhibitory concentration (MIC) assays with Vibrio harveyi, Staphylococcus aureus and Escherichia coli	Sargassum polycystum	Crude fucoidan of S. polycystum extracted three times with 1 N HCl, 95°C, 12 h	Fucoidans at 12 mg/mL exhibited antibacterial activity on V. harveyi (gram negative), S. aureus (gram positive), E. coli (gram negative) with zones of inhibition of 13, 10 and 9 mm, and MICs of 12, 12 and 6 mg/mL, respectively. The McFarland No 0.5 turbidity standard was used to prepare bacterial inocula to calculate MIC values
Anti-prion encephalopathy disease activity: in vivo assay with 5–8 week old Tg7 mice.	C. okamuranus	Sample 1: 87.8% fucoidan, 42.6 kDa; Sample 2: 87.1% fucoidan, 140.4 kDa	Dietary fucoidan extracts (at 2.5%, 5%, and 10% feed) delayed the onset of disease in mice enterally infected with scrapie when fed for 6 days post-inoculation, but not when given pre-inoculation
Induction of osteoblastic cell MG-63 (ATCC: CRL-1427™) differentiation: in vitro assay of bone health	U. pinnatifida fucoidan extract (Haewon Biotech Inc, Seoul, Korea)	Fucoidan composed of 62.12% polysaccharide and 34.20% sulfate	Fucoidan extract at 10–250 µg/mL increased MG-63 alkaline phosphatase (ALP) activity and osteocalcin (OC) secretion as phenotypic markers of early-stage human osteoblastic cell differentiation. Fucoidan extract at 100 µg/mL increased cell hydroxyapatite as an indicator of potential bone mineralization
γ-Radiation protective effect: in vitro, human monoblastic leukemia cells; U937 cells; and in vivo, male Balb/c mice (SLC)	Fucoidan (Heawon Biotech, Inc., Korea)	Fucoidan (85% sulfated polysaccharide), composed of 59% monosaccharides (27.5% fucose), 26.3% sulfate, 14.7% ash	Fucoidan at 1, 10, and 100 µg/mL increased U937 cell survival rates from 48% after 8 Gy irradiation, to 53%–83% (2×10^5 cells/mL) survival. Fucoidan treatment (100 mg/kg) protected against reductions in peripheral blood thrombocyte counts of Balb/c mice treated with 8 Gy radiation

(continued)

Table 24.3 (continued) Biological Activities of Fucoidans: In Vitro and Animal Model Studies

Biological Activity[a]	Source of Fucoidan	Structural Information[b]	Observed Effects
Inhibitory effects against myotoxic phospholipases A$_2$ (PLA$_2$s) from venoms of four crotaline snakes: in vitro: murine C2C12 skeletal muscle myoblasts (ATCC-1772™); in vivo: CD-1 mice	F. vesiculosus	Fucoidan (Sigma-Aldrich); 135 kDa MW	Fucoidan pre-treatment inhibited cytotoxicity of myotoxic PLA$_2$s to C2C12 skeletal muscle myoblasts varying from 50%–65% to 100% inhibition at molar ratios of 0.25:1 (fucoidan:myotoxin). Inhibition of myotoxic PLA$_2$s in CD-1 mice treated with 50 μg of toxin and 90–270 μg fucoidan was demonstrated by reduced plasma creatine kinase levels as a measure of skeletal muscle damage

Sources: Fujimura, T. et al., *Biol. Pharm. Bull.*, 23, 291, 2000; Fitton, J.H. et al., *Cosmet. Toilet.*, 122, 55, 2007; Thomes, P. et al., *Phytomedicine*, 18, 52, 2010; Raghavendran, H.R.B. et al., *Int. Immunopharmacol.*, 11, 157, 2011; Zvyagintseva, T.N. et al., *Comp. Biochem. Physiol. C Toxicol. Pharmacol.*, 126, 209, 2000; Oehninger, S. et al., *J. Androl.*, 13, 519, 1992; Chotigeat, W. et al., *Aquaculture*, 233, 23, 2004; Doh-ura, K. et al., *Antimicrob. Agents Chemother.*, 51, 2274, 2007; Cho, Y.-S. et al., *Food Chem.*, 116, 990, 2009; Rhee, K.H. and Lee, K.H., *Arch. Pharm. Res.*, 34, 645, 2011; Angulo, Y. and Lomonte, B., *Biochem. Pharmacol.*, 66, 1993, 2003.

[a] RBCs, red blood cells.
[b] MW, molecular weight.

2009; Chattopadhyay et al., 2010; Kim et al., 2007; Wang et al., 2008, 2009; Xue et al., 2001; Zhang et al., 2003; **Table 24.3**).

Mechanistically, fucoidans appear to be capable of interacting with multiple molecular targets resulting in, for example, binding to, and modulation of, proangiogenic growth factor bioavailability (Boisson-Vidal et al., 2007); up-regulation of trans-membrane integrin expression of fibroblasts (Fujimura et al., 2000, 2002); binding or blocking of leukocyte adhesion molecules of the selectin family, in the inhibition of pneumococcal meningitis (Angstwurm et al., 1995; Teixeira and Hellewell, 1997); inhibition of leukocyte elastase activity to protect against extracellular matrix degradation in a variety of inflammatory diseases (e.g., chronic leg ulcers, periodontitis, and rheuma-toid arthritis; Senni et al., 2006); stimulation of hematopoietic progenitor stem cell mobilization associated with disrupting the interaction between the CXCR4+ cell surface receptor and its ligand, stromal-derived factor-1 (SDF-1; Irhimeh et al., 2007); as well as immune boosting effects including increased white blood cell counts (Angstwurm et al., 1995). Moreover, fucoidans have been reported to bind to a variety of proteins, including adhesion proteins (e.g., fibronectin, collagen, and laminin) involved in endothelial progenitor cell mobilization; coagulation proteases and protease inhibitors (e.g., anti-thrombin and heparin cofactor II); cell signaling molecules (e.g., cytokines and glycoproteins) involved in neoangiogenesis, as well as with different

Table 24.4 Biological Activities of Fucoidans: Human Clinical Trials

Biological Activity	Source of Fucoidan	Experimental Design[a]	Observed Effects
Anti-gastric ulcer effects	*C. okamuranus*	Subjects treated with 100 mg/day fucoidan or glucose (placebo) for 3 weeks. A randomized control trial, double-blind study, 33 patients (aged 28–58 years; 17 F and 16 M) treated with proton-pump inhibitor	Severity of ulcer grade in fucoidan-treated group was decreased in 94% (16/17) patients versus 37.5% (6/16) of placebo patients. In subjects with active, severe gastric ulcers, fucoidan treatment was associated with tissue healing and scarring, versus placebo with healing only
Anti-viral activity: Human T-lymphotropic virus type-1- (HTLV-1) associated neurological and T-cell leukemia	Fucoidan was obtained from Kanehide Bio Co., Ltd. (Okinawa, Japan)	Subjects treated with 6 g/day fucoidan for 6–13 months, or untreated controls. A single-centre, open-label trial with 23 patients (aged 38–75 years; 18 F and 5 M) diagnosed with HTLV-1 associated myelopathy/tropical spastic paraparesis	Fucoidan decreased the HTLV-1 proviral DNA (copies/100 cells) load by approx. 42.4% in subjects who completed the full course of therapy; reduction was maintained 6 months post-treatment. In vitro studies indicated that fucoidan might control cell-to-cell HTLV-1 infection
Decrease of urinary acidity in metabolic syndrome (hypertension, hyperlithuria, impaired glucose tolerance)	*C. okamuranus* (Okinawa-mozuku and thread-form mozuku)	Subjects were treated with 600 mg/day fucoidan for 7 days. Seaweed extract was incorporated into breakfast, lunch and dinner of 11 subjects with hypertension	After consuming fucoidan, subject urinary pH levels increased from approx. 5.9 to 6.2; this effect lasted for at least 24 h. Fucoidan is effective as an urinary alkalinizing agent
Increasing mobilization of hematopoietic progenitor stem cells (HPCs) in treatment of neoplasias	*U. pinnatifida*	Subjects were treated with 3 g of whole *Undaria* containing 10% (w/w) fucoidan or 3 g of 75% fucoidan three times daily for 12 days. Single-blind, randomized, placebo-controlled trial with 37 healthy, M and F non-smokers	The 75% fucoidan treatment increased peripheral blood CD34+ cells/μL, as well as CD34+ expression of CXCR4+ receptor from 0.75 cells/μL at base line to 1.65 cells/μL. These results may assist with autologous transplants of HPCs

(continued)

Table 24.4 (continued) Biological Activities of Fucoidans: Human Clinical Trials

Biological Activity	Source of Fucoidan	Experimental Design[a]	Observed Effects
Anticoagulant activity	U. pinnatifida	Subjects were treated with 3 g of 75% fucoidan capsules or 3 g guar gum (placebo) for 12 days. Single-blinded, clinical phase I/II trial with 20 healthy non-smokers (aged 23–58 years; M and F)	The 75% fucoidan treatment increased the activated partial thromboplastin time (aPTT) from approx. 28.4 to 34 s after 12 days; thrombin time (TT) decreased from approx. 18.6 to 17.6 s after 4 days and antithrombin-III increased from 113.5% at baseline to 117% after 4 days. The increase in TT indicates that fucoidan appears to interfere with the final stage of coagulation. [Fucoidan] in treated subjects' plasma was 13.1 mg/L
Anti-aging effects on facial skin	F. vesiculosus	0.2 mL gel containing 1% aqueous fucoidan extract was applied topically to one side of the face twice daily for 5 weeks. Double-blind test with 10 healthy F subjects (aged 23–36 years)	The 1% fucoidan gel decreased age-related cheek skin thickness after 5 weeks and increased skin elasticity compared to untreated control facial skin. Effects with gels containing 2%, 3%, and 5% fucoidan were not different so a 1% fucoidan gel is sufficient for beneficial effects to be observed

Sources: Juffrie, M. et al., *Indones. J. Biotechnol.*, 11, 908, 2006; Araya, N. et al., *Antivir. Ther.*, 16, 89, 2011; Hisatome, I. et al., Food/beverage and pharmaceutical composition for oral administration for improvement in acidic urine each comprising fucoidan as active ingredient, U.S. Patent 0048507 A1, http://www.google.com/patents/about/12_530_332_FOOD_BEVERAGE_AND_PHARMACEUTI.html?id=gUrNAAAAEBAJ, 2010; Irhimeh, M.R. et al., *Exp. Hematol.*, 35, 989, 2007; Irhimeh, M.R. et al., *Blood Coagul. Fibrinolysis*, 20, 607, 2009; Fujimura, T. et al., *J. Cosmet. Sci.*, 53, 1, 2002.

[a] F, female; M, male.

types of growth factors (e.g., fibroblast growth factors [FGF-2] and transforming growth factors [TGF-β1]; Boisson-Vidal et al., 2007; O'Leary et al., 2004). These biological activities are supported by reports of fucoidan absorption when administered intramuscularly (i.m.; Deux et al., 2002) or intraperitoneally (i.p.; Guimarães and Mourão, 1997), as well as internalization by smooth muscle (Deux et al., 2002) or hepatic cells (Nakazato et al., 2010) as discussed later, thus enabling the binding of fucoidans with cellular growth factors as well as with target cell signaling proteins.

Fucoidans are also potential inhibitors of human complement activation in vitro via both classical and alternate pathways (Tissot and Daniel, 2003; Zvyagintseva et al., 2000). For example, the highly sulfated α-L-fucan from *L. cichorioides* exhibited a promising alternative pathway of complement (APC) inhibition at very low concentration levels (0.5–0.7 mg/mL) compared to those of purified β-D-glucans and λ-carrageenan (**Table 24.3**; Zvyagintseva et al., 2000). Fucoidan (22.6 kDa) isolated from the acidic lysate of *Ascophyllum nodosum* was found to inhibit the classical C3 convertase formation and formation/function of the alternative C3 convertase complexes in whole serum in vitro (Blondin et al., 1994). The inhibition capacity of fucoidan in the formation of C3 convertase is intriguing, as it may prevent the production of pro-inflammatory proteins (i.e., anaphylatoxins; Tissot and Daniel, 2003). Since complement is an important part of the human immune system, these reports indicate that fucoidans may have potential as anti-complementary and anti-inflammatory agents.

24.3.1 Anticoagulant activities of fucoidans

The anticoagulant activities of marine macroalgae have been very well documented, with the first report describing an anticoagulant effect of *Iridaea laminaroides* (marine red macroalga) from Chargaff et al. (1936), followed by the work of Springer et al. (1957) on the anticoagulant activity of *F. vesiculosus* fucoidan. While the sulfated polysaccharides of marine red (i.e., agar [backbone of alternating units of 3,6-anhydro-α-L-galactopyranosyl-(1→4) and O-β-D-galactopyranosyl-(1→3)], carrageenan [1,3-linked β-D-galactose, and 1,4-linked α-D-galactose]) and green macroalgae (i.e., glucuronoxylorhamnans, glucuronoxylorhamnogalactans, or xyloarabinogalactans) have been reported to exhibit anticoagulant activity, the greatest activities have been reported with fucoidans from marine brown macroalgae (Shanmugam and Mody, 2000). To date, approximately 60 marine brown macroalgal species have been reported to exhibit blood anticoagulant properties (Shanmugam and Mody, 2000). The anticoagulant activities of fucoidans are most often evaluated using citrated, pooled human plasma and determination of the following parameters: activated partial thromboplastin time (APTT) using an ellagic acid + bovine phospholipid reagent with clotting induced with $CaCl_2$; prothrombin time (PT) using rabbit thromboplastin; and thrombin time (TT) using bovine thrombin, to determine fucoidan efficacy in controlling the intrinsic, extrinsic, and fibrinolytic pathways in the human blood coagulation cascade, respectively (Athukorala et al., 2006; Jung et al., 2007; Wijesinghe et al., 2011).

An enzymatic hydrolysate of *Ecklonia cava*, composed of fucose, glucosamine, galactose, and mannose (molar ratio of 0.81:0.01:0.16:0.01) with a total sulfate content of 0.95 (sulfate/sugar) and a minor amount of hexouronic acid exhibited a similar or slightly reduced APTT activity compared to commercial heparin at the concentration level of 0.7 μg/mL; however, at higher concentrations, the APTT activity of heparin increased more rapidly compared to that of the *E. cava* anticoagulant (ECA; Athukorala et al., 2006). In a direct binding assay, the binding affinity between ECA/antithrombin III (ATIII) and activated blood coagulation factors was in the order of VIIa (blood clotting factor (F), FVIIa) > Xa (target factor, FXa) > thrombin (FIIa); kinetic analyses determined ECA dissociation constant (K_d) values for FVIIa, FXa, and FIIa of 15.1, 45.1, and 65.0 nM, respectively (Jung et al., 2007) (**Figure 24.3**). Thus, ECA exhibited strong and selectively enhanced ATIII-mediated coagulation factor inhibition in both the extrinsic and common coagulation pathways (Jung et al., 2007). The simplified mechanism for the anticoagulant effect of ECA is shown in **Figure 24.3**. In follow-up to these in vitro studies, ECA, administered to rats at 50–300 μg/kg body wt. via the carotid artery, extended the coagulation time in a dose- and time-dependent manner in both the APTT and TT assays, particularly at 30 min after the initial sample treatment (Wijesinghe et al., 2011). In the tail bleeding assay, ECA treatment resulted in a dose-dependent prolongation of bleeding times from 960 to >1800 s at 50–300 μg ECA/kg body wt. compared to the saline control (900 s), similar to heparin at 50–100 μg/kg body wt. (Wijesinghe et al., 2011).

Fucoidan has been reported to exert its anticoagulant activity mainly via heparin cofactor II (HCII) and ATIII, as it has been demonstrated that the presence of fucoidan enhances the HCII-thrombin interaction more effectively (over 3500 fold) at fucoidan concentrations from 0.1 to 10 μg/mL, than ATIII-thrombin (285 fold) and ATIII-FXa (35 fold) interactions at fucoidan concentrations of 30 and 500 μg/mL, respectively (Church at al., 1989). In addition, fucoidans can interact directly with thrombin and other proteases as a part of its anticoagulant activity mechanism. Generally, fucoidans with a greater amount of sulfate and lower amount of hexouronic acids exhibited good

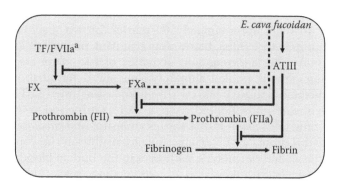

Figure 24.3 Anticoagulant mechanism of Ecklonia cava anticoagulant (ECA) in the human blood coagulation pathway. [a] TF, tissue factor; F, blood clotting factors; AT, antithrombin. The dotted line indicates weak-negative effects of ECA on the target factor FXa. (From Jung, W.-K. et al., J. Appl. Phycol., 19, 425, 2007.)

anticoagulant activity compared to those with higher uronic acid and lower sulfate contents (Shanmugam and Mody, 2000). In addition, the molecular weight (greatest activities between 50 and 100 kDa; lower activities >850 kDa), monosugar composition, stereochemical configuration, branching, and sulfate group position are important for the anticoagulant activity of fucoidans.

When the anticoagulant activity of *U. pinnatifida* fucoidan was tested in a single-blinded clinical phase I/II trial with 20 healthy human volunteers, fucoidan treatment enhanced APTT, TT, and PT values, with decreased ATIII time (Irhimeh et al., 2009; **Table 24.4**). Thus, the fucoidan exhibited low, but considerable, blood thinning properties by controlling the intrinsic coagulation pathway in subjects (Irhimeh et al., 2009). Due to its low clinical efficacy when consumed orally, fucoidans might not be able to be developed as an effective anticoagulant agent comparable to heparin; however, fucoidans could be used as a model or a probe to develop new anticoagulant drugs by studying its thrombin interactions.

24.3.2 Anticancer activities of fucoidans

While over 100 different types of cancer have been identified, there are but six essential alterations to cell physiology linked to malignant cell growth: (1) altered sensitivity to growth signals such as cell signaling pathways, phosphorylation of membrane receptors or growth factors; (2) insensitivity to antiproliferative signals which control progression through the cell cycle; (3) resistance to apoptosis (programmed cell death) through inactivation of tumor suppressor genes or transmission of anti-apoptotic signals; (4) activation of angiogenesis through increased gene transcription of vascular endothelial growth factor (VEGF) and down-regulation of proteases; (5) resistance to senescence associated with the normal erosion of chromosomal telomeres, through up-regulation of telomerase activity; and (6) resistance or inactivation of suppressors of tissue invasion and metastasis by epithelial cancers (Schumacher et al., 2011). Most often, at least a few of these aforementioned carcinogenic alterations are involved in the growth and proliferation of most human cancers. Therefore, cancer drugs that can interfere with at least one or several of the hallmarks of cancer cell growth are very important.

Reports suggest that fucoidans may control tumor growth through inhibition of growth signals (Boisson-Vidal et al., 2007), prevention of metastasis (Alekseyenko et al., 2007), suppression of angiogenesis (Koyanagi et al., 2003), induction of apoptosis (Athukorala et al., 2009; Park et al., 2011) and autophagy (type II programmed cell death; Park et al., 2011), and inhibition of cancer cell attachment (Rocha et al., 2001). Purified fucoidan from *E. cava* exhibited dose-dependent inhibition of the proliferation of several cancer cell lines (in descending order of efficacy): human leukemic monocyte lymphoma (U937) cells > promyelocytic leukemia (HL-60) cells > murine colon carcinoma (CT-26) cells > melanoma (B-16) cells (Athukorala et al., 2009). Apoptotic nuclear staining, indicative of chromatin condensation and/or fragmentation, in combination with cell cycle arrest in the sub-G1 phase as well as DNA fragmentation confirmed the induction of programmed cell death in U937 cells by *E. cava* fucoidan (Athukorala et al., 2009). Moreover, western blot analyses confirmed that the *E. cava* fucoidan increased U937 cell expression of caspase-7 and -8, increased poly ADP ribose polymerase (PARP) cleavage, and down-regulated the anti-apoptotic Bcl-xL protein expression. Similarly, commercially available fucoidan induced dose-dependent apoptosis in human gastric adenocarcinoma (AGS) cells, in association with down-regulation of anti-apoptotic Bcl-xL and Bcl-2 protein expression (Park et al., 2011). In addition, the increased Bax/Bcl-2 (and/or Bcl-xL) ratios and mitochondrial dysfunction as evidenced by loss of mitochondrial membrane potential were further indicators of apoptosis via the intrinsic pathway, which involves the proteins of the Bcl-2 family. These same workers also reported that fucoidan can inhibit the growth of AGS cells via autophagy (type II programmed cell death, or self-cannibalization) as confirmed by the increased conversion of the cytosolic protein LC3-I (18 kDa, nonlipidated) to LC3-II (16 kDa, lipidated form), as well as the dose-dependent increase in expression of the Beclin-1 protein levels (Park et al., 2011).

Fucoidan is able to bind to cell membranes, where it not only can stabilize the binding of growth factors to specific cell surface receptor proteins but also potentially increase local growth factor concentration in the vicinity of a receptor or act as a co-receptor for angiogenic growth factors (Boisson-Vidal et al., 2007). For example, fucoidan bound to the surface of human umbilical vein endothelial cells (HUVEC) can significantly suppress the binding of $VEGF_{165}$ to cell surface receptors, thereby down-regulating signal transduction in HUVECs (Koyanagi et al., 2003). Moreover, these workers reported that over-sulfated fucoidan exhibited a greater anti-angiogenic effect than native fucoidan. In mice, over-sulfated fucoidan (5 mg/kg body wt., i.v.) suppressed the neovascularization induced by implanted Sarcoma 180 cells, demonstrating the preventive effect of over-sulfated fucoidan on tumor-induced angiogenesis. Invasion of tumor cells through the basement membrane and adhesion to matrix proteins, such as laminin and vitronectin, are important in the progression of metastasis (Rocha et al., 2001). Therefore, fucoidans may be active in controlling cell adhesion and thereby preventing metastasis. The attachment of murine Lewis lung carcinoma cells (3LL) to laminin was prevented by fucoidan in a dose-dependent manner (Soeda et al., 1994). These workers hypothesized that fucoidans can selectively bind to the heparin-binding domain of laminin in the reconstituted basement membrane model, Matrigel™, controlling the attachment of 3LL cells to the membrane, followed by the suppression of the laminin-induced increase in extracellular urokinase-type plasminogen activator (u-PA).

As discussed earlier, the mechanisms through which fucoidans may exert anticarcinogenic effects are diverse, acting at many levels of regulation in the process of cellular growth, and consequently have potential as chemotherapeutic agents for multiple types of cancer cells. Moreover, fucoidans have been reported not to have any cytotoxic effects on the growth of normal cells (e.g., human lung fibroblasts, WI-38, Chinese hamster fibroblast cells, venous endothelial cells, and ECV-304; Athukorala et al., 2009; Jung et al., 2007; Park et al., 2011, respectively). Purified fucoidan preparations are almost odorless and colorless, which will allow incorporation into processed foods, functional foods, cosmeceuticals or pharmacognosic preparations. Due to the historical incorporation of marine brown macroalgae into the human diet (i.e., "Hijiki" (*Hizikia fusiformis*), "Wakame" (*U. pinnatifida*), "Makonbu" (*L. japonica*), "Hai dai" (*Laminaria* sp.), as well as the more recent functional food and nutraceutical industry developments outlined earlier, fucoidans have a very good ethno-medical history, which is very important when considering these extracts or preparations as potential cancer chemopreventive agents. Moreover, fucoidans are easily solubilized and stable (Fitton et al., 2007) and thereby have advantages over other natural anticarcinogenic agents such as curcumin, which is not water soluble and has a very short half-life due to the metabolic instability and poor absorptive characteristics of this bioactive constituent of *Curcuma longa* Linn (turmeric; Mimeault and Batra, 2011). Fucoidans can also be used as an adjuvant drug; thus, it is possible to obtain synergistic effects from fucoidans by using them in combination with other drugs. For example, when mice that were transplanted with Lewis lung adenocarcinoma cells were treated with both fucoidan and cyclophosphamide, significant antimetastatic effects were observed (Alekseyenko et al., 2007). Taken together, these reports indicate that fucoidans are important candidates for further pharmacognosic studies in the prevention or modulation of carcinogenesis and associated risk factors.

24.3.3 Antioxidant activities of fucoidans

Antioxidants are noted for acting to delay or prevent the initiation and/or propagation stages of lipid auto-oxidation and can thus behave as secondary antioxidants through chelation of transition metal ions or directly as singlet oxygen quenchers ($^1O_2$) and free radical scavengers or free radical chain inhibitors (i.e., e^- or H^+ donors; Yuan, 2007). Multiple in vitro and in vivo studies have demonstrated that fucoidans are effective in suppressing the harmful effects caused by reactive oxygen species (ROS) such as $^1O_2$, superoxide radical $\left(O_2^{-\bullet}\right)$, hydrogen peroxide ($H_2O_2$), and the highly reactive hydroxyl radical ($^\bullet OH$) as well as scavenging stable-free radicals (Athukorala et al. 2006; Kang et al., 2008; Kim et al., 2007; Thomes et al., 2010; Wang et al., 2009; Xue et al., 2001; Zhang et al., 2003). A number of studies have indicated that both crude and pure fucoidans can exert antioxidant activity through metal ion chelation (Wang et al., 2008, 2009); free radical scavenging (Kim et al., 2007; Zhang et al., 2003); reduced activities of hepatic oxidative enzymes when challenged with CCl_4 (e.g., glutamate oxaloacetate transaminase [GOT], glutamate pyruvate transaminase [GPT]; Kang et al., 2008); and, increased activities of hepatic and cardiac antioxidant enzymes (i.e., superoxide dismutase [SOD], catalase [CAT], glutathione peroxidise [GPx]), and small molecule antioxidants (i.e., glutathione [GSH]) in the presence of oxidative stress (Kang et al., 2008; Thomes et al., 2010). In addition, fucoidans are known to affect signal transduction and regulation by targeting molecular targets (e.g., nuclear factor-κB [NF-κB], matrix metalloproteinases [MMP]) that have a indirect connection with ROS generation via inflammation (Boisson-Vidal et al., 2007); therefore, a combination of mechanisms and kinetics is associated with the antioxidant activities of fucoidans.

Fucoidans have been reported to exhibit promising antioxidant activities compared to other sulfated and commercial polysaccharides; when fucoidan (containing fucose, xylose, galactose, sulfate), alginate, and laminaran fractions from the marine brown macroalga *Turbinaria conoides* were analyzed in a ferric reducing antioxidant (FRAP) assay, fucoidan showed the highest antioxidant activity at 4 and 30 min, followed by alginate and laminarin (Chattopadhyay et al., 2010). These workers also reported that fucoidan quenched the stable-free radical, 1,1-diphenyl-2-picrylhydrazyl (DPPH•) in a dose-dependent manner, with a scavenging capacity of 90% at 5 mg/mL. Similarly, when fucoidan (commercial, from *F. vesiculosus*) was compared to 66 carbohydrates, including chondroitin sulfate, agaro-oligosaccharide, D-glucuronic acid, and assorted mono-, di-, oligo-, and polysaccharides, fucoidan exhibited 30.1% quenching of DPPH• compared to the L-ascorbic acid control (Ajisaka et al., 2009). It is noteworthy that the antioxidant efficacies of fucoidans will vary greatly depending on the monosugar composition, degree of sulfation, and the antioxidant assay system employed. For example, two fucoidans isolated from *L. japonica*, a low-molecular-weight fraction (fucose > glucose > mannose) and a crude preparation (galactose > mannose > fucose), both exhibited strong inhibition of stable-free radical-induced human LDL oxidation (Xue et al., 2001). Interestingly, despite their similar chemical characteristics (i.e., sulfation, uronic acid, and amino sugar composition), these same sulfated polysaccharide fractions exhibited different efficacies against Cu^{2+}-induced LDL oxidation, with the former fraction conferring greater protection than the latter. Thus, the molecular mass of fucoidans strongly influences the antioxidant mechanisms of action through potential metal ion chelation as well as free radical quenching.

Several studies have reported that the in vivo antioxidant activities of fucoidans may exist in concert with other protective biological activities associated with oxidative stress. For example when fucoidan derived from *C. okamuranus* was administered to rats treated with isoproterenol to induce myocardial infarction, a significant reduction in the levels of serum and cardiac tissue lipid peroxidation, SOD, CAT, and GPX, and non-enzymatic antioxidants (GSH, α-tocopherol, and ascorbic acid) were observed in the fucoidan + isoproterenol versus those treated with isoproterenol alone (**Table 24.3**; Thomes et al., 2010). These workers demonstrated that fucoidan suppressed not only the biochemical and histological parameters of isoproterenol-induced oxidative stress in rats but also serum cholesterol, triacylglycerides, and LDL levels, thereby conferring protection against isoproterenol-induced myocardial infarction. Similarly, Kang et al. (2008) demonstrated a protective effect of *U. pinnatifida* fucoidan against CCl_4-induced oxidative stress in rats. Fucoidan (100 mg/kg body wt., i.p., 14 days) decreased the oxidative stress caused by CCl_4 (3.3 mL/kg body wt.; 1:1 ratio in olive oil) by reducing serum GOT, GPT, alkaline phosphatase (ALP), and lactate dehydrogenase (LDH) activities as well as hepatic lipid peroxidation, and increasing hepatic SOD, CAT, and GPx activities compared to the CCl_4-treated control animals. Thus, fucoidans prevented isoproterenol- and CCl_4-induced models of oxidative stress by scavenging damaging free radicals and strengthening the cellular oxidative system in animal models.

The presence of amino ($-NH_2$), carboxyl ($-COOH$), carbonyl ($-C=O$), and sulfonyl ($-SO_3H$) groups in carbohydrates is important for their antioxidant activities (Ajisaka et al., 2009; Wang et al., 2008, 2009; Xue et al., 2001). However, even if some polysaccharides possess some or all of these functional groups, antioxidant efficacies will vary (Ajisaka et al., 2009). The position or number of these groups in the polysaccharide structure is a key determinant of the structure–activity relationship for antioxidant activity (Ajisaka et al., 2009). As discussed earlier, polysaccharides with a sulfonyl group in their structure, such as fucoidans and chondroitin sulfate, exhibited a greater antioxidant activity than other polysaccharides. When Wang et al. (2009) studied the antioxidant efficacies of oversulfated, acetylated, or benzoylated *L. japonica* fucoidans, the benzoylated specimen exhibited the strongest $O_2^{-\cdot}$ and $\cdot OH$ radical scavenging activities, while the acetylated fucoidan exhibited the greatest activity against $\cdot OH$ and DPPH• radicals, as well as reducing power. Thus, the antioxidant activities of fucoidans are complex and highly dependent on the chemical groups and molecular weight of the sample. However, since fucoidans have been demonstrated to exhibit antioxidant activities in vitro and in vivo, there is potential for these biologically active compounds to have protective effects against oxidative stress to reduce chronic disease risk factors.

24.3.4 Cosmeceutical effects of fucoidans

Cosmeceuticals, which contain botanical or herbal constituents for dermatological applications (orally or topically), do not have a separate definition under the federal Food, Drug and Cosmetic Act (FD&C Act; Food and Drug Administration, 2012) of the United States, but rather encompass attributes of both cosmetics (i.e., are intended "to be rubbed, poured, sprinkled, or sprayed on or otherwise applied to the human body for cleansing, beautifying, promoting attractiveness, or altering the appearance") as well as drugs (i.e., are intended "for use in the cure, mitigation, treatment, or prevention of disease" and "articles (other than food) intended to affect the structure or any function of the body of man or other animals"; FD&C Act, section 201(i) and 201(g)(1); Food and Drug Administration, 2012). Therefore, they may not be reviewed for Food and Drug Administration (FDA) approval prior to being marketed, and thus, not subjected to the same requirements for demonstration of efficacy and safety (Allemann and Baumann, 2009). Fucoidans have generated considerable attention as an ingredient in cosmeceuticals, exhibiting activities such as diminution of the effects of skin aging, amelioration of skin sensitivities, and as an effective ingredient in hair-care products to reduce hair loss or assist with hair restoration (Mizutani et al., 2006). For example, fucoidan derived from *K. crassifolia* demonstrated promising hair-restoration effects after topical administration to shaved male C3H/He mice; animals exhibited 44.4%, 45.6%, and 26.7% trichogenous ratios (hair producing/shaved areas) after treatment with whole fucoidan, F-fucoidan (**Figure 24.2B**), and a placebo, respectively (Mizutani et al., 2006). In another trial, these same workers reported a greater induction of skin tissue glucose-6-phosphate dehydrogenase and alkaline phosphatase in mice treated with the F-fucoidan compared to the control group, suggesting a change in the hair-growth transition period from telogen (resting period of the hair follicle) to anagen (active growth phase of a hair follicle; Mizutani et al., 2006).

In related work, when a lotion containing *K. crassifolia* fucoidan was subjected to a blinded, clinical functional test by 25 healthy women, aged between 25 and 35 years, the perceived moistness, smoothness, and "liveliness" of the skin were increased compared to a placebo lotion, suggesting that the fucoidan-containing lotion had cosmeceutical effects (Mizutani et al., 2006). Fujimura et al. (2002) reported similar results with a gel containing 1% *F. vesiculosus* extract applied to one side of the faces of female volunteers (**Table 24.4**). These workers demonstrated that fucoidan-treated skin exhibited decreased thickness (by ultrasound) and increased elasticity compared to the untreated side. In previous work, Fujimura et al. (2000) reported that *F. vesiculosus* fucoidan promoted fibroblast-populated collagen gel contraction and relaxation time in a model of wound healing (**Table 24.3**). Moreover, when Sezer et al. (2008a,b) produced fucospheres (microspheres) and hydrogels from positively charged chitosan and negatively charged *F. vesiculosus* fucoidan and tested treatment efficacy on male New Zealand white rabbits with deep dermal burns with biopsy samples taken on the 7th, 14th, and 21st day after topical treatment, fucospheres (with high-surface charge and bioadhesive properties) showed promising healing effects after 21 days of treatment without any signs of edema, compared to a fucoidan solution, chitosan microspheres, or an untreated control wound. The fucosphere effects were associated with increased fibroblast migration and collagen synthesis with accelerated epithelial regeneration (Sezer et al., 2008b). Similarly, fucoidan-chitosan hydrogel-treated wounds did not exhibit signs of hemorrhage, edema, or inflammation, and in fact exhibited rapid epithelial regeneration and greater mononuclear leukocytes (MNLs) indicative of increased fibroblast and collagen production, as well as some hair regrowth, with wound healing completed by day 21 (Sezer et al., 2008a).

Related to wound repair, degradation of connective tissue is a hallmark of such inflammatory conditions as chronic leg ulcers, chronic wounds, rheumatoid arthritis, or periodontitis; thus, Senni et al. (2006) studied the ability of hydrolyzed *A. nodosum* fucoidan (55.7% L-fucose, 6.3% uronic acid, 29% sulfate; 16 kDa) to minimize human leukocyte elastase (HLE) to protect the dermal elastic fiber network (i.e., oxytalan, elaunin, and elastic fibers) against proteolytic degradation using human skin tissue sections. A fucoidan pre-treatment prior to the HLE reaction protected the skin elastic fiber network, demonstrating that fucoidan was capable of controlling important parameters involved in the breakdown of connective tissues in human skin. The three-dimensional structure and anionic charge density of fucoidan were thought to result in strong interactions and inhibition of HLE (Senni et al., 2006). These workers suggested that the inhibition of MMPs and serine proteases, as well as increased fibroblast number and matrix production induced by fucoidan treatment, may directly relate to the protective effects of fucoidans in models of inflammation.

Fucoidans have exhibited promising results in controlling UV-B induced MMP-1 expression, which is noted to play a role in the photo-aging of skin; as well, marked inhibition of extracellular signal regulated kinase (ERK) and slight inhibition of c-Jun N-terminal kinase (JNK) were observed in the presence of fucoidan (Moon et al., 2008, 2009). Thus, fucoidans appear to have promise as a cosmeceutical in the prevention and treatment of UV-B-induced skin photoaging (**Figure 24.4**). Based on recent literature, the UV-induced skin photoaging mechanism is shown in **Figure 24.4**, in that fucoidans can modulate both cellular as well as molecular responses in skin and thereby protect skin from photoaging. There are currently a number of fucoidan-containing algal extracts identified in the International Nomenclature of Cosmetic Ingredients (INCI) list along with recommended protective functions (Fitton et al., 2007; INCI Directory, 2012). For example, *L. japonica* (CAS number 223751-72-2; 92128-82-0) is identified for skin-protecting properties and potential applications in facial care, facial cleansing, body care, and baby care. Also, *F. vesiculosus* (CAS 84696-13-9; 100085-36-7) is listed for soothing, smoothing, emollient, and skin conditioning properties with potential applications in hair care, sun care, skin care, toiletries, and also in slimming applications. Similarly, *A. nodosum, U. pinnatifida, Dicksonia antarctica, Macrocystis pyrifera,* and *E. cava* are also listed in INCI for different skin care functions with similar end applications.

24.4 Absorption and bioavailability of fucoidans

A considerable amount of research has been conducted on the different biological activities of fucoidans, as discussed earlier. These studies have primarily made use of in vitro chemical assays, cell culture techniques, ex vivo tissue or LDL analyses, or in vivo rodent models. In addition, a few studies have

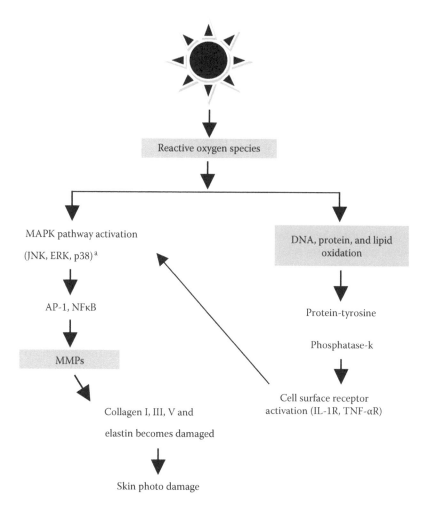

Figure 24.4 Simplified skin photodamaging mechanism by UV light. The shaded areas indicate where fucoidan modulates its biological activity to control skin photoaging. [a] MAPK, mitogen-activated protein kinases; JNK, jun amino-terminal kinase; ERK, extracellular-signal-regulated kinases; p38, p38 mitogen-activated protein kinase, AP-1, activator protein-1; NF-κB, nuclear factor-κB; MMPs, matrix metalloproteinases; IL-1R, interleukin-1 receptor; TNF-αR, tumor necrosis factor-α receptor. (Adapted from Sardy, M., Connect. Tissue, Res., 50, 132, 2009.)

been conducted with larger mammals including dogs (Prasad et al., 2008), rabbits (Deux et al., 2002), and baboons (Alwayn et al., 2000), as well as human subjects, to evaluate the topical cosmeceutical efficacy of fucoidans in vivo (Fujimura et al., 2002; Mizutani et al., 2006). However, pharmacokinetic data detailing the absorption, distribution, metabolism, and clearance/excretion of fucoidans, which are extremely important in the elucidation of the pharmacological effects of these polysaccharides, are very limited.

Guimarães and Mourão (1997) studied the pharmacokinetics (i.e., absorption into the circulation and urinary clearance versus molecular weight of fucoidan) of several polysaccharides including dextran sulfate, chondroitin-4-sulfate, chondroitin-6-sulfate, fucosylated chondroitin sulfate, sulfated L-galactans, and sulfated L-fucans after i.p. administration (65 mg/kg body wt.) to rats. Only small amounts of the fucoidans from *F. vesiculosus* (50 kDa) and *Laminaria brasiliensis* (150 kDa) were excreted in the urine at 12 h: 6.9% and 3.3% of the dose, respectively; whereas 44.9% of dextran sulfate (8 kDa) was excreted after 12 h. These results suggest that glomerular permeability and the limit of filtration are correlated with the molecular mass of the polysaccharide, in that the marine brown macroalgal fucoidans of higher molecular weight are cleared in the urine in only minor amounts due to their larger molecular mass (Guimarães and Mourão, 1997). These workers presumed that glomerular and vascular permeability are the factors that restrict the urinary secretion. Similarly, when a fluorolabeled, using 5-([4,6-dichlorotriazin-2-yl] amino) fluorescein, low-molecular-weight fucoidan (8 kDa) was administered (5 mg/kg body wt.) i.v. and i.m. to rats, 33.2% of the fucoidan was excreted in the urine (Deux et al., 2002). Thus, molecular weight and other structural characteristics (i.e., charge density and molecular structure) are important for the renal permeoselectivity of polysaccharides; lower-molecular-weight fucoidans result in greater urinary excretion than higher-molecular-weight fucoidans. In addition, Deux et al. (2002) reported an elimination half-life of approximately 56 min for fucoidan after i.v. administration and a constant plasma concentration of 10 μg/mL over 6 h after i.m. administration. When these workers studied the uptake of the fluorolabeled-fucoidan by cultured rabbit smooth muscle cells, a time-dependent accumulation of fucoidan was observed in the perinuclear region of the cells after 6 and 12 h of incubation; however, fucoidan was not detected inside the nucleus. Thus, low-molecular-weight fucoidan was able to migrate through cellular membranes, and thereby potentially bind with molecules in the intracellular signaling pathways (Deux et al., 2002), further supporting the observation that low-molecular-weight fucoidan was more highly localized in injured segments of the rabbit iliac arteries than in the normal arteries. Therefore, fucoidan appears to migrate preferentially into live cells. When Nakazato et al. (2010) stained hepatic cells of rats after administration of *N*-nitrosodiethylamine (DEN; to induce liver fibrosis) + high-molecular-weight fucoidan (41.1 kDa), or crude fucoidan, they were able to detect fucoidan localization in these cells. These authors also observed varying degrees of fucoidan staining between the two treatments: clearly visible staining in the high-molecular-weight fucoidan group, but only faint staining with the crude fucoidan; hence, it is plausible to assume that fucoidan can assimilate preferentially into cells.

Irhimeh et al. (2005) studied the bioavailability of fucoidan in healthy human volunteers. When a high-molecular-weight *U. pinnatifida* fucoidan (713 kDa; L-fucose linked 1→3 and 1→4, with sulfation at C-2 or C-4 and 29% of SO_4) was administered to volunteers, a considerable amount of free fucoidan was detected in plasma after 4, 8, and 12 days using a novel monoclonal antibody technique, but still small, at 0.6% of the oral dose. However, this study was not able to gather any information about the absorption of fucoidan by the human gut. The presence of 12–14 mg/L high-molecular-weight fucoidan in plasma over 4–12 days suggested that a small quantity of orally ingested fucoidan may cross the intestinal wall via endocytosis (Choi et al., 2005; Irimeh et al., 2005) similar to chondroitin sulfate. Interestingly, the plasma fucoidan concentrations did not change, even after 4 (11.9 mg/L), 8 (13.0 mg/L), or 12 (15.7 mg/mL) days. The antibody technique used in this study is in its infancy; however, as the first study of its kind for fucoidans, this methodology was able to provide some key information about the pharmacokinetics of fucoidan in humans.

While the low pH gastric conditions may result in limited hydrolysis of high-molecular-weight fucoidans, these polysaccharides are largely resistant to digestion by humans due to the absence of fucoidan digesting enzymes. A fermentation study conducted with purified L-fucose and fucoidan with human fecal bacteria (*Bacteroides* sp.) resulted in an absence of fermentation of the latter, but fermentation of the monosaccharide (Salyers et al., 1977). The bacteria exhibited no substrate reduction of the fucoidan, even after 24 h incubation time. The higher fucose (365 g/kg dry wt. of total fiber) and sulfate (221 g/kg) contents and the unique structure of the fucoidan in this study may have contributed to its greater resistance to enzymatic breakdown (Cherbut et al., 1991).

If fucoidans are administered by routes other than per os, including i.v., i.m., or i.p., the dose-dependence and efficacy will likely depend on the entrance rate of fucoidan into target tissues or organs. The distribution of fucoidans in body fluids and tissues may be influenced by the molecular weight, stereochemical configuration, branching and sulfate group positions, as well as on the monosugar residues and their arrangement. It remains unclear what fate awaits fucoidans after administration in vivo. Because fucoidans are negatively charged, binding to plasma proteins may easily occur in the circulation. If this is the case, are only unbound fucoidans biologically active? Thus, further elucidation of the digestion/fermentation, absorption, metabolism, and clearance/excretion of fucoidans is vital to understand the modes of action and efficacy of fucoidans as potential pharmacognosic agents.

24.5 Potential of fucoidans in pharmacognosy

Marine brown macroalgae containing considerable amounts of fucoidans, such as *U. pinnatifida* (Wakame, Sea Mustard), *H. fusiforme* (Hijiki), *L. japonica* (Makonbu, Sea Tangle), *C. okamuranus* (Mozuku), and *F. vesiculosus* (Bladder wrack) are widely consumed in Asia and have a long history of consumption (e.g., seasonings, condiments, and sea vegetables) and medicinal uses in Tonga, China, Korea, Japan, and Indonesia (Mower, 2010).

For example, Japanese consumption of marine macroalgae ranges between 0.4 and 29.2 g/day, with the people of Okinawa having the greatest intakes of macroalgae, thereby consuming a greater amount of fucoidan; on the other hand, most Western populations (i.e., North America and Europe) consume very low to zero marine macroalgae (Yuan, 2008). It is noteworthy that these consumption patterns are inversely related to chronic disease mortality rates in that the Japanese rates for coronary heart disease are 43 (males) and 22 (females)/100,000, approximately one quarter of the rates in North America; the pattern is similar for diet-related cancers such as breast and prostate cancers (Yuan, 2008). These epidemiological relationships may help to understand the potential protective, biological effects of fucoidans; however, the variety of bioactive compounds in marine macroalgae as well as the whole dietary pattern of the Japanese, and Okinawans in particular, must be considered as factors in the aforementioned observations.

The efficacy of fucoidans on gastric ulcers (Juffrie et al., 2006), anti-viral activity (Araya et al., 2011; Cooper et al., 2002), controlling urine acidity (Hisatome et al., 2010), stem cell modulation in the treatment of neoplasias (Irhimeh et al., 2007), anticoagulant activity (Irhimeh et al., 2009), as well as anti-aging of skin as a cosmeceutical (Fujimura et al., 2002) have been tested in human clinical trials in recent years (**Table 24.4**). The results of these studies are promising of the pharmacognosic potential of fucoidans; for example, when *C. okamuranus* fucoidan was administered to patients with active, severe, and moderate gastric ulcers, the grade of the ulcers, abdominal pain, and vomiting were reduced compared to the placebo group (**Table 24.4**; Juffrie et al., 2006). The fucoidans of *C. okamuranus* bind with proteins associated with gastric ulcers (i.e., gastric surface proteins), inhibit peptide activity, and improve the stability of basic fibroblast growth factor (bFGF) to promote ulcer healing (Shibata et al., 2000). Importantly, *C. okamuranus* fucoidan did not stimulate macrophage and polymorphonuclear leukocyte O_2^- generation, or tumor necrosis factor-α (TNFα) secretion by macrophages; thus, this fucoidan did not trigger any inflammatory effects (Shibata et al., 2000).

In a study evaluating the anti-viral efficacy of fucoidans, patients included seven cases of Herpes simplex virus type 1 (HSV-1); five cases of HSV-2; 3 of active Herpes zoster (chicken pox and shingles) and two of Epstein–Barr Virus (EBV); subjects were treated with fucoidan (one to four tablets containing 560 mg *U. pinnatifida* fucoidan) for a period ranging between 1 and 24 months (Cooper et al., 2002). These workers reported a reduction in the viral load at the end of the clinical trial without activating the host immune system. Moreover, this study demonstrated the inhibitory effects of fucoidan on the reactivation of HSV, and positive effects on active herpes infections (e.g., reduction in lesion severity, rapid lesion clearance, pain reduction, and clearance of EBV) indicative of the potential use of fucoidans as an anti-viral agent. Interestingly, there appears to be a difference in the incidence of HSV-1 reactivation, via asymptomatic shedding of the virus, between subjects in Japan and United States (Fitton, 2003; Okinaga, 2000). The lower HSV-1 isolation frequency from tears and saliva in healthy Japanese subjects compared to the United States may be related to the higher intake of fucoidan-rich marine brown macroalgae by the former population. Similarly, when patients diagnosed with human T-lymphotropic virus type-1-associated neurological and T-cell leukemia (HTLV-1) were treated with fucoidan, the HTLV-1 proviral DNA load was reduced by 42.4% after 13 months (**Table 24.4**; Araya et al., 2011). In other in vitro work, these workers reported that fucoidan may modulate cell-to-cell HTLV-1 infection.

When healthy volunteers were treated with *U. pinnatifida* fucoidan 3 times daily for 12 days, oral ingestion of fucoidan increased the CXCR4+ receptor hematopoietic progenitor stem cell (HPC) population in the peripheral blood (**Table 24.4**; Irhimeh et al., 2007). The ability to mobilize CD34+ HPCs and to increase CD34+ expression of CXCR4+, which is found on the surface of many different cell types that play a role in the human immune system, could be clinically valuable in the treatment of neoplastic diseases; for example, if fucoidan stimulates the mobilization of stem cells, these cells can replace dead cells, thereby enabling tissue and organ regeneration. The aforementioned clinical trial data (**Table 24.4**) provide evidence of the effect of fucoidans in reducing the risk factors of various disease states and potentially improving health; however, none of the aforementioned studies provided conclusive proof that a significant amount of intact or hydrolyzed fucoidan was absorbed and bioavailable in the circulation or tissues. In addition, most of the aforementioned studies have been conducted with small numbers of subjects (i.e., 11–33 male and female subjects); therefore, larger scale, prospective, randomized, double-blind and controlled clinical trials with ethnically diverse subjects including the Western population are needed to obtain a clear understanding about the efficacy of fucoidans. A number of clinical trials are currently being conducted to study the immunomodulatory activity and anti-inflammatory properties of fucoidans (Fitton et al., 2007). Because fucoidans have biological activities similar to those of glucosamine and chondroitin, clinical trials are currently ongoing to study fucoidans as a replacement for the aforementioned two nutraceutical products that are commercially obtained from animal connective tissues.

Marine brown macroalgae are considered a renewable bioresource, which are highly abundant and proliferate ubiquitously in the coastal regions of the world's oceans; therefore, fucoidans may be extracted from marine brown macroalgae at a low cost. As discussed earlier, fucoidans have promising antiviral activities; thus, if fucoidans can be developed as a safe natural drug, it may potentially replace or augment the high-priced anti-viral drugs currently on the market. For example, if at-risk populations are not able to afford anti-viral drugs that are available on the market, a diet supplemented with fucoidan-containing foods, nutraceuticals, or pharmacognosic agents might be a plausible goal.

Fucoidan-based functional foods, nutraceuticals, cosmeceuticals, or pharmacognosic agents could be manufactured in the form of beverages, capsules or tablets, or creams, lotions, or ointments, as formulations that consumers would be familiar with and find acceptable. Because fucoidans are water soluble, as well as tasteless and odorless, a variety of foods or drinks can be prepared with fucoidans, as discussed earlier (Mower, 2007, 2010). In addition, partially purified fucoidan contains a slight amount of protein, lipid, alginate, and laminarin with smaller amounts of other nutrients including minerals (i.e., sodium, magnesium, calcium, potassium, iron, phosphorous, and aluminum) and vitamins that may have potential health benefits (Mower, 2010; Oh, 2009; Shaklee et al., 2010). Therefore, fucoidans may be associated with more than one protective effect in functional foods, or as a nutraceutical, cosmeceutical, or pharmacognosic agent.

24.6 Conclusion

Taken together, the evidence discussed earlier indicates that marine brown macroalgal fucoidans have potential for further development as not only nutraceuticals and functional foods but also as cosmeceutical agents. However, due to the paucity of pharmacokinetic data on the digestion, absorption, metabolism, and clearance/excretion of fucoidans, the efficacy and mechanisms of action are difficult to interpret in the current literature. On the other hand, fucoidans that are well characterized, are uniform, and have defined structures, together with an understanding of the physical, chemical, biochemical, and biological properties will be extremely important for the development of standardized, commercial product under FDA and European pharmaceutical guidelines.

Acknowledgments

This work was funded by a Natural Sciences and Engineering Research Council of Canada (NSERC) Discovery Grant to YVY, NSERC Undergraduate Student Research Awards (USRAs), Ryerson University Postdoctoral Fellowship to YA, Ryerson Faculty of Community Services (FCS) Research Awards to YVY and FCS Travel Grants to YA. The authors thank Carmen Kwok and Susan Trang for their assistance with preparation of some of the figures, tables, and manuscript.

References

Ajisaka, K., S. Agawa, S. Nagumo et al. 2009. Evaluation and comparison of the antioxidative potency of various carbohydrates using different methods. *J. Agric. Food Chem.* 57:3102–3107.

Alekseyenko, T.V., S.Y. Zhanayeva, A.A. Venediktova et al. 2007. Antitumor and antimetastatic activity of fucoidan, a sulfated polysaccharide isolated from the Okhotsk Sea *Fucus evanescens* brown alga. *Bull. Exp. Biol. Med.* 143:730–732.

Allemann, I.B. and L. Baumann. 2009. Botanicals in skin care products. *Int. J. Dermatol.* 48:923–934.

Alves, A.P., B. Mulloy, J.A. Diniz, and P.A.S. Mourão. 1997. Sulfated polysaccharides from the egg jelly layer are species-specific inducers of acrosomal reaction in sperms of sea urchins. *J. Biol. Chem.* 272:6965–6971.

Alwayn, I.P., J.Z. Appel, C. Goepfert, L. Buhler, D.K. Cooper, and S.C. Robson. 2000. Inhibition of platelet aggregation in baboons: Therapeutic implications for xenotransplantation. *Xenotransplantation* 7:247–257.

Angstwurm, K., J.R. Weber, A. Segert et al. 1995. Fucoidin, a polysaccharide inhibiting leukocyte rolling, attenuates inflammatory responses in experimental pneumococcal meningitis in rats. *Neurosci. Lett.* 191:1–4.

Angulo, Y. and B. Lomonte. 2003. Inhibitory effect of fucoidan on the activities of crotaline snake venom myotoxic phospholipases A_2. *Biochem. Pharmacol.* 66:1993–2000.

Araya, N., K. Takahashi, T. Sato et al. 2011. Fucoidan therapy decreases the proviral load in patients with human T-lymphotropic virus type-1-associated neurological disease. *Antivir. Ther.* 16:89–98.

Athukorala, Y., G.N. Ahn, Y.-H. Jee et al. 2009. Antiproliferative activity of sulfated polysaccharide isolated from an enzymatic digest of *Ecklonia cava* on the U-937 cell line. *J. Appl. Phycol.* 21:307–314.

Athukorala, Y., W.-K. Jung, T. Vasanthan, and Y.-J. Jeon. 2006. An anticoagulative polysaccharide from an enzymatic hydrolysate of *Ecklonia cava*. *Carbohydr. Polym.* 66:184–191.

Baumann, R. and P. Rys. 1999. Metachromatic activity of betacyclodextrin sulfates as heparin mimics. *Int. J. Biol. Macromol.* 24:15–18.

Béress, A., O. Wassermann, T. Bruhn, and L. Béress. 1993. A new procedure for the isolation of anti-HIV compounds (polysaccharides and polyphenols) from the marine alga *Fucus vesiculosus*. *J. Nat. Prod.* 56:478–488.

Berteau, O. and B. Mulloy. 2003. Sulfated fucans, fresh perspectives: Structures, functions, and biological properties of sulfated fucans and an overview of enzymes active toward this class of polysaccharide. *Glycobiology* 13:29–40.

Bilan, M.I., A.A. Grachev, A.S. Shashkov, N.E. Nifantiev, and A.I. Usov. 2006. Structure of a fucoidan from the brown seaweed *Fucus serratus* L. *Carbohydr. Res.* 341:238–245.

Bilan, M.I., A.A. Grachev, N.E. Ustuzhanina, A.S. Shashkov, N.E. Nifantiev, and A.I. Usov. 2002. Structure of a fucoidan from the brown seaweed *Fucus evanescens* C. Ag. *Carbohydr. Res.* 337:719–730.

Bilan, M.I., A.A. Grachev, N.E. Ustuzhanina, A.S. Shashkov, N.E. Nifantiev, and A.I. Usov. 2004. A highly regular fraction of a fucoidan from the brown seaweed *Fucus distichus* L. *Carbohydr. Res.* 339:511–517.

Black, W.A.P. 1954. The seasonal variation in the combined L-fucose content of the common British Laminariaceae and Fucaceae. *J. Sci. Food Agri.* 5:445–448.

Blondin, C., E. Fischer, C. Boisson-Vidal, M.D. Kazatchkine, and J. Jozefonvicz. 1994. Inhibition of complement activation by natural sulfated polysaccharides (Fucans) from brown seaweeds. *Mol. Immunol.* 31:247–253.

Boisson-Vidal, C., F. Zemani, G. Caligiuri et al. 2007. Neoangiogenesis induced by progenitor endothelial cells: Effect of fucoidan from marine algae. *Cardiovasc. Hematol. Agents Med. Chem.* 5:67–77.

Chargaff, E.F., F.W. Bancroft., and M. Stanley-Brown. 1936. Studies on the chemistry of blood clotting by substances of high molecular weight. *J. Biol. Chem.* 115:155–161.

Chattopadhyay, N., T. Ghosh, S. Sinha, K. Chattopadhyay, P. Karmakar, and B. Ray. 2010. Polysaccharides from *Turbinaria conoides*: Structural features and antioxidant capacity. *Food Chem.* 118:823–829.

Cherbut, C., V. Salvador, J.-L. Barry, F. Doulay, and J. Delort-Laval. 1991. Dietary fibers effect on intestinal transit in man: Involvement of their physico-chemical and fermentative properties. *Food Hydrocoll.* 5:15–22.

Chevolot, L., A. Foucault, F. Chaubet et al. 1999. Further data on the structure of brown seaweed fucans: Relationships with anticoagulant activity. *Carbohydr. Res.* 319:154–165.

Chevolot, L., B. Mulloy, J. Ratiskol, A. Foucault, and S. Colliec-Jouault. 2001. A disaccharide repeat unit is the major structure in fucoidans from two species of brown algae. *Carbohydr. Res.* 330:529–535.

Chizhov, A.O., A. Dell, H.R. Morris et al. 1999. A study of fucoidan from brown seaweed *Chorda filum. Carbohydr. Res.* 320:108–119.

Cho, Y.-S., W.-K. Jung, J.-A. Kim, I.-W. Choi, and S.-K. Kim. 2009. Beneficial effects of fucoidan on osteoblastic MG-63 cell differentiation. *Food Chem.* 116:990–994.

Cho, M.L., B.-Y. Lee, and S.-G. You. 2011. Relationship between oversulfation and conformation of low and high molecular weight fucoidans and evaluation of their in vitro anticancer activity. *Molecules* 16:291–297.

Choi, E.-M., A.-J. Kim, Y.-O. Kim, and J.-K. Hwang. 2005. Immunomodulating activity of arabinogalactan and fucoidan *in vitro. J. Med. Food* 8:446–453.

Chotigeat, W., S. Tongsupa, K. Supamataya, and A. Phongdara. 2004. Effect of fucoidan on disease resistance of black tiger shrimp. *Aquaculture* 233:23–30.

Church, F.C., I.B. Meade, and R.E. Treanor. 1989. Antithrombin activity of fucoidan. *J. Biol. Chem.* 264:3618–3623.

Cooper, R., C. Dragar, K. Elliot, J.H. Fitton, J. Godwin, and K. Thompson. 2002. GFS, a preparation of Tasmanian *Undaria pinnatifida* is associated with healing and inhibition of reactivation of herpes. *BMC Complement. Altern. Med.* 2:1–7.

Deux, J.F., A. Meddahi-Pellé, A.F. Le Blanche et al. 2002. Low molecular weight fucoidan prevents neointimal hyperplasia in rabbit iliac artery in-stent restenosis model. *Arterioscler. Thromb. Vasc. Biol.* 22:1604–1609.

Doh-ura, K., T. Kuge, M. Uomoto, K. Nishizawa, Y. Kawasaki, and M. Iha. 2007. Prophylactic effect of dietary seaweed fucoidan against enteral prion infection. *Antimicrob. Agents Chemother.* 51:2274–2277.

Elkins, R.M.H. 2001. *Limu Moui: Prize Sea Plant of Tonga and the South Pacific.* Pleasant Grove, UT: Woodland Publishing.

Evans, L.V., M. Simpson, and M.E. Callow. 1973. Sulphated polysaccharide synthesis in brown algae. *Planta.* 110:237–252.

Food and Drug Administration. 2012. Cosmetics-compliance and regulatory information. http://www.fda.gov/cosmetics/guidancecomplianceregulatoryinformation/ucm074201.htm (accessed on June 18, 2012).

Fitton, J.H. 2003. Brown marine algae: A survey of therapeutic potentials. *Altern. Complement. Ther.* 9:29–33.

Fitton, J.H., M. Irhimeh, and N. Falk. 2007. Macroalgal fucoidan extracts: A new opportunity for marine cosmetics. *Cosmet. Toilet.* 122:55–64.

Fujii, M., D.-X. Hou, and M. Nakamizo. 2005. Fucoidan-based health food. U.S. Patent 0129708 A1. http://www.google.com/patents/about/10_735_958_Fucoidan_based_health_food.html?id=pkiVAAAAEBAJ (accessed on June 18, 2012).

Fujimura, T., K. Tsukahara, S. Moriwaki, T. Kitahara, T. Sano, and Y. Takema. 2002. Treatment of human skin with an extract of *Fucus vesiculosus* changes its thickness and mechanical properties. *J. Cosmet. Sci.* 53:1–9.

Fujimura, T., K. Tsukahara, S. Moriwaki, T. Kitahara, and Y. Takema. 2000. Effects of natural product extracts on contraction and mechanical properties of fibroblast populated collagen gel. *Biol. Pharm. Bull.* 23:291–297.

Guimarães, M.A.M. and P.A.S. Mourão. 1997. Urinary excretion of sulfated polysaccharides administered to Wistar rats suggests a renal permselectivity to these polymers based on molecular size. *Biochim. Biophys. Acta* 1335:161–172.

Hagiwara, H. 2010. Methods of extracting fucoidan. U.S. Patent 0056473 A1. http://www.google.com/patents/US20070087996 (accessed on June 18, 2012).

Hisatome, I., Y. Shirayoshi, H. Takeya, R. Teshima, and Y. Miki. 2010. Food/beverage and pharmaceutical composition for oral administration for improvement in acidic urine each comprising fucoidan as active ingredient. U.S. Patent 0048507 A1. http://www.google.com/patents/about/12_530_332_FOOD_BEVERAGE_AND_PHARMACEUTI.html?id=gUrNAAAAEBAJ (accessed on June 18, 2012).

Holtkamp, A.D., S. Kelly, R. Ulber, and S. Lang. 2009. Fucoidans and fucoidanases—Focus on techniques for molecular structure elucidation and modification of marine polysaccharides. *Appl. Microbiol. Biotechnol.* 82:1–11.

Honya, M., H. Mori, M. Anzai, Y. Araki, and K. Nisizawa. 1999. Monthly changes in the content of fucans, their constituent sugars and sulphate in cultured *Laminaria japonica*. *Hydrobiology* 398/399:411–416.

INCI Directory. 2012. International Nomenclature of Cosmetic Ingredients (INCI) Numbers. http://www.specialchem4cosmetics.com/services/inci/index.aspx (accessed on June 18, 2012).

Irhimeh, M.R., J.H. Fitton, and R.M. Lowenthal. 2007. Fucoidan ingestion increases the expression of CXCR4 on human CD34+ cells. *Exp. Hematol.* 35:989–994.

Irhimeh, M.R., J.H. Fitton, and R.M. Lowenthal. 2009. Pilot clinical study to evaluate the anticoagulant activity of fucoidan. *Blood Coagul. Fibrinolysis* 20:607–610.

Irhimeh, M.R., J.H. Fitton, R.M. Lowenthal, and P. Kongtawelert. 2005. A quantitative method to detect fucoidan in human plasma using a novel antibody. *Methods Find. Exp. Clin. Pharmacol.* 27:705–710.

Jiao, G., G.Yu, J. Zhang, and H.S. Ewart. 2011. Chemical structures and bioactivities of sulfated polysaccharides from marine algae. *Mar. Drugs* 9:196–223.

Juffrie, M., I. Rosalina, W. Damayanti, A. Djumhana, A. Rosalina, and H. Ahmad. 2006. The efficacy of fucoidan on gastric ulcer. *Indones. J. Biotechnol.* 11:908–913.

Jung, W.-K., Y. Athukoraka, Y.-J. Lee et al. 2007. Sulfated polysaccharide purified from *Ecklonia cava* accelerates antithrombin III-mediated plasma proteinase inhibition. *J. Appl. Phycol.* 19:425–430.

Kang, K.S., I.D. Kim, R.H. Kwon, and B.J. Ha. 2008. Undaria pinnatifida fucoidan extract protects against CCl4-induced oxidative stress. Biotechnol. Bioprocess Eng. 13:168–173.

Khan, M.N.A., M.-C. Lee, J.-Y. Kang, N.G. Park, H. Fujii, and Y.-K. Hong. 2008. Effects of the brown seaweed *Undaria pinnatifida* on erythematous inflammation assessed using digital photo analysis. *Phytother. Res.* 22:634–639.

Kim, S.H., D.S. Choi, Y. Athukorala, Y.-J. Jeon, M. Senevirathne, and C.K. Rha. 2007. Antioxidant activity of sulfated polysaccharides isolated from *Sargassum fulvellum*. *J. Food Sci. Nutr.* 12:65–73.

Koyanagi, S., N. Tanigawa, H. Nakagawa, S. Soeda, and H. Shimeno. 2003. Oversulfation of fucoidan enhances its anti-angiogenic and antitumor activities. *Biochem. Pharmacol.* 65:173–179.

Kusaykin, M., I. Bakunina, V. Sova et al. 2008. Structure, biological activity, and enzymatic transformation of fucoidans from the brown seaweeds. *Biotechnol. J.* 3:904–915.

Kylin, H. 1913. Zur biochemie der meersalgen. *Z. Physiol. Chem.* 83:171–197.

Laurienzo, P. 2010. Marine polysaccharides in pharmaceutical applications: An overview. *Mar. Drugs* 8:2435–2465.

Li, B., F. Lu, X. Wei, and R. Zhao. 2008. Fucoidan: Structure and bioactivity. *Molecules* 13:1671–1695.

Li. B., X.J. Wei, J.L. Sun, and S.Y. Xu. 2006. Structural investigation of a fucoidan containing a fucose-free core from the brown seaweed, *Hizikia fusiforme*. *Carbohydr. Res.* 341:1135–1146.

Mabeau, S. and B. Kloareg. 1987. Isolation and analysis of the cell walls of brown algae: *Fucus spiralis, F. ceranoides, F. vesiculosus, F. serratus, Bifurcaria* and *Laminaria digitata*. *J. Exp. Bot.* 38:1573–1580.

Mabeau, S., B. Kloareg, and J.P. Joseleau. 1990. Fractionation and analysis of fucans from brown algae. *Phytochemistry* 29:2441–2445.

Merkle, R.K. and I. Poppe. 1994. Carbohydrate composition analysis of glycoconjugates by gas-liquid chromatography/mass spectrometry. *Methods Enzymol.* 230:1–15.

Mian, A.J. and E. Percival. 1973. Carbohydrates of the brown seaweeds *Himanthalia lorea, Bifurcaria bifurcata*, and *Padina pavonia*: Part I. Extraction and fractionation. *Carbohydr. Res.* 26:133–146.

Mimeault, M. and S.K. Batra. 2011. Potential applications of curcumin and its novel synthetic analogs and nanotechnology-based formulations in cancer prevention and therapy. *Chin. Med.* 6:1–19.

Mizuno, M., Y. Nishitani, T. Tanoue et al. 2009. Quantification and localization of fucoidan in *Laminaria japonica* using novel antibody. *Biosci. Biotechnol. Biochem.* 73:335–338.

Mizutani, S., S. Deguchi, E. Kobayashi, E. Nishiyama, H. Sagawa, and I. Kato. 2006. Fucoidan-containing cosmetics. U.S. Patent 0093566. http://www.google.co.il/patents/about/7678368_Fucoidan_containing_cosmetics.html?id=nanNAAAAEBAJ (accessed on June 18, 2012).

Moon, H.J., S.H. Lee, M.J. Ku et al. 2009. Fucoidan inhibits UVB-induced MMP-1 promoter expression and down regulation of type I procollagen synthesis in human skin fibroblasts. *Eur. J. Dermatol.* 19:129–134.

Moon, H.J., S.R. Lee, S.N. Shim et al. 2008. Fucoidan inhibits UVB-induced MMP-1 expression in human skin fibroblasts. *Biol. Pharm. Bull.* 31:284–289.

Mower, T.E. 2007. Sports drink concentrate. U.S. Patent 0020358. http://www.google.com/patents/about/SPORTS_DRINK_CONCENTRATE.html?id=A7-XAAAAEBAJ (accessed on June 18, 2012).

Mower, T.E. 2010. Fucoidan compositions and methods for dietary and nutritional supplements. U.S. Patent 0330211 A1. http://www.google.com/patents/US7749545 (accessed on June 18, 2012).

Mulloy, B., A.C. Ribeiro, A.P. Alves, R.P. Vieira, and P.A.S. Mourão. 1994. Sulfated fucans from echinoderms have a regular tetrasaccharide repeating unit defined by specific patterns of sulfation at the *0*–2 and *0*–4 positions. *J. Biol. Chem.* 269:22113–22123.

Nagaoka, M., H. Shibata, I. Kimura-Takagi et al. 1999. Structural study of fucoidan from *Cladosiphon okamuranus* Tokida. *Glycoconj. J.* 16:19–26.

Nagumo, T. and T. Nishino. 1996. Fucan sulfates and their anticoagulant activities. In *Polysaccharides in Medicinal Applications*, ed. S. Dumitriu, pp. 545–574. New York: Marcel Dekker.

Nakazato, K., H. Takada, M. Iha, and T. Nagamine. 2010. Attenuation of N-nitrosodiethylamine-induced liver fibrosis by high-molecular-weight fucoidan derived from *Cladosiphon okamuranus*. *J. Gastroenterol. Hepatol.* 25:1692–1701.

Nishino, T., T. Nagumo, H. Kiyohara, and H. Yamada. 1991. Structural characterization of a new anticoagulant fucan sulfate from the brown seaweed *Ecklonia kurome*. *Carbohydr. Res.* 211:77–90.

Nishino, T., C. Nishioka, H. Ura, and T. Nagumo. 1994. Isolation and partial characterization of a novel amino sugar-containing fucan sulfate from commercial *Fucus vesiculosus* fucoidan. *Carbohydr. Res.* 255:213–224.

O'Leary, R., M. Rerek., and E.J. Wood. 2004. Fucoidan modulates the effects of transforming growth factor (TGF)-β1 on fibroblast proliferation and wound repopulation in in vitro models of dermal wound repair. *Biol. Pharm. Bull.* 27:266–270.

Oehninger, S., G.F. Clark, D. Fulgham et al. 1992. Effect of fucoidin on human sperm-zona pellucida interactions. *J. Androl.* 13:519–525.

Oh, K.D. 2009. Degradation of brown alga-derived fucoidan. U.S. Patent 0270607. http://www.patentstorm.us/applications/20090270607/description.html (accessed on June 18, 2012).

Okinaga, S. 2000. Shedding of *Herpes simplex* virus type-1 into tears and saliva in healthy Japanese adults. *Kurume Med. J.* 47:273–277.

Ozawa, T., J. Yamamoto, T. Yamagishi, N. Yamazaki, and M. Nishizawa. 2006. Two fucoidans in the holdfast of cultivated *Laminaria japonica*. *J. Nat. Med.* 60:236–239.

Park, H.S., G.-Y. Kim, T.-J. Nam, N.D. Kim, and Y.H. Choi. 2011. Antiproliferative activity of fucoidan was associated with the induction of apoptosis and autophagy in AGS gastric cancer cells. *J. Food Sci.* 76:77–83.

Ponce, N.M.A., C.A. Pujol, E.B. Damonte, M.L. Flores, and C.A. Stortz. 2003. Fucoidans from the brown seaweed *Adenocystis utricularis*: Extraction methods, antiviral activity and structural studies. *Carbohydr. Res.* 338:153–165.

Prasad, S., D. Lillicrap, A. Labelle et al. 2008. Efficacy and safety of a new-class hemostatic drug candidate, AV513, in dogs with hemophilia A. *Blood* 111:672–679.

Raghavendran, H.R.B., P. Srinivasan, and S. Rekha. 2011. Immunomodulatory activity of fucoidan against aspirin-induced gastric mucosal damage in rats. *Int. Immunopharmacol.* 11:157–163.

Ranatunga, 2007. Functional nutraceutical beverage: Youth juice. http://www.checkthescience.com/youthjuice.html (accessed on June 18, 2012).

Rhee, K.H. and K.H. Lee. 2011. Protective effects of fucoidan against γ-radiation-induced damage of blood cells. *Arch. Pharm. Res.* 34:645–651.

Rocha, H.A.O, C.R.C. Ranco, E.S. Trindade et al. 2001. A fucan from the brown seaweed *Spatoglossum schröederi* inhibits Chinese hamster ovary cell adhesion to several extracellular matrix proteins. *Braz. J. Med. Biol. Res.* 34:621–626.

Sakai, T., T. Kawai., and I. Kato. 2004. Isolation and characterization of a fucoidan-degrading marine bacteria strain and its fucoidanase. *Mar. Biotechnol.* 6:335–346.

Sakai, T., H. Kimura., and I. Kato. 2002. A marine strain of flavobacteriaceae utilizes brown seaweed fucoidan. *Mar. Biotechnol.* 4:399–404.

Sakai, T., H. Kimura., K. Kojima et al. 2003. Sulfated fucogalactan. U.S. Patent 6590097 B1. http://www.google.com/patents/US6590097

Salyers, A.A., J.R. Vercellotti, S.E.H. West, and T.D. Wilkins. 1977. Fermentation of mucin and plant polysaccharides by strains of *Bacteroides* from the human colon. *Appl. Environ. Microbiol.* 33:319–322.

Sardy, M. 2009. Role of matrix metalloproteinases in skin ageing. *Connect. Tissue Res.* 50: 132–138.

Schumacher, M., M. Kelkel, M. Dicato, and M. Diederich. 2011. Gold from the sea: Marine compounds as inhibitors of the hallmarks of cancer. *Biotechnol. Adv.* 29:531–547.

Senanayake, S.P.J.N., N. Ahmed, and J. Fichtali. 2011. Nutraceuticals and bioactives from marine algae. In *Handbook of Seafood Quality, Safety and Health Applications*, eds. C. Alasalvar, K. Miyashita, F. Shahidi, and U. Wanasundara, pp. 455–463. West Sussex, U.K.: Blackwell Publishing Ltd.

Senni, K., F. Gueniche, and A. Foucault-Bertaud et al. 2006. Fucoidan a sulfated polysaccharide from brown algae is a potent modulator of connective tissue proteolysis. *Arch. Biochem. Biophys.* 445:56–64.

Sezer, A.D., E. Cevher, F. Hatipoğlu, Z. Oğurtan, A.L. Baş, and J. Akbuğa. 2008a. Preparation of fucoidan-chitosan hydrogel and its application as burn healing accelerator on rabbits. *Biol. Pharm. Bull.* 31:2326–2333.

Sezer, A.D., E. Cevher, F. Hatipoğlu, Z. Oğurtan, A.L. Baş, and J. Akbuğa. 2008b. The use of fucosphere in the treatment of dermal burns in rabbits. *Eur. J. Pharm. Biopharm.* 69:189–198.

Shaklee, P.N., J. Bahr-Davidson, S. Prasad, and K. Johnson. 2010. Methods for fucoidan purification from seaweed extracts. U.S. Patent 0144667. http://www.google.com/patents/US20100144667 (accessed on June 18, 2012).

Shanmugam, M. and K.H. Mody. 2000. Heparinoid-active sulphated polysaccharides from marine algae as potential blood anticoagulant agent. *Curr. Sci.* 79:1672–1683.

Shibata, H., I. Kimura-Takagi, M. Nagaoka et al. 2000. Properties of fucoidan from *Cladosiphon okamuranus* tokida in gastric mucosal protection. *BioFactors* 11:235–245.

Skriptsova, A.V., N.M. Shevchenko, T.N. Zvyagintseva, and T.I. Imbs. 2010. Monthly changes in the content and monosaccharide composition of fucoidan from *Undaria pinnatifida* (Laminariales, Phaeophyta). *J. Appl. Phycol.* 22:79–86.

Soeda, S., S. Ishida, H. Shimeno, and A. Nagamatsu. 1994. Inhibitory effect of oversulfated fucoidan on invasion through reconstituted basement membrane by Murine Lewis lung carcinoma. *Jpn. J. Cancer Res.* 85:1144–1150.

Springer, G.F., H.A. Wurzel., G.M. Mcneal Jr., N.J. Ansell, and M.F. Doughty. 1957. Isolation of anticoagulant fractions from crude fucoidan. *Proc. Soc. Exp. Biol. Med.* 94:404–409.

Teixeira, M.M. and P.G. Hellewell. 1997. The effect of the selectin binding polysaccharide fucoidin on eosinophil recruitment *in vivo*. *Br. J. Pharmacol.* 120:1059–1066.

Thomes, P., M. Rajendran, B. Pasanban, and R. Rengasamy. 2010. Cardioprotective activity of *Cladosiphon okamuranus* fucoidan against isoproterenol induced myocardial infarction in rats. *Phytomedicine* 18:52–57.

Tissot, B. and R. Daniel. 2003. Biological properties of sulfated fucans: The potent inhibiting activity of algal fucoidan against the human complement system. *Glycobiology* 13:29–31.

Usov, A.I. and M.I. Bilan. 2009. Fucoidans-sulfated polysaccharides of brown algae. *Russ. Chem. Rev.* 78:785–799.

Usov, A.I., G.P. Smirnova, and N.G. Klochkova. 2001. Polysaccharides of algae: 55. Polysaccharide composition of several brown algae from Kamchatka. *Russ. J. Bioorg. Chem.* 27:395–399.

Usov, A.I., G.P. Smirnova, and N.G. Klochkova. 2005. Polysaccharides of algae: 58. The polysaccharide composition of the Pacific brown alga *Alaria fistulosa* P. et R. (Alariaceae, Laminariales). *Russ. Chem. Bull. Int. Ed.* 54:1282–1286.

Venugopal, V. 2011. Seaweed, microalgae, and their polysaccharides: Food applications. *In Marine Polysaccharides: Food Applications*, pp. 191–236. CRC Press, Boca Raton, FL.

Vilela-Silva, A.C.E.S., A.P. Alves, A.P. Valente, V.D. Vacquier, and P.A.S. Mourão. 1999. Structure of the sulfated α-L-fucan from the egg jelly coat of the sea urchin *Strongylocentrotus franciscanus*: Patterns of preferential 2-0- and 4-0-sulfation determine sperm cell recognition. *Glycobiology* 9:927–933.

Vilela-Silva, A.C.E.S., M.O. Castro, A.P. Valente, C.H. Biermann, and P.A.S. Mourão. 2002. Sulfated fucans from the egg jellies of the closely related sea urchins *Strongylocentrotus droebachiensis* and *Strongylocentrotus pallidus* ensure species-specific fertilization. *J. Biol. Chem.* 277:379–387.

Vitarich Laboratories, 2004. Limu plus vs. original limu. http://www.mlmwatchdog.com/ ComparisonLimuResearch.pdf (accessed on June 18, 2012).

Wang, J., L. Liu, Q. Zhang, Z. Zhang, H. Qi, and P. Li. 2009. Synthesized oversulphated, acetylated and benzoylated derivatives of fucoidan extracted from *Laminaria japonica* and their potential antioxidant activity *in vitro. Food Chem.* 114:1285–1290.

Wang, J., Q. Zhang, Z. Zhang, and Z. Li. 2008. Antioxidant activity of sulfated polysaccharide fractions extracted from *Laminaria japonica. Int. J. Biol. Macromol.* 42:127–132.

Wijesinghe, W.A.J.P., Y. Athukorala., and Y.-J. Jeon. 2011. The effect of anticoagulative sulfated polysaccharide purified from enzyme-assistant extract of a brown seaweed *Ecklonia cava* on Wistar rats. *Carbohydr. Polym.* 86:917–921.

Xue, C.-H., Y. Fang, H. Lin et al. 2001. Chemical characters and antioxidative properties of sulfated polysaccharides from *Laminaria japonica. J. Appl. Phycol.* 13:67–70.

Yuan, Y.V. 2007. Antioxidants from edible seaweeds. In *Antioxidant Measurement and Applications*, eds. F. Shahidi and C.-T. Ho, pp. 268–301. ACS symposium series (Vol. 956). Washington, DC: ACS.

Yuan, Y.V. 2008. Marine algal constituents. In *Marine Nutraceuticals and Functional Foods*, eds. C. Barrow, and F. Shahidi, pp. 259–296. Boca Raton, FL: CRC Press.

Zhang, Q., P. Yu, Z. Li, H. Zhang, Z. Xu, and P. Li. 2003. Antioxidant activities of sulfated polysaccharide fractions from *Porphyra haitanesis. J. Appl. Phycol.* 15:305–310.

Zvyagintseva, T.N., N.M. Shevchenko, I.V. Nazarova, A.S. Scobun, P.A. Luk'yanov, and L.A. Elyakova. 2000. Inhibition of complement activation by water-soluble polysaccharides of some far-eastern brown seaweeds. *Comp. Biochem. Physiol. C: Toxicol. Pharmacol.* 126:209–215.

Index

A

Abalone
  bioactive compounds
    anticancer activities, 65–66
    anticoagulant activities, 64–65
    antimicrobial and immune-
      modulatory activities, 66
    antioxidant activities, 63–64
    carotenoids, 63
    cosmeceutical activities, 66–67
  nutritional composition
    dietary recommendations, 59
    fatty acid composition of, 59
    fibril-forming collagens, 59
    isolation techniques, 59
    lipids, 60
    traditional medicine, nonedible
      abalone shell, 61
    viscera, 60
  value-added health-promoting
    products, 61–62
N-Acetyl-D-glucosamine (GlcNAc)
  endo-type chitinolytic enzyme,
    chitinase, 303–306
  enzymatic production, 309–311
  exo-type chitinolytic enzyme,
    NAHase, 306–308
  structure, 245–246, 248
Active peptides, *see* Bioactive peptides
Adipogenesis, 364–365
Allergic reaction, 159–162
Alzheimer's disease, 263–264
Angiotensin-I-converting enzyme
  (ACE), 52
Antiallergic activity, fucoidan, 138–139
Antibacterial effect, 234
Anticancer agent, uncultivated macroalgae
  brown algae, 171–175
  cuparene sesquiterpenes, 181–182
  green algae, 175–177
  herbal remedies, 181
  polysaccharides, 170
  red algae, 177–181
Anticoagulant activity, fucoidan
  blood coagulation disorders, 134
  Ecklonia kurome, 135
  heparin, 134
  molecular size, 135

Anti-neuroinflammatory activity
  Ecklonia cava, 22–23
  microglia, 22–23
  NSAIDs, 23–24
  pro-inflammatory mediators, 23
Antiobesity activity, 140–141
Antioxidant activity
  aminoderivatized chitosan
    preparation, 260–261
  enzymatic grafting, 262–263
  fucoidans, 139–140
  grafting reaction, 261–262
  protective effect, hydroxyl
    radical-induced DNA
    damage, 262–263
  ROS, 260
  SOD-1 and glutathione reductase
    (GSR), 262–263
Antioxidant ingredients
  bioactive peptides, 333–335
  bioprocessing and development,
    330–331
  carotenoids, 337–338
  chitooligosaccharide derivatives,
    331–333
  phlorotannins, 338–339
  sulfated polysaccharides
    (SPs), 335–337
Antituberculosis activity, 46–47
Apoptosis
  cholangiocarcinoma cell lines, 211
  intrinsic and extrinsic pathway, 210
  mitochondria, 212
  mitochondrial membrane potential
    (MMP), 210–211
  Wnt/β-catenin signaling
    molecules, 211
Appetite suppression, 157
Asthma, 106–107
Autoimmune diseases
  asthma, 106–107
  cytokines levels, IL-4, IL-5 and
    TNF-α, 108
  diabetes regulation
    obesity, 109
    phloroglucinol derivatives, 109
    plasma insulin and blood glucose
      levels, 109–110

Mitochondrial membrane potential
(MMP), 210–211
Mycobacterium tuberculosis, 46

N

Natural products, 187–190
Nutraceuticals, definition, 244

O

Obesity, *see* Body weight management
Omega-3 polyunsaturated fatty acids
(w3-PUFAs)
cancer
apoptosis, 210–212
autophagy-associated cancer cell
death, 212–213
cancer invasion and
metastasis, 216–217
checkpoint-controlled cell cycle, 209
chromatin modulation, 207–208
cyclin-CDK complex-controlled cell
cycle, 209
genome instability, 206
miRs, 208
tumor angiogenesis, 215–216
tumor-associated
inflammation, 214–215
dietary sources, 205
nomenclature and metabolic conversion
chemical structures of, 202
classification of, 201–202
cyclooxygenases (COXs), 203
enzymatic and nonenzymatic
oxidation of, 203–204
lipoxygenases (LOXs), 203
nonenzymatic oxidation of, 203
Oral care and hygiene, 237–239
Oxidative stress, 249, 258

P

Penaeus monodon, 315–316
Peptides, 194–195
Phlorotannins, 338–339
Pigments, microbial, 193
Polyunsaturated fatty acids (PUFA)
omega-3 fatty acids, 190–191
Thraustochytrids, 191
Probiotics, 195
Protein and bioactive peptide
microalgal species, 194
natural sources, 193–194
therapeutic activities, 194–195

R

Reactive oxygen species (ROS), 170, 258
Red algae
anti-inflammatory activity, 179
antioxidant properties, 178–179
antitumor activity, 179–180
cytotoxic acetogenins, 180–181
cytotoxic activity, 179
galactans and fucans, 177–178
secondary metabolites, 178
Rifampicin-induced hepatotoxicity,
chitosin, 259
Root canal treatment, 235–236
ROS, *see* Reactive oxygen species (ROS)

S

Sargassum ringgoldianum, 46–47
Sea lettuces
lipids and fatty acids, 9–10
minerals, 12
polysaccharides, 6–7
protein and amino acids, 7–9
Ulva sp.
U. conglobata, 7
U. fasciata, 11
U. lactuca, 7, 10
U. neumatoidea, 7
U. pertusa, 7
vitamins
vitamin B complex, 11
vitamin C, 11
vitamin E, 12
Seaweeds
anti-neuroinflammatory activity (*see*
Anti-neuroinflammatory activity)
antineurotoxic activity, 21–22
antioxidant activity
DPPH radical scavenging, 20
oxidative degradation, 21
oxidative stress, 20
phenolic compounds, 20–21
β-amyloid cleavage enzyme (BACE-1)
activity, 27
cancer, 18
cholinesterase inhibitory
activity, 24–26
farming, 170
fucoidan treatment, 27
metabolites, 19
neurite outgrowth, 26–27
neurodegenerative diseases, risk of, 19
neuronal death, inhibition of, 26
side effects, 18